The Flavonoids

The Flavonoids

Edited by

J. B. HARBORNE

Reader in Phytochemistry
University of Reading, U.K.

T. J. MABRY

Professor

and

HELGA MABRY

Research Assistant,
Department of Botany
University of Texas at Austin, U.S.A.

PART 2

(Chapters 12 to 20 and indices)

1975

ACADEMIC PRESS

New York   San Francisco
A subsidiary of Harcourt Brace Jovanovich, Publishers

Published in the United States
by Academic Press Inc.
111 Fifth Avenue
New York, New York 10003

First published 1975
by Chapman and Hall Ltd.
11 New Fetter Lane, London EC4P 4EE

© Chapman and Hall Ltd.

Typeset by Santype Ltd. (Coldtype Division)
Salisbury, Wiltshire
and printed in Great Britain by
Redwood Burn Limited
Trowbridge & Esher

ISBN 0-12-364602-4

LCCCN: 74-12868

Contents

Contributors

W. BARZ

Lehrstuhl für Biochemie der Pflanzen, Westfälische-Wilhelms-Universität, 44 Münster/Westf. Hindenburgplatz 55, Germany

B. A. BOHM

Department of Botany, University of British Columbia, Vancouver, Canada

M. L. BOUILLANT

Laboratory of Biological Chemistry, University of Lyon, 69 Villeurbanne, France

P. BRIDLE

Long Ashton Research Station, University of Bristol, U.K.

J. CHOPIN

Laboratory of Biological Chemistry, University of Lyon, 69 Villeurbanne, France

DERVILLA M. X. DONNELLY

Chemistry Department, University College, Belfield, Dublin 4, Ireland

L. FARKAS

Institut für Pharmazeutische Arzeimittelehre, 8-München-2, Karlstrasse 29, Germany

H. GEIGER

Fachgruppe 1, Universität Hohenheim, D-7000-Stuttgart-70, Windhalmweg 14, Germany

T. A. GEISSMAN

Department of Chemistry, University of California, Los Angeles, U.S.A.

O. R. GOTTLIEB — Instituto de Química, Universidade de São Paulo, Brazil

H. GRISEBACH — Biologisches Institut II der Universität, Schänzlestr. 9–11, D-7800 Freiburg/Br., Germany

H. HAHLBROCK — Biologisches Institut II der Universität, Schänzlestr. 9–11, D-7800 Freiburg/Br., Germany

J. B. HARBORNE — Plant Sciences Laboratories, The University, Reading, U.K.

E. HASLAM — Department of Chemistry, The University, Sheffield, U.K.

W. HÖSEL — Lehrstuhl für Biochemie der Pflanzen, Westfälische-Wilhelms-Universität, 44 Münster/Westf., Hindenburgplatz 55, Germany

T. J. MABRY — The Cell Research Institute and Department of Botany, University of Texas at Austin, Texas 78712, U.S.A.

K. R. MARKHAM — Chemistry Division, D.S.I.R., Petone, New Zealand

J. McCLURE — Department of Botany, Miami University, Oxford, Ohio, U.S.A.

C. QUINN — School of Botany, University of New South Wales, Kensington, NSW 2033, Australia

T. SWAIN — Biochemical Laboratory, Royal Botanic Gardens, Kew, Surrey, U.K.

C. F. TIMBERLAKE — Long Ashton Research Station, University of Bristol, Bristol, U.K.

K. VENKATARAMAN

National Chemical Laboratory, Poona 8, India

H. WAGNER

Institut für Pharmazeutische Arzeimittelhre, 8-München-2, Karlstrasse 29, Germany

CHRISTINE A. WILLIAMS

Plant Sciences Laboratories, The University, Reading, U.K.

E. WONG

Applied Biochemistry Division, D.S.I.R., Palmerston North, New Zealand

Preface

The flavonoids, one of the most numerous and widespread groups of natural constituents, are important to man not only because they contribute to plant colour but also because many members (e.g. coumestrol, phloridzin, rotenone) are physiologically active. Nearly two thousand substances have been described and as a group they are universally distributed among vascular plants. Although the anthocyanins have an undisputed function as plant pigments, the raison d'être for the more widely distributed colourless flavones and flavonols still remains a mystery. It is perhaps the challenge of discovering these yet undisclosed functions which has caused the considerable resurgence of interest in flavonoids during the last decade.

This book attempts to summarize progress that has been made in the study of these constituents since the first comprehensive monograph on the chemistry of the flavonoid compounds was published, under the editorship of T. A. Geissman, in 1962. The present volume is divided into three parts. The first section (Chapters 1—4) deals with advances in chemistry, the main emphasis being on spectral techniques to take into account the recent successful applications of NMR and mass spectral measurements to structural identifications. Recent developments in isolation techniques and in synthesis are also covered in this section. Advances in chemical knowledge of individual classes of flavonoid are mentioned *inter alia* in later chapters of the book.

The main section of the volume (Chapters 5—15) is concerned with providing a comprehensive account of the known structural variation among the fifteen classes of flavonoid compound. The need for a modern listing of all the known flavonoids has become more and more acute since so many new substances have been reported in recent years. Although the various contributors in this section have interpreted their tasks in different ways, this part of the book should provide, at least for the next decade, an indispensable guide to the chemical structures of flavonoid compounds.

One of the major changes of emphasis in the flavonoid field since 1962 has been away from purely chemical aspects towards bio-chemistry, systematic distribution and biological function. The last section (Chapters 16–20) therefore contains a summary of some of the more significant developments in these disciplines. Ten years ago, practically nothing was known of the enzymology of flavonoid biosynthesis; now, as is clear from Chapter 16, most of the enzymes involved have been detected and some of them have been fully characterised. Again, it is only recently that the turnover and metabolic fate of flavonoids have been seriously investigated and the results of the very latest experiments are included in Chapter 17. The final three essays cover more biological aspects and review what is known of the physiology, function, systematics and evolution of flavonoids.

The editors are grateful to their contributors, who have responded valiantly and courteously to all the demands made on them. They are particularly grateful to Professor T. A. Geissman for contributing the introduction. Finally, they thank the publishers for their guidance and assistance.

December 1974
Reading, U.K. and Austin, U.S.A.   The Editors.

The Flavonoids

PART 2

Chapter 12

C-Glycosylflavonoids

J. CHOPIN and M. L. BOUILLANT

12.1 Introduction

The rapid expansion of chromatographic analysis applied to plant extracts has disclosed the frequent occurrence of compounds which present the solubility and chromatographic properties of flavonoid glycosides, but which cannot be hydrolyzed even after prolonged treatment with acid, partial isomerization often taking place under these conditions. This resistance towards acid hydrolysis, the most distinctive feature of these compounds, results from the sugar being directly attached to the flavonoid nucleus by a carbon-carbon bond, and accounts for the difficulties encountered in the identification of the glycosyl residue. Furthermore, the anomalous results given by periodate oxidation led to conflicting assumptions about the structure of the side chain and the nature of the acid isomerization.

It was not until 1964 that these problems were resolved by NMR studies of vitexin and isovitexin (Horowitz and Gentili, 1964), orientin and iso-orientin (Koeppen, 1964), the 8- and 6- *C*-glucosyl derivatives of the flavones apigenin and luteolin respectively.

One year later the 6,8-di-*C*-glycosylluteolin structure of lucenin-1 (Seikel and Mabry, 1965) was established by NMR methods. It is now well demonstrated that in these natural products, the sugar is linked by its anomeric carbon atom to the aromatic A-ring, biogenetically derived from acetate, in the *ortho*-position to one or two phenolic hydroxyl groups. Although the term 'flavonoid *C*-glycosides' has been widely used in literature for these compounds, '*C*-glycosylflavonoids' seems to be preferable, owing to their 'aglycone' behaviour in hydrolytic processes.

Acid isomerization of 5-hydroxy-*C*-glycosylflavones mainly results from a Wessely-Moser rearrangement, involving opening of the pyrone ring followed by ring closure on either of the two phenolic hydroxyl groups *ortho* to the carbonyl group. A mixture of the two isomers is thus produced:

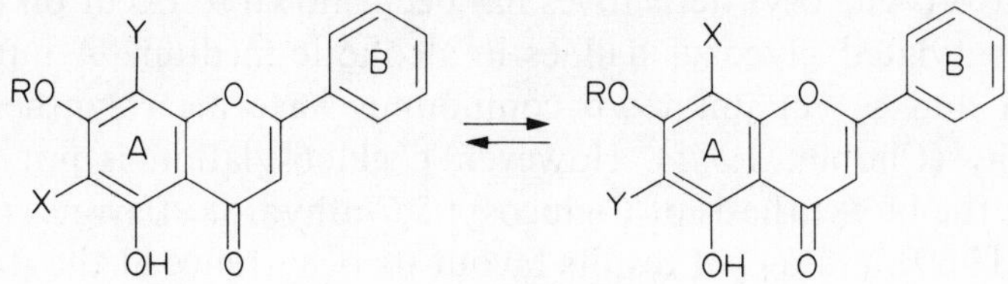

X = sugar, Y = H

The 6-isomers of the 8-*C*-glucosides vitexin and orientin were named saponaretin and homo-orientin before they were known to be interconvertible. The term iso has since been used in all other cases and should always be employed for the sake of uniformity. This chemical interconversion, which generally leads to an equilibrium mixture, does not find its equivalent *in vivo*. In *Lemna minor* for example, vitexin is converted to isovitexin, but the reverse reaction does not occur (Wallace and Mabry, 1970).

C-Glycosylflavonoids are rather widely distributed in the plant kingdom. They have now been found in dicotyledons, mono-cotyledons, ferns, mosses and green algae. All the genera (and the number of species in each genus) in which *C*-glycosylflavonoids have been characterized are listed in Table 12.1. Any part of the plant may be a source of *C*-glycosylflavonoids, but they are most frequently found in aerial parts. Systematic aspects of these compounds are discussed in Chapters 19 and 20.

Sometimes, a single *C*-glycosylflavonoid is the major constituent of the plant extract and can easily be isolated and thoroughly studied, but in most cases, small amounts of several *C*-glycosylflavo-noids co-occur with flavonoid *O*-glycosides and their separation only becomes possible when chromatographic techniques are used. The amounts available of a chromatographically pure compound are frequently so minute that only UV spectrometry and chromato-graphic tests can be employed for characterization. Fortunately, mass spectrometry requiring micro-amounts of substance is ex-tremely valuable in the field of *C*-glucosylflavonoids, which exhibit a characteristic fragmentation pattern (Prox, 1968). Circular dichroism now affords a new and very sensitive method of differentiation between 6- and 8-*C*-glycosylflavones (Gaffield and Horowitz, 1972) and much progress can be expected from the extended use of these techniques.

C-Glycosylation in the 6-position of 5,7-dihydroxyflavonoids and of their 8-*C*-glucosyl derivatives has been shown to occur on reaction with acetylated glycosyl halides in alcoholic medium. A number of natural and as yet unknown compounds have been synthesized in this way (Chopin, 1971). However, *C*-glucosylation is not the last step in the biosynthesis of *C*-glucosyl-5,7-dihydroxyflavones (Wallace *et al.*, 1969) and recent results favour its occurrence at the flavanone level (see Chapter 16).

Table 12.1   *Natural sources of C-glycosylflavonoids**

ALGAE	
Nitella (1)	Markham and Porter, 1969
BRYOPHYTA	
Mnium (1)	Melchert and Alston, 1965
Porella (Madotheca) (1)	Molisch, 1911; Nilsson, 1969, 1973; Tjukavkina *et al.,* 1970
Hymenophytum (1)	Markham *et al.,* 1969
Reboulia (1)	Markham *et al.,* 1972
Marchantia (2)	Markham and Porter, 1973
PTERIDOPHYTA	
PTERIDACEAE	
Sphenomeris (1)	Ueno *et al.,* 1963
CYATHEACEAE	
Cyathea (4)	Ueno *et al.,* 1963; Soeder and Babb, 1972
SPERMATOPHYTA	
PTERIDOSPERMAE	
CYCADACEAE	
Dioon (1)	Carson and Wallace, 1972
GYMNOSPERMAE	
PINACEAE	
Larix (1)	Niemann and Bekooy, 1971
ANGIOSPERMAE	
FAGALES	
FAGACEAE	
Nothofagus (1)	Hillis and Inoue, 1967
URTICALES	
ULMACEAE	
Zelkowa (1)	Funaoka, 1956, 1957; Funaoka and Tanaka, 1957 a, b Hillis and Horn, 1966
Trema (1)	Oelrichs *et al.,* 1968
MORACEAE	
Humulus (1)	Aritomi, 1962
POLYGONALES	
POLYGONACEAE	
Polygonum (1)	Hörhammer *et al.,* 1958, 1959b
Rumex (1)	Aritomi *et al.,* 1965
Fagopyrum (1)	Margna *et al.,* 1967
CENTROSPERMALES	
CHENOPODIACEAE	
Beta (1)	Gardner *et al.,* 1967

* The number of species in which *C*-glycosylflavonoids have been found is enclosed in parentheses.

Continued

Table 12.1 – *Continued*

CARYOPHYLLACEAE
Saponaria (1) — Barger, 1906
Silene (3) — Plouvier, 1967; Litvinenko and Darmograi, 1968; Darmograi and Litvinenko, 1971
Gypsophila (47) — Darmograi *et al.*, 1968, 1969; Litvinenko *et al.*, 1969b
Vaccaria (1) — Litvinenko *et al.*, 1967
Spergularia (1) — Zoll, 1972
Stellaria (1) — Zoll and Nouvel, 1974

RANALES
NYMPHAEACEAE
Cabomba (1) — Boutard, 1972
Nymphaea (1) — Taku *et al.*, 1970
RANUNCULACEAE
Adonis (1) — Hörhammer *et al.*, 1960; Chernobai *et al.*, 1968; Drozd *et al.*, 1971
Trollius (1) — Sachs, 1963
Ranunculus (3) — Drozd and Litvinenko, 1969; Drozd *et al.*, 1969 a, b
Thalictrum (3) — Wagner *et al.*, 1971; Kuczynski and Chyczewski, 1971

LAURACEAE
Beilschmiedia (1) — Harborne and Méndez, 1969

RHOEDALES
CRUCIFERAE
Alliaria (1) — Paris and Delaveau, 1962

ROSALES
ROSACEAE
Crataegus (26) — Fiedler, 1955; Geissman and Fiedler, 1956; Mrugasiewicz, 1963; Fizel, 1965; Lewak, 1966; Batyuk and Chernobrovaya, 1966; Batyuk *et al.*, 1966, 1969

LEGUMINOSAE
Aspalathus (1) — Koeppen *et al.*, 1962; Dahlgren, 1963; Koeppen and Roux, 1965b
Castanospermum (1) — Simes, 1949; Eade *et al.*, 1962, 1966
Cassia (2) — Anton and Duquenois, 1968; Trajman, 1972
Crotalaria (7) — Subramanian and Nagarajan, 1967 a, b, c, 1969, 1970
Harborne, 1969
Cytisus (6) — Stenhouse, 1851; Perkin, 1899; Mascré and Paris, 1937 a, b, c; Paris, 1957; Paris and Stambouli, 1961; Hörhammer *et al.*, 1962; Chopin *et al.*, 1965; Harborne, 1969; Paris and Brun-Bousquet, 1971

Continued

Table 12.1—*Continued*

Chamaecytisus (7)	Harborne, 1969
Teline (1)	Harborne, 1969
Ulex (5)	Harborne, 1969
Stauracanthos (1)	Harborne, 1969
Lygos (2)	Harborne, 1969
Calycotome, (2)	Harborne, 1969
Argyrolobium (2)	Harborne, 1969
Lupinus (15)	Harborne, 1969
Coronilla (1)	Sherwood *et al.,* 1973
Sophora (3)	Markham, 1973
Tamarindus (1)	Bhatia *et al.,* 1964, 1966a; Lewis and Neelakantan, 1962, 1964
Spartium (1)	Hörhammer *et al.,* 1959a
Genista (?)	Plouvier, 1967; Harborne, 1969
Desmodium (2)	Aritomi and Kawasaki, 1968; Chernobrovaya *et al.,* 1970
Glycyrrhiza (2)	Litvinenko and Kovalev, 1967; Litvinenko and Grankina, 1970; Litvinenko and Nadezhina, 1972
Pueraria (2)	Shibata *et al.,* 1959; Murakami *et al.,* 1960; Bhutani *et al.,* 1969
Lespedeza (2)	Paris and Charles, 1962; Paris and Etchepare, 1964; Glyzin *et al.,* 1970
	Wagner *et al.,* 1972e
Parkinsonia (1)	Bhatia *et al.,* 1966b
Trigonella (2)	Adamska and Lutomski, 1971; Wagner *et al.,* 1973; Seshadri *et al.,* 1972
Dalbergia (1)	Narayanan and Seshadri, 1971
Psoralea (30)	Ockendon *et al.,* 1966
GERANIALES	
OXALIDACEAE	
Oxalis (2)	Shimokoriyama and Geissman, 1962
LINACEAE	
Linum (2)	Volk and Sinn, 1968; Ibrahim, 1969; Ibrahim and Shaw, 1970; Dubois and Mabry, 1971; Wagner *et al.,* 1972a
EUPHORBIACEAE	
Jatropha (2)	Subramanian *et al.,* 1971a, b
Hevea (1)	Subramanian *et al.,* 1971b
Croton (1)	Wagner *et al.,* 1970
RUTALES	
RUTACEAE	
Citrus (2)	Chopin *et al.,* 1964; Horowitz and Gentili, 1966; Gentili and Horowitz, 1968
Teclea (1)	Paris and Etchepare, 1968

Continued

Table 12.1—*Continued*

SIMARUBACEAE
Ailanthus (1)   Kapoor *et al.*, 1971

SAPINDALES
ACERACEAE
Acer (1)   Aritomi, 1963

RHAMNALES
VITACEAE
Vitis (1)   Wagner *et al.*, 1967

MALVALES
MALVACEAE
Hibiscus (1)   Nakaoki, 1944

PARIETALES
THEACEAE
Thea (1)   Sakamoto, 1967, 1969, 1970
PASSIFLORAE
Passiflora (4)   Schilcher, 1968; Lutomski and Adamska,
                         1968; Glotzbach and Rimpler, 1968;
                         Poethke *et al.*, 1970

VIOLACEAE
Viola (1)   Hörhammer et *al.*, 1965; Rosprim, 1966;
                         Düll, 1970; Wagner *et al.*, 1972g

CUCURBITALES
CUCURBITACEAE
Bryonia (1)   Paris *et al.*, 1966; Leiba *et al.*, 1968
Cucumis (1)   Monties, 1971

MYRTALES
LYTHRACEAE
Lythrum (1)   Paris (M), 1968
MYRTACEAE
Eucalyptus (1)   Hillis and Carle, 1963
COMBRETACEAE
Combretum (1)   Jentzsch *et al.*, 1962

UMBELLIFLORAE
UMBELLIFERAE
Anethum (1)   Dranik, 1970; Harborne and Williams,1972
Opoponax (1)   Crowden *et al.*, 1969
Laretia (1)   Crowden *et al.*, 1969
Cryptotaenia (1)   Crowden *et al.*, 1969

ERICALES
ERICACEAE
Vaccinium (1)   Yasue *et al.*, 1965

PLUMBAGINALES
PLUMBAGINACEAE
Limonium   Asen and Plimmer, 1972

Continued

Table 12.1—*Continued*

GENTIANALES
  GENTIANACEAE
    Swertia (6)   Asahina *et al.*, 1942; Komatsu and
                              Tomimori, 1966; Komatsu *et al.*, 1967a, b
                              1968a; Tomimori and Komatsu, 1969
    Gentiana (8)   Lebreton and Dangy-Kaye, 1973; Bellmann
                              and Jacot-Guillarmod, 1973

RUBIALES
  DIPSACACEAE
    Dipsacus (2)   Plouvier, 1966; Zemtsova and Bandyukova,
                              1972
    Scabiosa (1)   Plouvier, 1966
    Cephalaria (1)   Plouvier, 1967; Bouillant *et al.*, 1972

TUBIFLORAE
  POLEMONIACEAE
    Phlox (1)   Mabry *et al.*, 1971
  VERBENACEAE
    Vitex (6)   Perkin, 1898, 1900; Rao and Venkateswarlu,
                              1956, 1962; Evans *et al.*, 1957; Briggs and
                              Cambie, 1958; Seikel *et al.*, 1959; Hänsel
                              and Rimpler, 1963; Seikel and Mabry, 1965;
                              Hänsel *et al.*, 1965; Rao, 1965
  SCROPHULARIACEAE
    Gratiola (1)   Litvinenko *et al.*, 1969a; Borodin *et al.*, 1970

SYNANTHERALES
  COMPOSITAE
    Eupatorium (1)   Düll, 1970; Wagner *et al.*, 1972b
    Tragopogon (5)   Kroschewsky *et al.*, 1969
    Helenium (3)   Wagner *et al.*, 1972d, f
    Gaillardia (1)   Wagner *et al.*, 1972c
    Helichrysum (1)   Rimpler *et al.*, 1963
    Artemisia (1)   Chumbalov and Fadeeva, 1970
    Staehelina (1)   Raynaud *et al.*, 1971
    Catananche (1)   Proliac, 1973
    Centaurea (1)   Asen and Jurd, 1967; Asen, 1967
    Senecio (10)   Glennie *et al.*, 1971

ALISMATALES
  ALISMATACEAE
    Alisma (1)   Boutard, 1972; Boutard *et al.*, 1973
    Sagittaria (2)   Boutard, 1972; Boutard *et al.*, 1973
    Echinodorus (1)   Boutard, 1972; Boutard *et al.*, 1973
  BUTOMACEAE
    Butomus (1)   Boutard, 1972; Boutard *et al.*, 1973
    Limnocharis (1)   Boutard, 1972; Boutard *et al.*, 1973
  HYDROCHARITACEAE
    Elodea (1)   Boutard, 1972; Boutard *et al.*, 1973

Continued

Table 12.1 —*Continued*

JUNCAGINACEAE
Triglochin (1)   Boutard, 1972; Boutard *et al.,* 1973

POTAMOGETONALES
POTAMOGETONACEAE
Potamogeton (5)   Boutard, 1972; Boutard *et al.,* 1972, 1973

PIPERALES
PIPERACEAE
Piper (2)   Boutard, 1972

ARALES
ARACEAE
Arum (1)   Phouphas, 1956; Boutard, 1972
Biarum (1)   Phouphas, 1956
LEMNACEAE
Lemna (8)   McClure, 1964; Tikhonov *et al.,* 1965; Wallace and Alston, 1966; McClure and Alston, 1966; Alston, 1966, 1968; Wallace *et al.,* 1969; Wallace and Mabry, 1970

Spirodela (4)   Jurd *et al.,* 1957; Tikhonov *et al.,* 1966; McClure and Alston, 1966; Alston, 1966, 1968; McClure, 1968; Reznik and Menschik, 1969

Wolffia (2)   Alston, 1966, 1968; McClure and Alston, 1966

PALMALES
PALMAE†   Williams *et al.,* 1971, 1973

COMMELINALES
COMMELINACEAE
Commelina (?)   Takeda *et al.,* 1966; Komatsu *et al.,* 1968b

GRAMINALES
GRAMINEAE‡   Seikel and Bushnell, 1959; Seikel *et al.,* 1962
Hordeum (1)   Seikel and Geissman, 1957
Avena (7)   Harborne and Hall, 1964
Triticum (7)   King, 1962; Harborne and Hall, 1964; Julian *et al.,* 1971
Agrostis (1)   Harborne and Hall, 1964
Ōryza (1)   Harborne and Hall, 1964
Poa (1)   Harborne and Hall, 1964
Stipa (2)   Harborne and Hall, 1964; Saleh *et al.,* 1971
Briza (17)   Harborne and Hall, 1964; Williams and Murray, 1972
Cathestecum (1)   Crawford and Lankow, 1972

Continued

Table 12.1—*Continued*

CYPERALES	
CYPERACEAE (34)§	Harborne, 1971; Kukkonen, 1971
LILIALES	
IRIDACEAE	
Iris (6)	Hirose *et al.*, 1962; Kawase and Yagishita, 1968; Kawase, 1968; Asen *et al.*, 1970; Carter, 1968

† Present in 84% of 125 spp. from 70 genera surveyed.
‡ Present in 90% of 193 spp. from 104 genera surveyed (C. A. Williams, unpublished results).
§ Present in 55% of 62 spp. from 11 genera surveyed.

The identification of the glycosyl residue is the most difficult problem in *C*-glycosylflavonoid chemistry and synthetic products have been used for determining the influence of this sugar on spectral and chromatographic properties. As expected, these properties are very similar when the only difference lies in the stereochemistry of the sugar moiety and much care is needed in the interpretation of chromatographic data. In the last ten years, the field of *C*-glycosylflavonoids has been rapidly growing and this progress has been frequently reviewed, see Hörhammer and Wagner (1961), Paris (1962), Seikel (1963), Haynes (1963, 1965), Chopin (1966), Wagner (1966), Bhatia and Seshadri (1967), Alston (1968) and Chopin (1971).

12.2 Naturally occurring *C*-glycosylflavonoids

The natural *C*-glycosylflavonoids known today can be divided in two groups: the unhydrolyzable 'aglycones' (mono and di-*C*-glycosylflavonoids) listed in Table 12.4 and their hydrolyzable derivatives (*O*-glycosides and *O*-acylated derivatives) lised in Table 12.5.

12.2.1 Mono-*C*-glycosylflavonoids

Until now, only five types of mono-*C*-glycosylflavonoids have been well defined: *C*-glycosylflavones, *C*-glycosylflavonols, *C*-glycosylflavanones, *C*-glycosylisoflavones, and *C*-glycosyldihydrochalcones. *C*-Glycosylflavones are by far the most important group. All these flavonoids have a phloroglucinol A-ring, except bayin (a flavone) and puerarin (an isoflavone) in which the A-ring derives from resorcinol. Only two of them are C-β-D-xylopyranosylflavonoids, all others being C-β-D-glucopyranosylflavonoids. A- and B-ring substituents in naturally occurring mono-C-β-D-glucopyranosylflavonoids are listed in Table 12.2. Closely related with C-glycosylflavonoids are the C-glucosylxanthones mangiferin and isomangiferin (Aritomi and Kawasaki, 1970) and the C-glucosylchromone aloesin (Haynes *et al.*, 1970).

12.2.1.1 Mono-C-glycosylflavones

Bayin excepted, all members of this group are 6- or 8-C-glycosyl-flavones deriving from apigenin (5,7,4'-trihydroxyflavone), luteolin (5,7,3',4'-tetrahydroxyflavone) or their methyl ethers.

Table 12.2   *Naturally occurring mono-C-β-D-glucopyranosyl flavonoids*

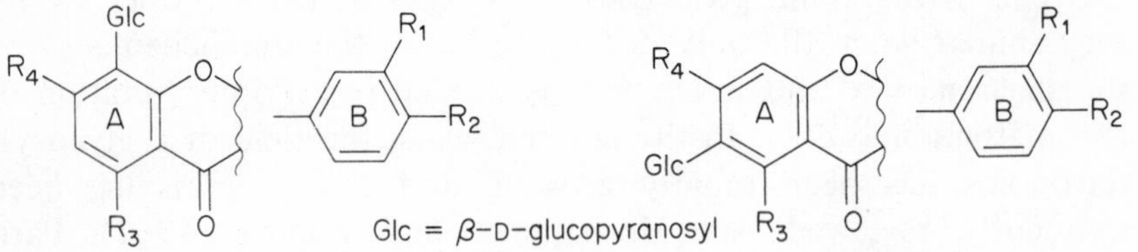

Glc = β-D-glucopyranosyl

8-isomer	R_1	R_2	R_3	R_4	6-isomer
		Flavones			
Bayin	H	OH	H	OH	–
Vitexin	H	OH	OH	OH	Isovitexin
Isoswertisin	H	OH	OH	OMe	Swertisin
Cytisoside	H	OMe	OH	OH	–
–	H	OMe	OH	OMe	Embigenin
Orientin	OH	OH	OH	OH	Iso-orientin
–	OH	OH	OH	OMe	Swertiajaponin
Parkinsonin A	OH	OH	OMe	OH	–
Scoparin	OMe	OH	OH	OH	Isoscoparin
8-C-β-D-glucopyranosyl diosmetin	OH	OMe	OH	OH	6-C-β-D-glucopyranosyl diosmetin
	OMe	OMe	OH	OMe	6-C-β-D-glucopyranosyl 7,3',4'-tri-O-methylluteolin
		Flavonols			
	H	OH	OH	OMe	Keyakinin
		Flavanones			
Isohemiphloin	H	OH	OH	OH	Hemiphloin
		Isoflavones			
	H	OH	H	OH	Puerarin
Dalpanitin	OMe	OH	OH	OH	–
		Dihydrochalcones			
Aspalathin	OH	OH	OH	OH	Aspalathin

Vitexin (8-C-β-D-glucopyranosylapigenin), the best known C-glycosylflavonoid, was first isolated from the wood of *Vitex lucens* (Verbenaceae) by Perkin (1898) and was considered to be an unusual type of apigenin glucoside. The extensive chemical studies of Evans *et al.* (1957) demonstrated the 8-position of the side chain and led them to propose a 2,5-anhydro hexityl structure (1) for the latter, periodate oxidation of vitexin giving, without any loss of carbon atoms, dehydrosecovitexin, a dialdehyde for which the hemiketal-structure (2) was deduced from the results of acidic methanolysis.

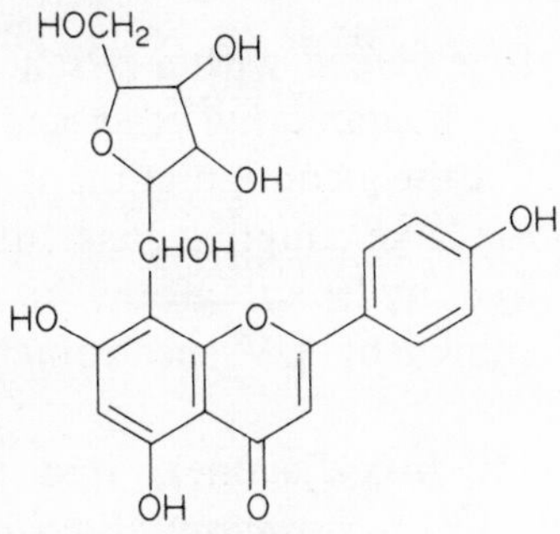

(1) 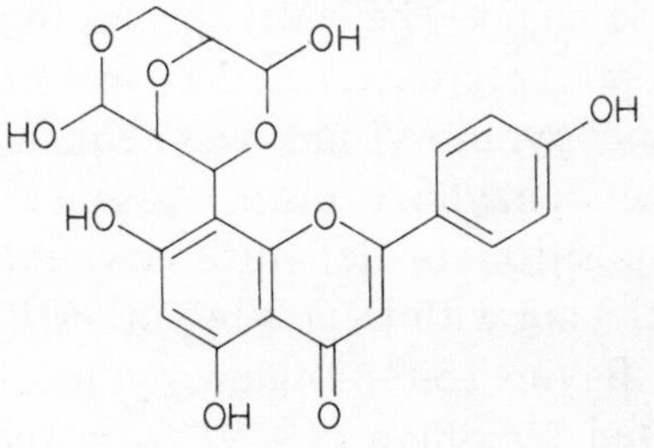 (2)

These conclusions were disputed by Rao and Venkateswarlu (1962), since their periodate oxidation results suggested a pyranosyl structure (3) for the side chain. Independently, a new formula (4) of dehydrosecovitexin was proposed by Dean (1963) starting from the same hypothesis.

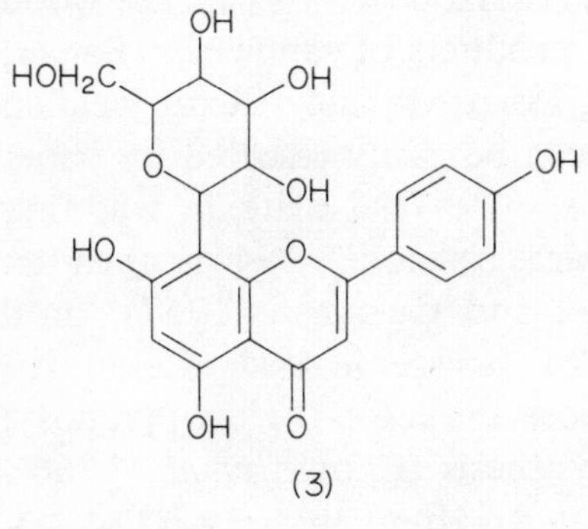

(3) 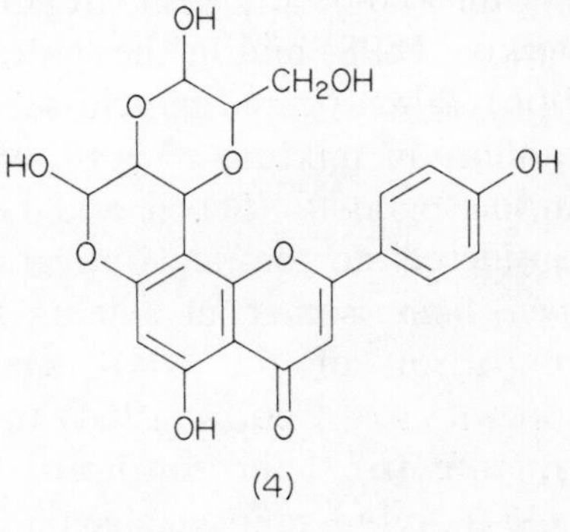 (4)

Eventually, the complete structure (5) of vitexin was elucidated by Horowitz and Gentili (1964). D-Glucose and arabinose were identified by paper chromatography as products of vitexin ozonolysis. Periodate consumption and formic acid production in the

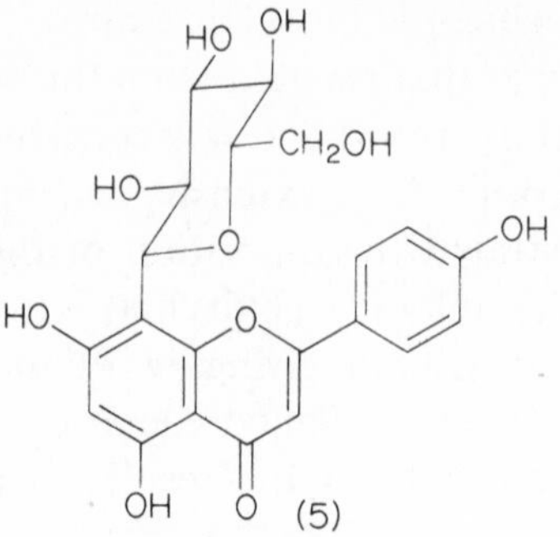

(5)

oxidation of vitexin 5,7,4'-trimethyl ether agreed with a pyranosyl side chain. The NMR spectra of vitexin, its heptamethyl ether and its heptaacetate excluded Evans' formula and the β-configuration of the glucopyranosyl link was deduced from the large coupling constant of the benzylic proton. Absence of a free hydroxyl group in the 7-position of dehydrosecovitexin was shown by UV spectrometry, agreeing with Dean's hypothesis.

Bayin (8-C-β-D-glucopyranosyl-7,4'-dihydroxyflavone), first iso-lated by Simes (1949) from the wood of *Castanospermum australe* (Leguminosae), has been thoroughly studied by Eade *et al.* (1962, 1966) and shown to be 5-deoxyvitexin. Hindered rotation about the C-1″–C-8 bond was evidenced by the splitting of certain bands in the NMR spectrum of bayin acetate at low temperatures (Eade *et al.*, 1965).

Isovitexin (saponaretin, homovitexin) (6-C-β-D-glucopyranosyl-apigenin) co-occurs with vitexin in the extracts of *Vitex lucens* wood (Perkin, 1898) and in the hydrolysis products of saponarin (Barger, 1906). When heated in acid solution, isovitexin and vitexin yield an equilibrium mixture of both, which can be easily resolved by paper chromatography (Seikel and Geissman, 1957). Isovitexin was first considered to be an optical isomer (Nakaoki, 1944), then an open-chain isomer of vitexin (Seikel and Geissman, 1957), until comparison of the NMR spectra of isovitexin and vitexin led Horowitz and Gentili (1964) to propose the 6-C-β-D-glucopyranosyl structure (6), later confirmed by synthesis (Chopin *et al.*, 1966). 6-C-β-D-Xylopyranosylapigenin, isolated after hydrolysis of an O-rhamnoside extracted from *Phlox drummondii* (Polemoniaceae) (Mabry *et al.*, 1971), was shown to be identical with the previously synthesized compound (Chopin and Bouillant, 1970a).

Avroside, isoavroside, neoavroside and isoneoavroside, found by

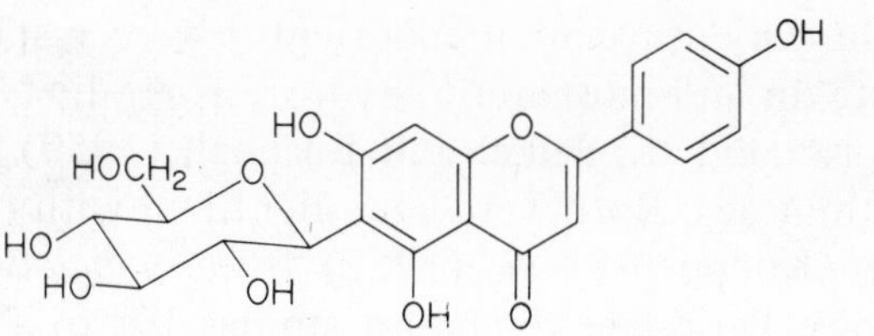

(6)

Borodin *et al.*, (1970) in *Gratiola officinalis*, are considered to be rotational isomers of 6-*C*-β and 6-*C*-α-D-glucopyranosylapigenin. Likewise, neovitexin and isoneovitexin, found in *Gypsophila pauli*, should be rotational isomers of 8-*C*-α-D-glucopyranosylapigenin (Litvinenko and Borodin, 1970). This is not in agreement with the fact that no rotational isomerism could be observed in the NMR spectra of isovitexin hexaacetate and vitexin tetraacetate, which are more hindered than the free compounds (Eade *et al.*, 1965).

Swertisin (6-*C*-β-D-glucopyranosylgenkwanin) from *Swertia japonica* (Gentianaceae) was extensively studied by Komatsu and Tomimori (1966) and shown to be 7-methylisovitexin. Its acid isomerization led to isoswertisin (7-methylvitexin), later identified (UV and chromatography) as a component of the blue pigment of *Centaurea cyanus* (Compositae) by Asen and Jurd (1967).* Cytoside (8-*C*-β-D-glucopyranosylacacetin), isolated from *Cytisus laburnum* (Leguminosae) by Paris (1957), was shown to be a *C*-glycosylacacetin by Paris and Stambouli (1961) and 4'-methylvitexin by Chopin *et al.* (1965). Its acid isomerization product, isocytisoside, not yet found in nature, was identified by comparison with synthetic 6-*C*-β-D-glucopyranosylacacetin (Chopin *et al.*, 1966). A different 8-*C*-glyco-sylacacetin was isolated, after hydrolysis, from *Reboulia hemi-spherica* (Bryophyta) by Markham *et al.* (1972). Embigenin (6-*C*-β-D-glucopyranosyl-7,4'-dimethylapigenin) was isolated by Kawase and Yagishita (1968) after hydrolysis of embinin, a dirhamnoside first extracted from *Iris tectorum* (Iridaceae) by Hirose *et al.* (1962). Its structure was established unambiguously as 7,4'-dimethylisovitexin.

Orientin (8-*C*-β-D-glucopyranosyl-luteolin) was found in *Polygonum orientale* (Polygonaceae) by Hörhammer *et al.* (1958), who showed it to be a luteolin glycoside, isomerized by acid into

*This identification has since been withdrawn (see Asen and Horowitz, 1974).

homoorientin. Orientin and homoorientin were respectively identified with lutexin and lutonaretin, hydrolysis products of lutonarin, a O-glucoside isolated by Seikel and Bushnell (1959) from *Hordeum vulgare* (Gramineae). Both were later identified with two compounds extracted by Koeppen *et al.* (1962) from *Aspalathus acuminatus* (Leguminosae). Periodate oxidation studies led to a hexopyranosyl structure of the side chain in both isomers (Koeppen, 1962; Hänsel and Rimpler, 1963; Sachs, 1963). The tetramethyl ethers yielded glucose and arabinose by ferric chloride oxidation in both cases, but different products by periodate oxidation. Thus, the glucopyranosyl side chain was not linked to the same position of the flavonoid skeleton and NMR spectra agreed with the 6-position in homoorientin and the 8-position in orientin. Moreover, comparison of the behaviour of the tri- and tetramethyl ethers towards acid showed that a free 5-hydroxyl group was needed for isomerization to take place (Koeppen, 1964; Koeppen and Roux, 1965a). The 6-C-β-D-glucopyranosyl-luteolin structure of homoorientin was later confirmed by synthesis (Chopin *et al.*, 1966). Epiorientin, isolated from *Parkinsonia aculeata* (Leguminosae) by Bhatia *et al.* (1966b), could not be distinguished from orientin by chromatography or periodate oxidation. Melting points and optical rotations of the free compounds and their tetramethyl ethers being different, the structure of 8-C-α-D-glucopyranosyl-luteolin was proposed for epiorientin.

Parkinsonins A (5-methylorientin) and B (5,7-dimethylepiorientin) from the same source were related by methylation with orientin and epiorientin respectively, the positions of methyl groups being assigned on the basis of UV spectral shifts. Two isomeric C-glycosyl-luteolins, chromatographically different from orientin and homoorientin, were found by Williams and Murray (1972) in *Briza media* (Gramineae). 6-C-β-D-xylopyranosyl-luteolin was obtained by hydrolysis of its O-rhamnoside occurring with the corresponding apigenin derivative in *Phlox drummondii* (Mabry *et al.*, 1971), and identified by comparison with the synthetic compound (Chopin and Bouillant, 1970a).

Swertiajaponin from *Swertia japonica* was unambiguously shown to be 7-methylhomoorientin (Komatsu et al., 1967b). Acid isomerization yielded isoswertiajaponin (7-methylorientin), not yet found in nature. Scoparin (8-C-β-D-glucopyranosylchrysoeriol), first isolated from *Sarothamnus scoparius* (Leguminosae) by Stenhouse (1851) was studied by Perkin (1899), Mascré and Paris (1937a,b,c), Paris and Stambouli (1961), Hörhammer *et al.* (1962, 1965) and shown to be 3'-methylorientin (Rosprim, 1966). Isoscoparin was prepared by acid isomerization of scoparin and identified with synthetic 6-*C*-β-D-glucopyranosylchrysoeriol (Chopin *et al.*, 1968a). It was later isolated from *Potamogeton natans* (Potamogetonaceae) and unambiguously characterized for the first time by Boutard *et al.* (1972).

6-*C*-β-D-glucopyranosyldiosmetin and 8-*C*-β-D-glucopyranosyl-diosmetin were isolated from *Citrus limon* (Rutaceae) by Gentili and Horowitz (1968), their structures determined from NMR and other spectral data, and confirmed by synthesis of the first compound (Chopin *et al.*, 1968b). 6-*C*-β-D-glucopyranosyl-7,3',4'-trimethyl-luteolin (7,3'4'-trimethylhomoorientin) was identified by Wagner *et al.* (1972a) as the 'aglycone' of linosids A and B from *Linum maritimum* (Linaceae).

12.2.1.2 Mono-*C*-glycosylflavonols

Keyakinin and keyakinol were isolated from *Zelkova serrata* (Ulmaceae) by Funaoka and Tanaka (1957a,b) and considered to be 6-*C*-pentosylrhamnocitrin and dihydrorhamnocitrin respectively. Keyakinin was reinvestigated by Hillis and Horn (1966) and the 6-*C*-β-D-glucopyranosylrhamnocitrin structure proposed on the basis of NMR spectra. Small amounts of another compound, keyakinin B were evidenced by paper chromatography, UV and R_f data suggesting a 6-*C*-glycosylrhamnetin structure.

12.2.1.3 Mono-*C*-glycosylflavanones

Hemiphloin (6-*C*-β-D-glucopyranosylnaringenin) and isohemiphloin (8-*C*-β-D-glucopyranosylnaringenin) were found by Hillis and Carle (1963) in *Eucalyptus hemiphloia* (Myrtaceae). Hemiphloin was shown to be dihydroisovitexin by iodine oxidation, and this structure was confirmed by NMR (Hillis and Horn, 1965) and synthesis (Chopin and Durix, 1966). Isomerization of these flavanones takes place in alkaline as well as in acid medium.

12.2.1.4 Mono-*C*-glycosylisoflavones

Puerarin (8-*C*-β-D-glucopyranosyldaidzein), isolated from *Pueraria thunbergiana* (Leguminosae) by Shibata *et al.* (1959), presents the same oxygenation pattern as bayin. Its structure was fully established by Murakami *et al.* (1960) and its β-configuration by NMR (Hillis and Horn, 1965).

12.2.1.5 Mono-*C*-glycosyldihydrochalcones

Aspalathin (3'-*C*-β-D-glucopyranosyl-3,4,2',4',6'-pentahydroxydihydrochalcone), isolated from *Aspalathus acuminatus* by Koeppen *et al.* (1962), was first considered to be a *C*-glycosyleriodictyol. Its structure was later revised by Koeppen and Roux (1965b). Konnanin and nothofagin, found in *Nothofagus fusca* (Fagaceae) by Hillis and Inoue (1967), are supposed to be *C*-glycosyl derivatives of 3,4,2',4',6'-pentahydroxy and 4,2',4',6'-tetrahydroxydihydrochalcones respectively.

12.2.2 Di-*C*-glycosylflavonoids

All known di-*C*-glycosylflavonoids have a phloroglucinol A ring. Only one di-*C*-glycosylisoflavone has been described, all other members of this group being di-*C*-glycosylflavones.

A 6,8-di-*C*-glycosylflavone structure was postulated by Seikel (1963) in order to explain the chromatographic behaviour of two groups of compound occurring in *Vitex lucens* extracts, the lucenins derived from luteolin and the vicenins derived from apigenin. Lucenins and vicenins migrated more rapidly than mono-*C*-glycosyl derivatives of luteolin and apigenin in strongly aqueous solvents and gave no sugar when heated with acid. Lucenin-1 and lucenin-3, vicenin-1 and vicenin-3 isomerized under these conditions, whereas lucenin-2 and vicenin-2 did not. It was therefore suggested that the latter two should be 6,8-di-*C*-glycosylflavones in which the glycosyl residues are identical, and the former 6,8-di-*C*-glycosylflavones with different glycosyl residues. However, each of the odd-numbered lucenins and vicenins yielded a mixture of four interconvertible products by acid treatment, instead of the two expected from a simple Wessely-Moser isomerization, but this could be explained by a reversible change at the sugar level (Seikel *et al.*, 1966).

In fact, this explanation has not yet received any experimental proof, owing to the lack of material. Lucenin-1 only could be isolated in sufficient amounts for NMR spectroscopy, and its 6,8-di-*C*-glycosyl-luteolin structure was thus established (Seikel and Mabry, 1965), but the nature of the glycosyl residues remain unknown. The difficulties encountered in the structural elucidation of mono-C-glycosylflavonoids are much greater in the case of di-C-glycosylflavonoids and, although a number of the latter have been found in nature, very few of them can be submitted to the chemical oxidations needed for sugar identification.

Ferric chloride oxidation of violanthin, isolated from *Viola tricolor* (Violaceae) by Hörhammer *et al.* (1965), yielded glucose, arabinose and rhamnose. From these and other chemical and physical data, Wagner *et al.* (1972g) proposed for violanthin a 6-C-β-D-glucopyranosyl 8-C-α-L-rhamnopyranosylapigenin structure, which was confirmed by chromatographic identification of isoviolanthin with synthetic 6-C-α-L-rhamnopyranosylvitexin (Biol and Chopin, 1972).

Paniculatin, an isoflavone from *Dalbergia paniculata* (Leguminosae), gave glucose and arabinose by ferric chloride oxidation and ozonolysis and was shown to be 6,8-di-C-β-D-glucopyranosylgenistein (Narayanan and Seshadri, 1971). 6,8-Di-C-β-D-glucopyranosylacacetin was isolated from *Trigonella corniculata* (Leguminosae) by Seshadri *et al.* (1972) and its structure established by similar methods. It has not yet been compared with the already known synthetic compound (Chopin and Bouillant, 1970b), but melting points are in agreement.

The other compounds listed in Table 12.3 were identified by comparison with synthetic products. A C-glycosylapigenin isolated from lemon peel extracts by Chopin *et al.* (1964), chromatographically identical with vicenin-2 (Seikel *et al.*, 1966), was shown by NMR to be a 6,8-di-C-glycosylapigenin and identified (IR) with synthetic 6,8-di-C-β-D-glucopyranosylapigenin (Chopin *et al.*, 1969). Non-crystalline vicenin-1 and vicenin-3 from *Vitex lucens* have been found to be chromatographically indistinguishable from synthetic crystalline 6-C-β-D-xylopyranosyl-8-C-β-D-glucopyranosylapigenin and its isomer respectively. By acid isomerization, the synthetic compounds gave the same chromatographic patterns as the natural ones (Bouillant and Chopin, 1971). Crystalline lucenin-1 from *Vitex lucens* has been identified (IR) with synthetic 6-C-β-D-xylopyranosyl-8-C-β-D-glucopyranosyl-luteolin, the isomer of which was chromatographically identical with lucenin-3. As before, the natural and synthetic compounds showed the same behaviour in acid isomerization (Bouillant and Chopin, 1972). By analogy with vicenin-2, lucenin-2 is probably 6,8-di-C-β-D-glucopyranosyl-luteolin.

Beside vicenins and lucenins, several other di-C-glycosyl derivatives

Table 12.3   *Natural di-C-glycosylflavonoids of known structure*

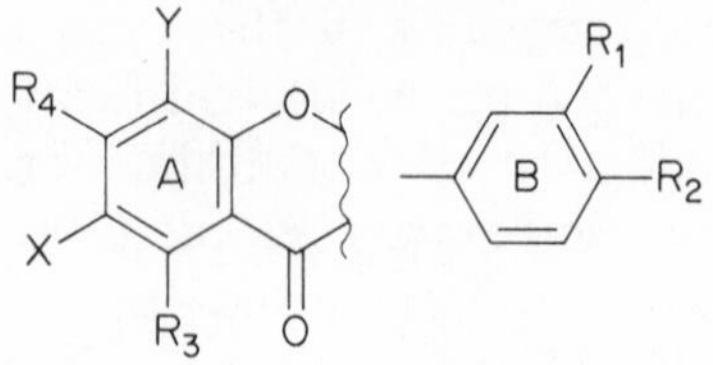

| | **Flavones** | | | | | |
	R_1	R_2	R_3	R_4	X	Y
Vicenin-1	H	OH	OH	OH	Xyl	Glc
Vicenin-2	H	OH	OH	OH	Glc	Glc
Vicenin-3	H	OH	OH	OH	Glc	Xyl
Violanthin	H	OH	OH	OH	Glc	Rha
Schaftoside	H	OH	OH	OH	Glc	Ara
Lucenin-1	OH	OH	OH	OH	Xyl	Glc
Lucenin-3	OH	OH	OH	OH	Glc	Xyl
	Isoflavones					
Paniculatin	H	OH	OH	OH	Glc	Glc

Glc = β-D-glucopyranosyl; Xyl = β-D-xylopyranosyl; Rha = α-L-rhamnopyranosyl.
Ara = α-L-arabinopyranosyl.

of apigenin and luteolin of unknown structure have been reported
(see Table 12.4). Schaftoside, isolated from *Silene schafta* (Caryophyllaceae) by Plouvier (1967), is a 6,8-di-C-glycosylapigenin with
hexosyl and pentosyl side chains (Chopin *et al.*, unpublished studies).
A di-C-rhamnosylapigenin from *Senecio* (Compositae) (Glennie *et al.*,
1971) has been compared with the synthetic product and found to
be different (Chopin, unpublished results). A di-C-glycosyl-
chrysoeriol structure has been proposed for compounds present in
Mnium affine (Melchert and Alston, 1965) and *Psoralea* (Ockendon
et al., 1966). The di-C-glycosyldiosmetin isolated from lemon peel
extracts by Chopin *et al.* (1964) and Gentili and Horowitz (1968) is
probably a 6,8-di-C-β-D-glucopyranosyl derivative, like its apigenin
companion.

12.2.3 O-Glycosides and acylated derivatives

C-Glycosylflavones and C-glycosylisoflavones are often found in
nature as O-glycosides and several examples of acylated derivatives
have been reported. Two types of O-glycosides have been described:
x'-O-glycosides in which the sugar is linked to a phenolic hydroxyl

Table 12.4   *Naturally occurring C-glycosylflavonoids*

Compounds	Natural sources	References
7,4'-Dihydroxyflavone derivatives		
Bayin: 5-deoxy vitexin = 8-*C*-β-D-gluco- pyranosyl-7,4'-dihydroxyflavone m.p. 220°; [α]$_D$ −1° (EtOH) hexaacetate: 128−9°; −74°5 (EtOH) di-*O*-methyl ether: 253°; −11°5 (MeOH)	*Castanospermum australe* (Leguminosae)	Simes, 1949; Eade *et al.*, 1962; Eade *et al.*, 1965 (NMR); Eade *et al.*, 1966; Hillis and Horn, 1965, (NMR); Eade and McDonald, 1973 (synthesis)
5,7,4'-Trihydroxyflavone (apigenin) derivatives		
Isovitexin (Saponaretin): 6-*C*-β-D-gluco- pyranosylapigenin m.p. 246−7°; [α]$_D$ + 16° (EtOH) heptaacetate: 248° hexaacetate: 169−71° tri-*O*-methyl ether: 308−12°	*Vitex lucens* (Verbenaceae) many other sources	Perkin, 1898; Barger, 1906; Nakaoki, 1944; Seikel and Geissman, 1957; Briggs and Cambie, 1958; Seikel *et al.*, 1959; Cambie, 1959; Bate-Smith and Swain, 1960; Horowitz and Gentili, 1964; Hillis and Horn, 1965 (NMR); Chopin *et al.*, 1966 (synthesis); Rosprim, 1966; Prox, 1968 (MS); Mabry *et al.*, 1970 (NMR)
Vitexin: 8-*C*-β-D-glucopyranosylapigenin m.p. 264−5°; [α]$_D$ −14° (pyridine) heptaacetate: 257−8°; −73° (acetone) hexaacetate: 157−9° pentaacetate: 146−7°; −4°5 (acetone) hexa-*O*-methyl ether: 205°; + 13°5 (MeOH) penta-*O*-methyl ether: 220° tri-*O*-methyl ether: 288° tetra-*O*-acetyl tri-*O*-methyl ether: 202°; −10° (acetone)	*Vitex lucens* (Verbenaceae) many other sources	Perkin, 1898, 1900; Barger, 1906; Nakaoki, 1944; Evans *et al.*, 1957; Briggs and Cambie, 1958; Hörhammer *et al.*, 1958, 1959b; Rao and Venkateswarlu, 1962; Dean, 1963; Horowitz and Gentili, 1964; Rosprim, 1966; Prox, 1968 (MS); Mabry *et al.*, 1970 (NMR)
6-*C*-β-D-xylopyranosylapigenin m.p. 229−31°	*Phlox drummondii* (Polemoniaceae) (from *O*-rhamnoside)	Mabry *et al.*, 1971; Chopin and Bouillant, 1970a (synthesis)

Continued

Table 12.4—*Continued*

Compounds	Natural sources	References
Other mono-C-glycosylapigenins		
Avrosid: syn 6-C-β-D-glucopyranosylapigenin	⎫	Borodin et al., 1970; Litvinenko and Borodin, 1970
Isoavrosid: anti 6-C-β-D-glucopyranosylapigenin	⎬ Gratiola officinalis (Scrophulariaceae)	
Neoavrosid: syn 6-C-α-D-glucopyranosylapigenin		
Isoneoavrosid: anti 6-C-α-D-glucopyranosyl- apigenin	⎭	
Neovitexin: syn 8-C-α-D-glucopyranosylapigenin	⎫ Gypsophila pauli (Caryophyllaceae)	Litvinenko and Borodin, 1970
Isoneovitexin: anti 8-C-α-D-glucopyranosyl- apigenin	⎭	
6,8-di-C-β-D-glucopyranosylapigenin m.p. 233–6°; [α]D +35° (H₂O)	Citrus limon (Rutaceae)	Chopin et al., 1964; Chopin and Bouillant, 1967 (synthesis); Gentili and Horowitz, 1968; Chopin et al., 1969 (synthesis)
	Spergularia (Arenaria) rubra (Caryophylla- ceae)	Zoll, 1972
	Viola tricolor (Violaceae)	Wagner et al., 1972b, g
Vicenin-2 (same as above ?)	Eupatorium serotinum (Compositae)	
	Vitex lucens (Verbenaceae)	Bouillant and Chopin, 1971; Markham, 1973;
	Triticum aestivum (Gramineae)	Seikel et al., 1966; Julian et al., 1971;
	Tragopogon spp. (Compositae)	Kroschewsky et al., 1969; Dubois and
	Linum usitatissimum (Linaceae)	Mabry, 1971; Wagner et al., 1970
	Croton zambezicus (Euphorbiaceae)	
	Sophora spp. (Leguminosae)	
Vicenin-1 = (?) 6-C-β-D-xylopyranosyl- 8-C-β-D-glucopyranosylapigenin m.p. (synthetic) 234–6°	Vitex lucens (Verbenaceae) Tragopogon spp. (Compositae) Linum usitatissimum (Linaceae) Nitella hookeri (Chlorophyta-F. Characeae)	Bouillant and Chopin, 1971 (synthesis); Seikel et al., 1966; Kroschewsky et al., 1969; Dubois and Mabry, 1971; Markham and Porter, 1969
Vicenin-3 = (?) 6-C-β-D-glucopyranosyl- 8-C-β-D-xylopyranosylapigenin	Vitex lucens (Verbenaceae)	Seikel et al., 1966; Bouillant and Chopin, 1971 (synthesis)
Violanthin: 6-C-β-D-glucopyranosyl- 8-C-α-L-rhamnopyranosylapigenin m.p. 228–30°; [α]D −31° (pyridine) decaacetate: 241°; +21° (CHCl₃) tri-O-methyl ether: 277–80°	Viola tricolor (Violaceae)	Hörhammer et al., 1965; Rosprim, 1966; Düll, 1970; Wagner et al., 1972g; Biol and Chopin, 1972; Biol, 1973

Compound	Source (Family)	References
Schaftoside: 6,8-(hexosyl, pentosyl)apigenin m.p.: 224–6°; [α]$_D$ + 71° (pyridine); + 110° (H$_2$O)	*Silene schafta* (Caryophyllaceae) *Catananche cerulea* (Compositae)	Plouvier, 1967; Chopin and Bouillant (unpublished); Biol, 1973; Proliac, 1973
Isoschaftoside: 6,8-(pentosyl, hexosyl)apigenin	*Catananche cerulea* (Compositae)	Proliac, 1973

Other di-C-glycosylapigenins

Compound	Source (Family)	References
Vicenin-4	*Vitex lucens* (Verbenaceae)	Seikel et al., 1966.
'Di-C-rhamnosylapigenin' (?)	*Senecio* spp. (Compositae)	Glennie et al., 1971; Biol, 1973
Vicenins H$_1$ and H$_2$	*Hymenophytum flabellatum* (Bryophyta)	Markham et al., 1969
Di-C-glycosylapigenins IIIa and IIIb	*Thea sinensis* (Theaceae)	Sakamoto, 1967, 1970
Flavonoid B m.p. 222–5° dec. acetates: 245° and 185–90° tri-O-methyl ether: 213°	*Triticum* (Wheat germ)(Gramineae)	King, 1962; Harborne and Hall, 1964; Biol, 1973
Vicenin	*Mnium affine* (Bryophyta)	Melchert and Alston, 1965
Vicenins	*Psoralea* spp. (Leguminosae)	Ockendon et al., 1966
Vicenins	*Lemna* and *Wolffia* spp. (Lemnaceae)	McClure and Alston, 1966
Vicenins	*Iris chrysophylla* (Iridaceae)	Carter, 1968
Vicenin	*Porella platyphylla* (Bryophyta)	Nilsson, 1973
Flavonoid glycoside A 6-β, 8-α and 6-α, 8-β-di-C-glucopyranosyl apigenins (?)	*Silene conica* (Caryophyllaceae)	Litvinenko and Darmograi, 1968
Di-C-glycosylapigenins	*Artemisia transiliensis* (Compositae)	Chumbalov and Fadeeva, 1970
Vicenin	*Viola tricolor* (Violaceae)	Düll, 1970; Wagner et al., 1972g
	Cathestecum prostratum (Gramineae)	Crawford and Lankow, 1972

5-Hydroxy-7-methoxy-4'-hydroxyflavone (genkwanin, 7-O-methylapigenin) derivatives

Compound	Source (Family)	References
Swertisin: 6-C-β-D-glucopyranosyl genkwanin m.p. 243° dec.; [α]$_D$ −10° (pyridine) hexaacetate: 155–8° pentaacetate: 186–7°	*Swertia japonica* (Gentianaceae)	Nakaoki, 1927; Asahina et al., 1942; Komatsu and Tomimori, 1966; Komatsu et al., 1967a
	Swertia spp. (Gentianaceae)	Komatsu et al., 1968a, b; Takeda et al., 1966
	Commelina spp. (Commelinaceae)	Aritomi and Kawasaki, 1968
	Desmodium caudatum (Leguminosae)	Kroschewsky et al., 1969
	Tragopogon spp. (Compositae)	Wagner et al., 1972c, f
	Gaillardia pulchella (Compositae)	
	Helenium brevifolium (Compositae)	

Continued

653

Table 12.4—Continued

Compounds	Natural sources	References
Isoswertisin: 8-C-β-D-glucopyranosyl genkwanin m.p. 295° dec. hexaacetate: 134–6°	*Iris tingitana* hybrid (Prof. Blaauw) (Iridaceae) *Iris nertshinskia* (Iridaceae) *Centaurea cyanus* (Compositae) *Centaurea cyanus* (Compositae) (from blue pigment)	Asen *et al.*, 1970 Kawase, 1968 Asen and Jurd, 1967; Asen, 1967 Asen and Jurd, 1967; Komatsu, *et al.*, 1967a Asen, 1967
5,7-Dihydroxy 4'-methoxyflavone (acacetin) derivatives		
Cytisoside: 8-C-β-D-glucopyranosylacacetin m.p. 235–6°; [α]D –22° (pyridine) hexaacetate (two forms): 170–4° and 217–22°; –93° (pyridine)	*Cytisus laburnum* (Leguminosae)	Paris, 1957; Paris and Stambouli, 1961; Stambouli and Paris, 1961; Hörhammer *et al.*, 1962; Sachs, 1963; Chopin *et al.*, 1965, 1966; Chopin and Bouillant (unpublished) (MS, NMR, . . .)
	Trema aspera (Ulmaceae) (from *O*-glucoside acetates)	Oelrichs *et al.*, 1968
8-*C-glycosylacacetin*	*Reboulia hemispherica* (Bryophyta) (from *O*-glycoside)	Markham *et al.*, 1972
6,8-di-C-β-D-glucopyranosylacacetin m.p. 245–6° X'' monoacetate: 196–8°	*Trigonella corniculata* (Leguminosae)	Chopin and Bouillant, 1970b (synthesis); Seshadri *et al.*, 1972
5-Hydroxy-7,4'-dimethoxyflavone(7,4'-di-O-methylapigenin) derivatives		
Embigenin: 6-C-β-D-glucopyranosyl 7,4'-di-*O*-methylapigenin m.p. 236–8° undecaacetate: 189–90° acetate: 217°	*Iris tectorum* (Iridaceae) (from di-*O*-rhamnoside) *Iris germanica* (Iridaceae) (from di-*O*-rhamnoside)	Hirose *et al.*, 1962 Kawase and Yagishita, 1968
5,7,3',4'-Tetrahydroxyflavone (luteolin) derivatives		
Isoorientin (homoorientin): 6-C-β-D-gluco-pyranosyl-luteolin	*Polygonum orientale* (Polygonaceae) many other sources	Hörhammer *et al.* 1958, 1959b; Seikel and Bushnell, 1959; Koeppen, 1962, 1964; Koeppen

654

Compound and physical data	Source	Reference
m.p. 238–9°; $[\alpha]_D$ +31° (pyridine) acetate: 140–4°; +27° (acetone) 7,3',4'-tri-O-methyl ether: 272°; −10° (pyridine) tetra-O-methyl ether: 267–8°; +38° (aq. 50% acetone) tetra-O-acetyl tetra-O-methyl ether: 128–30°; 0° (acetone)		et al., 1962; Hänsel and Rimpler, 1963; Koeppen and Roux, 1965a; Chopin et al., 1966 (synthesis)
6-C-β-D-xylopyranosyl-luteolin m.p. 241–2°	*Phlox drummondii* (Polemoniaceae) (from O-rhamnoside;	Mabry et al., 1971; Chopin and Bouillant, 1970a (synthesis)
6-C-glycosyl-luteolin	*Briza media* (Gramineae)	Williams and Murray, 1972
Orientin: 8-C-β-D-glucopyranosyl-luteolin m.p. 265–7° dec.; $[\alpha]_D$ +18° (pyridine) octaacetate: 196–7°; −54° (acetone) 7,3',4'-tri-O-methyl ether: 271–3°; −38° (pyridine) tetra-O-methyl ether: 273–6° dec.; +51° (aq. 50% acetone) tetra-O-acetyl tetra-O-methyl ether: 253–4° dec.; 0° (acetone)	*Polygonum orientale* (Polygonaceae) many other sources	Hörhammer et al., 1958, 1959a, b; Seikel and Bushnell, 1959; Koeppen et al., 1962; Koeppen, 1962, 1964; Sachs, 1963; Koeppen and Roux, 1965a, b
Epiorientin: 8-C-α-D-glucopyranosyl-luteolin (?) m.p. > 300°; $[\alpha]_D$ 0° (pyridine) octaacetate: 191–2°; −63° (pyridine) tetra-O-methyl ether: 259°; −40° (pyridine)	*Parkinsonia aculeata* (Leguminosae)	Bhatia et al., 1966b
8-C-glycosyl-luteolin	*Briza media* (Gramineae) *Vitex lucens* (Verbenaceae)	Williams and Murray, 1972; Seikel, 1963; Seikel and Mabry, 1965; Seikel et al., 1966
Lucenin-2: 6,8-di-C-glucosyl-luteolin (?)	*Tragopogon* spp. (Compositae) *Linum usitatissimum* (Linaceae) *Spergularia (Arenaria) rubra* (Caryophyllaceae)	Kroschewsky et al., 1969; Dubois and Mabry, 1971; Zoll, 1972;
Lucenin-1: 6-C-β-D-xylopyranosyl 8-C-β-D-glucopyranosyl-luteolin m.p. (synthetic) 205° dec.	*Sophora* spp. (Leguminosae) *Vitex lucens* (Verbenaceae)	Markham, 1973; Seikel, 1963; Seikel and Mabry, 1965; Seikel et al., 1966; Bouillant and Chopin, 1972 (synthesis)

Continued

655

Table 12.4—*Continued*

Compounds	Natural sources	References
	Triticum dicoccum (Gramineae)	Harborne and Hall, 1964
	Triticum aestivum (Gramineae)	Julian *et al.*, 1971
	Tragopogon spp. (Compositae)	Kroschewsky *et al.* 1969
	Linum usitatissimum (Linaceae)	Dubois and Mabry, 1971
Lucenin-3 = (?) 6-*C*-β-D-glucopyranosyl-8-*C*-β-D-xylopyranosyl luteolin	*Vitex lucens* (Verbenaceae)	Seikel, 1963; Seikel and Mabry, 1965; Seikel *et al.*, 1966; Bouillant and Chopin, 1972
	Triticum dicoccum (Gramineae)	Harborne and Hall, 1964
	Triticum aestivum (Gramineae)	Julian *et al.*, 1971
Other di-C-glycosyl-luteolins		
Lucenin-4	*Vitex lucens* (Verbenaceae)	Seikel and Mabry, 1965; Seikel *et al.*, 1966
Compounds 1, 1a and 3	*Nitella hookeri* (Chlorophyta-F. Characeae)	Markham and Porter, 1969
Lucenins	*Psoralea* spp. (Leguminosae)	Ockendon *et al.*, 1966
Lucenins	*Lemna* spp. (Lemnaceae)	McClure and Alston, 1966
Glycosid EG₂	*Viola tricolor* (Violaceae)	Rosprim, 1966
Lucenins	*Iris chrysophylla* (Iridaceae)	Carter, 1968
5-Methoxy-7,3',4'-trihydroxyflavone (5-O-methyl-luteolin) derivatives		
Parkinsonin A (5-O-methylorientin): 5-*O*-methyl 8-*C*-β-D-glucopyranosyl-luteolin m.p. 224°; [α]D −32° (pyridine) hexaacetate: 144–5°; +26° (pyridine) tri-*O*-methyl ether: 273–4°; +51° (pyridine)	*Parkinsonia aculeata* (Leguminosae)	Bhatia *et al.*, 1966b
5,3',4'-Trihydroxy-7-methoxyflavone (7-O-methyl-luteolin) derivatives		
Swertiajaponin (7-O-methylisoorientin): 7-*O*-methyl 6-*C*-β-D-glucopyranosyl-luteolin m.p. 265° dec.; [α]D −3° (pyridine) heptaacetate: 156–60°	*Swertia japonica* (Gentianaceae)	Komatsu *et al.*, 1967b, 1968a; Komatsu and Tomimori, 1966
	Tragopogon spp. (Compositae)	Kroschewsky *et al.*, 1969
	Iris tingitana hybrid (Prof. Blaauw) (Iridaceae)	Asen *et al.*, 1970
	Iris nertshinskia (Iridaceae)	Kawase, 1968
	Cephalaria leucantha (Dipsacaceae)	Plouvier, 1967; Bouillant *et al.*, 1972

5,7-Dimethoxy 3',4'-dihydroxyflavone (5,7-di-O-methyl-luteolin) derivatives

Parkinsonin B (5-7-di-O-methylepiorientin);   Parkinsonia aculeata (Leguminosae)   Bhatia et al., 1966b
5,7-di-O-methyl 8-C-α-D-glucopyranosyl-luteolin (?)
m.p. > 300°; [α]D 0° (pyridine)
di-O-methyl ether: 258–9°; –40° (pyridine)

5,7,4'-Trihydroxy 3'-methoxyflavone (chrysoeriol) derivatives

Isoscoparin: 6-C-β-D-glucopyranosylchrysoeriol   Hordeum vulgare (Gramineae)   Seikel et al., 1962; Chopin et al., 1968a
m.p. 206–10° dec.; [α]D +20° (MeOH)   (from 7-O-glucoside)   (synthesis)
                                                  Lemna minor (Lemnaceae)   Wallace et al., 1969
                                                  Potamogeton natans (Potamogetonaceae) Boutard et al., 1972

Scoparin: 8-C-β-D-glucopyranosylchrysoeriol   Sarothamnus (Cytisus) scoparius   Perkin, 1899; Stenhouse, 1851; Mascré and
m.p. 250°; [α]D 0° (MeOH)   (Leguminosae)   Paris, 1937a, b, c; Paris and Stambouli, 1961;
heptaacetate: 240–1°   Stambouli and Paris, 1961; Seikel et al., 1962;
                                                                                        Hörhammer et al., 1962
                                                  Mnium affine (Bryophyta)   Melchert and Alston, 1965
                                                  Psoralea spp. (Leguminosae)   Ockendon et al., 1966

6,8-di-C-glycosylchrysoeriol   Mnium affine (Bryophyta)   Melchert and Alston, 1965
                                                  Psoralea spp. (Leguminosae)   Ockendon et al., 1966

5,7,3'-Trihydroxy 4'-methoxyflavone (diosmetin) derivatives

6-C-β-D-glucopyranosyldiosmetin   Citrus limon (Rutaceae)   Gentili and Horowitz, 1968; Chopin et al., 1968b
m.p. 243–5°   (synthesis)
8-C-β-D-glucopyranosyldiosmetin   Citrus limon and C. sinensis (Rutaceae) Gentili and Horowitz, 1968
m.p. 267–8°
6,8-di-C-glycosyldiosmetin   Citrus limon (Rutaceae)   Chopin et al., 1964; Gentili et Horowitz, 1968

5-Hydroxy 7,3',4'-trimethoxyflavone (7,3',4'-tri-O-methyl-luteolin) derivatives

6-C-β-D-glucopyranosyl 7,3',4'-tri-O-   Linum maritimum (Linaceae)   Volk and Sinn, 1968; Koeppen and Roux,1965a
methylluteolin   (from O-rhamnoside: Linosid B)   (synthesis); Wagner et al., 1972a
m.p. 253–5°

Continued

657

Table 12.4—Continued

Compounds	Natural sources	References
5,4'-Dihydroxy 7-methoxyflavonol (7-O-methylkaempferol) derivatives		
Keyakinin: 6-*C*-β-D-glucopyranosyl-7-*O*-methylkaempferol m.p. 233–4° heptaacetate: 235–6° hexaacetate: 132–4°	*Zelkova serrata* (Ulmaceae)	Funaoka, 1956, 1957; Funaoka and Tanaka, 1957a, b; Hillis and Horn, 1966
5,3',4'-Trihydroxy 7-methoxyflavonol (7-O-methylquercetin) derivatives		
Keyakinin B: 6-*C*-glycosylquercetin	*Zelkova serrata* (Ulmaceae)	Hillis and Horn, 1966
5,7,4'-Trihydroxyflavanone (naringenin) derivatives		
Hemiphloin: 6-*C*-β-D-glucopyranosylnaringenin (two forms) m.p. I 210–11°, II 228–9° [α]_D I +41° (aq. 50% acetone) II +39° (aq. 50% acetone) heptaacetate: 146–7°; –36° (acetone) hexaacetate: 144–5°	*Eucalyptus hemiphloia* (Myrtaceae)	Hillis and Carle, 1963; Hillis and Horn, 1965; Chopin and Durix, 1966 (synthesis)
Isohemiphloin: 8-*C*-β-D-glucopyranosyl-naringenin [α]_D – 12° (aq. 50% acetone) heptaacetate: m.p. 223–4°; –85° (acetone)	*Eucalyptus hemiphloia* (Myrtaceae)	Hillis and Carle, 1963; Hillis and Horn, 1965
7,4'-Dihydroxyisoflavone (daidzein) derivatives		
Puerarin: 8-*C*-β-D-glucopyranosyldaidzein m.p. 187° dec.; hexaacetate: 128–9° di-*O*-methyl ether: 156°	*Pueraria thunbergiana* (Leguminosae)	Murakami *et al.*, 1960; Hillis and Horn, 1965; Shibata *et al.*, 1959
4',6''-diacetate: 250–1°; [α]_D +18° (MeOH)	*Pueraria tuberosa* (Leguminosae) (and 4',6''-diacetate)	Bhutani *et al.*, 1969

5,7,4'-Trihydroxyisoflavone (genistein) derivatives

Paniculatin: 6,8-di-C-β-D-glucopyranosyl genistein
m.p. 225–7°
acetate: 161–2°
tri-O-methyl ether; 209–10°

Dalbergia paniculata (Leguminosae) — Narayanan and Seshadri, 1971

5,7,3',4'-Tetrahydroxyisoflavone (orobol) derivatives

Dalpanitin: 8-C-β-D-glucopyranosyl-3'-O-methyl orobol
m.p. 213–4° (dec.); $[\alpha]_D + 35°$ (EtOH)
heptaacetate: 130–1°

Dalbergia paniculata (Leguminosae) — Adinarayana and Rao, 1972

2',4',6',4-Tetrahydroxydihydrochalcone derivatives

Nothofagin: C-glycosyl-2',4',6',4-tetrahydroxy-dihydrochalcone

Nothofagus fusca (Fagaceae) — Hillis and Inoue, 1967

2',4',6',3,4,-Pentahydroxydihydrochalcone derivatives

Aspalathin: 3'-C-β-D-glucopyranosyl 2',4',6',3,4,-pentahydroxydihydrochalcone
nonaacetate m.p.: 152–4°
α-bromo derivative: 185°

Aspalathus linearis (acuminatus) (Leguminosae) — Dahlgren, 1963; Koeppen and Roux, 1965b; Koeppen *et al.*, 1962

Konnanin: C-glycosyl 2',4',6',3,4-pentahydroxy dihydrochalcone

Nothofagus fusca (Fagaceae) — Hillis and Inoue, 1967

659

Table 12.5   *Naturally occurring C-glycosylflavonoid O-glycosides and acylated derivatives*

Compounds	Natural sources	References
Isovitexin (6-C-β-D-glucopyranosylapigenin) derivatives		
Saponarin: 7-O-glucosylisovitexin m.p. 235–6°; [α]D−79° (pyridine) decaacetate: 195–6° ˢubacetate: 274.5−275.5°	Saponaria officinalis (Caryophyllaceae)	Barger, 1906; Rosprim, 1966; Mabry et al., 1970 (NMR)
	Hordeum vulgare (Gramineae)	Seikel and Geissman, 1957; Seikel et al., 1962
	Hibiscus syriacus (Malvaceae)	Nakaoki, 1944
	Spirodela oligorrhiza (Lemnaceae)	Jurd et al., 1957
	Triticum monococcum (Gramineae)	Harborne and Hall, 1964
	Dipsacus and Scabiosa spp. (Dipsacaceae)	Plouvier, 1966
	Thea sinensis (Theaceae)	Sakamoto, 1969
	Arum italicum and Biarum tenuifolium (Araceae)	Phouphas, 1956
	Passiflora spp. (Passifloraceae)	Schilcher, 1968; Glotzbach and Rimpler, 1968
	Porella (Madotheca) platyphylla (Bryophyta)	Molisch, 1911; Nilsson, 1969; Tjukavkina et al., 1970
Isosaponarin: 4'-O-glucosylisovitexin m.p.; [α]D (from Vaccaria segetalis) 236–7°; −92° (MeOH) decaacetate: 190–2°; −51° (CHCl₃)	Spirodela oligorrhiza (Lemnaceae)	Jurd et al., 1957
	Spirodela intermedia (Lemnaceae)	McClure and Alston, 1966
	Lemna minor (Lemnaceae)	McClure and Alston, 1966
	Psoralea spp. (Leguminosae)	Ockendon et al.,1966
	Vaccaria segetalis (Caryophyllaceae)	Litvinenko et al., 1967
	Gypsophila spp. (Caryophyllaceae)	Darmograi et al., 1968; Litvinenko et al., 1969
	Briza media (Gramineae)	Williams and Murray, 1972
X''-O-arabinosylisovitexin	Coronilla varia (Leguminosae)	Sherwood et al., 1973
X''-O-xylosylisovitexin	Avena spp. (Gramineae)	Harborne and Hall, 1964
X''-O-rhamnosylisovitexin	Tragopogon dubius (Compositae)	Kroschewsky et al., 1969
5-O-rhamnosylisovitexin m.p. 221–3°	Sophora microphylla (Leguminosae)	Markham and Godley, 1972
	Thalictrum rugosum (Ranunculaceae)	Kuczynski and Chyczewski, 1971
O-Gentiobiosylisovitexin	Gentiana lutea (Gentianaceae)	Bellman and Jacot-Guillarmod, 1973
Other O-glycosylisovitexins	Lemna minor (Lemnaceae)	McClure and Alston, 1966
	Cucumis melo (Cucurbitaceae)	Monties, 1971

6-C-β-D-Xylopyranosyl apigenin derivatives

X''-O-rhamnosyl 6-C-β-D-xylopyranosyl-
apigenin   *Phlox drummondii* (Polemoniaceae)   Mabry et al., 1971
m.p. 210°

Vitexin (8-C-β-D-glucopyranosylapigenin) derivatives

7-O-glucosylvitexin   *Trigonella foenumgraecum* (Leguminosae)   Adamska and Lutomski, 1971
m.p. 240–3°
4'-O-glucosylvitexin   *Briza media* (Gramineae)   Williams and Murray, 1972
4'-O-rhamnosylvitexin   *Crataegus oxyacantha* (Rosaceae)   Fiedler, 1955; Geissman and Fiedler, 1956;
m.p. 215°; [α]D −66° (EtOH)   Lewak, 1966
−90° (pyridine)   *Crataegus* spp. (Rosaceae)   Mrugasiewicz, 1963

acetate: 139–45°
benzoate: 161°
4''-O-rhamnosylvitexin   *Crataegus curvisepala* (Rosaceae)   Batyuk and Chernobrovaya, 1966; Batyuk et al.,
m.p. 214–5°; [α]D −35° (MeOH)   (from O-acetyl derivative: cratenacin)   1969
acetate: 148–51°
Acetyl 4'-O-rhamnosylvitexin   *Crataegus monogyna* (Rosaceae)   Fizel, 1965
m.p. 248–50°; [α]D −61° (EtOH)
−74° (pyridine)

acetate: 139–45°
Cratenacin: 6''-O-acetyl 4''-O-   *Crataegus curvisepala* (Rosaceae)   Batyuk and Chernobrovaya, 1966; Batyuk et al.,
rhamnosylvitexin   1966, 1969
m.p. 254°; [α]D −29° (MeOH)
acetate: 148–51°
4'-O-rhamnoglucosylvitexin   *Crataegus oxyacantha* (Rosaceae)   Lewak, 1966
X''-O-rhamnosylvitexin   *Avena* spp. (Gramineae)   Harborne and Hall, 1964
                                              Sophora microphylla (Leguminosae)   Markham and Godley, 1972
                                              Vitex lucens (Verbenaceae)   Seikel et al., 1959; Seikel et al., 1966; Horowitz
                                                                                         and Gentili 1966
2''-O-β-D-xylopyranosylvitexin   *Citrus sinensis* (Rutaceae)   Horowitz and Gentili, 1964, 1966; Gentili and
m.p. 210°   Horowitz, 1968
nonaacetate: 221°   *Larix laricina* (Pinaceae)   Niemann and Bekooy, 1971
                                              (from 7-O-glucoside)

                                              Continued

661

Table 12.5–*Continued*

Compounds	Natural sources	References
Katchimoside: 6''-O-xylosylvitexin m.p. 200°	*Crotalaria* spp. (Leguminosae) *Gypsophila paniculata* (Caryophyllaceae)	Subramanian and Nagarajan, 1967a,b,c; Darmograi *et al.*, 1968
O-xylosylvitexin	*Gypsophyila* spp. *Beta vulgaris* (Chenopodiaceae) (from O-glucoside)	Litvinenko *et al.*, 1969b Gardner *et al.*, 1967
2''-O-p-hydroxybenzoylvitexin m.p. 204°	*Vitex lucens* (Verbenaceae)	Horowitz and Gentili, 1966
7-O-glucosylxylosylvitexin	*Larix laricina* (Pinaceae)	Niemann and Bekooy, 1971
Other O-glycosyl C-glycosylapigenins		
7-O-glycosyl-C-glycosylapigenin	*Cassia occidentalis* (Leguminosae)	Anton and Duquenois, 1968
Alliaroside m.p. 260°–2°; [α]D –66° (pyridine) acetate: 194–6°	*Alliaria officinalis* (Cruciferae) *Bryonia dioica* (Cucurbitaceae)	Paris and Delaveau, 1962 Paris *et al.*, 1966; Leiba *et al.*, 1968
Petrocomoside: 7-O-glucosylisoavroside	*Petrocoma (Silene) hoefftiana* (Caryophyllaceae)	Darmograi and Litvinenko, 1971
O-glycosyl 6,8-di-C-glycosylapigenins		
O-glycosylvicenin	*Psoralea* spp. (Leguminosae) *Mnium affine* (Bryophyta)	Ockendon *et al.*, 1966 Melchert and Alston, 1965
7-O-rhamnosylglucosylvicenin	*Linum usitatissimum* (Linaceae)	Ibrahim, 1969; Ibrahim and Shaw, 1970
5-O-glucosyl 7-O-rhamnosyl-vicenin	*Linum usitatissimum*	Ibrahim, 1969. Ibrahim and Shaw, 1970
Flavonoid A: sinapic ester of flavonoid B (6,8-di-C-glycosylapigenin) m.p. 199° dec. acetate: 200–225°	*Triticum* (Gramineae)	King, 1962
Swertisin (6-C-β-D-glucopyranosylgenkwanin) (7-O-methylisovitexin) derivatives		
Flavocommelinin: 4'-O-glucosylswertisin *Flavoayamenin*: X''-O-glucosylswertisin m.p. 258–9°	*Commelina* spp. (Commelinaceae) *Iris nertshinskia* (Iridaceae)	Takeda *et al.*, 1966; Komatsu *et al.*, 1968b Kawase, 1968

662

Compound	Source	Reference
X''-O-xylosylswertisin	Iris tingitana hybrid (Prof. Blaauw) (Iridaceae)	Asen, 1970
Isoswertisin (8-C-β-D-glucopyranosylgenkwanin) derivatives		
4'-O-glucosylisoswertisin	Triticum aestivum (Gramineae)	Julian et al., 1971
Cytisoside (8-C-β-D-glucopyranosylacacetin) derivatives		
7-O-glucosylcytisoside m.p. 223°–6°	Trema aspera (Ulmaceae) (from O-acetyl derivatives)	Oelrichs et al., 1968
Tremasperin: mixed mono and di-O-acetyl 7-O-glucosylcytisoside m.p. nonaacetate 155–8°	Trema aspera	Oelrichs et al., 1968
O-glycosyl 8-C-glycosylacacetin		
Embigenin (6-C-β-D-glucopyranosyl 7,4'-di-O-methylapigenin) derivatives	Reboulia hemispherica (Bryophyta)	Markham et al., 1972
Embinin: x''-O-rhamnosylembigenin m.p. 181–2°	Iris tectorum (Iridaceae)	Hirose et al., 1962
	Iris germanica (Iridaceae)	Kawase and Yagishita, 1968
Isoorientin (6-C-β-D-glucopyranosyl-luteolin) derivatives		
Lutonarin: 7-O-glucosylisoorientin m.p. 235° dec.	Hordeum vulgare (Gramineae)	Seikel and Bushnell, 1959; Seikel et al., 1962
	Psoralea spp. (Leguminosae)	Ockendon et al., 1966
	Lemna minor (Lemnaceae)	McClure and Alston, 1966
	Spirodela spp. (Lemnaceae)	McClure and Alston, 1966
	Triticum aestivum (Gramineae)	Julian et al., 1971
	Coronilla varia (Leguminosae)	Sherwood et al., 1973
4'-O-glucosylisoorientin	Briza media (Gramineae)	Williams and Murray, 1972
	Coronilla varia (Leguminosae)	Sherwood et al., 1973
Wyomin: 7-O-rutinosylisoorientin acetate: m.p. 144–5°	Triticum aestivum (Gramineae)	Julian et al., 1971
2''-O-Rhamnosylisoorientin	Triticum dicoccum (Gramineae)	Harborne and Hall, 1964
	Coronilla varia (Leguminosae)	Sherwood et al., 1973
Other O-glycosylisoorientins		
Isolutonarin = (?) 4'-O-glycosyl-isoorientin	Psoralea spp. (Leguminosae)	Ockendon et al., 1966

Continued

Table 12.5—*Continued*

Compounds	Natural sources	References
7-*O*-glycosylisoorientins E and F	*Linum usitatissimum* (Linaceae)	Ibrahim, 1969; Ibrahim and Shaw, 1970
X''-*O*-glucosylisoorientin	*Stipa lemonii* (Gramineae)	Saleh *et al.*, 1971
	Gentiana lutea (Gentianaceae)	Bellmann and Jacot-Guillarmod, 1973
Melosides L, a and l	*Cucumis melo* (Cucurbitaceae)	Monties, 1971
6-*C*-β-D-Xylopyranosyl-luteolin derivatives		
X''-*O*-rhamnosyl 6-*C*-β-D-xylopyranosyl-luteolin	*Phlox drummondii* (Polemoniaceae)	Mabry *et al.*, 1971
m.p. 198°		
Orientin (8-C-β-D-glucopyranosyl-luteolin) derivatives		
4'-*O*-glucosylorientin	*Briza media* (Gramineae)	Williams and Murray, 1972
2'' (?)-*O*-xylosylorientin	*Vitex lucens* (Verbenaceae)	Seikel *et al.*, 1966
X''-*O*-xylosylorientin	*Ranunculus lingua* (Ranunculaceae)	Drozd *et al.*, 1969a
Adonivernith: x''-*O*-xylosylorientin	*Adonis vernalis* (Ranunculaceae)	Hörhammer *et al.*, 1960; Rosprim, 1966;
m.p. 205°		Chernobai *et al.*, 1968
acetate: 226–8°	*Gypsophila* spp. (Caryophyllaceae)	Darmograi *et al.*, 1968; Litvinenko *et al.*, 1969b
7-*O*-rhamnosylorientin	*Linum usitatissimum* (Linaceae)	Ibrahim and Shaw, 1970
Lucenin (6,8-di-C-glycosyl-luteolin) derivatives		
7-*O*-rhamnosyl-lucenin	*Linum usitatissimum* (Linaceae)	Ibrahim, 1969; Ibrahim and Shaw, 1970
X''-*O*-glycosyl-lucenin	*Psoralea* spp. (Leguminosae)	Ockendon *et al.*, 1966
Isoscoparin (6-C-β-D-glucopyranosylchrysoeriol) derivatives		
Lutonarin 3'-methyl ether =	*Hordeum vulgare* (Gramineae)	Seikel *et al.*, 1962
7-*O*-glucosylisoscoparin		
m.p. 190–5° dec.		
Scoparin (8-C-β-D-glucopyranosylchrysoeriol) derivatives		
O-glycosylscoparin	*Mnium affine* (Bryophyta)	Melchert and Alston, 1965

664

6,8-Di-C-glycosylchrysoeriol derivatives

O-glycosyl-6,8-di-C-glycosyl-
chrysoeriol

7,3',4'-Tri-O-methylisoorientin derivatives

Linoside B: 6″ or 2″ O-α-L-rhamnosyl-
7,3',4'-trimethyliso-orientin
m.p. 179–82°

Linoside A: 6″ or 2″ O-α-L-
(O-acetylrhamnosyl)-7,3',4'-trimethyl-
iso-orientin
m.p. 174–5°
acetate: 135–9°

Puerarin (8-C-β-D-glucopyranosyldaidzein) derivatives

Diacetylpuerarin: 4',6″-di-O-
acetyl 8-C-β-D-glucopyranosyl-
daidzein
m.p. 250–1°; $[\alpha]_D$ +18° (MeOH)
O-methyl ether: 142–4°
O-xylosylpuerarin

Mnium affine (Bryophyta)	Melchert and Alston, 1965
Psoralea spp. (Leguminosae)	Ockendon et al., 1966
Linum maritimum (Linaceae)	Volk and Sinn, 1968; Wagner et al., 1972a
Linum maritimum (Linaceae)	Volk and Sinn, 1968; Wagner et al., 1972a
Pueraria tuberosa (Leguminosae)	Bhutani et al., 1969
Pueraria thunbergiana (Leguminosae)	Shibata et al., 1959; Murakami et al., 1960

group of the flavonoid, i.e. C-glycosyl flavonoid O-glycosides; and x″-O-glycosides in which the sugar is linked to an hydroxyl group of the glycosyl side chain, i.e. C-diglycosylflavonoids. The former, unlike the latter, exhibit a change in UV spectral shifts after hydrolysis, and this property has been widely used for elucidating the linkage position on the flavonoid. Isomerization of the C-glycosylflavonoid occurs during acid hydrolysis and kinetic studies or enzymatic hydrolysis are needed to identify the true 'aglycone'.

Saponarin (saponaretin 7-O-glucoside) thus gives glucose, saponaretin and vitexin on acid hydrolysis. However saponaretin, appearing first in the mixture, is the true 'aglycone' (Seikel and Geissman, 1957). This was confirmed by methylation of saponarin (precluding isomerization), followed by hydrolysis and remethylation, yielding saponaretin trimethyl ether (Rosprim, 1966). The fact that the 7-position is occupied by a glucose residue is clear from the UV data.

7-O-glucosides of vitexin, cytisoside (as acetyl derivatives = tremasperin), isoavroside (petrocomoside), homoorientin (lutonarin), isoscoparin (lutonarin 3′-methyl ether), 7-O-rhamnosides of orientin and lucenin, 7-O-rhamnoglucosides of vicenin and homoorientin (wyomin), as well as 4′-O-glucosides of saponaretin (isosaponarin), vitexin, swertisin (flavocommelinin), isoswertisin, homoorientin and orientin, 4′-rhamnoside and 4′-rhamnoglucoside of vitexin, have been reported. Only two 5-O-glycosides are known: a 5-O-rhamnoside of isovitexin and a 5-O-glucoside-7-O-rhamnoside of vicenin.

C-diglycosyl derivatives have the same UV spectrum as the corresponding C-monoglycosides. Methylation of the phenolic hydroxyl groups followed by acid hydrolysis yields a product, the UV spectrum of which is not shifted in the presence of alkaline reagents. On this basis, several C-diglycosylflavones were found in Gramineae by Harborne and Hall (1964) and a number of such compounds are now known. However, identification of the glycosidic linkage raises difficult problems which have not been resolved, except in a few cases. It was possible to identify the 2″-O-β-D-xylopyranosylvitexin from *Vitex lucens* and *Citrus sinensis* (Horowitz and Gentili, 1966), owing to the characteristic shielding of the 2″-O-acetyl or 2″-O-methyl groups in NMR spectra of the corresponding derivatives of 8-C-β-D-glucopyranosylflavones. Likewise, the apparent absence of the 2″-O-acetyl resonance in the NMR spectrum of adonivernith acetate given by Rosprim (1966) implies that adonivernith isolated from *Adonis vernalis* by Hörhammer *et al.*(1960) is 2″-*O*-xylo-pyranosylorientin.

The structure of cratenacin, isolated from *Crataegus curvisepala,* has been established as *O*-acetyl 4″-*O*-rhamnosylvitexin by Batyuk and Chernobrovaya (1966). Periodate oxidation of deacetylcra-tenacin [found to be identical with Fiedler's (1965) vitexin 4′-*O*-rhamnoside!], followed by nitric acid oxidation yielded tartaric acid, from which the 4″-position of the glycosidic bond was deduced. x″-*O*-glucosides of swertisin (flavoayamenin), homoorientin and lucenin, x″-*O*-xylosides of saponaretin, vitexin, swertisin and orientin, x″-*O*-rhamnosides of vitexin, 6-xylosylapigenin, embigenin (embinin), 6-xylosylluteolin and 7,3′,4′-trimethylhomoorientin (linoside B), and x″-*O*-arabinoside of saponaretin have been reported, in which the position of the glycosidic bond is still unknown.

O-Acylated derivatives of mono- and di-*C*-glycosylflavonoids, *C*-glycosylflavonoid *O*-glycosides or *C*-diglycosylflavones have been reported. *O*-Acetyl groups were found in puerarin 4′,6″-diacetate, di-*C*-glucosylacacetin acetate, tremasperin and linoside A, an *O*-*p*-hydroxybenzoyl group in 2″-*O*-*p*-hydroxybenzoylvitexin, and an *O*-sinapyl group in flavonoid A (from wheat germ).

12.3 Synthesis of *C*-glycosylflavonoids

When the structures of vitexin and orientin were elucidated, it became evident that *C*-glycosylflavonoids could result from the attack of an activated sugar by the carbanionic form of a phenolate anion. The extensive studies of Seshadri and co-workers on the nuclear methyla-tion of polyhydroxyflavonoids had previously shown that *C*-methyla-tion requires free hydroxyl groups in the 5- and 7-positions and takes place exclusively in the 6-position, when methyl iodide is used as the alkylating agent, potassium hydroxide or sodium methoxide as the base, and water or alcohol as the solvent. *O*-Methylation also occurs and the reaction mixture contains *O*- and *C*-methyl derivatives. When 2,3,4,6-tetraacetyl-α-D-glucopyranosyl bromide and a trace of sodium iodide are used instead of methyl iodide, solvolysis is the main reaction; but small amounts of *C*-glucosyl derivatives are formed, beside O-glucosides (deacetylation taking place by saponifi-cation or transesterification). After acid hydrolysis of the latter, the 6-*C*-glucosyl-5,7-dihydroxyflavonoids can be isolated pure by chro-

matography, but the yields are less than 1% (Chopin, 1971). The
corresponding 8-*C*-glucosyl derivatives resulting from acid isomerization can be characterized by their chromatographic properties and
UV spectral data. The 6-*C*-β-D-glucopyranosyl structure of the
synthetic compounds so obtained was proved either by chromatographic and spectral (UV, IR, MS) identification with known natural
products (or their 6-isomers), or by NMR and mass spectra for
unknown compounds.

   C-Glycosylation of some 5,7-dihydroxyflavones was then attempted with other acetohalogenoses in order to ascertain the
influence of different sugars on the chromatographic and spectral
properties of *C*-glycosylflavones. Reaction of 2,3,4-triacetyl-α-D-
xylopyranosyl bromide with chrysin, acacetin, apigenin and luteolin
led to the corresponding 6-*C*-β-D-xylopyranosylflavones (from NMR
and MS data), and the latter two could be identified with the
hydrolysis products of two natural compounds isolated from *Phlox
drummondii* flowers (Mabry *et al.*, 1971). Likewise, reaction of
2,3,4,6-tetraacetyl-α-D-galactopyranosyl bromide with apigenin and
acacetin, of 2,3,4-triacetyl-β-L-arabinopyranosyl bromide and
2,3,4-triacetyl-α-L-rhamnopyranosyl bromide with apigenin led to
products considered to be the corresponding 6-*C*-β-D-galacto-
pyranosyl, 6-*C*-α-L-arabinopyranosyl and 6-*C*-α-L-rhamnopyranosyl-
flavones respectively.

   Meanwhile, it was shown that 6,8-di-*C*-glycosyl-5,7-dihydroxy-
flavones could be obtained in the same way from natural 8-*C*-gluco-
syl-5,7-dihydroxyflavones. Thus, reaction of 2,3,4,6-tetraacetyl-α-D-
glucopyranosyl bromide with cytisoside and vitexin led to the
6,8-di-*C*-β-D-glucopyranosyl derivatives of acacetin and apigenin
respectively, the latter being identified with a natural product from
Citrus limon; the former being different from a substance in
Trigonella corniculata. 6-*C*-β-D-Xylopyranosyl-8-*C*-β-D-glucopyrano-
sylapigenin and 6-*C*-β-D-glucopyranosyl-8-*C*-β-D-xylopyranosyl-
apigenin, chromatographically identical with natural vicenins 1 and
3 respectively, were obtained from vitexin and 2,3,4-triacetyl
α-D-xylopyranosyl bromide. By contrast, 6-*C*-a-L-arabinopyranosyl
8-*C*-β-D-glucopyranosylapigenin (from vitexin and 2,3,4-triacetyl β-L-
arabinopyranosyl bromide) has different chromatographic properties.
Lucenins 1 and 3 were similarly undistinguishable from 6-*C*-β-D-
xylopyranosyl-8-*C*-β-D-glucopyranosyl-luteolin   and   6-*C*-β-D-gluco-

pyranosyl-8-C-β-D-xylopyranosyl-luteolin, prepared by C-xylosylation of an orientin/isoorientin mixture.

Reaction of 2,3,4-triacetyl α-L-rhamnopyranosyl bromide with vitexin led to 6-C-α-L-rhamnopyranosyl-8-C-β-D-glucopyranosylapigenin, chromatographically identical with isoviolanthin, the isomerization product of natural violanthin, confirming the structure proposed by Wagner et al. (1972g) for the latter. 6-C-β-D-galactopyranosyl-8-C-β-D-glucopyranosyl apigenin was also prepared from vitexin and 2,3,4,6-tetraacetyl α-D-galactopyranosyl bromide. While 6,8-disubstitution was observed in the C-xylosylation of acacetin, apigenin or luteolin, and in the C-rhamnosylation of apigenin, the 6,8-di-C-xylosylflavones and the 6,8-di-C-rhamnosylapigenin formed can only be characterized by chromatographic and UV spectral data.

All synthetic C-glycosylflavonoids obtained as crystalline and chromatographically pure compounds are reported in Table 12.6. The sugar residues of these synthetic compounds are illustrated in Fig. 12.1.

A total synthesis of 7,4'-di-O-methylbayin was recently performed by Eade and McDonald (1973) (Fig. 12.2). 2,6-dimethoxyphenylmagnesium bromide (7) was condensed with α-acetochloroglucose (8) to give a mixture of α- and β-D-glucopyranosyl-2,6-dimethoxybenzene (9). The α-isomer rearranged by heating with acid to the β-isomer. The tetraacetate of the latter (10) was acetylated by acetyl chloride and aluminium chloride in ether at room temperature and the resulting o-hydroxyacetophenone (11) was condensed with anisaldehyde to give the chalcone (12), the selenium dioxide oxidation of which led to 7,4'-di-O-methylbayin (13).

12.4 Identification of C-glycosylflavonoids

12.4.1 Chemical methods

Acid treatment of a C-glycosylflavonoid does not yield any flavonoid aglycone under the conditions (Harborne, 1965) in which all types of flavonoid O-glycosides are hydrolyzed, and this reaction is always used for their characterization. Several C-glycosylflavonoid 'aglycones' may appear as a result of isomerization. Alkali fusion of the 'aglycone' leads to a polyphenol and a substituted benzoic acid respectively derived from the A- and B-rings of the flavonoid.

Table 12.6 *Synthetic C-glycosylflavonoids*

6-C-glycosylflavones

6-C-β-D-glucopyranosylchrysin	Chopin *et al.*, 1970a
6-C-β-D-glucopyranosylapigenin (isovitexin)	Chopin *et al.*, 1966
6-C-β-D-glucopyranosylacacetin (isocytisoside)	Chopin *et al.*, 1965, 1966
6-C-β-D-glucopyranosyl-luteolin (iso-orientin)	Chopin *et al.*, 1966
6-C-β-D-glucopyranosylchrysoeriol (isoscoparin)	Chopin *et al.*, 1968a
6-C-β-D-glucopyranosyldiosmetin *(Citrus)*	Chopin *et al.*, 1968b
6-C-β-D-galactopyranosylapigenin	} Chopin *et al.*, 1971
6-C-β-D-galactopyranosylacacetin	
6-C-β-D-xylopyranosylchrysin	Chopin *et al.*, 1970a
6-C-β-D-xylopyranosylapigenin *(Phlox)*	
6-C-β-D-xylopyranosylacacetin	} Chopin and Bouillant, 1970a
6-C-β-D-xylopyranosyl-luteolin *(Phlox)*	
6-C-α-L-arabinopyranosylapigenin	Chopin *et al.*, 1972
6-C-α-L-rhamnopyranosylapigenin	Chopin and Biol, 1972

6-C-glycosylflavanones

6-C-β-D-glucopyranosylnaringenin (hemiphloin)	Chopin and Durix, 1966

6-C-glycosylflavonols

6-C-β-D-glucopyranosylgalangin	
6-C-β-D-glucopyranosylkaempferol	} Chopin *et al.*, 1970b
6-C-β-D-glucopyranosylquercetin	

6,8-di-C-glycosylflavones

6,8-di-C-β-D-glucopyranosylapigenin (vicenin-2)	Chopin *et al.*, 1969
6,8-di-C-β-D-glucopyranosylacacetin	Chopin and Bouillant, 1970b
6-C-β-D-xylopyranosyl 8-C-β-D-glucopyranosylapigenin (vicenin-1)	
6-C-β-D-glucopyranosyl 8-C-β-D-xylopyranosylapigenin (vicenin-3)	} Bouillant and Chopin, 1971
6-C-β-D-xylopyranosyl 8-C-β-D-glucopyranosylacacetin	Chopin and Bouillant, 1970b
6-C-α-L-rhamnopyranosyl 8-C-β-D-glucopyranosylapigenin (isoviolanthin)	Biol and Chopin, 1972
6-C-β-D-galactopyranosyl 8-C-β-D-glucopyranosylapigenin	} Biol, 1973
6-C-α-L-arabinopyranosyl 8-C-β-D-glucopyranosylapigenin	
6-C-β-D-xylopyranosyl 8-C-β-D-glucopyranosyl-luteolin (lucenin-1)	
6-C-β-D-glucopyranosyl 8-C-β-D-xylopyranosyl-luteolin (lucenin-3)	} Bouillant and Chopin, 1972

Heating with hydriodic acid in phenol yields the parent poly-hydroxyflavonoid by demethylation and cleavage of the glycosyl bond.

Selective methylation of phenolic hydroxyl groups is best obtained with excess diazomethane. Alkaline degradation of the products with barium hydroxide leads to the *C*-glycosyl-*o*-hydroxy-acetophenone (or benzoic acid) derived from the A-ring without cleavage of the glycosyl bond. Permethylation requires the use of the

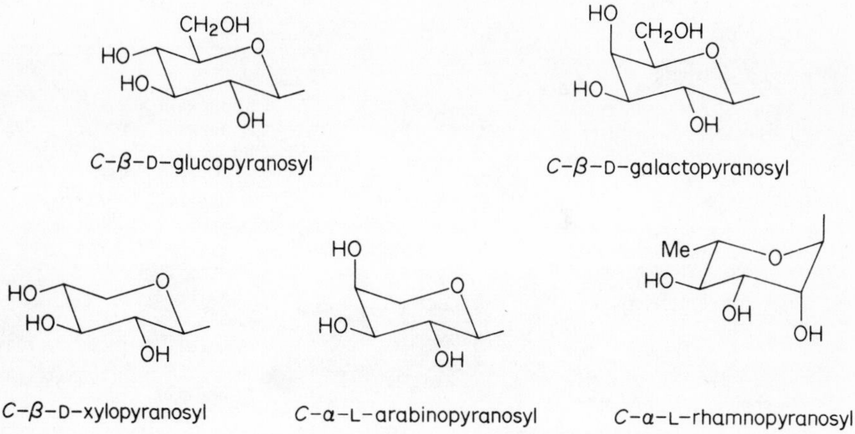

Figure 12.1 *C*-glycosyl residue in synthetic *C*-glycosylapigenins.

classical methods of carbohydrate chemistry. Peracetylation is much easier though some difficulties are encountered with the chelated and sterically hindered 5-hydroxyl group in 6-*C*-glycosyl and 6,8-di-*C*-glycosyl compounds. Sulphuric acid as catalyst has been recommended by Wagner (1966).

Periodate oxidation has been widely used in the study of the glycosyl side chain. Normal results are obtained only after methylation of the phenolic hydroxyl groups. However, direct periodate oxidation, followed by borohydride reduction and hydrolysis (Viscontini *et al.*, 1955) normally yields glycerol from *C*-glucopyranosylflavonoids and propylene glycol from *C*-rhamnopyranosylflavonoids.

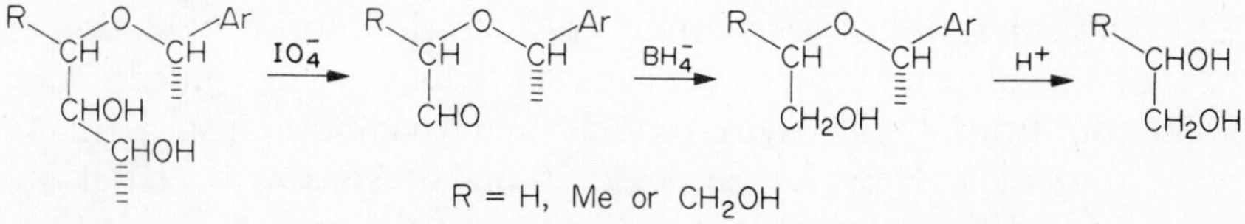

R = H, Me or CH_2OH

Periodic acid oxidation in strongly acidic medium with titration of the formic acid produced has been systematically used by Hörhammer, Wagner and co-workers.

The position of the glycosyl residue on the flavonoid can be ascertained by lead tetraacetate or periodic oxidation of the side chain, leading to *C*-formylflavonoids or derivatives identified by comparison with synthetic products, eventually after oxidation or reduction of the formyl group.

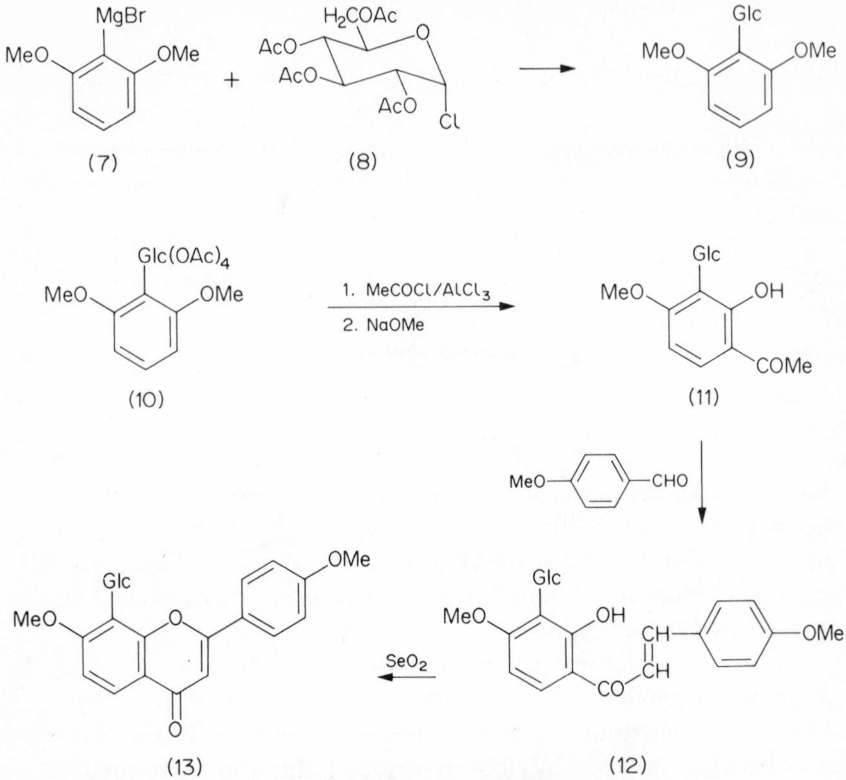

Figure 12.2 Synthesis of 7,4'-di-O-methylbayin.

The main problem remains: the identification of the sugar residue. The only solution has been found in the ferric chloride oxidation or ozonolysis of the C-glycosylflavonoid which yields glucose and arabinose from C-glucosylflavonoids and rhamnose from C-rhamnosylflavonoids. However, at least 20 mg of starting material are needed for chromatographic identification of the sugar formed.

12.4.2 Physical methods

12.4.2.1 UV spectrometry

The UV spectra of flavonoids are not appreciably affected by the introduction of one or two C-glycosyl substituents on the A-ring, and the characteristic shifts observed in the presence of aluminium chloride, aluminium chloride-hydrochloric acid, sodium acetate, sodium acetate-boric acid, sodium ethylate or sodium hydroxide, remain the same (cf. Chapter 2). UV spectral methods are always used in the identification of the flavonoid moiety and they are often the only criterion when insufficient amounts of material are available for chemical studies.

Distinction between C-diglycosylflavonoids and C-glycosylflavonoid O-glycosides can be made on a microscale (Harborne and Hall, 1964). After complete methylation of phenolic hydroxyl groups (ensured by the absence of shift in alkaline medium) and acid hydrolysis, the absence of an alkaline shift will indicate that the starting material was a C-diglycosylflavonoid.

12.4.2.2 IR spectrometry

IR spectrometry can be employed with very small amounts of material and, when combined with chromatographic comparison, is the best method of identification with a reference compound. However, IR spectra in the solid state (in KBr pellets) can vary according to the crystalline state of the specimens, which should be the same for the unknown and the reference product.

12.4.2.3 NMR spectrometry

NMR spectrometry has played a fundamental role in the structural analysis of C-glycosylflavonoids. Owing to the different chemical shifts of the H-3, H-6 and H-8 protons in 5,7-dihydroxyflavones, position 8 was assigned to the glycosyl residue in vitexin and orientin, and position 6 in isovitexin and homoorientin, leading to the assumption that their acid isomerization results from a Wessely-Moser rearrangement (Horowitz and Gentili, 1964; Koeppen, 1964).

This hypothesis was shown to be valid by acid isomerization of 6-methyl and 8-methylapigenins under the same experimental conditions (Chopin and Chadenson, 1966), and was conclusively proved by the identification of 3-methyl-6-hydroxy-2,4-dimethoxy-acetophenone as a reduction product of the C-formylacetophenone obtained from swertisin(7-methylisovitexin) (Komatsu *et al.,* 1967a).

At the present time, therefore, attribution of the 6- and 8-position to the glycosyl residues in di-C-glycosylflavonoids is based on the absence of H-6 and H-8 protons in their NMR spectra, as first shown for lucenin-l (Seikel and Mabry, 1965).

The unusually high field position (δ 1.70–1.83) of one O-acetyl

signal is a characteristic feature in the NMR spectra of acetylated
C-β-D-glucopyranosyl flavonoids. This signal is assigned to the equa-
torial 2″-acetyl, since it can be expected to be influenced by the
magnetic anisotropy of the aromatic A-ring (Hillis and Horn, 1965).
An NMR method for differentiating 6- and 8-C-glucosyl isomers lies
in the chemical shift difference exhibited by 2″- and 6″-acetyl
groups. In 8-C-glucosylflavones, both the 2″-acetyl (δ 1.70–1.73)
and the 6″-acetyl (δ 1.90–1.95) occur at a higher field than they do
in 6-C-glucosylflavones: δ 1.77–1.83 for the 2″-acetyl and
δ 1.98–2.04 for the 6″-acetyl (Gentili and Horowitz, 1968). The
greater shielding of the 2″- and 6″-acetyl groups in 8-C-glucosyl-
flavones has been explained on the assumption that the 2″-acetyl is
shielded not only by the A-ring of the flavone, but to some extent by
the B-ring, and that the 6″-acetyl lies in the diamagnetic region of
the B-ring. There is a smaller difference in the signal position of the
3″-acetyl group, which is likely to be shielded by the B-ring in
8-C-glucosylflavones.

Restricted rotation about the C-8–C-1″ bond as a result of steric
hindrance of the 2″- and 6″-acetyl groups with both the 7-acetyl
group and the phenyl B-ring was deduced by Eade *et al.* (1965) from
the temperature-dependent variations in the NMR spectra of bayin
and vitexin acetates. The existence of two conformers is evidenced
by the splitting or broadening of the signals from some of the
aromatic and acetyl protons. The effect disappears when the phenyl
group is attached to the C-ring at the 3-position as in the 8-C-gluco-
sylisoflavone puerarin, when the 7-hydroxyl group is free as in
vitexin tetraacetate, and when the sugar residue is attached at the
6-position as in saponaretin hexaacetate.

The β-configuration of the glucopyranosyl radical in vitexin and
isovitexin (Horowitz and Gentili, 1964), orientin and homoorientin
(Koeppen, 1964) was deduced from the large coupling constant
(10 Hz) of the benzylic H-1″ proton, and has since been found in all
other C-glucosylflavonoids.

12.4.2.4 Mass spectrometry

The usefulness of mass spectrometry in the characterization and
structure elucidation of C-glycosylflavonoids was shown by Prox
(1968), who interpreted the fragmentation patterns of 6- and
8-C-glucosylflavones related to apigenin and luteolin. Unlike flavone

O-glucosides which eliminate the sugar moiety and give only the mass spectra of the aglycones, C-glucosylflavones show initial fragmentation of the glucose moiety. The molecular ion M is absent or very weak and the higher mass peaks C, C', C'' result from successive losses of three molecules of water. The base peak is not the flavone radical ion C_1, but the characteristic benzyl ion A, accompanied by the radical ion A' (A + 1) (Fig. 12.3). The benzyl ion A gives rise in a retro-Diels-Alder process to the ion A_1 with the A-ring of the flavone and the radical ion A_3 with the B-ring, affording information about the substituents present on these rings. On the other hand, a distinction can be made between 6- and 8-C-glucosylflavones, due to the much greater intensity of the A' fragment in the 6-substituted flavones and of the B fragment in those that are 8-substituted.

The fragmentation patterns of 6-C-β-D-glucopyranosyl flavanones (hemiphloin) or flavonols containing a phloroglucinol-type A-ring remain the same as those of the corresponding 6-C-β-D-glucopyranosylflavones, but the A_3 fragments are characteristic of the flavonoid type. In the mass spectra of 6-C-pentopyranosylflavones, the C peak (M-H_2O) remains higher than C'(M-2 H_2O), but C''(M-3 H_2O) becomes negligible and a new peak M-47 appears. In the spectrum of one 6-C-rhamnopyranosylflavone, the C'-peak is higher than C; the M-47 peak is important and C'' very weak.

Only a few MS data are available in the field of di-C-glycosylflavonoids. The first published mass spectrum was that of violanthin, 6-C-glucosyl-8-C-rhamnosylapigenin, (Wagner *et al.*, 1972g), in which the highest mass peak was M-3 H_2O, followed by M-4 H_2O, M-5 H_2O and M-6 H_2O. The main peak m/e 295 has been interpreted as the bis-methylene flavone ion (see Fig. 12.3) from which the retro-Diels-Alder fragment m/e 177 is formed. A similar fragmentation pattern was found in the mass spectrum of 6-C-xylosyl-8-C-glucosylapigenin, the highest mass peak remaining M-3 H_2O (Bouillant and Chopin, 1971).

A complete spectrum was obtained with schaftoside, a C-pentosyl-C-hexosylapigenin, the highest mass peak being M-H_2O, followed by M-2,3,4,5 and 6 H_2O (Chopin, unpublished results). Many peaks can be interpreted on the basis of Prox's scheme (Biol, 1973). However, other di-C-glycosylflavones have not given satisfactory spectra under the same conditions. The use of acetyl or trimethylsilyl derivatives will probably afford a solution of this problem.

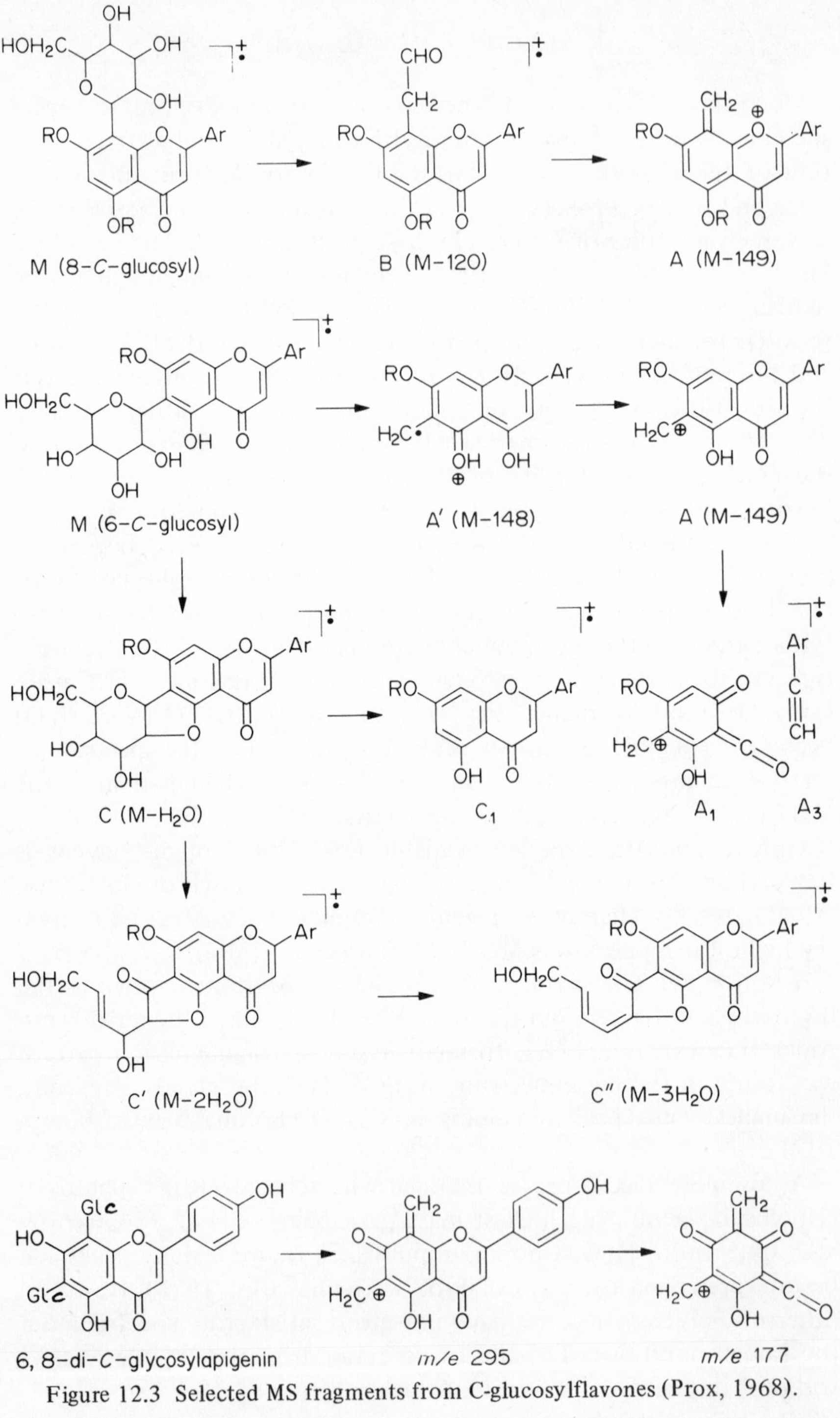

Figure 12.3 Selected MS fragments from C-glucosylflavones (Prox, 1968).

Mass spectrometry has been applied to the tetraacetates of tri-O-methylvitexin and isovitexin by Aritomi et al. (1970). Here again, a distinction can be made between the 6- and 8-C-glucosyl-flavones, the main peak being M-59 with the former and the molecular ion with the latter.

12.4.2.5 Optical methods

Generally, 6-C-β-D-glucopyranosylflavones and their acetates have a more positive optical rotation than their 8-isomers (Hillis and Horn, 1965). An interesting chiroptical differentiation of 6- and 8-C-glucosylflavones results from circular dichroism studies which have shown that a positive Cotton effect at 250–275 nm indicates that the glycosyl residue is linked to C-6 while a negative Cotton effect at 250–275 nm indicates it is linked to C-8 (Gaffield and Horowitz, 1972).

12.4.3 Comparison with known compounds

When sufficient amounts of substance are available, the classical methods of comparison can be used: m.p. and m.m.p., UV, IR, NMR and mass spectra, optical rotations of the compounds themselves or of their acetyl or methyl derivatives, along with chromatography. However, when only small amounts are available, paper or thin layer co-chromatography has been widely used alone for identification of C-glycosylflavonoids and it is therefore of importance to test the validity of these methods with synthetic compounds. Gas chromatography of trimethylsilyl ethers can be applied to C-glycosyl-flavonoids (Paris and Paris, 1966) but it seems to have met with limited success.

12.4.3.1 Comparison of mono-C-glycosylapigenins

Chromatography Fig. 12.5 shows the paper chromatographic behaviour of the known mono-C-glycopyranosylapigenins in BAW/15% AcOH, the most widely used two-dimensional system for the characterization of C-glycosylflavones. The following conclusions can be drawn: (1) whatever the sugar, 6-C-glycopyranosylapigenins migrate faster than their 8-isomers in both solvent systems, but R_f differences decrease markedly in the order hexosyl > pentosyl > rhamnosyl; (2) 6- and 8-C-rhamnosyl, 6-C-pentosyl-and 6-C-hexosylapigenins are well separated, but 6-C-glucosyl-and 6-C-xylosylapigenins cannot be easily distinguished from 6-C-galactosyl-

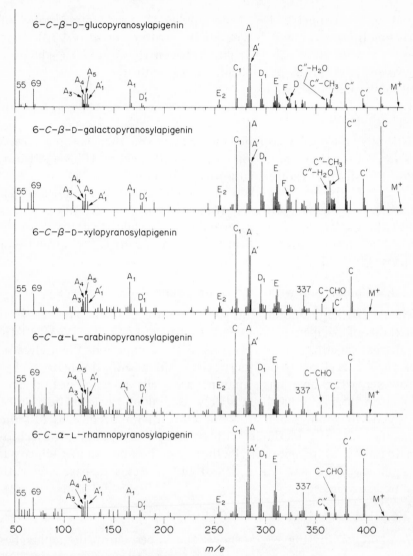

Figure 12.4 Mass spectra of 6-*C*-glycosylapigenins.

and 6-*C*-arabinosylapigenins respectively; (3) the situation is even
more unfavourable in the case of 8-*C*-hexosyl and 8-*C*-pentosyl-
apigenins, which are difficult to separate.

TLC on freshly activated silicagel G (Merck) in the solvent system
ethyl acetate-pyridine-water-methanol (80 : 20 : 10 : 5) (APWM)
allows a good separation of the pairs of 6-*C*-hexosyl- and 6-*C*-

pentosylapigenins (Table 12.7). Under the same conditions, 8-C-glycosylapigenins migrate faster than their 6-isomers; the pair of 8-C-hexosylapigenins is well resolved, but 8-C-pentosylapigenins are not clearly separated.

UV spectra As expected, no significant differences are observed in the UV spectrum and spectral shifts of apigenin and those of C-glycosylapigenins.

IR spectra Each of the five studied 6-C-glycosylapigenins exhibits a characteristic IR spectrum (in KBr pellets); the most pronounced difference in absorption bands are in the region 1000–1200 cm^{-1}.

Mass spectra (Fig. 12.4) The fragmentation patterns of 6-C-galacto-pyranosyl- and 6-C-arabinopyranosylapigenins cannot be distin-guished from those of 6-C-glucopyranosyl- and 6-C-xylopyranosyl-apigenins respectively. Characteristic features of the 6-C-pentosyl- and 6-C-rhamnosylapigenins are the presence of significant peaks at M-(H_2O + CHO) and m/e 337, and the absence or weakness of the peak M-3 H_2O.

The fragmentation pattern remains the same for all 6-C-glyco-pyranosylapigenins below m/e 320. However, the apigenin ion is

Table 12.7   *TLC of C-glycosylapigenins*

R_f value on activated silica gel G Merck in EtOAc-pyridine-H_2O-MeOH
(80 : 20 : 10 : 5)

Mono-C-glycosylapigenins

Compound	R_f	Compound	R_f
6-C-glucopyranosyl	0.59	8-C-glucopyranosyl	0.71
6-C-galactopyranosyl	0.37	8-C-galactopyranosyl	0.59
6-C-xylopyranosyl	0.72	8-C-xylopyranosyl	0.73
6-C-arabinopyranosyl	0.61	8-C-arabinopyranosyl	0.68
6-C-rhamnopyranosyl	0.78	8-C-rhamnopyranosyl	0.72

Di-C-glycosylapigenins

Compound	R_f
6,8-di-C-glucopyranosyl	0.13
6-C-galactopyranosyl 8-C-glucopyranosyl	0.11
6-C-xylopyranosyl 8-C-glucopyranosyl	0.26
6-C-arabinopyranosyl 8-C-glucopyranosyl	0.15
6-C-rhamnopyranosyl 8-C-glucopyranosyl	0.26

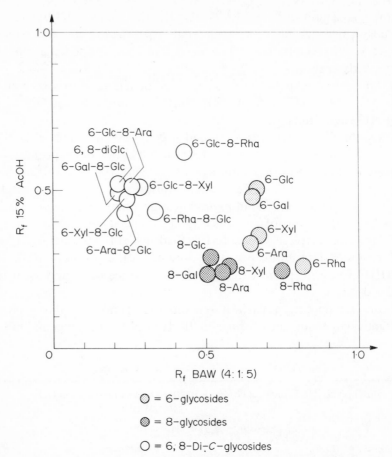

Figure 12.5. Two-dimensional PC of C-glycopyranosylapigenins (Whatman paper No. 1) (Glc = β-D-glucopyranosyl; Gal = β-D-galactopyranosyl; Xyl = β-D-xylopyranosyl; Ara = α-L-arabinopyranosyl; Rha = α-L-rhamnopyranosyl).

slightly more abundant than the benzylic ion for 6-C-pentosyl-apigenins, since the sugar-flavone bond is more unstable towards acids than in 6-C-hexosyl or rhamnosylapigenins.

Acid isomerization. Two dimensional chromatography in BAW/15% BAW/15% AcOH of the mixture obtained by refluxing a solution of 6-C-glycosylapigenin in 4N HCl–MeOH (1 : 1) shows that small amounts of apigenin are produced from the C-pentosyl derivatives exclusively. Moreover, although 6-C-glucosylapigenin only gives the two spots expected from Wessely-Moser isomerization, the other

6-*C*-glycosylapigenins produce more complex mixtures, with several weaker spots besides the starting product and its 8-isomer. For instance, 6-*C*-xylopyranosylapigenin gives two secondary spots with higher R_f in BAW than the 6- and 8-isomers.

12.4.3.2 Comparison of 6-*C*-glycosyl 8-*C*-glucosylapigenins

Chromatography. As shown in Fig. 12.5, 6-*C*-glycopyranosyl 8-*C*-glucopyranosylapigenins have much lower R_f values in BAW than the 6-*C*-glycosylapigenins, and the R_f increase in 15% AcOH brought by the 8-*C*-glucosyl residue is much larger for the 6-*C*-rhamnosyl and 6-*C*-pentosyl than for the 6-*C*-hexosylvitexins. Accordingly, 6-*C*-rhamnosylvitexin alone is well separated from the four others which are strongly overlapping, 6-*C*-hexosylvitexins migrating only slightly faster than 6-*C*-pentosylvitexins. On activated silica gel G (Merck) in APWM (80 : 20 : 10 : 5), migration of 6-*C*-glycosylvitexins on TLC is much lower than that of 6-*C*-glycosylapigenins. 6-*C*-xylosyl and 6-*C*-arabinosylvitexins are well separated, but 6-*C*-glucosyl and 6-*C*-galactosylvitexins are not. 6-*C*-Xylosyl and 6-*C*-rhamnosylvitexins migrate together.

UV spectra. The UV spectra and spectral shifts of 6-*C*-glycosyl-vitexins remain the same as those of apigenin.

IR spectra. 6-*C*-glycosylvitexins are not so well differentiated as 6-*C*-glycosylapigenins by their IR spectrum (in KBr pellets), except between 1000 and 1200 cm^{-1}.

Mass spectra. Only in some cases have been mass spectra obtainable from 6,8-di-*C*-glycosylflavonoids. 6-*C*-Rhamnosylvitexin exhibits the same peaks as violanthin up to m/e 427, but the higher mass peaks are not observed. Fragmentation patterns of 6-*C*-xylosylvitexin and violanthin are closely similar, the highest mass peak being M-3 H_2O, followed by M-4 H_2O, M-5 H_2O, M-6 H_2O, and the main peak the benzylic ion m/e 295, characteristic of 6,8-di-*C*-glycosylapigenins.

Acid isomerization. Wessely-Moser isomerization can be expected for unsymmetrical 6,8-di-*C*-glycosyl 5,7-dihydroxyflavones. Indeed the symmetrical 6-*C*-glucosylvitexin is not affected by heating with acids, whereas 6-*C*-rhamnosylvitexin gives rise to one new spot, migrating faster in 15% AcOH, and indistinguishable from natural violanthin. Several spots are formed from 6-*C*-xylosylvitexin, one of them showing the slightly higher R_f in 2% AcOH and BAW expected for the 6-*C*-glucosyl isomer, and indistinguishable from both natural vicenin-3 and the second compound isolated in the C-xylosylation of vitexin. Two other spots of lower R_f in 2% AcOH compare well with vicenins-4 and 5. Likewise, 6-C-arabinosylvitexin gives rise to a spot which could be considered as the 6-C-glucosyl isomer, and another with lower R_f in 15% AcOH. From 6-C-galactosylvitexin appears only spots of lower R_f in 15% AcOH, but it is likely that the 6-C-glucosyl isomer is indistinguishable from the starting material.

12.4.3.3 Comparison of C-glucosyl and C-xylosylflavones

Mono-C-glycosyl flavones. The synthesis of C-xylopyranosyl derivatives of chrysin, acacetin, apigenin and luteolin has allowed the study of the respective influences of the sugar and flavone moieties on chromatographic properties, by comparison with the corresponding C-glucopyranosylflavones. On paper chromatography, whatever the flavone, 6-C-xylosylflavones, like 6-C-glucosylflavones, migrate faster than the 8-isomers in 15% AcOH and BAW. Methoxylation or hydroxylation in the 2-phenyl ring of the flavone reduces the migration rate in 15% AcOH for both C-xylosyl and C-glucosyl-flavones. For a given flavone, R_f in 15% AcOH are in the order: 6-C-glucosyl > 6-C-xylosyl > 8-C-glucosyl ≥ 8-C-xylosyl, the difference being much smaller for the last two, and the ratio R_f 6-C-xylosyl/R_f 6-C-glucosyl decreases from 0.86 for chrysin to 0.78 for acacetin, 0.70 for apigenin and 0.65 for luteolin derivatives. Thus flavone hydroxylation has a greater effect on C-xylosyl- than on C-glucosylflavones, as expected from the number of hydroxyl groups in the sugar moiety. However, in BAW, 6- and 8-C-xylosyl-flavones have nearly the same migration rate as the corresponding 6- and 8-C-glucosylflavones respectively.

On silica gel TLC, in APWM (80 : 12 : 10 : 5), for a given flavone R_f values are in the reverse order: 8-C-xylosyl > 8-C-glucosyl ≥ 6-C-xylosyl > 6-C-glucosyl. Comparison of IR spectra (KBr) show that 6-C-glucosylflavones absorb more strongly at 9.25 and 9.8 μ than 6-C-xylosylflavones, which in turn more strongly absorb between 9.4 and 9.6 μ than 6-C-glucosylflavones. All the 6-C-xylopyranosyl-flavones exhibit in their mass spectra the characteristic peaks M-47 and M-65, the M-3 H_2O peak being absent or very weak.

Di-C-glucosylflavones. As previously mentioned, small amounts of 6,8-di-C-xylopyranosyl derivatives have been obtained in the C-xylosylation of 5,7-dihydroxyflavones, and 6-C-xylopyranosyl 8-C-

glucopyranosyl acacetin, apigenin and luteolin have been prepared by *C*-xylosylation of cytisoside, vitexin and orientin respectively.

Replacement of one or two *C*-glucosyl by *C*-xylosyl residues leads, as expected, to a decrease of the migration rate in 15% AcOH on paper chromatography, in the order 6,8-di-*C*-glucosyl = 6-*C*-glucosyl 8-*C*-xylosyl > 6-*C*-xylosyl 8-*C*-glucosyl > 6,8-di-*C*-xylosyl flavone, but differences are much smaller than between 6-*C*-glucosyl and 6-*C*-xylosyl flavones. On silica gel TLC, in APWM (80 : 12 : 10 : 5), mobilities are in the reverse order 6,8-di-*C*-glucosyl < 6-*C*-glucosyl 8-*C*-xylosyl < 6-*C*-xylosyl 8-*C*-glucosyl < 6,8-di-*C*-xylosyl flavone.

12.4.3.4 Conclusion

The above results show that much care is needed in the chromatographic identification of the sugar residues in *C*-glycosylflavonoids. For example, mono-*C*-glycosylapigenins, *C*-hexosyl and *C*-pentosylapigenins can easily be recognized by the chromatographic behaviour of the 6-isomers on paper. However, TLC is needed for differentiating isomeric 6-*C*-hexosyl or 6-*C*-pentosylapigenins. On the other hand, di-*C*-hexosylapigenins are much more difficult to distinguish by purely chromatographic means from di-*C*-glycosylapigenins containing both pentosyl and hexosyl residues, whereas di-*C*-pentosyl and di-*C*-rhamnosylapigenins remain clearly separated from the others.

Acknowledgements

The authors are grateful to Professors H. Wagner, J. W. Wallace, R. A. Eade, P. Lebreton, A. M. Debelmas, Drs A. Zoll, J. Raynaud, H. Combier and A. Proliac for supplying papers and information prior to publication, and to Professor T. J. Mabry for numerous NMR spectra.

References

Adamska, M. and Lutomski, J. (1971), *Planta med.* **20**, 224.
Adinarayana, D. and Rao, J. R. (1972), *Tetrahedron* **28**, 5377.
Alston, R. E. (1966), In *Comparative Phytochemistry* (ed. T. Swain), p. 40. Academic Press, London and New York.
Alston, R. E. (1968), In *Recent Advances in Phytochemistry* (eds. T. J. Mabry, R. E. Alston and V. C. Runeckles), p. 305. Appleton Century Crofts, New York.

Anton, R. and Duquenois, P. (1968), *Ann. pharm. franc.* **26**, 673.

Aritomi, M. (1962), *J. pharm. Soc. Japan* **82**, 1331.

Aritomi, M. (1963), *J. pharm. Soc. Japan* **83**, 737.

Aritomi, M and Kawasaki, T. (1968), *Chem. pharm. Bull.* **16**, 1842.

Aritomi, M. and Kawasaki, T. (1970), *Chem. pharm. Bull.* **18**, 2327.

Aritomi, M., Kiyota, I. and Mazaki, T. (1965), *Chem. pharm. Bull.* **13**, 1470.

Aritomi, M., Komori, T. and Kawasaki, T. (1970), *Liebigs Ann. Chem.* **734**, 91.

Asahina, Y., Asano, J. and Uyeno, Y. (1942), *J. pharm. Soc. Japan* **62**, 22.

Asen, S. (1967), *Am. Soc. hort. Sci.* **91**, 653.

Asen, S. and Horowitz, R. M. (1974), *Phytochemistry* **13**, 1219.

Asen, S. and Jurd, L. (1967), *Phytochemistry* **6**, 577.

Asen, S. and Plimmer, J. R. (1972), *Phytochemistry* **11**, 2601.   [9, 619.

Asen, S., Stewart, R. N., Norris, K. H. and Massie, D. R. (1970), *Phytochemistry*

Barger, G. (1906), *J. chem. Soc.* **89**, 1210.

Bate-Smith, E. C. and Swain, T. (1960), *Chem. Ind.* 1132.

Batyuk, V. S. and Chernobrovaya, N. V. (1966), *Khim. Prir. Soedin.* **2**, 90.

Batyuk, V. S., Chernobrovaya, N. V. and Kolesnikov, D. G. (1969), *Khim. Prir. Soedin.* **5**, 234.

Batyuk, V. S., Chernobrovaya, N. V. and Prokopenko, A. P. (1966), *Khim. Prir. Soedin.* **2**, 288.

Bellmann, G. and Jacot-Guillarmod, A. (1973), *Helv. chim. Acta* **56**, 284.

Bhatia, V. K., Gupta, S. R. and Seshadri, T. R. (1964), *Curr. Sci.* **33**, 581.

Bhatia, V. K., Gupta, S. R. and Seshadri, T. R. (1966a), *Phytochemistry* **5**, 177.

Bhatia, V. K., Gupta, S. R. and Seshadri, T. R. (1966b), *Tetrahedron* **22**, 1147.

Bhatia, V. K. and Seshadri, T. R. (1967), *Curr. Sci.* **36**, 111.

Bhutani, S. P., Chibber, S. S. and Seshadri, T. R. (1969), *Indian J. Chem.* **7**, 210.

Biol, M. C. (1973), Thèse de Doctorat de Spécialité, Université Claude Bernard, Lyon, $n°$ 225.

Biol, M. C. and Chopin, J. (1972), *C. r. Acad. Sci.*, Ser. C **275**, 1523.

Borodin, L. I., Litvinenko, V. I. and Kurinnaya, N. V. (1970), *Khim. Prir. Soedin.* **6**, 19.

Bouillant, M. L. and Chopin, J. (1971), *C. r. Acad. Sci.*, Ser. C **273**, 1759.

Bouillant, M. L. and Chopin, J. (1972), *C. r. Acad. Sci.*, Ser. C **274**, 193.

Bouillant, M. L., Chopin, J. and Plouvier, V. (1972), *Phytochemistry* **11**, 1858.

Boutard, B. (1972), Thèse de Doctorat de Spécialité, Université Claude Bernard, Lyon, n° 136.

Boutard, B., Bouillant, M. L., Chopin, J. and Lebreton, Ph. (1972), *C. r. Acad. Sci.*, Ser. D **274**, 1099.

Boutard, B., Bouillant, M. L., Lebreton, Ph. and Chopin, J. (1973), *Biochem. Syst.*, **1**, 133.

Briggs, L. H. and Cambie, R. C. (1958), *Tetrahedron* **3**, 269.

Cambie, R. C. (1959), *Chem. Ind.* 87.

Carson, J. L. and Wallace, J. W. (1972), *Phytochemistry* **11**, 842.

Carter, L. (1968). Cited by Alston, R. E. In *Recent Advances in Phytochemistry*, (eds. T. J. Mabry, R. E. Alston and V. C. Runeckles) p. 312. Appleton Century Crofts, New York.

Chernobai, V. T., Komissarenko, N. F. and Litvinenko, V. I. (1968), *Khim. Prir. Soedin.* **4**, 51.

Chernobrovaya, N. V., Komissarenko, N. F., Batyuk, V. S. and Kolesnikov, D. G. (1970), *Khim. Prir. Soedin.* **5**, 634.

Chopin, J. (1966). In *Actualités de Phytochimie Fondamentale*, (ed. C. Mentzer), p. 44. Masson et Cie, Paris.

Chopin, J. (1971). In *Pharmacognosy and Phytochemistry* (eds. H. Wagner and L. Hörhammer), p. 111. Sprinter-Verlag, Heidelberg, Berlin, New York.

Chopin, J. and Biol, M. C. (1972), *C. r. Acad. Sci.*, Ser. C **275**, 1435.

Chopin, J., Biol, M. C. and Bouillant, M. L. (1972), *C. r. Acad. Sci.*, Ser. C **274**,

Chopin, J., Biol, M. C. and Bouillant, M. L. (1972), *C. r. Acad. Sci.*, Ser. C, **274**, 1840.

Chopin, J. and Bouillant, M. L. (1967), *C. r. Acad. Sci.*, Ser. C **264**, 1875.

Chopin, J. and Bouillant, M. L. (1970a), *C. r. Acad. Sci.*, Ser. C **270**, 331.

Chopin, J. and Bouillant, M. L. (1970b), *C. r. Acad. Sci.*, Ser. C **270**, 222.

Chopin, J., Bouillant, M. L. and Biol, M. C. (1971), *C. r. Acad. Sci.*, Ser. C **273**, 1262.

Chopin, J., Bouillant, M. L. and Durix, A. (1965), *C. r. Acad. Sci.* **260**, 4850.

Chopin, J., Bouillant, M. L. and Durix, A. (1970a), *C. r. Acad. Sci.*, Ser. C **270**, 69.

Chopin, J., Bouillant, M. L. and Wallach, J. (1968b), *C. r. Acad. Sci.*, Ser. C **267**, 1722.

Chopin, J. and Chadenson, M. (1966), *C. r. Acad. Sci.*, Ser. C **262**, 662.

Chopin, J., Chadenson, M., Hauteville, A. and Hauteville, M. (1970b), *C. r. Acad. Sci.*, Ser. C **270**, 733.

Chopin, J. and Durix, A. (1966), *C. r. Acad. Sci.*, Ser. C **263**, 951.

Chopin, J., Durix, A. and Bouillant, M. L. (1966), *Tetrahedron Letters* **31**, 3657.

Chopin, J., Durix, A. and Bouillant, M. L. (1968a), *C. r. Acad. Sci.*, Ser. C **266**, 1334.

Chopin, J., Roux, B., Bouillant, M. L., Durix, A., D'Arcy, A., Mabry, T. J. and Yoshioka, Y. H. (1969), *C. r. Acad. Sci.*, Ser. C **268**, 980.

Chopin, J., Roux, B. and Durix, A. (1964), *C. r. Acad. Sci.* **259**, 3111.

Chumbalov, T. K. and Fadeeva, O. V. (1970), *Khim. Prir. Soedin.* **6**, 364.

Crawford, D. J. and Lankow, D. L. (1972), *Phytochemistry* **11**, 2571.

Crowden, R. K., Harborne, J. B. and Heywood, V. H. (1969), *Phytochemistry* **8**, 1963.

Dahlgren, R. (1963), *Opera Botanika* **9**, 1.

Darmograi, V. N., Krivenchuk, P. E. and Litvinenko, V. I. (1969), *Farmatsiya* **18**, 30.

Darmograi, V. N. and Litvinenko, V. I. (1971), *Farm. Zh. (Kiev)* **26**, 70.

Darmograi, V. N., Litvinenko, V. I. and Krivenchuk, P. E. (1968), *Khim. Prir. Soedin.* **4**, 248.

Dean, F. M. (1963), *Naturally Occurring Oxygen Ring Compounds*, p. 308. Butterworths, London.

Dranik, L. I. (1970), *Khim. Prir. Soedin.* **2**, 268.

Drozd, G. A., Koreshchuk, K. E., Khapugina, L. L. and Miroschnikov, E. V. (1971), *Khim. Prir. Soedin.* **7**, 526.

Drozd, G. A., Koreshchuk, K. E. and Litvinenko, V. I. (1969a), *Farm. Zh. (Kiev)* **24**, 56.

Drozd, G. A., Koreshchuk, K. E. and Litvinenko, V. I. (1969b), *Khim. Prir. Soedin.* **5**, 180.

Drozd, G. A. and Litvinenko, V. I. (1969), *Khim. Prir. Soedin.* **5**, 180.

Dubois, J. and Mabry, T. J. (1971), *Phytochemistry* **10**, 2839.

Düll, P. (1970). Ph.D. dissertation, Ludwigs-Maximilians Universität, München.

Durix, A. (1967), Thèse de Doctorat de Spécialité, Université Claude Bernard, Lyon, n°250.

Eade, R. A., Hillis, W. E., Horn, D. H. S., and Simes, J. J. H. (1965), *Aust. J. Chem.* **18**, 715.

Eade, R. A. and McDonald, F. J. (1973), personal communication.

Eade, R. A., Salasoo, I. and Simes, J. J. H. (1962), *Chem. Ind.* 1720.

Eade, R. A., Salasoo, I. and Simes, J. J. H. (1966), *Aust. J. Chem.* **19**, 1717.

Evans, W. H., McGookin, A., Jurd, L., Robertson, A. and Williamson, W. R. N. (1957), *J. chem. Soc.* 3510.

Fiedler, U. (1955), *Arzneim. Forsch.* **5**, 609.

Fizel, J. (1965), *Arzneim. Forsch.* **15**, 1417.

Funaoka, K. (1956), *Chem. Abstr.* **50**, 14729.

Funaoka, K. (1957), *Mokuzai Gakkaishi* **3**, 218.

Funaoka, K. and Tanaka, M. (1957a), *Mokuzai Gakkaishi* **3**, 173.

Funaoka, K. and Tanaka, M. (1957b), *Mokuzai Gakkaishi* **3**, 144.

Gaffield, W. and Horowitz, R. M. (1972), *Chem. Comm.* 648.

Gardner, R. L., Kerst, A. F., Wilson, D. M. and Payne, M. G. (1967), *Phytochemistry* **6**, 417.

Geissman, T. and Fiedler, U. (1956), *Naturwissenschaften* **43**, 226.

Gentili, B. and Horowitz, R. M. (1968), *J. org. Chem.* **33**, 1571.

Glennie, C. W., Harborne, J. B., Rowley, G. D. and Marchant, C. J. (1971), *Phytochemistry* **10**, 2413.

Glotzbach, B. and Rimpler, H. (1968), *Planta med.* **16**, 1.

Glyzin, V. I., Bankoski, A. I., Zhurba, O. V. and Shejchenko, V. I. (1970), *Khim. Prir. Soedin.* **4**, 473.

Hänsel, R., Leuckert, Ch., Rimpler, H. and Schaaf, D. (1965), *Phytochemistry* **4**, 19.

Hänsel, R. and Rimpler, H. (1963), *Arch. Pharm.* **296**, 598.

Harborne, J. B. (1965), *Phytochemistry* **4**, 107.

Harborne, J. B. (1969), *Phytochemistry* **8**, 1449.

Harborne, J. B. (1971), *Phytochemistry* **10**, 1569.

Harborne, J. B. and Hall, E. (1964), *Phytochemistry* **3**, 421.

Harborne, J. B. and Mendez, J. (1969), *Phytochemistry* **8**, 763.

Harborne, J. B. and Williams, C. A. (1972), *Phytochemistry* **11**, 1741.

Haynes, L. J. (1963), *Adv. Carbohyd. Chem.* **18**, 227.

Haynes, L. J. (1965), *Adv. Carbohyd. Chem.* **20**, 357.

Haynes, L. J., Holdsworth, D. K. and Russel, R. (1970), *J. chem. Soc.* 2581.

Hillis, W. E. and Carle, A. (1963), *Aust. J. Chem.* **16**, 147.

Hillis, W. E. and Horn, D. H. S. (1965), *Aust. J. Chem.* **18**, 531.

Hillis, W. E. and Horn, D. H. S. (1966), *Aust. J. Chem.* **19**, 705.

Hillis, W. E. and Inoue, T. (1967), *Phytochemistry* **6**, 59.

Hirose, Y., Hayashi, K. A., Wagishi, E. and Shibata, B. (1962), *Kumamoto pharm. Bull.* **5**, 48.

Hörhammer, L. and Wagner, H. (1961), In *Recent Developments in the Chemistry of Natural Phenolic Compounds* (ed. W. D. Ollis), p. 185. Pergamon Press, Oxford, London, New York, Paris.

Hörhammer, L., Wagner, H. and Beyersdorff, P. (1962), *Naturwissenschaften* **49**, 392.

Hörhammer, L., Wagner, H. and Dhingra, H. A. (1959a), *Arch. Pharm.* **292**, 83.

Hörhammer, L., Wagner, H. and Gloggengiesser, F. (1958), *Arch. Pharm.* **291**, 126.

Hörhammer, L., Wagner, H. and Leeb, W. (1960), *Arch. Pharm.* **65**, 264.

Hörhammer, L., Wagner, H., Nieschlag, H. and Wildi, G. (1959b), *Arch. Pharm.* **292**, 380.

Hörhammer, L., Wagner, H., Rosprim, L., Mabry, T. J. and Rösler, H. (1965), *Tetrahedron Letters* **22**, 1707.

Horowitz, R. M. and Gentili, B. (1964), *Chem. Ind.* 498.

Horowitz, R. M. and Gentili, B. (1966), *Chem. Ind.* 625.

Ibrahim, R. K. (1969), *Biochim. Biophys. Acta* **192**, 549.

Ibrahim, R. K. and Shaw, M. A. (1970), *Phytochemistry* **9**, 1855.

Jentzsch, K., Spiegel, P. and Fuchs, L. (1962), *Planta med.* **10**, 1.

Julian, E. A., Johnson, G., Johnson, D. K. and Donnelly, B. J. (1971), *Phytochemistry* **10**, 3185.

Jurd, L., Geissman, T. A. and Seikel, M. K. (1957), *Arch. Biochem. Biophys.* **67**, 284.

Kapoor, S. K., Ahmad, P. I. and Zaman, A. (1971), *Phytochemistry* **10**, 3333.

Kawase, A. (1968), *Agric. biol. Chem.* **32**, 1028.

Kawase, A. and Yagishita, K. (1968), *Agric. biol. Chem.* **32**, 537.

King, H. G. C. (1962), *J. Fd. Sci.* **27**, 446.

Koeppen, B. H. (1962), *Chem. Ind.* 2145.

Koeppen, B. H. (1964), *Ztsch. Naturf.* **19b**, 173.

Koeppen, B. H. and Roux, D. G. (1965a), *Biochem. J.* **97**, 444.

Koeppen, B. H. and Roux, D. G. (1965b), *Tetrahedron Letters* 3497.

Koeppen, B. H., Smit, C. J. B. and Roux, D. G. (1962), *Biochem. J.* **83**, 507.

Komatsu, M. and Tomimori, T. (1966), *Tetrahedron Letters* 1611.

Komatsu, M., Tomimori, T. and Ito, M. (1967a), *Chem. pharm. Bull.* **15**, 263.

Komatsu, M., Tomimori, T. and Makiguchi, Y. (1967b), *Chem. pharm. Bull.* **15**, 1567.

Komatsu, M., Tomimori, T., Makiguchi, Y. and Asano, K. (1968a), *Yakugaku Zasshi* **88**, 832.

Komatsu, M., Tomimori, T., Takeda, K. and Hayashi, K. (1968b), *Chem. pharm. Bull.* **16**, 1413.

Kroschewsky, J. R., Mabry, T. J., Markham, K. R. and Alston, R. E. (1969), *Phytochemistry* **8**, 1495.

Kuczynski, L. and Chyczewski, T. (1971), *Diss. Pharm. Pharmacol.* **23**, 519.

Kukkonen, I. (1971), *Mitt. bot. Staatssamml (München)* **10**, 622.
Lebreton, Ph. and Dangye-Caye, M. P. (1973), *Plant. med. Phytother.* VII (2), 87.
Leiba, S., Delaveau, P. and Paris, R. R. (1968), *Plant. med. Phytother.* II (2), 81.
Lewak, S. (1966), *Rocz. Chem.* **40**, 445.
Lewis, Y. S. and Neelakantan, S. (1962), *Curr. Sci.* **31**, 508.
Lewis, Y. S. and Neelakantan, S. (1964), *Curr. Sci.* **33**, 460.
Litvinenko, V. I., Amanmuradov, K. and Abubakirov, N. K. (1967), *Khim. Prir. Soedin.* **3**, 159.
Litvinenko, V. I. and Borodin, L. I. (1970), *Farm. Zh. (Kiev)* **25**, 84.
Litvinenko, V. I., Borodin, L. I. and Kurinnaya, N. V. (1969a), *Khim. Prir. Soedin.* **5**, 328.
Litvinenko, V. I. and Darmograi, V. N. (1968), *Dokl. Akad. Nauk. Ukrajin, RSRB,* **30**, 639.
Litvinenko, V. I., Darmograi, V. N., Krivenchuk, P. E. and Zoz, I. G. (1969b), *Rast. Resur.* **5**, 369.
Litvinenko, V. I. and Grankina, V. P. (1970), *Rast. Resur.* **6**, 395.
Litvinenko, V. I. and Kovalev, I. P. (1967), *Khim. Prir. Soedin.* **3**, 56.
Litvinenko, V. I. and Nadezhina, T. P. (1972), *Rast. Resur.* **8**, 35.
Lutomski, J. and Adamska, M. (1968), *Herba Polon.* **14**, 249.
Mabry, T. J., Markham, K. R. and Thomas, M. B. (1970), *The Systematic Identification of Flavonoids*, Springer-Verlag, Berlin, Heidelberg, New York.
Mabry, T. J., Yoshioka, H., Sutherland, S., Woodland, S., Rahman, W., Ilyas, M., Usmani, J. N., Hameed, N., Chopin, J. and Bouillant, M. L. (1971), *Phytochemistry* **10**, 677.
Margna, U., Hallop, L., Margna, E. and Tohver, M. (1967), *Biochim. Biophys. Acta* **136**, 396.
Markham, K. R. (1973), *Phytochemistry* **12**, 1091.
Markham, K. R. and Godley, E. J. (1972) *New Zeal. J. Bot.* **10**, 627.
Markham, K. R., Mabry, T. J. and Averett, J. E. (1972), *Phytochemistry* **11**, 2875.
Markham, K. R. and Porter, L. J. (1969). *Phytochemistry* **8**, 1777.
Markham, K. R. and Porter, L. J. (1973), *Phytochemistry* **12**, 2007.
Markham, K. R., Porter, L. J. and Brehm, B. G. (1969), *Phytochemistry* **8**, 2193.
Mascré, M. and Paris, R. R. (1937a), *Bull. Soc. pharmacol.* **44**, 401.
Mascré, M. and Paris, R. R. (1937b), *C. r. Acad. Sci.* **204**, 1270.
Mascré, M. and Paris, R. R. (1937c), *C. r. Acad. Sci.* **204**, 1581.
McClure, J. W. (1964), Ph.D. dissertation, The University of Texas, Austin, USA.
McClure, J. W. (1968), *Pl. Physiol.* **43**, 193.
McClure, J. W. and Alston, R. E. (1966), *Am. J. Bot.* **53**, 849.
Melchert, T. E. and Alston, R. E. (1965), *Science* **150**, 1170.
Molisch, H. (1911), *Ber. dt. bot. Ges.* **29**, 486.
Monties, B. (1971), IInd International Congress on Photosynthesis, Stresa.
Mrugasiewicz, K. (1963), *Biul. Inst. Roslin. Leczniczych.* **9**, 1.
Murakami, T., Nishikawa, Y. and Ando, T. (1960), *Chem. pharm. Bull.* **8**, 688.

Nakaoki, T. (1927), *J. pharm. Soc. Japan* **47**, 144.

Nakaoki, T. (1944), *J. pharm. Soc. Japan* **64**, 57.

Narayanan, V. and Seshadri, T. R. (1971), *Indian J. Chem.* **9**, 14.

Niemann, G. J. and Bekooy, R. (1971), *Phytochemistry* **10**, 893.

Nilsson, E. (1969), *Acta chem. scand.* **23**, 2910.

Nilsson, E. (1973), *Phytochemistry* **12**, 722.

Ockendon, D. J., Alston, R. E. and Naifeh, K. (1966), *Phytochemistry* **5**, 601.

Oelrichs, P., Marshall, J. T. B. and Williams, D. H. (1968), *J. chem. Soc.* 941.

Paris, M. (1968), *Plant. med. Phytother.* III (1), 32.

Paris, R. R. (1957), *C. r. Acad. Sci.* **245**, 443.

Paris, R. R. (1962), *Pharm. Acta Helv.* **37**, 336.

Paris, R. R. and Brun-Bousquet, M. (1971), *C. r. Acad. Sci.*, Ser. D **273**, 1116.

Paris, R. R. and Charles, A. (1962), *C. r. Acad. Sci.* **254**, 352.

Paris, R. R. and Delaveau, P. (1962), *C. r. Acad. Sci.* **254**, 928.

Paris, R. R., Delaveau, P. G. and Leiba, S. (1966), *C. r. Acad. Sci.*, Ser. D **262**, 1372.

Paris, R. R. and Etchepare, S. (1964), *C. r. Acad. Sci.* **258**, 6003.

Paris, R. R. and Etchepare, S. (1968), *Ann. pharm. franc.* **26**, 51.

Paris, R. R. and Paris, M. (1966), *C. r. Acad. Sci.*, Ser. D **263**, 792.

Paris, R. R. and Stambouli, A. (1961), *C. r. Acad. Sci.* **252**, 1659.

Perkin, J. (1898), *J. chem. Soc.* **73**, 1019.

Perkin, A. G. (1899), *Proc. chem. Soc.* **15**, 123.

Perkin, J. (1900), *J. chem. Soc.* **77**, 422.

Phouphas, C. (1956), *C. r. Acad. Sci.* **242**, 1641.

Plouvier, V. (1966), *C. r. Acad. Sci.*, Ser. D **262**, 1368.

Plouvier, V. (1967), *C. r. Acad. Sci.*, Ser. D **265**, 516.

Poethke, W., Schwarz, C. and Gerlach, H. (1970), *Planta med.* **19**, 177.

Proliac, A. (1973), Ph.D. thesis, Claude Bernard University, Lyon, France.

Prox, A. (1968), *Tetrahedron* **24**, 3697.

Rao, C. B. and Venkateswarlu, V. (1956), *Curr. Sci.* **25**, 328.

Rao, C. B. and Venkateswarlu, V. (1962), *J. sci. ind. Res. India* **21B**, 313.

Rao, D. S. (1965), *Naturwissenschaften* **52**, 262.

Raynaud, J., Gorunovic, M. and Lebreton, Ph. (1971), *Planta med.* **20**, 199.

Reznik, H. and Menschick, R. (1969), *Ztsch. Pflanzenphysiol.* **61**, 348.

Rimpler, H., Langhammer, L. and Frenzel, H. J. (1963), *Planta med.* **11**, 325.

Rosprim, L. (1966), Dissertation, Ludwigs-Maximilians-Universität, München.

Sachs, A. (1963), Dissertation, Ludwigs-Maximilians-Universität, München.

Sakamoto, Y. (1967), *Agr. biol. Chem.* **31**, 1029.

Sakamoto, Y. (1969), *Agr. biol. Chem.* **33**, 959.

Sakamoto, Y. (1970), *Agr. biol. Chem.* **34**, 919.

Saleh, N. A. M., Bohm, B. A. and Maze, J. R. (1971), *Phytochemistry* **10**, 490.

Schilcher, H. (1968), *Ztsch. Naturf.* **23b**, 1393.

Seikel, M. K. (1963), Proceeding of a Symposium of the Plant Phenolics Group of North America (University of Toronto, Toronto, Canada), (ed. V. C. Runeckles), p. 19.

Seikel, M. K. and Bushnell, A. J. (1959), *J. org. Chem.* **24**, 1995.

Seikel, M. K., Bushnell, A. J. and Birzgalis, R. (1962), *Arch. Biochem. Biophys.* **99**, 451.

Seikel, M. K., Chow, J. H. S. and Feldman, L. (1966), *Phytochemistry* **5**, 439.

Seikel, M. K. and Geissman, T. A. (1957), *Arch. Biochem. Biophys.* **71**, 17.

Seikel, M. K., Holder, D. J. and Birzgalis, R. (1959), *Arch. Biochem. Biophys.* **85**, 272.

Seikel, M. K. and Mabry, T. J. (1965), *Tetrahedron Letters* 1105.

Seshadri, T. R., Sood, A. R. and Varshney, I. P. (1972), *Indian J. Chem.* **10**, 26.

Sherwood, R. T., Shamma, M., Moniot, J. L. and Kroschewsky, K. R. (1973), *Phytochemistry* **12**, 2275.

Shibata, S., Murakami, T., Nishikawa, Y. and Harada, M. (1959), *Chem. pharm. Bull.* **7**, 134.

Shimokoriyama, M. and Geissman, T. A. (1962). In *Chemistry of Natural and Synthetic Colouring Matters* (eds. T. Gore *et al.*), p. 255. Academic Press, New York and London.

Simes, J. J. H. (1949), M.S. thesis, University of Sydney, Australia.

Soeder, R. W. and Babb, M. S. (1972), *Phytochemistry* **11**, 3079.

Stambouli, A. and Paris, R. R. (1961), *Ann. pharm. franc.* **19**, 435.

Stenhouse, J. (1851), *Ann. Chem.* **78**, 15.

Subramanian, S. S. and Nagarajan, S. (1967a), *Curr. Sci.* **36**, 364.

Subramanian, S. S. and Nagarajan, S. (1967b), *Curr. Sci.* **36**, 403.

Subramanian, S. S. and Nagarajan, S. (1967c), *Indian J. Pharm.* **29**, 311.

Subramanian, S. S. and Nagarajan, S. (1969), *Curr. Sci.* **38**, 65.

Subramanian, S. S. and Nagarajan, S. (1970), *Phytochemistry* **9**, 2581.

Subramanian, S. S., Nagarajan, S. and Sulochana, N. (1971a), *Phytochemistry* **10**, 1690.

Subramanian, S. S., Nagarajan S. and Sulochana, N. (1971b), *Phytochemistry* **10**, 2548.

Takeda, K., Mitsui, S. and Hayashi, K. (1966), *Bot. Mag. (Tokyo)* **79**, 578.

Taku, E. P., Tikhonov, A. I. and Litvinenko, V. I. (1970), *Khim. Prir. Soedin.* **6**, 629.

Tikhonov, O. I., Krivenchuk, P. E. and Litvenenko, V. I. (1966), *Farm. Zh. (Kiev)* **21**, 40.

Tikhonov, O. I., Krivenchuk, P. E., Litvinenko, V. I. and Kovalov, I. P. (1965), *Farm. Zh. (Kiev)* **20**, 53.

Tjukavkina, N. A., Benesova, V. and Herout, V. (1970), *Coll. Czech. Chem. Commun.* **35**, 1306.

Tomimori, T. and Komatsu, M. (1969), *Yakugaku Zasshi* **89**, 1276.

Trajman, G. (1972), Thèse Université, Grenoble.

Ueno, A., Oguri, N., Hori, K., Saiki, Y. and Harada, T. (1963), *Yakugaku Zasshi* **83**, 420.

Viscontini, M., Hoch, D. and Karrer, P. (1955), *Helv. chim. Acta* **38**, 642.

Volk, O. and Sinn, M. (1968), *Ztsch. Naturf.* **23b**, 1017.

Wagner, H. (1966). In *Comparative Phytochemistry* (ed. T. Swain), p. 309. Academic Press, London and New York.

Wagner, H. Budweg, W., Iyengar, M. A., Volk, O. and Sinn, M. (1972a), *Ztsch. Naturf.* **27b**, 809.

Wagner, H., Hörhammer, L. and Kiraly, I. C. (1970), *Phytochemistry* 9, 897.

Wagner, H., Iyengar, M. A. and Beal, J. L. (1971), *Phytochemistry* 10, 2553.

Wagner, H., Iyengar, M. A., Düll, P. and Herz, W. (1972b), *Phytochemistry* 11, 1506.

Wagner, H., Iyengar, M. A. and Herz, W. (1972c), *Phytochemistry* 11, 851.

Wagner, H., Iyengar, M. A. and Herz, W. (1972d), *Phytochemistry* 11, 446.

Wagner, H., Iyengar, M. A. and Hörhammer, L. (1972e), *Phytochemistry* 11, 1518.

Wagner, H., Iyengar, M. A. and Hörhammer, L. (1973), *Phytochemistry* 12, 2548.

Wagner, H., Iyengar, M. A., Hörhammer, L. and Herz, W. (1972f), *Phytochemistry* 11, 1857.

Wagner, H., Patel, J., Hörhammer, L., Yap, F. and Reichardt, A. (1967), *Ztsch. Naturf.* 22b, 988.

Wagner, H., Rosprim, L. and Düll, P. (1972g), *Ztsch. Naturf.* 27b, 954.

Wallace, J. W. and Alston, R. E. (1966), *Pl. Cell Physiol.* 7, 699.

Wallace, J. W. and Mabry, T. J. (1970), *Phytochemistry* 9, 2133.

Wallace, J. W., Mabry, T. J. and Alston, R. E. (1969), *Phytochemistry* 8, 93.

Williams, C. A., Harborne, J. B. and Clifford, H. T. (1971), *Phytochemistry* 10, 1059.

Williams, C. A., Harborne, J. B. and Clifford, H. T. (1973), *Phytochemistry* 12, 2417.

Williams, C. A. and Murray, B. G. (1972), *Phytochemistry* 11, 2507.

Yasue, M., Itaya, M., Oshima, H. and Funahashi, S. (1965), *Yakugaka Zasshi* 85, 533.

Zemtsova, G. N. and Bandyu Kova, V. A. (1972), *Khim. Prir. Soedin.* 5, 678.

Zoll, A. (1972), Thèse de Doctorat en Pharmacie, Université scientifique et medicale de Grenoble, no. 1.

Zoll, A. and Nouvel, G. (1974), *Plant. med. Phytother.* 8, 134.

Chapter 13

Biflavonoids

HANS GEIGER and CHRISTOPHER QUINN

13.1 Introduction

Harborne (1967) could still define biflavonyls as 'dimers' of apigenin and as such they were well distinguished from other flavonoid 'dimers' like the proanthocyanidins, theaflavins and dracorubin. During the past six years, however, a wealth of biflavonoids has been discovered, which differ from the 'classical biflavones' (amento-, hinoki, cupressu- and agathis-flavone) not only in the hydroxylation pattern of the aromatic rings, but also in the oxidation level of the central heterocycle (see Locksley, 1973).

Despite the range of biflavonoids now known, their formation in all cases may be explained in terms of oxidative coupling of two chalkone units and subsequent modification of the central C_3-units. If an electron is abstracted from the C-4 anion of naringenino-chalkone, a radical is formed which may be represented by the canonical formulae R_α, R_β and R_γ:

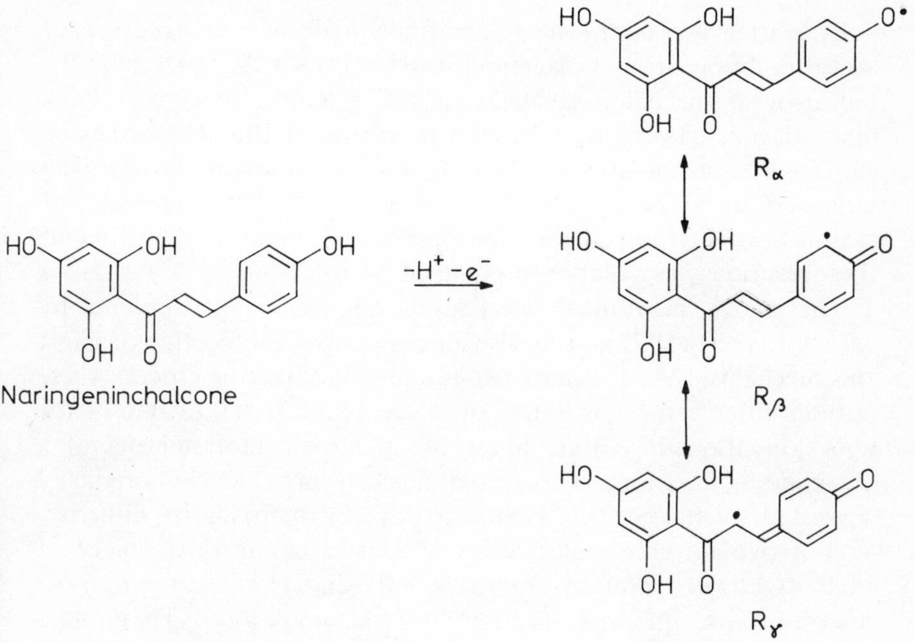

Naringeninchalcone

$-H^+ - e^-$

R_α

R_β

R_γ

Abstraction of an electron from the C-4' anion of the same chalkone yields another radical, which again may be represented by several canonical formulae. In the present context, only the canonical formula R_δ is important.

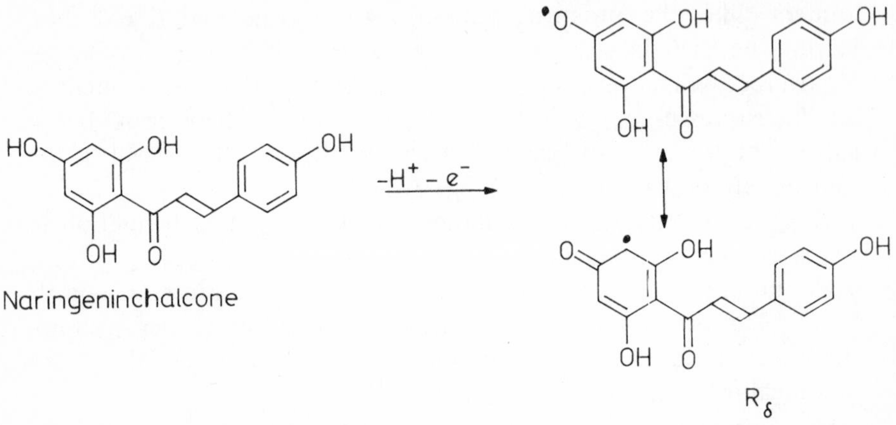

Naringeninchalcone

R_δ

By pairing two of the above mentioned radicals, the precursors of all naturally occurring biflavonoids can be formed: $R_\alpha + R_\beta$ gives the precursor of the ochnaflavone group; $R_\alpha + R_\delta$ the precursor of the hinokiflavone group; $R_\beta + R_\delta$ the precursor of the amentoflavone and robustaflavone groups; $R_\gamma + R_\delta$ the precursor of the *Garcinia* biflavonoids; and $R_\delta + R_\delta$ the precursor of the cupressu- and agathis-flavone group. It must be emphasized however, that none of these reactions has so far been observed *in vitro*.

The above mentioned mechanism has been put forward by Jackson *et al.* (1971), who also discuss possible objections against this mechanism. As an alternative to the radical pairing process, these authors discuss the possibility of electrophilic attack of one of the above mentioned radicals upon the phloroglucinol nucleus of a chalcone or flavanone. This second mechanism, which was originally suggested by Baker *et al.* (1963) to explain the oxidative dimerization of two apigenin molecules, would also account for the fact that in most known naturally occurring biflavonoids at least one 6- or 8-position is involved in the interflavonyl link. These two mechanisms, although speculative, provide a basis for grouping the various biflavonoids treated in this chapter into one 'natural family' of plant products.

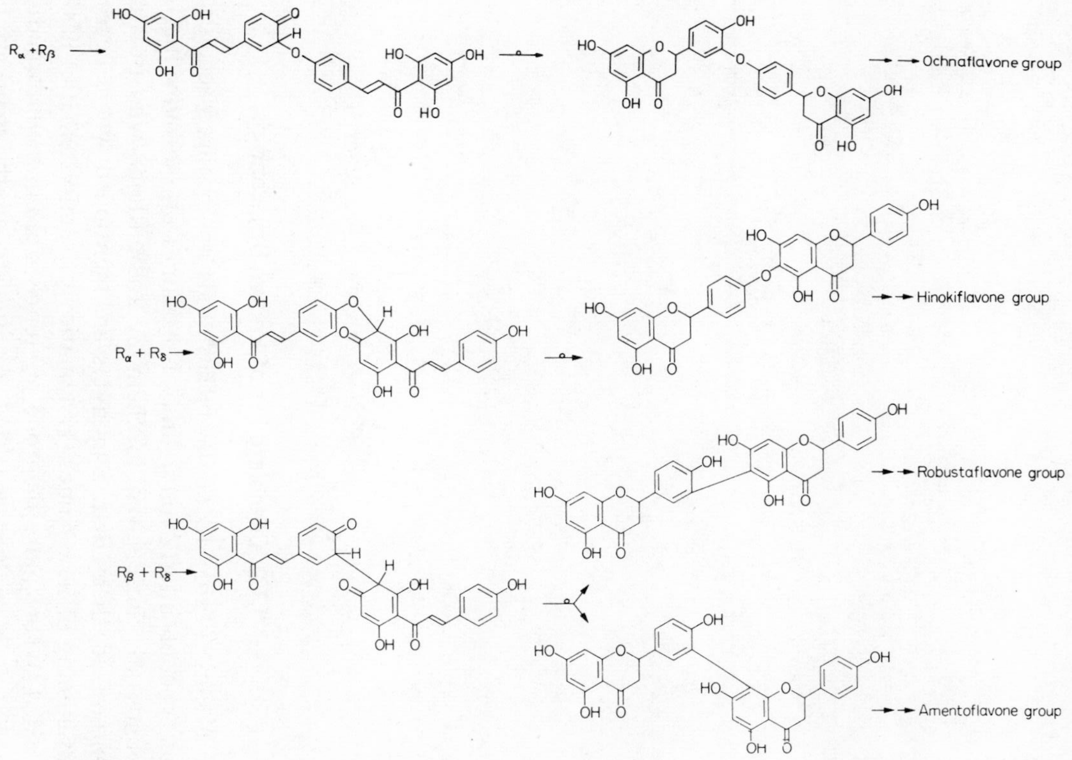

Ochnaflavone group

Hinokiflavone group

Robustaflavone group

Amentoflavone group

$R_\alpha + R_\beta$

$R_\alpha + R_\delta$

$R_\beta + R_\delta$

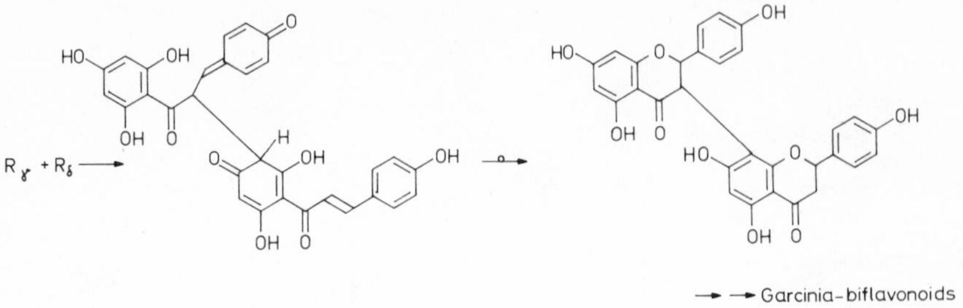

Garcinia-biflavonoids

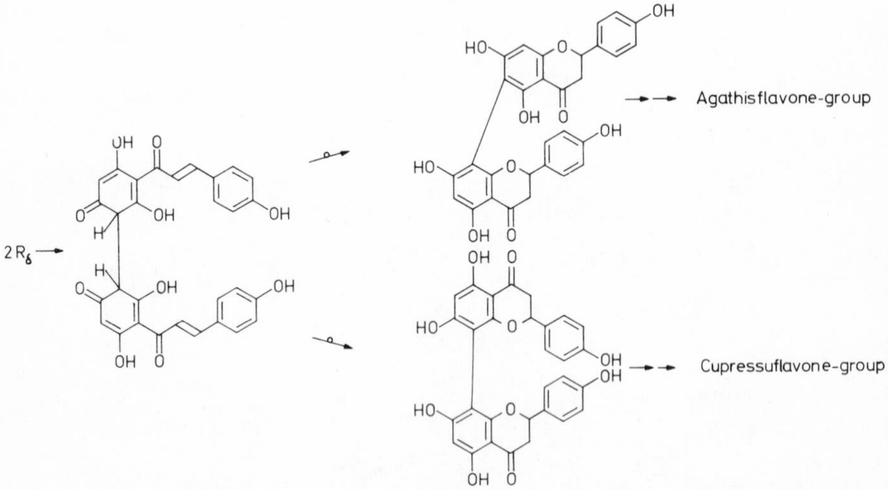

Agathisflavone-group

Cupressuflavone-group

13.2 Complete list of known biflavonoids

In this list (Table 13.1) the biflavonoids are arranged according to the type of interflavonyl link. The names of the individual compounds are those given to them by the authors, who isolated or synthesized them first. Synonyms are mentioned where necessary, but coining of new names (systematic* or semisystematic) has been avoided. Where more than one reference is given, the information to be found in each is specified. Comments on the main groups of biflavonoids follow.

*For a tentative nomenclature of biflavonoids see: Jackson *et al.*, 1971, p. 3794.

Table 13.1

The agathisflavone group. Biflavones as well as flavanonyl-flavones and biflavones have been reported in this group. The fully aromatic members of this group occur as optically active atropisomers. Agathisflavone hexamethyl ether has been synthesized unequivocally.

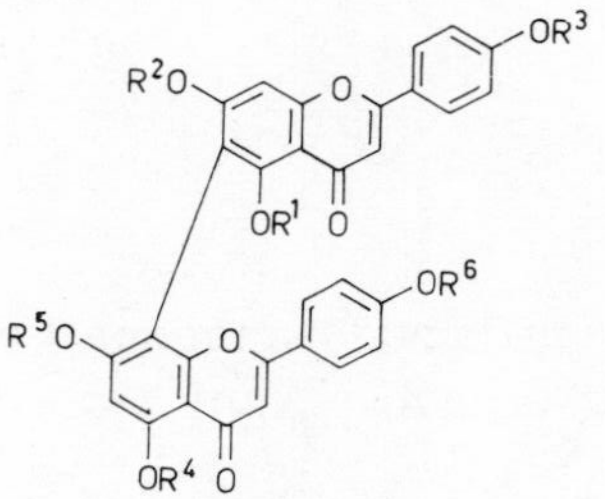

	R^1	R^2	R^3	R^4	R^5	R^6	Leading References
(1) agathisflavone	H	H	H	H	H	H	Khan *et al.*, 1972 (isolated as the hexamethyl ether).
(2) 7-*O*-methylagathisflavone (= WA-I)	H	ME	H	H	H	H	Pelter *et al.*, 1969c.
(3) 7,7''-di-*O*-methylagathisflavone	H	Me	H	H	Me	H	Khan *et al.*, 1972.
(4) 7,4'''-di-*O*-methylagathisflavone (= WA-VII)	H	Me	H	H	H	Me	Pelter *et al.*, 1969c.
(5) agathisflavone tetramethyl ether	H	Me	Me	H	Me	Me	cf. Handa *et al.*, 1971a.
(6) agathisflavone hexamethyl ether	Me	Me	Me	Me	Me	Me	Pelter *et al.*, 1969c (struct., NMR, UV, opt. rot.); Pelter *et al.*, 1971c (synth.); Moriyama *et al.*, 1972 (synth.).

697

Table 13.1–*Continued*

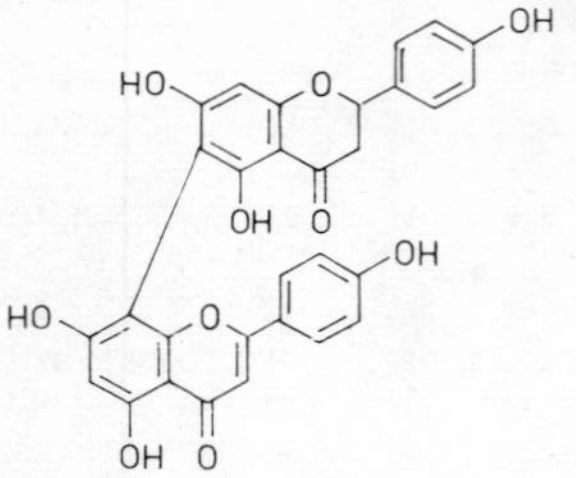

(6a) rhusflavone
Chen *et al.*, 1974.

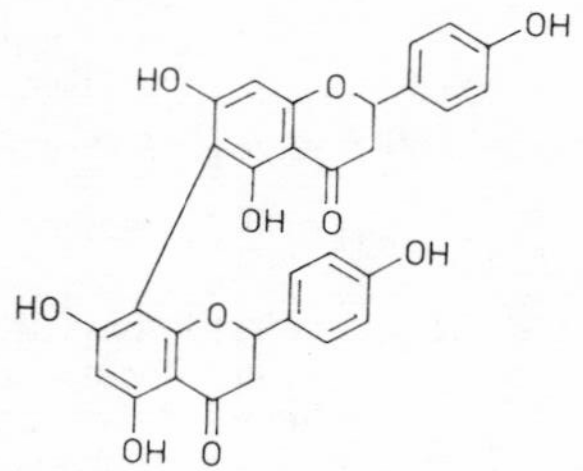

(6b) rhusflavanone
Lin and Chen, 1973.

The cupressuflavone group. In this group only fully aromatic compounds have been reported. Optically active atropisomers have been reported for all members. The structure of cupressuflavone hexamethyl ether has been proved by complete synthesis.

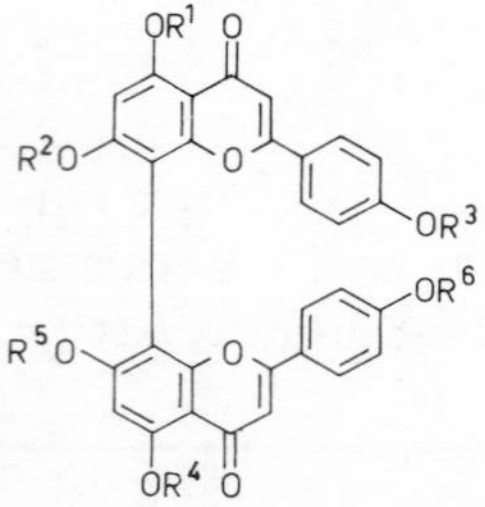

	R¹	R²	R³	R⁴	R⁵	R⁶	Leading References
(7) cupressuflavone	H	H	H	H	H	H	Murti *et al.*, 1967 (isoln. struct., UV, IR, NMR synth.); Pelter *et al.*, 1971c (opt. rot.); Ahmad and Razaq, 1971 (synth.).
(8) 7-*O*-methylcupressuflavone	H	Me	H	H	H	H	Khan *et al.*, 1972.
(9) 7,7''-di-*O*-methylcupressuflavone	H	Me	H	H	Me	H	Khan *et al.*, 1972 and refs. quoted therein; Natarajan *et al.*, 1970a (synth.).
(10) 4',7''-cupressuflavone trimethyl ether	H	Me	Me	H	Me	H	Khan *et al.*, 1971a.
(11) cupressuflavone tetramethyl ether (= Wb1 = AC3)	H	Me	Me	H	Me	Me	Murti *et al.*, 1967 (synth.); Rahman and Bhatnagar, 1968 (isoln., UV, IR, NMR); Ilyas *et al.*, 1968 (isoln., UV, NMR, MS, opt. rot.)
(11a) cupressuflavone pentamethyl ether	H	Me	Me	Me	Me	Me	Natarajan *et al.*, 1970a (synth.).
(12) cupressuflavone hexamethyl ether (= '8,8''-biflavone compound')	Me	Me	Me	Me	Me	Me	Nakazawa, 1962 (synth.); Murti *et al.*, 1967 (NMR, IR); Natarajan *et al.*, 1969 (MS); Ahmad and Razaq, 1971 (synth.); Moriyama *et al.*, 1972 (synth.).

The amentoflavone group. In this large group not only fully aromatic biflavones but also flavanone-flavones and biflavanones are found. The most abundant of these types are the biflavones, some of which have been isolated as optically active atropisomers. Ginkgetin and isoginkgetin were the first biflavones whose structures were elucidated and confirmed by complete synthesis.

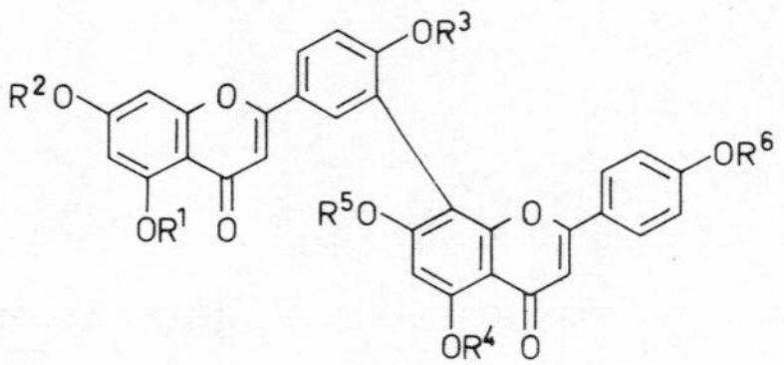

	R¹	R²	R³	R⁴	R⁵	R⁶	Leading References
(13) Amentoflavone	H	H	H	H	H	H	Hörhammer et al., 1965; Pelter et al., 1970 (opt. rot).
(14) sequoiaflavone	H	Me	H	H	H	H	Miura and Kawano, 1968b.
(15) bilobetin	H	H	Me	H	H	H	Baker et al., 1963.
(16) sotetsuflavone	H	H	H	H	Me	H	earlier reports (cf. refs. in Baker et al., 1963) are doubtful, but it occurs in *Araucaria* spp. (Kawano, private communication).
(17) podocarpusflavone A	H	H	H	H	H	Me	Miura et al., 1969 (isoln. struct); Chexal et al., 1970b (opt. rot).
(18) ginkgetin	H	Me	Me	H	H	H	Baker and Ollis, 1961 and refs. quoted therein; Baker et al., 1963; Nakazawa and Ito, 1963 (synth.).
(19) podocarpusflavone B (= putraflavone)	H	Me	H	H	H	Me	Miura et al., 1969; Garg and Mitra, 1971.
(20) amentoflavone 4′,7″-dimethyl ether	H	H	Me	H	Me	H	Khan et al., 1971a.
(21) isoginkgetin	H	H	Me	H	H	Me	Baker and Ollis, 1961 and refs. quoted therein; Baker et al., 1963.
(22) amentoflavone 7″,4‴-dimethyl ether	H	H	H	H	Me	Me	Beckmann et al., 1971.
(23) sciadopitysin	H	Me	Me	H	H	Me	Baker and Ollis 1961 and refs. quoted therein; Baker et al., 1963.
(24) heveaflavone	H	Me	H	H	Me	Me	Madhav, 1969; c. Chandramouli et al., 1971.
(25) kayaflavone	H	H	Me	H	Me	Me	Baker and Ollis, 1961 and refs. quoted therein; Baker et al., 1963.
(26) amentoflavone 7,7″,4′,4‴-tetra-methyl ether (= W13)	H	Me	Me	H	Me	Me	Nakazawa, 1962 (synth.); Hodges, 1965 (isoln); Pelter et al., 1969a (opt. rot.).
(27) amentoflavone hexamethyl ether (= dioonflavone)	Me	Me	Me	Me	Me	Me	Nakazawa, 1962 (synth.); Natarajan et al., 1969 (MS) Dossaji et al., 1973 (detectn.).

(28) 5',8''-biluteolin

Nilsson, 1973.

	R^1	R^2	R^3	R^4	R^5	R^6	Leading References
(29) 2,3-dihydroamentoflavone	H	H	H	H	H	H	Geiger and de Groot-Pfleiderer, 1971.
(30) 2,3-dihydroamentoflavone 7'',4'''- dimethyl ether	H	H	H	H	Me	Me	
(31) 2,3-dihydrosciadopitysin	H	Me	H	Me	Me	Me	Beckmann *et al.*, 1971
(32) 2,3-dihydroamentoflavone hexamethyl ether	Me	Me	Me	Me	Me	Me	

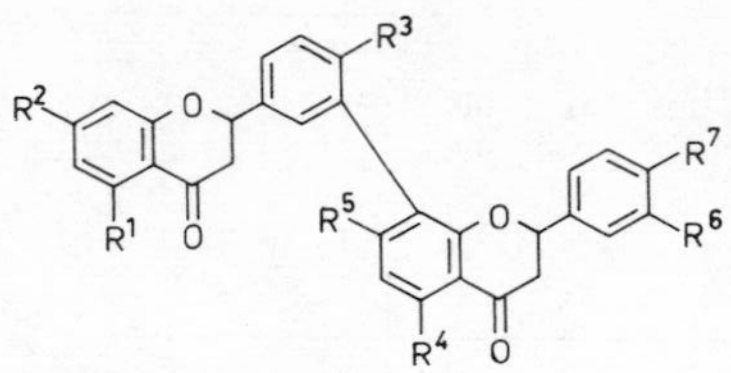

(33a) biflavanone A
(33b) biflavanone B
(33c) biflavanone C

$R^1=R^2=R^3=R^4=R^5=R^6=R^7=OH$
$R^1=R^2=R^3=R^4=R^5=R^7=OH; R^6=H$
$R^2=R^3=R^5=R^7=OH; R^1=R^4=R^6=H$

Rao *et al.*, 1973 (structures deduced from NMR of methylation products, compounds as such not yet isolated).

The robustaflavone group. Before robustaflavone had been isolated from a natural source, its hexamethyl ether had been prepared by Wessely-Moser-rearrangement of amentoflavone hexamethyl ether (27). So far only one member of this group has been found to occur naturally.

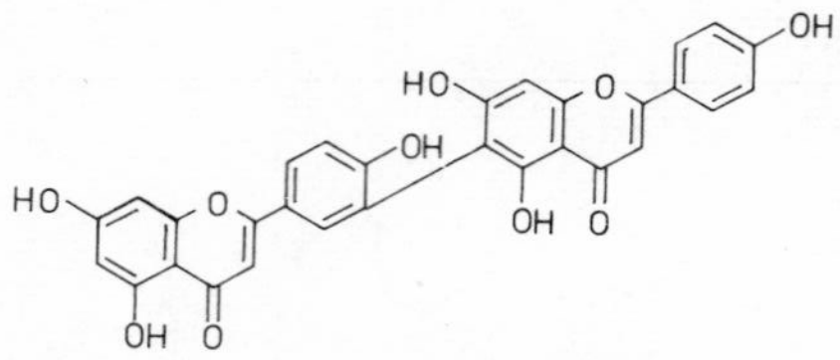

(33d) robustaflavone

Chexal *et al.*, 1970a (synth.); Varshney *et al.* 1973b (isoln. struct.)

The garcinia biflavonoids. This is the only group in which biflavonoid glycosides have been found. Fully aromatic members of this group have not been found naturally, and it is doubtful whether they are ever likely to be found, since it would need a special enzyme to dehydrogenate the heterocycle bearing the interflavonyl link. The structures of fukugetin and saharanflavone have been confirmed by complete synthesis. Since all members contain at least one asymmetric carbon atom, they are all optically active. Racemates have also been reported.

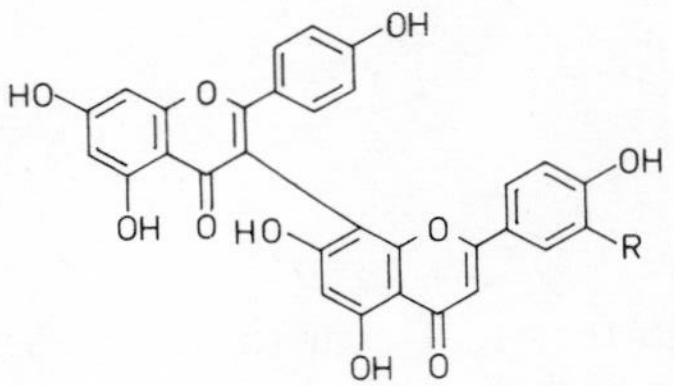

		R
(34)	bisdehydro-GB1a	H
(35)	saharanflavone	OH

Leading References

Jackson *et al.*, 1971 (synth.); Pelter *et al.*, 1971a (synth. of hexamethyl ether).
Pelter *et al.*, 1971a (synth. of heptamethyl ether); Ikeshiro and Konoshima 1972 (synth. of heptamethyl ether).

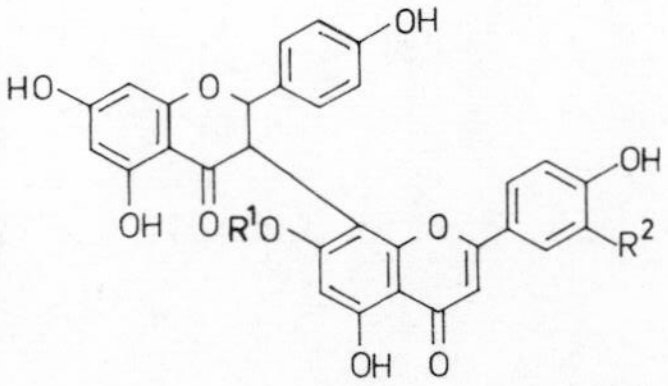

	R^4	R^2	Leading References
(36) volkensiflavone (= BGH-III = talbotiflavone)	H	H	Herbin et al., 1970 (isoln., struct., NMR, MS); Joshi et al., 1970 (isoln., NMR, MS); Pelter et al., 1971a (isoln., NMR). Konoshima et al., 1970.
(37) spicatiside	β–D-glc	H	Perkin and Phipps, 1904 (isoln.); Karanjgaokar et al., 1967 (isoln., struct.); Konoshima et al., 1969 (struct., opt. rot., IR, NMR, MS; Pelter et al., 1971a (isoln., NMR); Ikeshiro and Konoshima, 1972 (synth. of heptamethyl ether).
(38) fukugetin (= BGH-II = morelloflavone)	H	OH	
(39) 3-O-methylfukugetin	H	OMe	Konoshima et al., 1969.
(40) fukugiside	β–D-glc	OH	Konoshima and Ikeshiro, 1970.

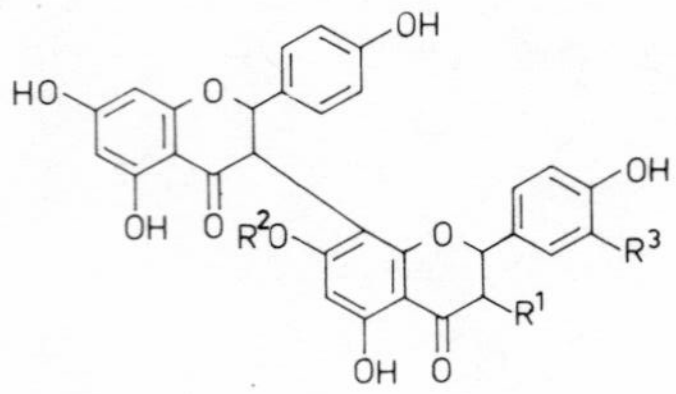

	R^1	R^2	R^3	Leading References
(41) G B 1a	H	H	H	Jackson et al., 1971.
(42) G B 1	OH	H	H	Jackson et al., 1971.
(43) G B 2a	H	H	OH	Jackson et al., 1971.
(44) G B 2	OH	H	OH	Jackson et al., 1971.
(45) Xanthochymusside	OH	β–D–glc	OH	Konoshima et al., 1970.

704

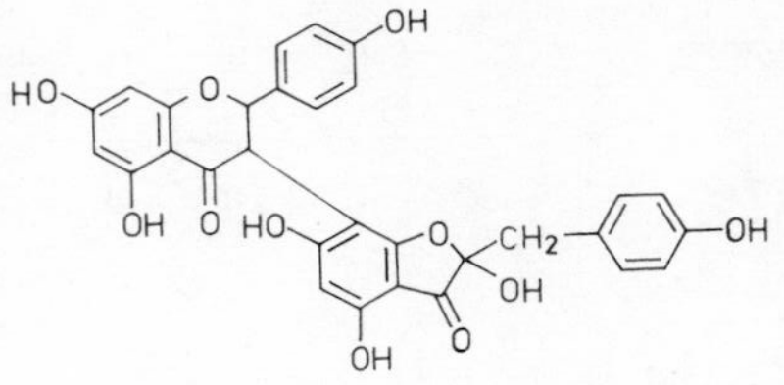

(46) zeyherin

du Volsteed and Roux, 1971.

The hinokiflavone group. In contrast to most other biflavonoids, the interflavonyl link here is not a C—C bond but a diaryl ether link. Besides the fully aromatic compounds, one flavanone-flavone is known. When Fukui and Kawano (1959) worked on the chemistry of hinokiflavone, they put forward two alternative structures (47) and (62). The second (62) was the preferred one until Nakazawa (1968) proved (47) to be the correct structure by complete synthesis. This has since been confirmed by NMR measurements (Pelter *et al.*, 1969b). Hence in all early publications where (62) is used as the structure of hinokiflavone, it has to be replaced by (47). Other features of the molecules such as the position of methoxyl groups remain unaffected. In the following list only the correct structures are given.

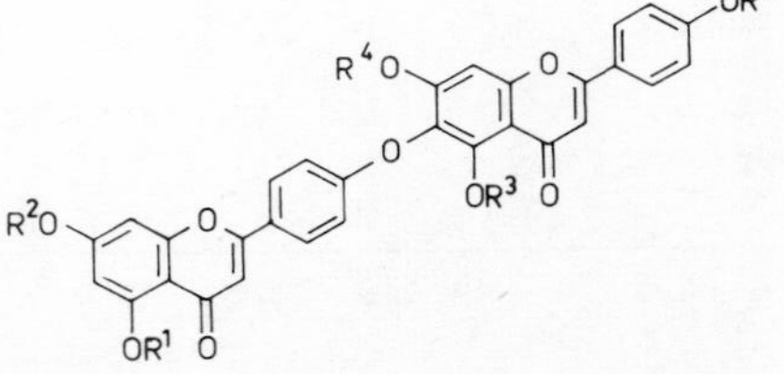

		R¹	R²	R³	R⁴	R⁵	Leading References
(47)	hinokiflavone	H	H	H	H	H	Kariyone and Sawada, 1958 (isoln.); Nakazawa, 1968 (synth. struct.).
(48)	neocryptomerin	H	Me	H	H	H	Miura and Kawano, 1968c (synth.); Miura et al., 1969 (isoln.).
(49)	isocryptomerin	H	H	H	Me	H	Muira and Kawano, 1967 (synth.); Miura et al., 1968 (isoln.).
(50)	cryptomerin A	H	H	H	H	Me	Miura et al., 1966.
(51)	chamaecyparin	H	Me	H	Me	H	Miura and Kawano, 1968c (synth.); Miura, 1967a; Miura et al., 1968 (isoln.).
(52)	hinokiflavone 7,4'''-dimethyl ether	H	Me	H	H	Me	Miura and Kawano, 1968a.
(53)	cryptomerin B	H	H	H	Me	Me	Miura et al., 1966.
(54)	hinokiflavone 7,7'',4'''-trimethyl ether	H	Me	H	Me	Me	Kariyone and Fukui, 1960 (synth.); Miura and Kawano, 1968a (isoln.).
(55)	hinokiflavone pentamethyl ether	Me	Me	Me	Me	Me	Nakazawa, 1968 (synth. struct.); Pelter et al., 1969b (NMR); Natarajan et al., 1969 (MS).

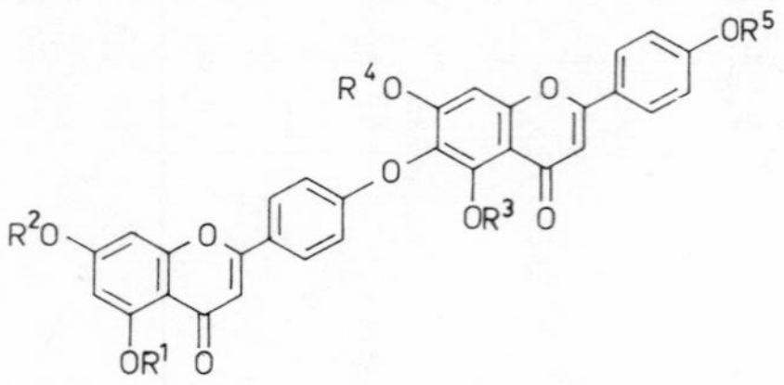

		R¹	R²	R³	R⁴	R⁵	Leading References
(56)	2,3-dihydrohinokiflavone	H	H	H	H	H	Beckmann et al., 1971.
(57)	2,3-dihydrohinokiflavone pentamethyl ether	Me	Me	Me	Me	Me	

The ochnaflavone group. Members of this group have been recently isolated from *Ochna squarrosa* L. (Ochnaceae). The structure of the basic skeleton has been confirmed by a complete synthesis of ochnaflavone pentamethyl ether. This type constitutes the first example of a naturally occurring biflavonoid in which neither of the A-rings is involved in the interflavonyl link. So far, only fully aromatic members of this group are known.

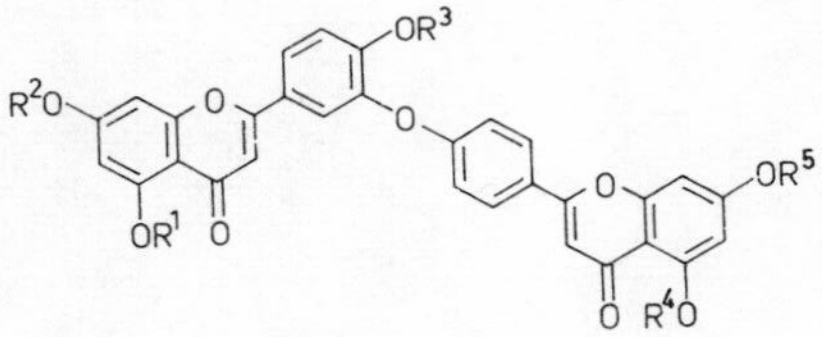

	R^1	R^2	R^3	R^4	R^5
(57a) ochnaflavone (= OS I)	H	H	H	H	H
(57b) ochnaflavone 4'-methyl ether (= OS II)	H	H	Me	H	H
(57c) ochnaflavone 4', (7 or 7'')-dimethyl ether (= OS III)	H	Me	Me or Me	H	H
	H	H	Me	Me	H
(57d) ochnaflavone pentamethyl ether	Me	Me	Me	Me	Me

Okigawa *et al.*, 1973.

Table 13.1—*Continued*

Miscellaneous synthetic biflavonoids. This comprises several bi-flavonoids which might occur naturally, since their interflavonyl links are situated between atoms which take part in the interflavonyl link of naturally occurring biflavonoids. Only the basic compounds are given, irrespective of whether this or a derivative has actually been obtained.

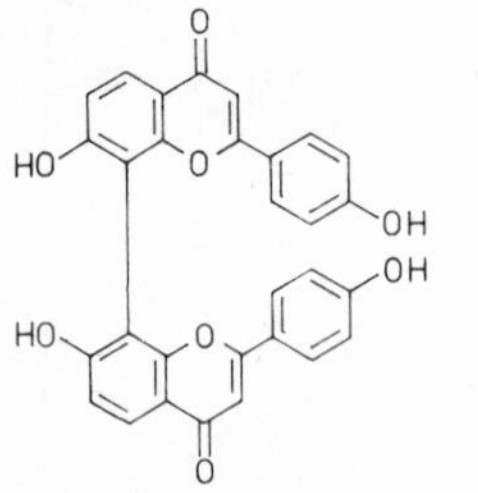

(58) 5,5'-dideoxycupressuflavone
Chandramouli *et al.*, 1972.

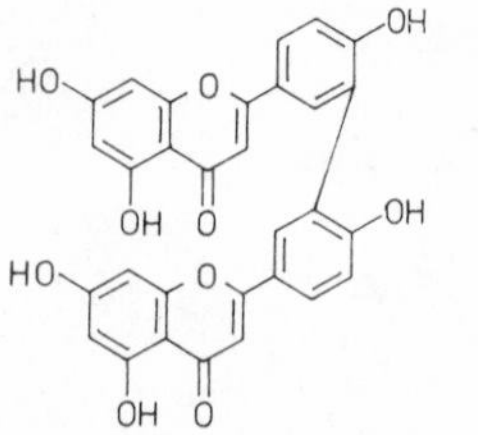

(59) 3',3'''-biapigenin
Nakazawa, 1962 (synth. of hexamethyl ether)

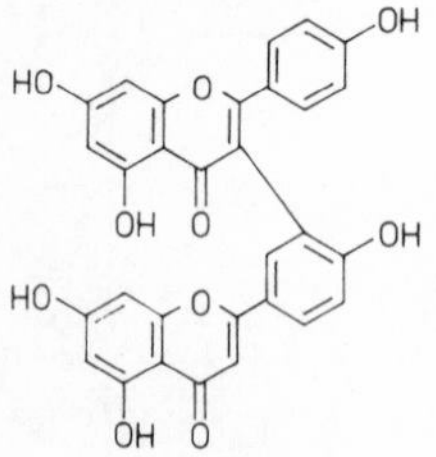

(60) 3,3'''-biapigenin
Molyneux *et al.*, 1970.

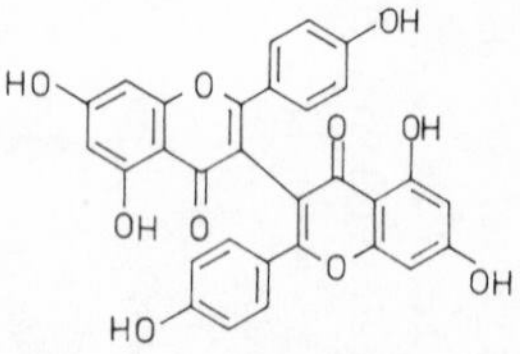

(61) 3,3''-biapigenin
Molyneux *et al.*, 1970.

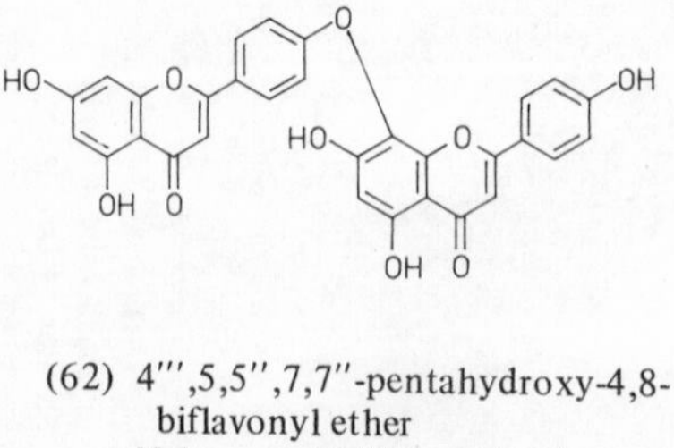

(62) 4''',5,5'',7,7''-pentahydroxy-4,8-biflavonyl ether
Nakazawa 1968 (synth. of pentamethyl ether)

13.3 Detection and isolation of biflavonoids

13.3.1 Extraction

The solvent used for the extraction of biflavonoids depends largely upon the nature of the biflavonoids and the nature of the plant material from which they have to be extracted, since all biflavonoids are more or less strongly absorbed in the tissue in which they occur. The following statements are, unless otherwise stated, based on our own experience with biflavonoids of the hinoki- and amentoflavone groups.

Saturated aliphatic hydrocarbons are the only solvents in which biflavonoids are insoluble and so they can be used for preliminary defatting. In some instances, benzene has extracted appreciable amounts of biflavonyl trimethyl ethers, e.g., sciadopitysin. Methylene chloride is still more effective. Trichlorethylene has been used for the extraction of kayaflavone (Kawano, 1961), and chloroform has been used for the extraction of *Garcinia* biflavonoids (cf. Herbin *et al.*, 1970; Jackson *et al.*, 1971). Ether extraction in most cases yields all the biflavonoids present, although the unmethylated types are sometimes extracted rather slowly. In this case acetone, ethyl methyl ketone, methanol or ethanol with or without addition of some water may be used. The only drawback of these excellent solvents is that they extract large amounts of other substances along with the biflavonoids.

Since many biflavonoids are only sparingly soluble in the solvents used for their extraction, in many instances the extracts, after reduction of their volume, deposit crude crystals of biflavonoids, which may be filtered off. The mother liquors, however, may still contain considerable amounts of biflavonoids, which may be recovered by a suitable combination of the methods described below.

Biflavonoids and other acidic material may be separated from water-immiscible solvents by extraction with dilute sodium hydroxide or sodium carbonate and subsequent acidification. This method may, however, lead to some racemization with biflavonoids containing a flavanone moiety. Distribution between two immiscible solvents is another means of preliminary purification: e.g., water-soluble material can be removed by distribution between ethyl methyl ketone and water, whereas lipids may be removed by distribution between dimethylformamide and light petroleum.

Finally column and preparative thin layer chromatography (see below) may be used not only for resolving biflavone mixtures, but also for the elimination of by-products. The result of all operations described above has, of course, to be followed by TLC.

13.3.2 Thin layer chromatography

Thin layer chromatography (TLC) has now almost completely replaced paper chromatography for the separation of biflavones. Silica gel, cellulose and polyamides have been used as TLC supports. TLC separations, especially on silica gel, are affected by so many known and unknown factors that systems giving excellent results in the hands of one worker may prove unsatisfactory in the hands of others. Therefore, current techniques are treated here in general terms without mention of specific separations, examples of which can be found in nearly every paper dealing with biflavonoids.

With silica gel as support, the solvents used can be divided into two categories. The solvents of the first category are composed of 50–90% benzene or toluene, 0–10% formic or acetic acid and 5–40% of one of the following solvents: pyridine, ethyl formate, ethyl acetate, dioxane, dimethylformamide, methanol or mixtures thereof. The two most widely used systems of this type, toluene/ethyl formate/formic acid (5 : 4 : 1) and benzene/pyridine/formic acid (36 : 9 : 5), were originally introduced for monoflavonoids (Stahl and Schorn, 1961; Hörhammer et al., 1964). In the paper of Chexal et al. (1970a), R_f values of a large number of biflavonoids on silica gel are listed. The second category of solvents comprises chloroform and its mixtures with up to 50% ethyl acetate or up to 15% methanol. Since commercial chloroform contains about 1% of ethanol as stabilizer, chloroform from the bottle will give higher R_f values than purified chloroform. Solvents of this type have been used, for example, by Herbin et al. (1970), Beckmann et al. (1971) and Rao et al. (1973).

On cellulose plates n-butanol saturated with 2N ammonia has been successfully used as solvent (Harborne, 1967; Harborne and Quinn, unpublished). With polyamides (polyamide 6, 6.6 or 11), mixtures of nitromethane and methanol between 3 : 4 and 7 : 3 and methanol/acetic acid (9 : 1) are used (c.f. Beckmann et al., 1971). In most TLC systems, separation of biflavonoids into groups having the same degree of methylation normally offers no problem, whereas the separation within these groups is usually poor. The system cellulose and n-butanol/ammonia is especially effective for the separation of isomeric methyl ethers, since it responds to the differences in acidity of the remaining free hydroxyls (cf. Baker et al., 1963).

Two-dimensional TLC itself has so far not been used with biflavonoids, but the combination of preparative TLC or column chromatography and analytical TLC is widely used. The 'multiple elimination' TLC technique of Van Sumere et al. (1968a, b) might also be useful with biflavonoids. Preparative TLC (Halpaap, 1963) on silica gel layers is an excellent means for the isolation and purification of weighable amounts of biflavonoids, especially the more highly methylated ones. For the separation of small amounts, the same solvents may be used as in analytical TLC, but if larger amounts (e.g., 100 mg on a 2 mm layer on a 20 x 20 cm^2 plate) are to be separated, the developing mixture should be such that it dissolves the substances to be separated; otherwise streaking may take place. For examples of preparative TLC separations, see Khan et al., 1972; Beckmann et al., 1971 and Jackson et al., 1971.

Visualization of biflavonoids is effected in the same way as for monoflavonoids. Biflavones with free hydroxyl groups in the 5-positions appear in UV light as dark spots quenching any fluorescence of the layer; and, after spraying with ferric chloride (Kawano et al., 1964, c.f. Miura et al., 1969), aluminium chloride, diphenylboric acid β-aminoethyl ester (Beckmann et al., 1971) or other substances with which they form chelates, they fluoresce yellow, green or brown. These colours are sufficiently different for it to be possible to distinguish between various biflavonoids, if they are on the same chromatogram, but they are not different enough to be described unequivocally. Biflavone permethyl ethers fluoresce brightly in UV light without any spray, but the colours are described quite subjectively (Chexal et al., 1970a; Natarajan et al., 1970b; Beckmann et al., 1971); determination of the fluorescence maxima would be very useful. Biflavonoids containing at least one flavanone moiety may be visualized as orange or red spots by the method of Horowitz (1957), namely spraying with a solution of sodium borohydride in isopropanol and subsequent exposure of the plate to HCl fumes (cf. Beckmann et al., 1971). This reaction may fail when analytical grade isopropanol is used; in this case, a small drop of acetone must be added to the spray solution (El-Dessouki and Geiger, unpublished).

13.3.3 Column chromatography

Columns of silica gel, magnesium silicate, polyamide, polyvinyl-pyrrolidone (polyclar AT^R) and sephadex LH 20^R have been used. Suitable eluents can be found by preliminary tests on thin layers, when the R_f values of the compounds to be separated should preferably not exceed 0.5, thus allowing fast travelling contaminants to appear in the fore run. The biflavonoids being separated must be reasonably soluble in the eluting solvent. When crude extracts are subjected to column chromatography, it is best to start eluting with a solvent that leaves the biflavonoids near the origin, thus eliminating as many contaminants as possible before the solvent that elutes the biflavonoids is applied.

The inorganic absorbents, silica gel and magnesium silicate, have been used for preliminary purifications (e.g., Garg and Mitra, 1971; Khan *et al.*, 1972; Rao *et al.*, 1973) as well as for actual separations (e.g., Natarajan *et al.*, 1970b; Joshi *et al.*, 1970). However, preparative TLC is usually preferred for separations with these absorbents because of its superior resolution. Good separations on polyamide and polyvinylpyrrolidone, two absorbents of very similar properties, have been reported (Beckmann *et al.*, 1971; Geiger and De Groot, 1971; Dossaji *et al.*, 1973) the eluents used being ethylene glycol monomethyl ether, aqueous acetone or chloroform/methanol/ethyl acetate (4 : 2 : 1). Sephadex LH 20^R is equally good for the separation of biflavonoids from each other as well as from high-molecular-weight contaminants that often interfere with the crystallization of these compounds. It is thus advantageously used in the final step of the purification. For monitoring the separation on columns of adsorbent, TLC is employed. Maximum information is obtained if a TLC system is used whose separation characteristics differ from those of the column. In this way, a 'two-dimensional' chromatogram is obtained.

13.3.4 Countercurrent distribution

Countercurrent distribution is an excellent means for the separation of very similar compounds. Since most commercial machines allow recycling, several thousand transfers are normally possible. In the biflavone field, two phase-systems have been described. The system ethyl methyl ketone/borate buffer was introduced by Baker *et al.* (1963) and has been used successfully several times since (Miura *et al.*, 1968; Khan *et al.*, 1972). Beckmann *et al.* (1971) used the phase pair aqueous dimethylformamide and ether. In Beckmann's system, the biflavonoids are separated into groups containing about the same number of methoxyl groups, whereas Baker's system also separates isomeric pairs that differ in the acidity of the free hydroxyl groups (e.g., ginkgetin and isoginkgetin).

13.4 Identification and determination of structure

13.4.1 Derivatives

Acetates, methyl ethers and deuteromethyl ethers are the derivatives usually prepared during identification and structural elucidation of biflavonoids. With biflavones, there is no difficulty in preparing these derivatives by standard methods. If, however, a flavanone moiety is present in the molecule, the heterocycle of this moiety, under the influence of bases, may be cleaved and the corresponding chalkone formed. The conditions under which this occurs vary from one compound to another; for details, reference must be made to the original literature (Beckmann *et al.*, 1971; Jackson *et al.*, 1971; Joshi *et al.*, 1970; Konoshima and Ikeshiro, 1970; Pelter *et al.*, 1971a; Rao *et al.*, 1973; see also Asahina and Inubuse, 1928; and Dean, 1963).

Deuteromethyl ethers are prepared in the same way as the methyl ethers, using the deutero analogue of the methylating agent. A very economic method uses deuterodiazomethane, which can easily be obtained by base catalysed D/H-exchange between diazomethane and deuterium oxide (van der Merwe *et al.*, 1964; Khan *et al.*, 1972).

13.4.2 Identification with known compounds

If an authentic sample is at hand, identification may be achieved by any of the usual methods (co-chromatography, mixed melting points of the substance itself and its derivatives, UV, IR, NMR and mass spectrometry). At least two independent methods should be used. Apart from the facts that R_f values and UV spectra are in many cases very similar and that a number of biflavonoids melt with decomposition, another pitfall is that some biflavonoids may form solvates with more than one solvent. This may lead to some confusion with NMR and especially IR spectra (Geiger, unpublished).

If no authentic sample is available, the physical and chemical properties of the substance must be compared with those reported in the literature. If no good NMR and MS data are reported, this may mean that the structure must be proved again, as in the original determination. In the case of the many partial methyl ethers, the type to which they belong is determined by preparing the permethyl ethers, which can be readily identified by TLC (Natarajan *et al.*, 1970b). The methylation pattern is then determined by suitable fragmentation, as described in the next section.

13.4.3 Determination of structure

This section is mainly confined to classical chemical and physical methods. NMR and MS methods, which are often superior, are dealt with briefly here, since they are covered comprehensively in other chapters. Even in the first steps of any structure elucidation, the elemental analysis and molecular weight determination, classical methods are inferior because of the tendency of biflavonoids and their derivatives to form rather stable solvates. The molecular weights and empirical formulae of the permethyl ethers are more easily obtained by high resolution mass spectrometry. The number of methoxyl groups can be learned either from the NMR spectrum (integral of the methyl protons) or from the mass spectra of the deuteriomethylated and methylated compounds.

Nevertheless, with the exception of the cupressu- and agathis-flavone type, the determinations of the basic skeleton of all types are mainly based on chemical facts. The extensive work of Japanese and British researchers that led to the structures of the amento- and hinokiflavone groups has been comprehensively summarized by Baker and Ollis (1961) and Baker *et al.* (1963). The chemical proofs of the structures of di- and tetrahydro-biflavonoids of these groups have been described by Beckmann *et al.* (1971) and Rao *et al.* (1973). The chemical work on *Garcinia* biflavonoids is found in the papers of Karanjgaokar *et al.* (1967), Konoshima *et al.* (1969), Jackson *et al.* (1971) and Pelter *et al.* (1971a). For chemical work on cupressu- and agathisflavone, see Murti *et al.* (1967) and Pelter *et al.* (1971c). The main procedures used are alkaline and alkaline oxi-dative degradations, by which biflavonoids are cleaved analogously to the corresponding monoflavonoids (see Venkataraman, 1962, and the papers quoted above). Alkaline degradation of biflavonoids thus yields three types of fragments. Two of them are derived from the aromatic rings not involved in the biflavonoid formation, whilst the third type is derived from the two aromatic rings connected by the interflavonyl link, and thus is important for establishing the position of this link.

The only structural feature of biflavonoids for which chemical degradation cannot supply the answer is the question whether in a given case C-6 or C-8 is involved in the interflavonyl link. This question can be dealt with either by complete synthesis (see next paragraph) or by NMR. The NMR technique used for this purpose involves the benzene induced methoxyl shifts, which occur only if one position *ortho* to a given methyl group is unsubstituted. For more details, see Chapter 2 and especially Pelter *et al.* (1969b, c and 1971a). The most unequivocal method of distinguishing between the two possibilities is that used by Chandramouli *et al.* (1972) to establish the location of the interflavonyl link in cupressuflavone (7). The complete methyl ether of the biflavone is selectively de-methylated in the 5- and 5′-positions, tosylated and finally reduced to yield the complete methyl ether of the corresponding 5,5′-dideoxybiflavone (58), the NMR spectrum of which reveals clearly whether there are protons *ortho* to the newly introduced protons in the 5- and 5′-positions.

A full account of the MS work on biflavonoids is found in Chapter 3. UV spectra, which are very useful in the monoflavone series (cf. Mabry *et al.,* 1970), are of limited value with biflavonoids, because in this case there are two independent chromophores, which respond independently to various shift reagents. Only where both moieties are alike, can conclusions be drawn (cf. Baker *et al.,* 1963).

Although many biflavonoids are known to be optically active, nothing is known about their absolute configuration. Since to our knowledge biflavonoids are the only natural products which show atropisomerism, it would be desirable to know more about this. In the case of biflavanones which have additional centres of chirality, a number of diastereomers are possible, but certain configurations seem to be preferred; otherwise, the existence in such cases of racemates (e.g., ± fukugetin) would be surprising.

13.4.4 Synthesis and transformations

Complete syntheses have been performed of amentoflavone hexa-methyl ether (Nakazawa, 1962), ginkgetin (Nakazawa and Ito, 1963), hinokiflavone pentamethyl ether (Nakazawa, 1968; Natarajan *et al.*, 1968 and 1970c), cupressuflavone hexamethyl ether (Nakazawa, 1962; Murti *et al.*, 1967; Ahmad and Razag, 1971; Moriyama *et al.*, 1972), agathisflavone hexamethyl ether (Pelter *et al.*, 1971c; Moriyama *et al.*, 1972) the heptamethyl ethers of fukugetin and saharanflavone (Ikeshiro and Konoshima, 1972), as well as ochnaflavone pentamethyl ether (Okigawa *et al.*, 1973).

Of more practical interest are some partial syntheses performed by modification of existing biflavonoids. Three types of reactions are important in this context: Wessely-Moser skeletal rearrangement, partial or total demethylation, and dehydrogenation of flavanone to flavone units. The Wessely-Moser rearrangement of 5-hydroxy-flavones is the result of an opening and reclosure of the heterocycle in such a way that the hetero-oxygen atom of the starting material becomes the 5-oxygen of the product and vice versa. In consequence, the 6- and 8-positions of the product are interchanged as compared with the starting material. This reaction is catalysed by acids; usually, hydrogen iodide in acetic anhydride at the boiling point is used. A typical example of this reaction is the interconversion of cupressu- and agathisflavone:

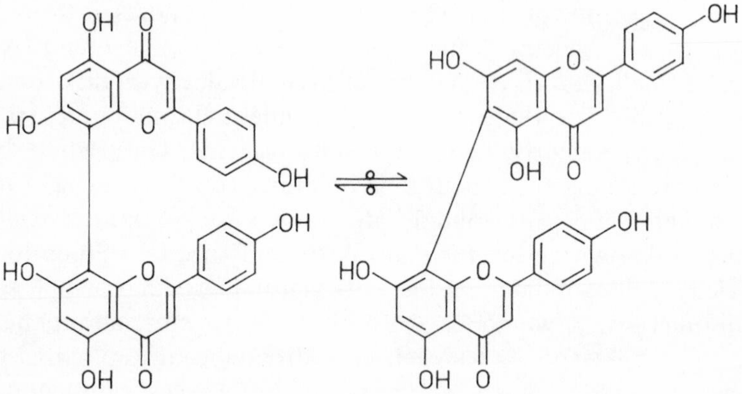

In this case the rearrangement leads to an equilibrium mixture (Pelter *et al.*, 1971c). For further examples, see Nakazawa (1968) and Chexal *et al.* (1970a, reference 20).

Partial demethylation is a very useful means for obtaining some partial methyl ethers starting from natural or synthetic more highly methylated products. The ease with which demethylation occurs depends upon the position of the methoxyl group, decreasing in the sequence $5 > 4' > 7$. For selective demethylation in the 5-position, anhydrous aluminium chloride in nitrobenzene has been used (Nakazawa, 1962). Partial demethylations leaving the 7-positions unaffected have been produced by hydrogen iodide in phenol containing acetic anhydride at reflux temperature (Miura *et al.*, 1966; Miura and Kawano, 1968b, c), or pyridinium chloride (Beckmann *et al.*, 1971). For complete demethylation, the use of anhydrous aluminium chloride in benzene (Murti *et al.*, 1967) and boron trichloride (Ahmad and Razag, 1971) has been reported. If only partial demethylation is wanted, it is in any case advisable to follow the reaction by TLC and to stop when the desired product has reached a maximum concentration. Although it is claimed that Wessely-Moser rearrangement does not occur during demethylation, it is advisable to check this point by remethylation to the full methyl ether of the starting product.

Two methods have been used to dehydrogenate flavanone units of biflavonoids to flavones. Iodine/alkali acetate in acetic acid has been used for compounds with free hydroxyl groups (Herbin *et al.*, 1970; Pelter *et al.*, 1971a) and chloranil in xylene has been employed for the dehydrogenation of permethyl ethers (Beckmann *et al.*, 1971).

13.5 Taxonomic significance of biflavonoids

13.5.1 Validity of the data

Much of the available data on the natural occurrence of biflavones comes from the work of chemists who have had no immediate thought of its application to taxonomy. The result is that it is often difficult to link some observations to a particular species with any degree of certainty. Several authors have drawn attention to this problem, and pointed out the need to cite authorities for any botanical names and also to cite voucher specimens of the material used. While the omission of this information in the early work on biflavones is understandable, the continuation of the practice up to the present day is to be deplored. There is little value in the assertion by one recent author that 'all specimens were correctly identified before the work was started'. For taxonomic purposes, it is essential that the sense in which a botanical name is being applied is made known. For example, *Podocarpus falcatus* R. Br., *P falcata* Hort. and *P. falcata* A. Cunningham are three quite different *Podocarpus* species. It is also essential to be able to refer back to the material, not only in cases where misidentification is suspected, but also where intraspecific variation is detected or where a new species is subsequently recognized. It is to be hoped that editors of phytochemical journals will soon come to insist on the inclusion of these details (cf. Notes to Contributors in the journal *Phytochemistry*).

In compiling the tables of natural distribution of biflavonyls, authorities are given wherever it is possible to assign them with any degree of certainty. It should be borne in mind, however, that in some cases these may not be the correct authorities for the name as applied by the original authors. Where no authority is given in the tables, there is confusion as to which is the correct one. The nomenclature of Dallimore and Jackson (1966) is used for the Coniferales and Taxales. Where work has been reported under a synonym, these are given as footnotes to the table.

Since the last useful review* of the field by Murti (1967), the range of compounds isolated and the range of taxa screened have been greatly extended; the conclusions drawn by Murti on the data then available are in many cases no longer justified. In addition, many of the earlier records, such as those by Sawada (1958), though extremely important at the time, are now of questionable value because of the limitations of the techniques used. A re-examination of some species has already shown that only the major chromatographic component was isolated by him in most instances. Some errors were also made in identification. Compare, for example, the results of Sawada with those of Khan *et al.* (1971c) for *Cephalotaxus harringtonia* var. *drupacea* (Sieb. & Zucc.) Koidzumi; the former recorded kayaflavone only, whereas the latter authors correct this to sciadopitysin and add ginkgetin, sequoiflavone and a trace of the parent compound, amentoflavone. In constructing the tables of natural distribution (Tables 13.2 to 13.5), therefore, reports by early workers quoted in Baker and Ollis (1961) that have not been confirmed or clearly refuted by later workers are represented by an open circle. In most cases these are in agreement with the results of later

*But see Kasakov *et al.* (1972).

Table 13.2   The distribution of biflavonoids in the Pteridophyta, Cycadales and Ginkgoales

Biflavonoid	13	14	15	16	18	21	23	25	26	27	29	47	49	56	References
PSILOTALES															
Psilotum triquetrum Swartz	•											•			Voirin & Lebreton, 1966.
SELAGINELLALES															
Selaginella nipponica Franch. & Savat.		•													Okigawa *et al.*, 1971.
S. pachystachys Koidz.		•													Okigawa *et al.*, 1971.
S. tamariscina (Beauv.) Spring.		•											•		Okigawa *et al.*, 1971; Hsu, 1959.
CYCADALES															
Cycadaceae															
Cycas revoluta Thunb.	•			?							•	•		•	Geiger & de Groot, 1971; Handa *et al.*, 1971b; Kawano & Yamada, 1960.
Zamiaceae															
Zamia angustifolia Jacq.	•	•	•	•		•	•	•	•	•					Handa *et al.*, 1971b.
Dioon edule Lindl.	•	•	•	•		•	•	•	•	•					Dossaji *et al.*, 1973.
D. spinulosum Dyer ex Eichl.	•	•	•	•		•	•	•							Dossaji *et al.*, 1973.
D. imbricata Dyer	•	•	•	•		•	•	•							Dossaji *et al.*, 1973.
D. mejea Stand. & L. C. Williams	•	•	•	•		•	•	•							Dossaji *et al.*, 1973.
D. purpusii Rose	•	•	•	•		•	•	•							Dossaji *et al.*, 1973.
D. san blas (authority ?)				•		•	•								Dossaji *et al.*, 1973.
GINKGOALES															
Ginkgo biloba L.					•	•	•								Baker *et al.*, 1963; Handa *et al.*, 1971b; Miura *et al.*, 1969.

Table 13.3   The distribution of biflavonoids in the araucariaceae and cupressaceae

Biflavonoid	1	2	3	4	7	8	9	10	11	13	15	16	20	23	25	26	47	49	50	51	References
AURACARIACEAE																					
Agathis dammara (Lamb.) Rich.[a]	●	●			●	●	●	●	●	●							●				Khan et al., 1971b, 1972; Mashima et al., 1970.
A. palmerstoni F. Muell.	●	●			●	?															Khan et al., 1971b; Mashima et al., 1970; Pelter et al., 1969c.
Araucaria bidwillii Hook.	●				●	●	?	●	●	●	●						●				Khan et al., 1971b; 1970; Mashima et al., 1970.
A. columnaris (Forst.) Hook.[b]						?	●	●	●						●	●	●				Ilyas et al., 1968; Khan et al., 1971b; 1972; Pelter et al., 1969a.
A. cunninghamii D. Don						?	●	●	●				●		●	●	●				Ilyas et al., 1968; Khan et al., 1971a, 1971b, 1972; Rahman & Bhatnagar, 1968.
CUPRESSACEAE																					
CUPRESSOIDEAE																					
Cupresseae																					
Cupressus arizonica Green				●	●	●											○				Miura & Kawano, 1968a; Sawada, 1958.
C. funebris Endl.				●	●	●				●							●	●	●		Natarajan et al., 1970b; Pelter et al., 1971c.
C. goveniana Gord.				●	●	●				●							●	●			Miura & Kawano, 1968a; Natarajan et al., 1970b.
C. lusitanica Miller				●	●	●				●							●	●			Natarajan et al., 1970b.
C. sempervirens var. sempervirens L.				●	●	●				●							●				Murti et al., 1964; Natarajan et al., 1970b.
var. stricta Ait.																					Natarajan et al., 1970b; Pelter et al., 1971c.
C. torulosa D. Don				●	●	●				●							●				Murti et al., 1964; Natarajan et al., 1970b.
Chamaecyparis obtusa (Sieb. & Zucc.) Endl.																	○	●	●		Miura, 1967b; Miura & Kawano, 1968a; Sawada, 1958.
cv. 'Breviramea'													○				●	●	●		Miura & Kawano, 1968a; Miura et al., 1968; Sawada, 1958.
C. pisifera (Sieb. & Zucc.) Endl.																	●	●	●		Miura & Kawano, 1968a.
cv. 'Filifera'																	○		●		Sawada, 1958.
cv. 'Squarrosa'																	●	●	●	●	Miura & Kawano, 1968a.

Thujopsideae										
Thuja occidentalis L.							○			Sawada, 1958.
T. orientalis L.[d]			●				●			Miura & Kawano, 1968a; Pelter *et al.*, 1970.
T. plicata D. Don			●				●			Rahman *et al.*, 1972.
T. standishii (Gord.) Carr.							●			Miura & Kawana, 1968a.
Thujopsis dolabrata (L. f.) Sieb. & Zucc.				○	○		●	●	●	Miura & Kawano, 1968a; Sawada, 1958.
Calocedrus decurrens (Torrey) Florin[e]							○			Sawada, 1958.
C. formosana (Florin) Florin[f]							○			Sawada, 1958.
Junipereae										
Juniperus chinensis L.[g]			●				●			Pelter *et al.*, 1971b.
cv. 'Sargentii'[h]						○	○			Sawada, 1958.
cv. 'Kaizuku'							●	●		Miura & Kawano, 1968a.
J. conferta Parl.						○	○			Sawada, 1958.
J. horizontalis Moench.	●		●		●		●			Hameed *et al.*, 1973b.
J. procumbens (Endl.) Miquel.						○	○			Sawada, 1958.
J. recurva Buch.-Ham.	●	●	●				●			Hameed *et al.*, 1973b.
CALLITROIDEAE										
Actinostrobeae										
Callitris columellaris F. Muell.[i]							○			Sawada, 1958.

a = *Agathis alba* Foxworthy; b = *Araucaria cooki* R. Br. ex D. Don; c = *Cupressus glauca* Lamarck; d = *Biota orientalis* Endl.; e = *Libocedrus decurrens* Torrey; f = *Libocedrus formosana* Florin; g = *Sabina chinensis* Antoine ?; h = *Sabina sargentii* Nakai ?; i = *Callitris glauca* R. Br. ex R. T. Baker & H. G. Smith.

Table 13.4   The distribution of biflavonoids in the Coniferales and Taxales

Biflavonoid	13	14	15	16	17	18	19	21	22	23	25	26	30	31	47	48	49	50	52	53	References
CONIFERALES																					
Podocarpaceae																					
Podocarpus gracilior Pilger	●							●													Chexal *et al.*, 1970b.
P. macrophyllus (Thunb.) D. Don[a]	●				●		●								●	●					Miura *et al.*, 1968, 1969.
P. montanus (Willd.) Loddiges[b]		●	●		●																Hameed *et al.*, 1973b.
P. nagi (Thunb.) Makino	●			●	●			●	●		●				●			●			Miura *et al.*, 1968, 1969.
Dacrydium cupressinum Sol.												●									Hodges, 1965.
Taxodiaceae																					
Cryptomeria japonica (L. f.) D. Don	●	●							●	●			●	●	●		●	●		●	Kawano *et al.*, 1964; Miura *et al.*, 1966.
cv. 'Zindai'	●	●											●	●	●		●	●		●	Miura *et al.*, 1968a.
cv. 'Araucarioides'			○	○					○		○				○						Sawada, 1958.
Cunninghamia lanceolata (Lamb.) Hook. f.			○	○							○				○						Sawada, 1958.
C. konishii Hayata[c]																					Miura & Kawano, 1968b.
Glyptostrobus lineatus (Poiret) Druce[d]																●	●		?		Miura & Kawano, 1968a.
Metasequoia glyptostroboides Hu & Cheng	●		?							●			●	●			●				Beckman *et al.*, 1971.
Sequoia sempervirens (D. Don) Endl.	●	●							●	●	○				●						Miura *et al.*, 1968b; Sawada, 1958.
Sequoiadendron giganteum (Lindl.) Buchholz	●					●									●					●	Geiger & Buck, 1973.
Taiwania cryptomerioides Hayata		●													○						Sawada, 1958.
Taxodium distichum (L.) Rich.	●			●					●	●					●		●	●			Geiger & de Groot, 1973.
T. mucronatum Tenore										●					●			●			Pelter *et al.*, 1969b.
Sciadopitys verticillata (Thunb.) Sieb. & Zucc.										○											Sawada, 1958.
Cephalotaxaceae																					
Cephalotaxus harringtonia var. *drupacea* (Sieb. & Zucc.) Koidzumi[e]	●					●				●											Khan *et al.*, 1971c.

TAXALES

Taxaceae

Amentotaxus formosana Li   Hörhammer *et al.*, 1966.

Taxus baccata L.   Di Modica *et al.*, 1960.

T. cuspidata Sieb. & Zucc.   Sawada, 1958.

T. floriana Chapman   Sawada, 1958.

Torreya nucifera L.   Sawada, 1958.

a = *Podocarpus chinensis* Sweet ?; b = *Podocarpus taxifolius* Kunth.; c = *Cunninghamia lanceolata* var. *konishii* Fujita ?; d = *Glyptostrobus pensilis* (Staunton) K. Koch; e = *Cephalotaxus drupaceae* Sieb. & Zucc.

Table 13.5   The distribution of biflavonoids in the angiosperms

Biflavonoid	7	13	17	19	24	33a	33b	33c	36	37	38	39	40	41	42	43	44	45	46	47	63	64	65	References
ANACARDIACEAE																								
*Rhus succedanea**	●																			●				Lin & Chen, 1973; Chen et al., 1974.
Semecarpus anacardium L. f.						●	●	●																Rao et al., 1973.
CAPRIFOLIACEAE																								
Viburnum prunifolium L.	●																							Horhammer et al., 1966.
CASUARINACEAE																								
Casuarina cunninghamiana Miq.				●																				Natarajan et al., 1971b.
C. stricta Ait.																			○					Sawada, 1958.
C. junghuhniana Miq.																			●					Natarajan et al., 1971b.
C. suberosa Otto & Dietr.																			●					Natarajan et al., 1971b.
EUPHORBIACEAE																								
Hevea brasiliensis Mull. Arg.					●																			Chandramouli et al., 1971; Madhav, 1969.
Putranjiva roxburghii Wall.		●	●																					Garg & Mitra, 1971; Varshney et al., 1973.
GUTTIFERAE																								
Allanblackia floribunda Oliv.									●		●													Locksley & Murray, 1971.
Calophyllum inophyllum L.	●																							Desai et al., 1967.
Garcinia buchananii Baker														●	●	●	●							Jackson et al., 1971.
G. eugeniifolia Wall.														●	●	●	●							Jackson et al., 1971.
G. linii Chang.														●	●	●	●							Konoshima et al., 1970.
G. livingstonei T. Anders.			●							●	●													Petter et al., 1971a.
G. morella									●	●	●													Karanjgaokar et al., 1967.
G. multiflora Champ. ex Benth.														●	●	●	●							Konoshima et al., 1970.
G. spicata Hook. f.									●	●	●	●	●	●	●	●								Konoshima et al., 1969, 1970; Konoshia & Ikeshiro, 1970.

G. talboti Raiz.	• •			Joshi et al., 1970.
G. volkensii Engl.				Herbin et al., 1970.
G. xanthochymus Roxb.				Konoshima et al., 1970.
OCHNACEAE				
Ochna squarrosa L.		• • •		Okigawa & Kawano, 1973.
RHAMNACEAE				
Phyllogeiton zeyheri (Sond.)			•	du Volsteed & Roux, 1971.
Suesseng.				

*also contains biflavonoids 1, 6a and 6b.

workers for closely related species, and so may be accepted as correct though perhaps incomplete. In other cases, where they diverge from later results (e.g., records of sotetsuflavone (16) in two species of the Cupressaceae – see Table 13.3), these early results must be treated cautiously. In either case, it is important that they be checked.

Many recent reports still include unidentified partial methyl ethers; this is because of the great difficulty in separating and identifying some of the complex mixtures encountered. Thus, many of the records given in the tables are incomplete, even for those species that have been intensively studied (e.g. Khan *et al.*, 1971b). Therefore, despite the great increase over the last six years in our knowledge of the natural distribution of biflavonyls, we are still likely to draw incorrect biochemical distinctions between different taxa. This is particularly true at the species level. In many instances differences between the known biflavonyl patterns of two closely related species is due mainly to different workers being responsible for the data. The genus *Garcinia* provides a good example of this. There seems little doubt that *G. eugeniifolia* and *G. buchananii* will prove to contain some of the flavanone-flavones (36–40) which are present in all other species of the genus, just as *G. morella* will probably be found to possess the biflavanones (41–44) which were unknown at the time the work on this species was done.

That differences do occur between the biflavonyl patterns of closely related species is certain, but where a range of species from the one genus has been screened by the same workers, there is generally a high degree of uniformity. Differences between species tend to involve the position or degree of methylation of the parent compounds. On the other hand, differences in biflavonyl groups are generally restricted to the sub-family level or above. For example, in the recent survey of *Dioon* (Cycadales, Table 13.2), Dossaji *et al.* (1973) recorded amentoflavone (13) as the major biflavone in all six species. In addition, all contained lesser amounts of the same four partial methyl ethers (15, 18, 23, 26). Only two compounds were inconsistently distributed: amentoflavone 7-methyl ether (14) is in all but *D. san blas (sic)*, and the fully methylated amentoflavone (dioonflavone-27) is in only *D. edule* and *D. spinulosum*.

That more basic differences may occasionally separate closely related species is evident from the report by Natarajan *et al.* (1970b) that *Cupressus sempervirens* is distinguished from all other species of the genus so far examined by the absence of the hinokiflavone group (Table 13.3). However, in most other cases where such differences between the biflavonyl patterns of closely related species appear in the tables, they are unlikely to be borne out by a careful reinvestigation.

In spite of the limitations of the present data, the biflavone pattern shows promise of being very useful at the specific level, and may help to sort out some of the generic and subgeneric problems in the conifers in particular. The large amount of work involved in screening the widest possible range of species in such groups would certainly seem to be well justified.

13.5.2 General distribution

At an early stage in the study of biflavonyls it became apparent that while they were characteristic of the gymnosperms with the exception of the Pinaceae and Gnetales, they were also present in some angiosperms (*Casuarina stricta* Ait.; Sawada, 1958) and some of the more primitive vascular plants (*Selaginella tamariscina*; Hsu, 1959). This broad pattern of distribution has been little changed by subsequent work. Biflavonyls have now been isolated from four of the five living orders of the gymnosperms and from every family of the Coniferales with the exception of the Pinaceae. Although the absence of biflavonyls from the Gnetales still rests upon a published report on only one species (*Ephedra gerardiana* Wall; Sawada, 1958), a recent unpublished screening of members of all three families by Swain and co-workers (pers. comm.) has confirmed this absence. The inclusion of the Gnetales in the gymnosperms is, of course, strongly debated, and some workers prefer to place them as a separate group, the Chlamydosperms, distinct from both the gymnosperms and angiosperms. The absence of biflavones, which are such a characteristic feature of all other orders of the gymnosperms, certainly sets the Gnetales apart as a very distinct order.

13.5.3 Non-seed plants

The early report of amentoflavone (13) in one species of *Selaginella* by Hsu (1959) has been confirmed and extended to two other species by Okigawa *et al.* (1971) (see Table 13.1). *S. tamariscina* also possesses hinokiflavone (47) and two of its dimethyl ethers (49 and unidentified: Okigawa *et al.*, 1971). No biflavone has been reported from any member of the related order of lycopods, Lycopodiales, nor has one been found in the ferns. Voirin and Lebreton (1966) reported both amentoflavone and hinokiflavone in *Psilotum tri-quetrum* (Psilotales), a member of the order that shows one of the most primitive organizations of any living vascular plant. It would be very interesting to know whether *Tmesipteris*, the only other genus in this order, also produces biflavones.

The only record of a biflavonyl in a non-vascular plant is the recent isolation of biluteolin (28) from the moss *Dicranum scoparium* Hedw. by Nilsson (1973).

13.5.4 Gymnosperms

Cycadales. This order includes nine living genera, and is regarded as a group of very ancient gymnosperms. According to Johnson (1959), it comprises three families. The first of these, the Cycadaceae, includes the genus *Cycas* alone, of which there are 15 species. According to recent work by Handa *et al.* (1971a) and Geiger and de Groot (1971), amentoflavone (13) and hinokiflavone (47) are present in the leaves of *Cycas revoluta* (Table 13.2). In addition, the latter authors record 2,3-dihydro derivatives of both compounds (29, 56). The presence of sotetsuflavone (16) has been in dispute since it was first reported for the species by Kawano and Yamada (1960), and to date the matter has not been satisfactorily resolved.

There are seven genera in the Zamiaceae, the largest being *Zamia* with 30 species. Only one species of this genus has been examined for biflavones (Handa *et al.*, 1971a; see Table 13.2). *Dioon*, on the other hand, has been extensively surveyed by Dossaji *et al.* (1973). A highly uniform pattern emerges from this work, and according to the latter author, unpublished work on other genera in the family agrees with this pattern. The absence of hinokiflavone and its derivatives clearly distinguishes the Zamiaceae from the Cycadaceae, though it is highly desirable that a wider range of species be examined, particularly in the latter family.

The third family, the Stangeriaceae, includes only one species, *Stangeria paradoxa* T. Moore from Natal. This species possesses by far the greatest number of primitive and fern-like features of any of the cycads (Sporne, 1965; p. 117). It would therefore be of particular interest to see whether it shows any distinctive features in its biflavone pattern.

Ginkgoales. This monotypic order was the source of ginkgetin (18), the first biflavone to be identified. Subsequently several other members of the amentoflavone group (15, 21, 23) have been isolated from *Ginkgo biloba* (Table 13.2), but no other biflavonyl group has been detected.

Araucariaceae. This family includes two genera, namely *Agathis* (21 species) and *Araucaria* (14 species). Five species have been investigated for the biflavone content of their leaves (Table 13.3). In addition to the definite records included in this table, further incompletely identified partial methyl ethers have been isolated from each species (Khan *et al.*, 1971b). Thus all five species are known to contain members of the amentoflavone, hinokiflavone and cupressuflavone group, as well as a group based on agathisflavone (1) which is unique to the family. The biflavone pattern of this family is therefore the most complex yet known. The occurrence of highly methylated derivatives of cupressuflavone is also a distinguishing feature of the family. There is only one record of such a derivative outside the family; that being 7,7''-dimethylcupressuflavone found recently in *Juniperus recurva* by Hameed *et al.* (1973b).

The Araucariaceae have long been regarded as a very distinctive family of conifers because of their primitive embryology and wood anatomy, highly specialised cone and leaf morphology, and their distinctive chromosome number of $n = 13$. The complexity and distinctiveness of their biflavone pattern reinforces this view.

Cupressaceae. According to Dallimore and Jackson (1966) there are eighteen genera in the family, seven of which are monotypic. Only three of the remaining genera are large: *Juniperus* with 70 species, *Cupressus* with 20 species and *Callitris* with 15 species. The most recent classification of the family (Li, 1953) recognizes two sub-families: the Callitroideae, which is largely of southern distribution, and the Cupressoideae, which is confined to the northern hemisphere. Each sub-family is further divided into three tribes.

Recent reports are available on the biflavones of five genera, and there are early records for two more (Table 13.3). Apart from the report of hinokiflavone in *Callitris glauca* R. Br. (*C. columellaris* F. Muell.) by Sawada (1958), all species studied belong to the Cupressoideae. It will be particularly interesting to see if further studies of the Callitroideae produce any biochemical data to support this basic division of the family, which appears well founded on morphological and geographical criteria.

Those northern species studied have been drawn almost equally from the three tribes (Table 13.3). The most significant feature is the occurrence of cupressuflavone (7), along with amentoflavone (13) and/or hinokiflavone (47) and their methyl ethers in both *Cupressus* (Natarajan *et al.*, 1970b) and more recently in *Juniperus* (Hameed *et al.*, 1973a). This similarity in biflavones clearly supports Li's grouping of *Juniperus* and *Cupressus* within the one sub-family, as compared with Pilger's (1926) separation of them into different subfamilies. It is possible that a reinvestigation of the remaining species of *Juniperus*, and of the other genera in the tribe, will show cupressuflavone to be typical of the Junipereae.

On present evidence there is a marked contrast between the biflavone pattern of *Cupressus* and that of *Chamaecyparis*, the only other genus of the Cupresseae so far investigated. Miura and Kawano (1968a) have recorded only hinokiflavone and its methyl ethers in *Chamaecyparis obtusa* and *C. pisifera*; Sawada's early record of sotetsuflavone (16) in *C. obtusa* cv. Breviramea, is most probably in error. It does not seem possible that both amentoflavone and cupressuflavone could have escaped detection by Miura and Kawano, so the tribe has a discontinuous biflavone pattern. An examination of the remaining four species of *Chamaecyparis* and of the monotypic *Fokienia* is obviously desirable to confirm this biochemical discontinuity within the Cupresseae.

Cupressuflavone appears to be absent from the tribe Thujopsideae. Amentoflavone (13) and hinokiflavone (47) were both recorded in two species of *Thuja* examined by Pelter *et al.* (1970), although the former was not found in *Thuja orientalis* L. by Natarajan *et al.* (1970a, *ut Biota orientalis* Endl.) or in its 'var. *mexican* Dümmer, Dallimore and Jackson (1966, p. 617) state that *Thuja orientalis* var. *mexicana* Dümmer is most probably synonymous with *Juniperus flaccida* var. *poblana* Martinez, but all recent reports on the genus *Juniperus* have also included amentoflavone. Pelter's record is therefore accepted for *Thuja orientalis* L., and the taxa examined by Natarajan must remain in doubt. The fact that both hinokiflavone and amentoflavone occur in *Thuja plicata* (Rahman *et al.*, 1972), which has been confirmed by one of us (C.Q.), supports the contention that both biflavone groups are typically present in this genus.

Miura and Kawano (1968a) record only hinokiflavone and its methyl ethers in the leaves of *Thujopsis dolabrata*, but make no reference to the earlier record of amentoflavone derivatives (16, 23) in this species by Sawada (1958). The only other records for this tribe are of hinokiflavone in *Calocedrus* (Sawada, 1958, *ut Libocedrus formosana* Florin and *L. decurrens* Torrey), but a reinvestigation of this genus may reveal a wider spectrum of compounds as has been the case with most other genera.

Despite the fragmentary nature of the records, then, the biflavone patterns in the Cupressaceae are of considerable interest from the taxonomic standpoint, and the accumulation of further data is highly desirable so that the biochemical discontinuities can be more accurately defined.

Podocarpaceae. This family is distinguished by having an almost completely southern distribution. There are seven genera, of which the largest are *Podocarpus* (*ca.* 100 species) and *Dacrydium* (23 species). The remainder contain between one and five species. In view of recent suggestions that both *Podocarpus* and *Dacrydium* are unnatural genera (Tengner, 1965; de Laubenfels, 1969; Quinn, 1970), the biflavone pattern in this family is of particular interest to taxonomists since it may help in forming a new set of generic boundaries. Apart from the report by Hodges (1965) of amentoflavone tetramethyl ether as the main biflavonyl in the leaves of *Dacrydium cupressinum*, the only data available for the family are for four species, each belonging to a different section of *Podocarpus* (Table 13.4). Cambie and James (1967) reported amentoflavone derivatives in the leaves of several unnamed members of the family, but were unable to locate any derivatives of hinokiflavone or cuppressuflavone. Miura *et al.* (1968, 1969) and Hameed *et al.* (1973b), however, have found hinokiflavone in two species of *Podocarpus*, and unpublished work by one of us (C.Q.) has shown it to be present in two Australian species of the genus. It is probable, therefore, that this family will prove to have both amentoflavone and hinokiflavone derivatives in its basic biflavonyl pattern. Podocarpusflavone A (17), which has only been found in one other conifer species, is present in every species of *Podocarpus* so far investigated. Whether this compound proves to be a useful taxonomic marker at either the generic or family level will rest on the results of a much wider survey of the family, which is at present in hand.

Taxodiaceae. This family, which is a much reduced relict of a formerly important group of forest trees, is represented today by only ten genera, the largest of which include only three species (*Taxodium* and *Arthrotaxis*). Recent reports on the biflavones are available for seven of these general (Table 13.4). *Taxodium, Metasequoia* and *Sequoiadendron* possess both amentoflavone and hinokiflavone derivatives. The two dihydro-derivatives of amento-flavone (30, 31) found in *Metasequoia* by Beckmann *et al.* (1971) are the only instances of such compounds in the Coniferales. Recent reports on *Cryptomeria* have described a range of hinokiflavone biflavonyls (47, 50, 53), but make no mention of the amentoflavone derivatives (16, 23, 25) reported by Sawada (1958). Sawada (l.c.) also reported hinokiflavone (47) and kayaflavone (25) from *Cunninghamia* and *Sequoia*, while Miura and Kawano (1968b) isolated sequoia-flavone (14) from both genera. According to Miura and Kawano (1968a) *Glyptostrobus* contains only hinokiflavone derivatives.

The only records for the remaining two genera, *Taiwania* and *Sciadopitys*, are Sawada's report of hinokiflavone (47) from the former and sciadopitysin (23) from the latter. No examination has been made of *Arthrotaxis*, the sole outlier of the family in the southern hemisphere, but this genus seems to be a typical member of the family on cytological as well as embryological and morphological grounds. *Sciadopitys*, on the other hand, is atypical of the family in both its embryogeny and cytology (Sporne, 1965, p. 143), and there have been proposals to remove it to a family of its own. The absence of the hinokiflavone group from the biflavonyl pattern of the genus is further evidence to support this move (Baker and Ollis, 1961, p. 183), though in view of the many changes made to Sawada's results for other genera in the family, it is important that the biflavonyl pattern of *Sciadopitys* should be checked to confirm this distinction.

Cephalotaxaceae. This monogeneric family has been found to contain only amentoflavone (13) and its methyl ethers (14, 18, 23) (Table 13.4). This is the only occurrence of ginkgetin (18) apart from its original isolation from *Ginkgo biloba*. The absence of all but the amentoflavone group makes this one of the simplest biflavonyl patterns in the Coniferales.

Taxales. Apart from the record by Kariyone (see Hörhammer *et al.*, 1967) of amentoflavone in the leaves of *Amentotaxus formosana*, there has been no additional information on the biflavones of this family since the review by Baker and Ollis (1961). Judging from recent work on other families, further studies are likely to add greatly to the range of biflavones recorded in the Taxaceae. On present records there is nothing in the biflavone pattern to support the removal of the family from the Coniferales to a separate order, the Taxales (Florin, 1948), unless it is the absence of any but the amentoflavone series, which could be regarded as a primitive pattern (see Section 13.6.2).

13.5.5 Angiosperms

Biflavonoids have now been recorded in members of seven angiosperm families (Table 13.5). They appear to be widespread in two of these. Four of the seven species so far examined from the mono-generic family Casuarinaceae have proved to possess biflavones (Natarajan *et al.*, 1971b). *Casuarina cunninghamiana* was previously reported to be without biflavones (Bate-Smith, 1962), but the former workers found that while this is true of the Australian material, material cultivated in India possesses cupressuflavone rather than the hinokiflavone found in the other species. A more extensive survey is needed to determine the limits of this variation in the family.

The Guttiferae has proved to be extremely rich in biflavonyls. While the amentoflavone group has been recorded from the leaves of two species (*Garcinia livingstonei*, Pelter *et al.*, 1971a; *Calophyllum inophyllum*, Desai *et al.*, 1967), the most characteristic feature of the family is the 3-8 linked *Garcinia* biflavonyl group. A range of these compounds has been isolated from bark (Konoshima and Ikeshiro, 1970), roots (Joshi *et al.*, 1970) or heartwood (Jackson *et al.*, 1971; Karanjgaokar *et al.*, 1967; Locksley and Murray, 1971; Pelter *et al.*, 1971a) of ten species of *Garcinia* and one species of *Allanblackia*. More recently, Okigawa *et al.* (1973) have found a series of 3'-4' linked biflavone ethers based on ochnaflavone (63) in the leaves of *Ochna squarrosa* belonging to a family, Ochnaceae, which is regarded as being closely related to the Guttiferae.

There are reports of biflavonyls in isolated species of four other families. In the Caprifoliaceae, Hörhammer *et al.* (1966) reported amentoflavone in the bark of one species of *Viburnum*, but failed to find any biflavones in twenty-one other species examined. The amentoflavone group have also been isolated from the leaves of two species in the Euphorbiaceae: *Hevea brasiliensis* (Madhav, 1969;

Chandramouli *et al.*, 1971) and *Putranjiva roxburghii* (Garg and Mitra, 1971; Varshney *et al.*, 1973). Another record comes from the Rhamnaceae, where the heartwood of *Phyllogeiton zeyheri* has been found to contain an unusual biflavonyl, zeyherin (46), (du Volsteedt and Roux, 1971). Finally, Rao *et al.* (1973) record three related biflavones (33a, b, c) in *Semecarpus anacardium* and Chen *et al.* (1974) five compounds (1, 6a, 6b, 13 and 47) in the drupes of *Rhus succedanea* (both of the Anacardiaceae).

These seven families are not regarded as forming a closely related group by either Takhtajan (1969) or Cronquist (1968) in their recent classifications of the angiosperms.

13.6 Evolutionary aspects

13.6.1 Production of biflavonyls

The occurrence of hinokiflavone in *Casuarina stricta* was seen by Baker and Ollis (1961, p. 183) as a retention of a primitive gymnospermous character in what he regarded as one of the most primitive angiosperms. Quite apart from the fact that the primitiveness of *Casuarina* is not tenable within current concepts, biflavonyls are now known to occur in six other widely separated angiosperm families, some of which are clearly very specialized. This raises the question of whether the occurrence of biflavonyls in angiosperms is in fact due to the retention of a primitive biochemical feature, or whether the ability to produce biflavonyls has evolved separately in each case. If the latter is the case, then the usefulness of these compounds as taxonomic markers must be greatly diminished. The biflavonyls present in four of the seven families (Anacardiaceae, Caprifoliaceae, Casuarinaceae and Euphorbiaceae) are either of the amentoflavone or hinokiflavone group. These are the most widely distributed groups of biflavonyls in the gymnosperms, and the same biflavonyl groups also occur in some primitive land plants. This suggests that in these angiosperm families at least, the presence of biflavones may be regarded as the retention of a more primitive biochemistry, since the repeated independent evolution of biflavonyls would be expected to produce a much greater variety of structures.

While the most common biflavones in the Guttiferae are of the *Garcinia* group, amentoflavone has been isolated from the leaves of

Calophyllum inophyllum (Desai *et al.*, 1967) and amentoflavone
and its 7-methyl ether from the leaves of *Garcinia livingstonei* (Pelter
et al., 1971a). It is possible that this group of biflavones may be
much more widespread in the family than present records indicate,
particularly since most effort appears to have been directed to
locating the novel biflavonyls from the wood and bark. But the
occurrence of amentoflavone in two species is in itself enough to
indicate that the Guttiferae may have been derived from an ancestral
stock in which biflavones of the amentoflavone group were typically
present, and the ability to form the *Garcinia* biflavonyls arose after
the differentiation of the family. In the case of *Phyllogeiton zeyheri*
(Rhamnaceae), however, it is quite possible that the ability to form
the unusual biflavonyl, zeyherin (46), has been evolved independently. In either case, this must be regarded as a biochemical
specialization. A better understanding of the problem must await a
more extensive survey of the families.

The complete absence of biflavonyls from the Pinaceae was seen
by Baker and Ollis (1961) as support for the separation of the family
from the Coniferales into an order of its own, the Pinales. The
absence of biflavones from the family cannot, however, be viewed as
a primitive condition when the universal occurrance of these
compounds throughout the gymnosperms is taken into account. It is
clear that the ancestral gymnosperm must have been capable of
forming biflavonones. While this ability has been retained or even
elaborated during the evolution of all other branches, it was lost
early in the evolution of the Pinaceae. Such a feature, while it is
evidence of the phylogenetic unity of the family, no more destroys
the unity of the order Coniferales than does the occurrence of a new
group of biflavones in the Araucariaceae. There are many other
features (e.g., embryology, cone morphology) that bind all members
of the order together.

The absence of biflavonyls from all orders of ferns, which has
recently been confirmed by studies by Swain and Cooper-Driver on a
range of material drawn from the Marattiales, Ophioglossales and
Filicales (pers. comm.), indicates that while more primitive vascular
plants apparently produced biflavonyls, this feature was lost early in
the evolutionary path leading to these orders of ferns. This biochemical distinction between the seed plants and the ferns is further
evidence for regarding the latter as a side line in the evolution of

vascular plants, and supports the view that the progenitor of the seed plants should be looked for among the less specialized orders of the pteridophytes.

The occurrence of the unique biflavonyl, biluteolin (28), in one species of moss (*Dicranum scoparium*) is probably an example of an independently evolved feature, and cannot be seen as evidence of any relationship between this moss and those vascular plants containing biflavones.

Thus, with the exceptions mentioned above, it is reasonable to regard the ability to synthesize biflavonyls as a monophyletic character. The actual types of biflavonyls formed have, of course, been elaborated independently during the subsequent evolution of some families, while in others the ability to form biflavonyls has been completely lost (e.g., most of the angiosperms).

13.6.2 Primitive and advanced biflavonyls

The amentoflavone group is present in every plant family containing biflavonyls with the exception of the Casuarinaceae, Ochnaceae and Rhamnaceae. The hinokiflavone group is only slightly less widespread, having been reported from the Psilotales, Selaginellales, one family of the Cycadales, four families from the Coniferales, and one angiosperm family (Casuarinaceae). With such widely separated taxa possessing the same biflavones, the ability to form them must have been elaborated at an early stage in the evolution of vascular plants. These, then, must surely represent the most primitive of the groups of biflavonyls.

The agathisflavone group is clearly a much more recent specialization that occurred after the differentiation of the Araucariaceae, since it does not occur anywhere outside the family.

The cupressuflavone group has a curious distribution. The fact that it occurs in only two coniferous families seems to indicate that the formation of these biflavones requires biochemical pathways that were only evolved after the differentiation of the Coniferales. In fact it could be used to argue that these two families are more closely related to one another than they are to any of the other families in the order. The occurrence of cupressuflavone in one species of *Casuarina*, however, presents a problem. Even those that still regard the Casuarinaceae as a primitive angiosperm family are unlikely to contemplate a direct link with either the Araucariaceae or Cupressaceae (but see Greguss, 1955). It is much more likely that the ability to form cupressuflavone is a specialization that has arisen independently in the Coniferales and the Casuarinaceae.

The *Garcinia* group of biflavonyls, being found only within one family of angiosperms (Guttiferae), must also be seen as a type that has arisen relatively recently, and thus is a more advanced biflavonyl. The occurrence of biflavonyl glycosides, which has also only been reported in the Guttiferae, must also be a highly specialized feature.

13.6.3 Conclusion

The formation of biflavonyls appears to be a feature that was developed early in the evolution of vascular plants. The most primitive types of biflavonyls appear to be of the amentoflavone and hinokiflavone groups, while the agathisflavone, cupressuflavone and Garcinia groups are subsequent specializations. It is possible that the other unusual biflavonyls are the result of independent lines of evolution. The ability to form biflavones has been secondarily lost in the evolutionary pathways leading to the Pinaceae, Gnetales and the majority of the angiosperms, and probably also to the ferns and Lycopodiales.

References

Ahmad, S. and Razaq, S. (1971), *Tetrahedron Letters*, 4633.

Asahina, Y. and Inubuse, M. (1928), *Ber. dt. Chem. Ges.* **61**, 1514.

Baker, W. and Ollis, W. D. (1961), In *Recent Developments in the Chemistry of Natural Phenolic Compounds* (W. D. Ollis ed.), p. 152. Pergamon Press, Oxford.

Baker, W., Finch, A. C. M., Ollis, W. D. and Robinson, K. W. (1963), *J. Chem. Soc.*, 1477.

Bate-Smith, E. C. (1962), *J. Linn. Soc., London* **58**, 95.

Beckmann, S., Geiger, H. and de Groot-Pfleiderer, W. (1971), *Phytochemistry* **10**, 2465.

Cambie, R. C. and James, M. A. (1967), *New Zealand J. Sci.* **10**, 918.

Chandramouli, N., Natarajan, S., Murti, V. V. S. and Seshadri, T. R. (1971), *Indian J. Chem.* **9**, 895.

Chandramouli, N., Murti, V. V. S., Natarajan, S. and Seshadri, T. R. (1972), *Indian J. Chem.* **10**, 1115.

Chen, F.-C., Lin, Y.-M. and Wu, J.-C. (1974), *Phytochemistry*, 1571.

Chexal, K. K., Handa, B. K., and Rahman, W. (1970a), *J. Chromatog.* **48**, 484.

Chexal, K. K., Handa, B. K., Rahman, W. and Kawano, N. (1970b), *Chem. and Ind.*, 28.

Cronquist, A. (1968) *The Evolution and Classification of Flowering Plants*, Nelson, London.

Dallimore, W. and Jackson, A. B. (1966), *A Handbook of Coniferae and Ginkgoaceae* 4th. Edn. (revised by S. G. Harrison). Arnold, London.

Dean, F. M. (1963) *Naturally Occurring Oxygen Ring Compounds*, Butterworths, London.

de Laubenfels, D. J. (1969), *J. Arnold Arbor.* **50**, 274.

Desai, P. D., Dutia, M. D., Ganguly, A. K., Govindachari, T. R., Joshi, B. S., Kamat, V. N., Prakash, D., Rane, D. F., Sathe, S. S. and Viswanathan, N. (1967), *Ind. J. Chem.* **5**, 523.

Di Modica, G., Rossi, P. F., Rivero, A. M. and Bortello, E. (1960), *Rend. Accad. Nazl. Lincei* **29**, 74.

Dossaji, S. F., Bell, E. A. and Wallace, J. W. (1973), *Phytochemistry* **12**, 371.

du R. Volsteedt, F. and Roux, D. G. (1971), *Tetrahedron Letters*, 1647.

Florin, R. (1948), *Bot. Gaz.* **110**, 31.

Fukui, Y. and Kawano, N. (1959), *J. Am. Chem. Soc.* **81**, 6331.

Garg, H. S. and Mitra, C. R. (1971), *Phytochemistry* **10**, 2787.

Geiger, H. and Buck, R. (1973), *Phytochemistry* **12**, 1176.

Geiger, H. and de Groot-Pfleiderer, W. (1971), *Phytochemistry* **10**, 1936.

Geiger, H. and de Groot-Pfleiderer, W. (1973), *Phytochemistry* **12**, 465.

Greguss, P. (1955), *Identification of Living Gymnosperms on the Basis of Xylotomy*, Akademiai Kiado, Budapest.

Halpaap, H. (1963), *Chemie-Ing.-Techn.* **35**, 488.

Hameed, N., Ilyas, M., Rahman, W., Okigawa, M. and Kawano, N. (1973a), *Phytochemistry* **12**, 1494.

Hameed, N., Ilyas, M., Rahman, W., Okigawa, M. and Kawano, N. (1973b), *Phytochemistry*, **12**, 1497.

Handa, B. K., Chexal, K. K., Mah, T. and Rahman, W. (1971a), *J. Indian Chem. Soc.* **48**, 177.

Handa, B. K., Chexal, K. K., Rahman, W., Okigawa, M. and Kawano, N. (1971b), *Phytochemistry* **10**, 436.

Harborne, J. B. (1967), *Comparative Biochemistry of the Flavonoids*, Academic Press, London, New York.

Herbin, G. A., Jackson, B., Locksley, H. D., Scheinmann, F. and Wolstenholme, W. A. (1970), *Phytochemistry* **9**, 221.

Hodges, R. (1965), *Aust. J. Chem.* **18**, 1491.

Hörhammer, L., Wagner, H. and Hein, K. (1964), *J. chromatog.* **13**, 235.

Hörhammer, L., Wagner, H. and Reinhardt, H. (1965), *Naturwissenschaften* **52**, 161.

Hörhammer, L., Wagner, H. and Reinhardt, H. (1966), *Bot. Mag. Tokyo* **79**, 510.

Hörhammer, L., Wagner, H. and Reinhardt, H. (1967), *Z. Naturforschg.* **22b**, 768.

Horowitz, R. M. (1957), *J. org. Chem.* **22**, 1733.

Hsu, H. Y. (1959), *Bull. Taiwan Prov. Hyg. Lab.* 1.

Ikeshiro, Y. and Konoshima, M. (1972), *Tetrahedron Letters*, 4383.

Ilyas, M., Usmani, J. N., Bhatnagar, S. P., Ilyas, M., Rahman, W. and Pelter, A. (1968), *Tetrahedron Letters*, 5515.

Jackson, B., Locksley, H. D., Scheinmann, F. and Wolstenholme, W. A. (1971), *J. chem. Soc.* C, 3791.

Johnson, L. H. (1959), *Proc. Linn. Soc. N.S.W.* **84**, 64.

Joshi, B. S., Kamat, V. N. and Viswanathan, N. (1970), *Phytochemistry* **9**, 881.

Karanjgaokar, C. G., Radhakrishnan, P. V. and Venkataraman, K. (1967), *Tetrahedron Letters*, 3195.

Kariyone, T. and Fukui, Y. (1960), *J. Pharm. Soc. Japan* **80**, 746.

Kariyone, T. and Sawada, J. (1958), *J. Pharm. Soc. Japan* **78**, 1020.

Kazakov, A. L., Bandyukova, V. A. and Shinkarenko, A. L. (1972), *Rast. Resur.* **8**, 140.

Kawano, N. (1961), *Chem. Pharm. Bull. (Tokyo)* **9**, 358.

Kawano, N., Miura, H. and Kikuchi, H. (1964), *J. Pharm. Soc. Japan* **84**, 469.

Kawano, N. and Yamada, M. (1960), *J. Pharm. Soc. Japan* **80**, 1576.

Khan, N. U., Ilyas, M., Rahman, W., Okigawa, M. and Kawano, N. (1970), *Tetrahedron Letters* 2941.

Khan, N. U., Ansari, W. H., Rahman, W., Okigawa, M. and Kawano, N. (1971a), *Chem. Pharm. Bull.* (Tokyo) **19**, 1500.

Khan, N. U., Ansari, W. H., Usmani, J. N., Ilyas, M. and Rahman, W. (1971b), *Phytochemistry* **10**, 2129.

Khan, N. U., Ilyas, M., Rahman, W., Okigawa, M. and Kawano, N. (1971c), *Phytochemistry* **10**, 2541.

Khan, N. U., Ilyas, M., Rahman, W., Mashima, T., Okigawa, M. and Kawano, N. (1972), *Tetrahedron* **28**, 5689.

Konoshima, M. and Ikeshiro, Y. (1970), *Tetrahedron Letters* 1717.

Konoshima, M., Ikeshiro, Y., Nishinaga, A., Matsuura, T., Kubota, T. and Sakamoto, H. (1969), *Tetrahedron Letters* 121.

Konoshima, M., Ikeshiro, Y., Miyahara, S. and Yen Kun Ying (1970), *Tetrahedron Letters* 4203.

Li, H.-L. (1953), *J. Arnold Arbor.* **34**, 17.

Lin, Y.-M. and Chen, F.-C. (1973), *Tetrahedron Letters* 4747.

Locksley, H. D. (1973), *Fortschr. Chem. org. Naturst.* **30**, 207.

Locksley, H. D. and Murray, I. G. (1971), *J. chem. Soc.* C 1332.

Mabry, T. J., Markham, K. R. and Thomas, M. B. (1970), *The Systematic Identification of Flavonoids*, Springer-Verlag, Berlin, Heidelberg, New York.

Madhav, R. (1969), *Tetrahedron Letters* 2017.

Mashima, T., Okigawa, M., Kawano, N., Khan, N. U. and Rahman, W. (1970), *Tetrahedron Letters* 2937.

Miura, H. (1967a), *J. Pharm. Soc. Japan* **87**, 7.

Miura, H. (1967b), *J. Pharm. Soc. Japan* **87**, 871.

Miura, H. and Kawano, N. (1967), *Chem. Pharm. Bull (Tokyo)* **15**, 232.

Miura, H. and Kawano, N. (1968a), *J. Pharm. Soc. Japan* **88**, 1459.

Miura, H. and Kawano, N. (1968b), *J. Pharm. Soc. Japan* **88**, 1489.

Miura, H. and Kawano, N. (1968c), *Chem. Pharm. Bull (Tokyo)* **16**, 1838.

Miura, H., Kawano, N. and Waiss, A. C. Jr. (1966), *Chem. Pharm. Bull (Tokyo)* **14**, 1404.

Miura, H., Kihara, T. and Kawano, N. (1968), *Tetrahedron Letters* 2339.

Miura, H., Kihara, T. and Kawano, N. (1969), *Chem. Pharm. Bull. (Tokyo)* **17**, 150.

Molyneux, R. J., Waiss, A. C. Jr. and Haddon, W. F. (1970), *Tetrahedron* **26**, 1409.

Moriyama, S., Okigawa, M. and Kawano, N. (1972), *Tetrahedron Letters* 2105.

Murti, V. V. S. (1967), *Bull. Nat. Inst. Sci. India* **34**, 161.

Murti, V. V. S., Raman, P. V. and Seshadri, T. R. (1964), *Tetrahedron Letters* 2995.

Murti, V. V. S., Raman, P. V. and Seshadri, T. R. (1967), *Tetrahedron* **23**, 397.

Nakazawa, K. (1962), *Chem. Pharm. Bull. (Tokyo)* **10**, 1032.

Nakazawa, K. (1968), *Chem. Pharm. Bull. (Tokyo)* **16**, 2503.

Nakazawa, K. and Ito, M. (1963), *Chem. Pharm. Bull (Tokyo)* **11**, 283.

Natarajan, S., Murti, V. V. S. and Seshadri, T. R. (1968), *Indian J. Chem.* **6**, 549.

Natarajan, S., Murti, V. V. S. and Seshadri, T. R. (1969), *Indian J. Chem.* **7**, 751.

Natarajan, S., Murti, V. V. S. and Seshadri, T. R. (1970a), *Indian J. Chem.* **8**, 113.

Natarajan, S., Murti, V. V. S. and Seshadri, T. R. (1970b), *Phytochemistry* **9**, 575.

Natarajan, S., Murti, V. V. S. and Seshadri, T. R. (1970c), *Indian J. Chem.* **8**, 116.

Natarajan, S., Murti, V. V. S. and Seshadri, T. R. (1971a), *Indian J. Chem.* **9**, 383.

Natarajan, S., Murti, V. V. S. and Seshadri, T. R. (1971b), *Phytochemistry* **10**, 1083.

Nilsson, E. (1973), *Chem. Scripta* **4**, 66.

Okigawa, M., Chiang Wu Hwa, Kawano, N. and Rahman, W. (1971), *Phytochemistry* **10**, 3286.

Okigawa, M., Kawano, N. Aqil, M., and Rahman, W. (1973), *Tetrahedron Letters* 2003.

Pelter, A., Warren, R., Ilyas, M., Usmani, J. N., Bhatnagar, S. P., Rizivi, R. H., Ilyas, M. and Rahman, W. (1969a), *Experientia* **25**, 350.

Pelter, A., Warren, R., Usmani, J. N., Ilyas, M. and Rahman, W. (1969b), *Tetrahedron Letters* 4259.

Pelter, A., Warren, R., Usmani, J. N., Rizvi, R. H., Ilyas, M. and Rahman, W. (1969c), *Experientia* **25**, 351.

Pelter, A., Warren, R., Hameed, N., Khan, N. U., Ilyas, M. and Rahman, W. (1970), *Phytochemistry* **9**, 1897.

Pelter, A., Warren, R., Chexal, K. K., Handa, B. K. and Rahman, W. (1971a), *Tetrahedron* **27**, 1625.

Pelter, A., Warren, R., Hameed, N., Ilyas, M. and Rahman, W. (1971b), *J. Indian Chem. Soc.* **48**, 204.

Pelter, A., Warren, R., Handa, B. K., Chexal, K. K. and Rahman, W. (1971c), *Indian J. Chem.* **9**, 98.

Perkin, A. G. and Phipps, A. (1904), *J. chem. Soc.* **85**, 56.

Pilger, R. (1926), In *Die Natürlichen Pflanzenfamilien* (A. Engler and K. Prantl eds.) 2nd. Edn. Verlag von Wilhelm Engelmann, Leipzig.

Quinn, C. J. (1970), *Proc. Linn. Soc. N.S.W.* **94**, 166.

Rahman, W. and Bhatnagar, S. P. (1968), *Tetrahedron Letters* 675.

Rahman, W., Hameed, N. and Ilyas, M. (1972), *J. Indian Chem. Soc.* **49**, 917.

Rao, N. S. P., Row, L. R. and Brown, R. T. (1973), *Phytochemistry* **12**, 671.

Sawada, T. (1958), *J. Pharm. Soc. Japan* **78**, 1023.

Sporne, K. R. (1965), *The Morphology of the Gymnosperms*, Hutchinson, London.

Stahl, E. and Schorn, P. J. (1961), *Hoppe-Seyler's Z. physiol. Chem.* **325**, 263.

Takhtajan, A. (1969), *Flowering Plants – Origin and Dispersal*, Oliver and Boyd, Edinburgh.

Tenger, J. (1967), *Bot. Notiser* **180**, 450.

Van der Merwe, K. J., Steyn, P. S. and Eggers, S. H. (1964), *Tetrahedron Letters* 3923.

Van Sumere, C. F., Cottenic, J. and Teuchy, H. (1968a), *Arch. Internat. Physiol. Bioch.* **76**, 967.

Van Sumere, C. F., Kint, J. and Cottenic, J. (1968b), *Arch. Internat. Physiol. Bioch.* **76**, 396.

Varshney, A. K., Aqil, M., Rahman, W., Okigawa, M. and Kawano, N. (1973a), *Phytochemistry* **12**, 1501.

Varshney, A. K., Rahman, W., Okigawa, M. and Kawano, N. (1973b), *Experientia* **29**, 784.

Venkataraman, K. (1962), In *The Chemistry of Flavonoid Compounds*. (T. A. Geissman, Ed.) Pergamon Press, London.

Voirin, B. and Lebreton, P. (1966), *Compt. Rend.* **262D**, 707.

Chapter 14

The Isoflavonoids

E. WONG

14.1 Introduction

Isoflavonoids differ from other flavonoid compounds in having as a basic structural feature the branched $C_6 C_3 C_6$ skeleton shown in (1). Within this group are included many classes of natural products. Isoflavones, isoflavanones, rotenoids, pterocarpans and coumestans are well established members of this group (Ollis, 1962), while others such as isoflavans, 3-aryl-4-hydroxycoumarins, coumarono-chromones, and hydroxy- and dehydro-variants of pterocarpans and rotenoids have only recently been reported as natural products. The skeletal structures of the various classes of isoflavonoid, arranged in order of their oxidation level, are given in Fig. 14.1. The structural variety displayed in the isoflavonoids is, in fact, greater than existing in the normal flavonoid series.

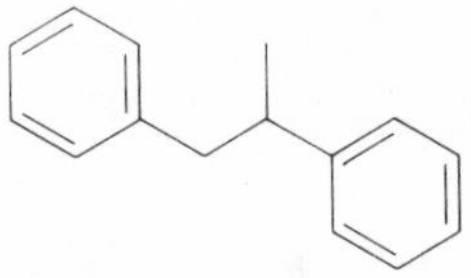

(1)

In the decade since the comprehensive reviews of Ollis (1962) and Dean (1963), the number of naturally occurring isoflavonoids has increased more than threefold. All known structures within each class of compound will be presented in the following sections; the total number of compounds (aglycones) covered is 180.

Recent developments in the structural study of isoflavonoids have centred largely on the application of spectroscopic methods. The main stages in the structural elucidation of an isoflavonoid are: recognition of the class to which the compound belongs, and determination of the nature and orientation of substituent groups in the aromatic rings. Both of these stages are well served by the application of UV, IR, NMR, and MS techniques. Application of physical methods often permits the selection of a minimal number of chemical reactions to provide for the maximum amount of structural information (Ollis, 1968), and many examples exist of complete structural elucidation of isoflavonoids without recourse to chemical degradation (e.g. Highet and Highet, 1967; Pelter and Amenechi, 1969; Burden *et al.*, 1972). In this chapter, only those aspects of structural elucidation will be emphasized which concern the characterization of the different classes of isoflavonoid.

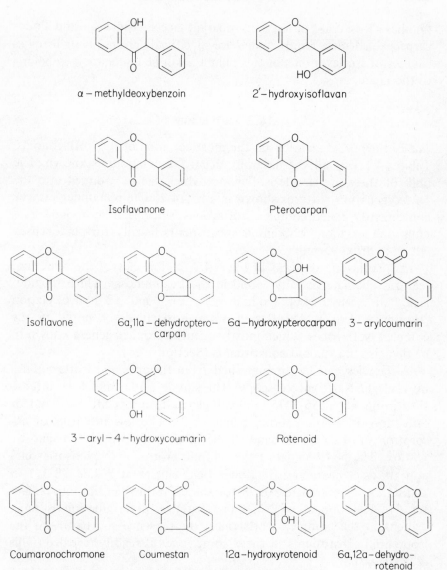

α – methyldeoxybenzoin

2′ – hydroxyisoflavan

Isoflavanone

Pterocarpan

Isoflavone

6a,11a – dehydroptero‑carpan

6a – hydroxypterocarpan

3 – arylcoumarin

3 – aryl – 4 – hydroxycoumarin

Rotenoid

Coumaronochromone

Coumestan

12a – hydroxyrotenoid

6a,12a – dehydro‑rotenoid

Figure 14.1 Skeletal structures for the isoflavonoid classes.

A distinctive feature of the isoflavonoids is their very limited taxonomic distribution. In contrast to the near ubiquitous occurrence in higher plants of other flavonoids, isoflavonoids are confined very largely to one group of plants, the sub-family Lotoideae of the Leguminosae (Jay *et al.*, 1971; Harborne, 1971), They also occur occasionally in the sub-family Caesalpinioideae and in a few other families (Rosaceae, Moraceae, Amaranthaceae, Iridaceae and Podo-carpaceae). Unless family names are specifically given, plants listed as sources of isoflavonoids in this chapter will be understood to belong to the Leguminosae.

14.2 Isoflavones

The isoflavones remain by far the most common of the isoflavonoids. Table 14.1 lists all the naturally occurring compounds known. The bulk of these are simple isoflavones which can be grouped into the 18 oxygenation patterns shown in Fig. 14.2. The remainder may be conveniently called complex isoflavones, since each compound contains one or more isoprenoid substituents usually further cyclised with a hydroxyl group (52)–(75).

The isoflavone patterns in Fig. 14.2 exhibit clearly the structural features characteristically found in the Leguminosae which include: lack of 5-hydroxyl, 6-hydroxylation and 2'-hydroxylation (Harborne, 1967). This last feature is particularly noted in the complex isoflavones, which mostly occur in tropical genera known to be rich also in rotenoid constituents (Section 14.4).

Isoflavones can be distinguished from flavones and isoflavanones by UV and NMR spectroscopy. The simple isoflavones have intense absorption at ~255–275 nm and generally a less intense band or inflection at ~310–330 nm (Chapter 2). The low intensity of absorption of the second band of isoflavones is a valuable diagnostic feature. The NMR signal of the olefinic proton at C-2 in isoflavones appears as a characteristic down field singlet at ~7.8δ (8.3δ) in DMSO) as compared to ~6.7δ for the C-3 proton in flavones (Chapter 2). The impact of NMR on structure determination is most evident in the complex isoflavones; the presence and nature of the isoprenoid substituents in these compounds is readily revealed (Ollis et al., 1967).

Mass spectrometry cannot distinguish between flavones and iso-flavones since fragmentation via the retro Diels-Alder process results in identical fragments for both classes of compounds. The presence of a 2'-methoxyl group in isoflavones, however, has been found to influence the fragmentation pattern profoundly. These compounds show a strong M-31 peak due to the loss of methoxyl; this probably occurs via ring closure as shown in Fig. 14.3 (Campbell et al., 1969).

Table 14.1 *Natural isoflavones*

Isoflavone (formula)	Plant sources	References*
Simple Isoflavones		
Daidzein (2)	*Glycine max*	
	Pueraria spp.	Murakami *et al.*, 1960.
	Baptisia spp.	Markham *et al.*, 1970.
	Trifolium spp.	Guggolz *et al.*, 1961.
	Machaerium villosum	Oliveira *et al.*, 1968.
	tribe Genisteae	Harborne, 1969.
	Dalbergia ecastophyllum	Matos *et al.*, 1970.
Formononetin (3)	*Cicer arietinum*	
	Baptisia spp.	Markham *et al.*, 1970.
	Trifolium spp.	Francis *et al.*, 1967.
	Dalbergia paniculata	Narayanan and Seshadri, 1971.
	Machaerium villosum	Oliveira *et al.*, 1968.
	Pterocarpus spp.	Seshadri, 1972.
	Pericopsis spp.	Imamura *et al.*, 1968.
	Castanospermum australe	Eade *et al.*, 1963.
	Diplotropis purpurea	Braz Filho *et al.*, 1973a.
	tribe Genisteae	Harborne, 1969.
	Glycyrrhiza glabra	Elgamal and Fayez, 1972.
Isoformononetin (4)	*Machaerium villosum*	Oliveira *et al.*, 1968.
Di-*O*-methyl- daidzein (5)	*Dalbergia violacea*	Ollis, 1966.
Durlettone (6)	*Millettia dura*	Ollis *et al.*, 1967.
7,3',4'- Trihydroxy- (7)	*Machaerium villosum*	Oliveira *et al.*, 1968.
7,4'-Dihydroxy- 3'-methoxy- (8)	*Machaerium villosum*	Oliveira *et al.*, 1968.
	Cyclolobium clausseni	Oliveira *et al.*, 1971a.
	Machaerium mucronulatum } *M. vestitum*	Gottlieb, 1973.
Calycosin (9)	*Baptisia* spp.	Markham *et al.*, 1970.
	Pterocarpus dalbergioides	Parthasarathy *et al.*, 1969.
Cladrin (10)	*Cladrastis lutea*	Shamma and Stiver, 1969.
Pseudobaptigenin (11)	*Baptisia* spp.	Markham *et al.*, 1970.
	Pterocarpus erinaceous	Bevan *et al.*, 1966.
	Maackia amurensis	Suginome and Kio, 1966.
	Dalbergia spruceana	Gottlieb, 1973.
Cabreuvin (12)	*Myrocarpus fastigiatus*	
	Myroxylon balsamum	
Maxima isoflavone B (13)	*Tephrosia maxima*	Kukla and Seshadri, 1962.
Baptigenin (14)	*Baptisia tinctoria* (as baptisin)	Farkas *et al.*, 1963.
Maxima isoflavone C (15)	*Tephrosia maxima*	
Genistein (16)	*Trifolium* spp.	Francis *et al.*, 1967.
	Cicer arietinum	
	Podocarpus spicatus (Podocarpaceae)	
	Baptisia spp.	Markham *et al.*, 1970.
	tribe Genisteae	Harborne, 1969.

Continued

Table 14.1—*Continued*

Isoflavone (formula)	Plant sources	References*
Biochanin A (17)	*Cicer arietinum*	
	Trifolium spp.	Francis *et al.*, 1967.
	Ferreirea spectabilis	
	Andira parviflora	Braz Filho *et al.*, 1973a.
	Baptisia spp.	Markham *et al.*, 1970.
	Dalbergia spp.	Narayanan and Seshadri, 1971.
Prunetin (18)	*Prunus puddum* (Rosaceae)	
	Pterocarpus angolensis	
	Dalbergia miscolobium	Gottlieb, 1973.
5-Methylgenistein (19)	tribe Genisteae	Harborne, 1969.
	Ormosia excelsa	Gottlieb and Rocha, 1972.
Orobol (20)	*Baptisia* spp.	Markham *et al.*, 1970.
Santal (21)	*Pterocarpus santalinus*	
	P. osum	Akisanya *et al.*, 1959.
	Baphia nitida	
Pratensein (22)	*Trifolium* spp.	Wong, 1963.
	Cicer arietinum	Wong *et al.*, 1965.
3'-Methylorobol (23)	*Dalbergia paniculata* (as 8-*C*-glucoside)	Adinarayana and Rao, 1972a.
	Thermopsis montana	Dement and Mabry, 1972.
	Dalbergia inundata	Almeida and Gottlieb, 1974.
Derrustone (24)	*Derris robusta*	East *et al.*, 1969.
Retusine (25)	*Dalbergia retusa*	Jurd *et al.*, 1972.
8-Methylretusine (26)	*Dalbergia retusa*	Jurd *et al.*, 1972.
	D. variabilis	Gottlieb, 1973.
Maxima isoflavone A (27)	*Tephrosia maxima*	Kukla and Seshadri, 1962.
6,7,4'-Trihydroxy- (28)	*Glycine max* (fermented)	Gyorgy *et al.*, 1964.
Glyciteine (29)	*Glycine max* (as glucoside)	Naim *et al.*, 1973.
Texasin (30)	*Baptisia* spp.	Markham *et al.*, 1970.
	Platymiscium praecox	Oliveira *et al.*, 1972.
Afrormosin (31)	*Afrormosia elata*	
	Baptisia spp.	Markham *et al.*, 1970.
	Pericopsis spp.	Imamura *et al.*, 1968.
	Pterodon apparicioi	Gottlieb, 1973.
	Myrocarpus fastigiatus *Myroxylon balsamum*	Harborne *et al.*, 1963.
	Amphimas pterocarpoides	Bevan *et al.*, 1966.
	Castanospermum australe	Eade *et al.*, 1963.
	Dalbergia riparia	Braz Filho *et al.*, 1973b.
Cladrastin (32)	*Cladrastis lutea*	Shamma and Stiver, 1969
Fujikinetin (33)	*Cladrastis platycarpa*	Imamura *et al.*, 1972.
	Dalbergia riparia	Braz Filho *et al.*, 1973b.
	Pterodon apparicioi	Gottlieb, 1973.
6,7,3',4'-Tetra-methoxy- (34)	*Pterodon pubescens*	Braz Filho *et al.*, 1971.
	Cordyla africana	Campbell *et al.*, 1969.

Continued

Table 14.1—*Continued*

Isoflavone (formula)	Plant sources	References*
6,7-Dimethoxy-3',4'-methylenedioxy- (35)	*Cordyla africana*	Campbell *et al.*, 1969.
6,7,2',3',4'-Penta-methoxy- (36)	*Pterodon pubescens*	Braz Filho *et al.*, 1971.
6,7,3'-Trimethoxy-4',5'-methylene-dioxy- (37)	*Cordyla africana*	Campbell *et al.*, 1969.
7-Hydroxy-6,2'-dimethoxy-4',5,-methylenedioxy- (38)	*Dalbergia paniculata* (as 7-glucoside)	Adinarayana and Rao, 1972a.
Milldurone (39)	*Cordyla africana*	Campbell *et al.*, 1969.
	Pterodon pubescens	Braz Filho *et al.*, 1971.
	Millettia dura	Ollis *et al.*, 1967.
6,7,2',4',5'-Penta-methoxy (40)	*Cordyla africana*	Campbell *et al.*, 1969.
	Pterodon pubescens	Gottlieb, 1973.
6-Hydroxy-genistein (41)	*Baptisia hirsuta*	Markham *et al.*, 1968.
Tectorigenin (42)	*Baptisia* spp.	Markham *et al.*, 1970.
	Dalbergia riparia	Braz Filho *et al.*, 1973b.
Muningin (43)	*Pterocarpus angolensis*	
Irisolidone (44)	*Iris nepalensis* (Iridaceae)	Prakash *et al.*, 1965.
7-Methyltectori-genin (45)	*Pterocarpus angolensis*	Morgan and Orsler, 1967.
	Dalbergia sissoo	Banerji *et al.*, 1963.
	D. spruceana	Gottlieb, 1973.
7,4'-Dimethyl-tectorigenin (46)	*Dalbergia sissoo*	Banerji *et al.*, 1965.
Irisolone (47)	*Iris nepalensis* (Iridaceae)	Gopinath *et al.*, 1961.
	Iris germanica	Dhar and Kalla, 1972.
Tlatlancuayin (48)	*Iresine celosioides* (Amaranthaceae)	
Podospicatin (49)	*Podocarpus spicatus* (Podocarpaceae)	
Irigenin (50)	*Iris* spp. (Iridaceae) (as 7-*O*-glucoside)	
Caviunin (51)	*Dalbergia* spp.	Braz Filho *et al.*, 1973b.
Complex Isoflavones		
Alpinumisoflavone (52)	*Laburnum alpinum*	Jackson *et al.*, 1971.
Parvisoflavone A (53)	*Poecilanthe parviflora*	Assumpcao and Gottlieb, 1973.
Parvisoflavone B (54)	*Poecilanthe parviflora*	Assumpcao and Gottlieb, 1973
Derrubone (55)	*Derris robusta*	East *et al.*, 1969.
Robustone (56)	*Derris robusta*	East *et al.*, 1969.
Robustone methyl ether (57)	*Derris robusta*	East *et al.*, 1969.
Dehydroneotenone (58)	*Pachyrrhizus erosus* *Neorautanenia pseudo-pachyrriza*	Crombie and Whiting, 1963.
	N. amboensis *N. edulis*	Brink *et al.*, 1965.

Continued

Table 14.1—*Continued*

Isoflavone (formula)	Plant sources	References*
Durmillone (59)	*Millettia dura*	Ollis *et al.*, 1967.
	M. ferruginea	Highet and Highet, 1967.
Jamaicin (60)	*Piscidia erythrina*	
Ichthynone (61)	*Piscidia erythrina*	Schwarz *et al.*, 1964.
Toxicarol iso-flavone (62)	*Derris malaccensis*	Harper and Underwood, 1965.
Ferrugone (63)	*Millettia ferruginea*	Highet and Highet, 1967.
Piscerythrone (64)	*Piscidia erythrina*	Falshaw *et al.*, 1966.
Piscidone (65)	*Piscidia erythrina*	Falshaw *et al.*, 1966.
Auriculin (66)	*Millettia auriculata*	Shabbir and Zaman, 1970.
Pomiferin (67)	*Maclura pomifera* (Moraceae)	
Osajin (68)	*Maclura pomifera* (Moraceae)	
Scandinone (69)	*Derris scandens*	Falshaw *et al.*, 1969. Pelter and Stainton, 1966.
Scandenone (70)	*Derris scandens*	Pelter and Stainton, 1966.
Auriculatin (71)	*Millettia auriculata*	Shabbir *et al.*, 1968.
Isoauriculatin (72)	*Millettia auriculata*	Shabbir and Zaman, 1970.
Chandalone (73)	*Derris scandens*	Falshaw *et al.*, 1969.
Munetone (74)	*Mundulea sericea*	Dyke *et al.*, 1963.
Mundulone (75)	*Mundulea sericea*	Dyke *et al.*, 1963.

*References to publications prior to 1962 are not given. These may be found in the earlier reviews of Ollis (1962) and Dean (1963).

Chemical methods for the structure determination of isoflavones are well documented (Ollis, 1962; Dean, 1963). Isoflavones are much more susceptible to alkali hydrolysis than flavones. The production of a deoxybenzoin (76) under mild conditions, together with formic acid, constitutes important evidence for an isoflavone structure (Fig. 14.4). Confirmation may further be obtained by resynthesis of the isoflavone from the deoxybenzoin with the appropriate reagent. Methods for the synthesis of isoflavones are discussed in Chapter 4. The majority of the natural simple isoflavones have been synthesized.

Glycosides of isoflavones have been known since a very early date (De Laire and Tieman, 1893). The majority of the natural glycosides currently known (Table 14.2), however, have been reported only in recent years, mainly as a result of more systematic analysis of plant extractives. Thus chromatographic surveys of *Baptisia* and *Thermopsis* species have led to the recognition of a host of new isoflavone glycosides (Markham *et al.*, 1970; Dement and Mabry, 1972). In some cases, e.g. in subterranean clover *(Trifolium subterraneum),*

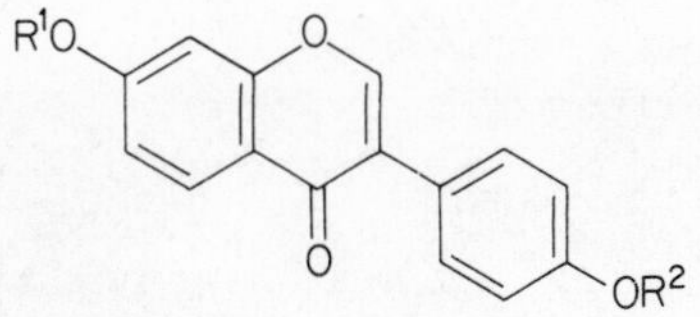

(2, $R^1 = R^2 = H$ Daidzein)
(3, $R^1 = H, R^2 = Me$ Formononetin)
(4, $R^1 = Me, R^2 = H$ Isoformononetin)
(5, $R^1 = R^2 = Me$)
(6, $R^1 = Me, R^2 = -CH_2CH=CMe_2$ Durlettone)

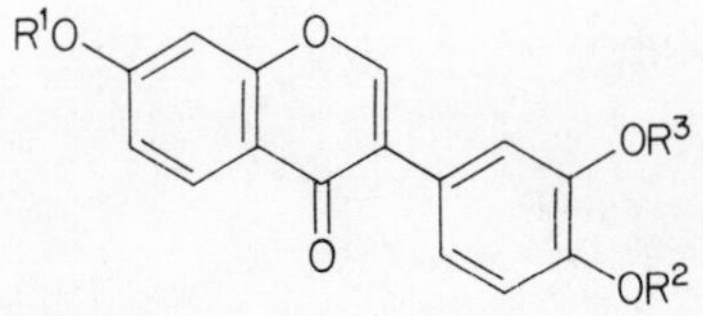

(7, $R^1 = R^2 = R^3 = H$)
(8, $R^1 = R^2 = H, R^3 = Me$)
(9, $R^1 = R^3 = H, R^2 = Me$ Calycosin)
(10, $R^1 = H, R^2 = R^3 = Me$ Cladrin)
(11, $R^1 = H, R^2 + R^3 = CH_2$ ψ-baptigenin)
(12, $R^1 = R^2 = R^3 = Me$ Cabreuvin)
(13, $R^1 = -CH_2CH=CHMe_2$
$R^2 + R^3 = CH_2$ Maxima isoflavone B)

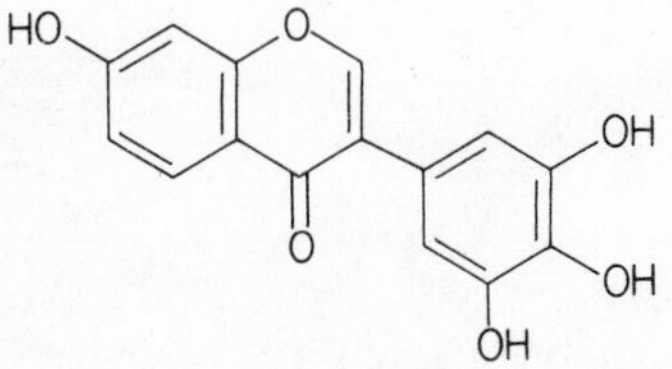

(14, Baptigenin)

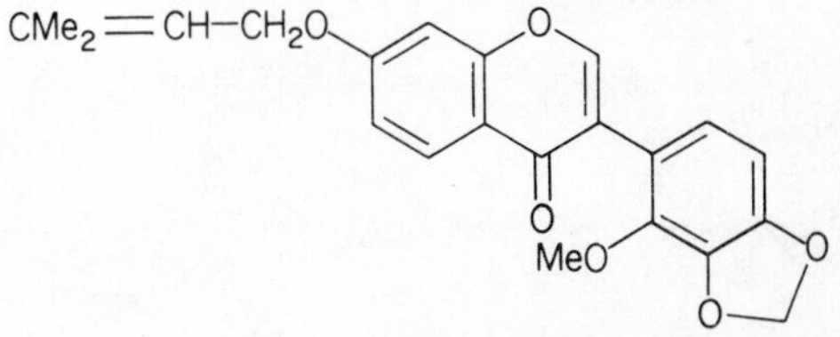

(15, Maxima isoflavone C)

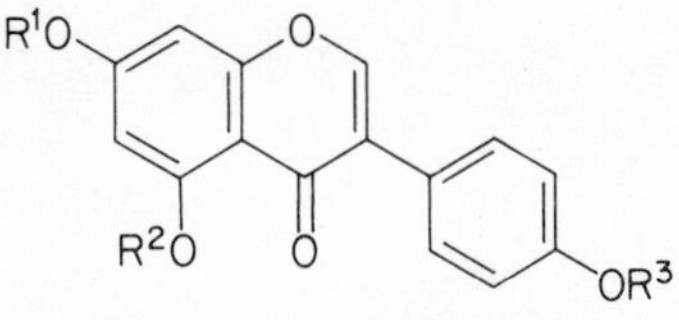

(16, $R^1 = R^2 = R^3 = H$ Genistein)
(17, $R^1 = R^2 = H, R^3 = Me$ Biochanin A)
(18, $R^1 = Me, R^2 = R^3 = H$ Prunetin)
(19, $R^1 = R^3 = H, R^2 = Me$ 5-methylgenistein)

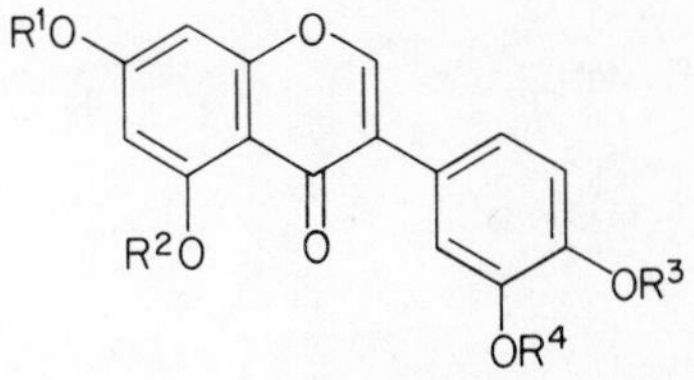

(20, $R^1 = R^2 = R^3 = R^4 = H$ Orobol)
(21, $R^1 = Me, R^2 = R^3 = R^4 = H$ Santal)
(22, $R^1 = R^2 = R^4 = H, R^3 = Me$ Pratensein)
(23, $R^1 = R^2 = R^3 = H, R^4 = Me$ 3'-methylorobol)
(24, $R^1 = R^2 = Me, R^3 + R^4 = CH_2$ Derrustone)

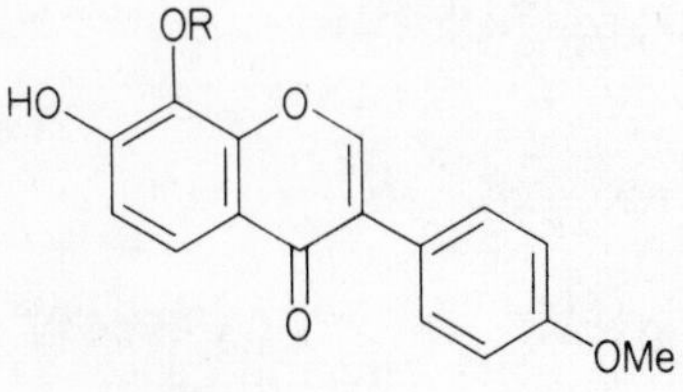

(25, $R = H$ Retusine)
(26, $R = Me$)

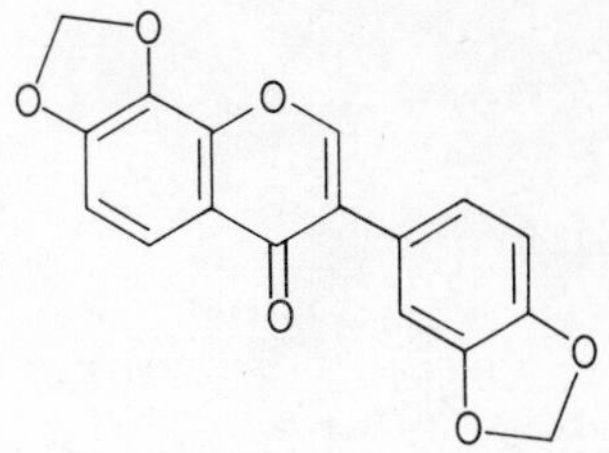

(27, Maxima isoflavone A)

Figure 14.2 Simple isoflavones.

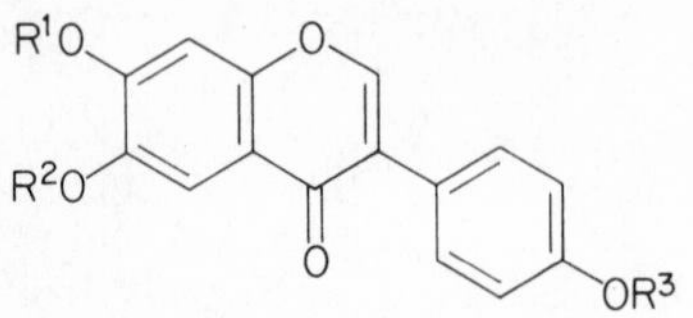

(28, $R^1=R^2=R^3=H$)
(29, $R^1=R^3=H, R^2=Me$ Glyciteine)
(30, $R^1=R^2=H, R^3=Me$ Texasin)
(31, $R^1=H, R^2=R^3=Me$ Afrormosin)

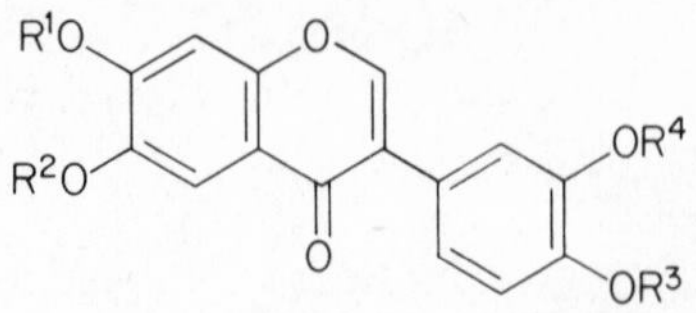

(32, $R^1=H, R^2=R^3=R^4=Me$ Cladrastin)
(33, $R^1=H, R^2=Me, R^3+R^4=CH_2$ Fujikinetin)
(34, $R^1=R^2=R^3=R^4=Me$)
(35, $R^1=R^2=Me, R^3+R^4=CH_2$)

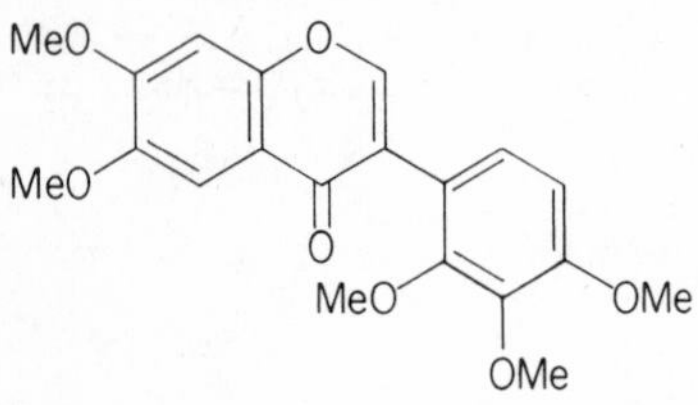

(36)

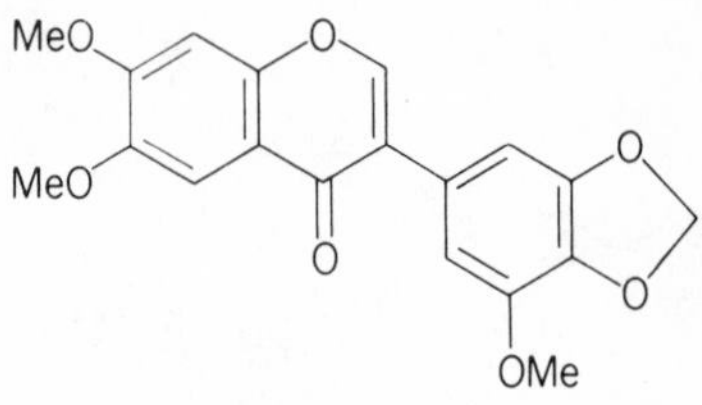

(37)

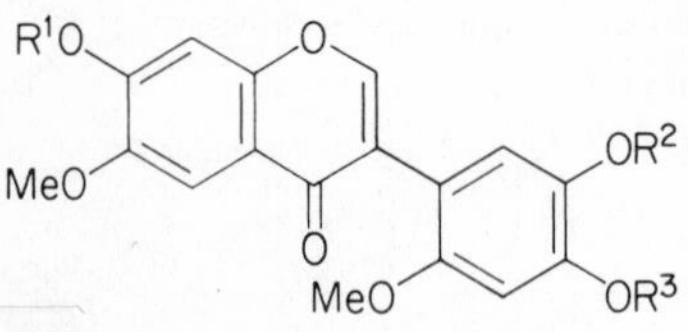

(38, $R^1=H, R^2+R^3=CH_2$)
(39, $R^1=Me, R^2+R^3=CH_2$ Milldurone)
(40, $R^1=R^2=R^3=Me$)

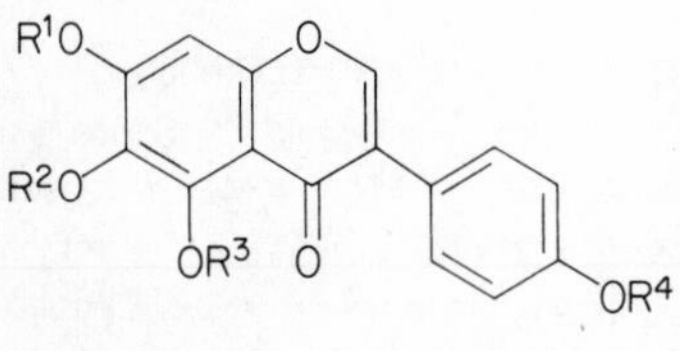

(41, $R^1=R^2=R^3=R^4=H$ 6-hydroxygenistein)
(42, $R^1=R^3=R^4=H, R^2=Me$ Tectorigenin)
(43, $R^1=R^3=Me, R^2=R^4=H$ Muningin)
(44, $R^1=R^3=H, R^2=R^4=Me$ Irisolidone)
(45, $R^1=R^2=Me, R^3=R^4=H$ 7-methyltectorigenin)
(46, $R^1=R^2=R^4=Me, R^3=H$ 7,4'-dimethyltectorigenin)
(47, $R^1+R^2=CH_2, R^3=Me, R^4=H$ Irisolone)

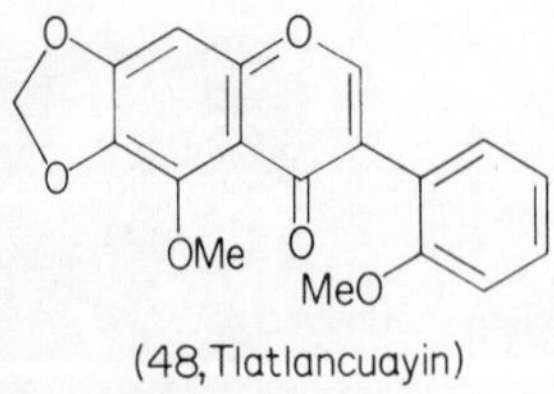

(48, Tlatlancuayin)

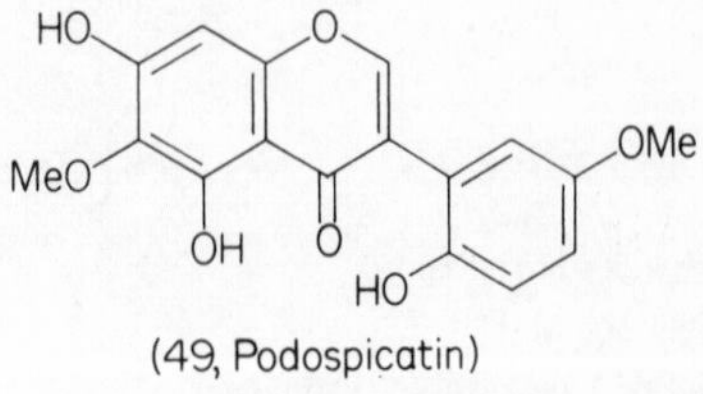

(49, Podospicatin)

Figure 14.2—*Continued*

752

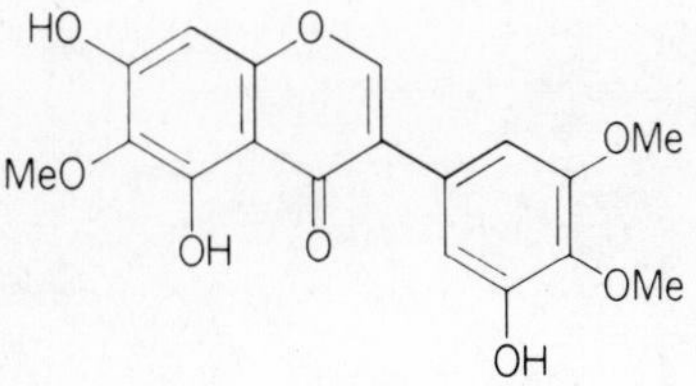

(50, Irigenin)

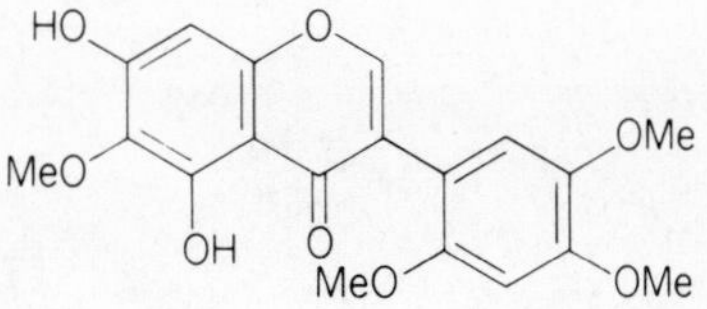

(51, Caviunin)

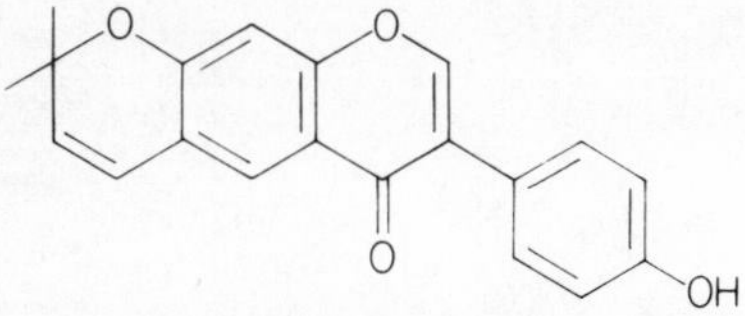

(52, Alpinumisoflavone)

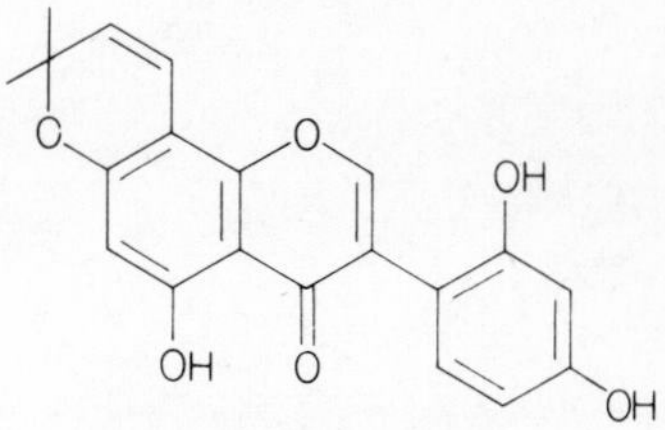

(53, Parvisoflavone A)

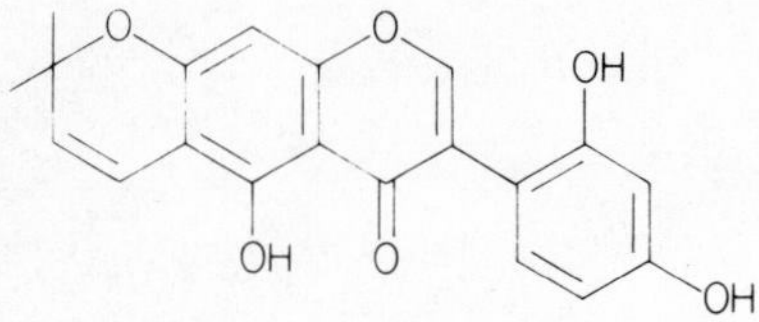

(54, Parvisoflavone B)

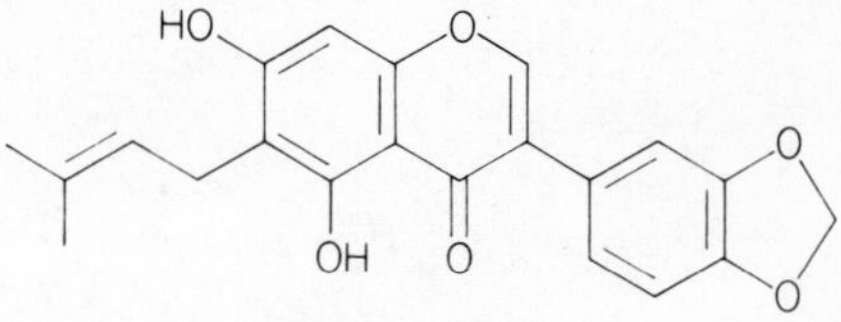

(55, Derrubone)

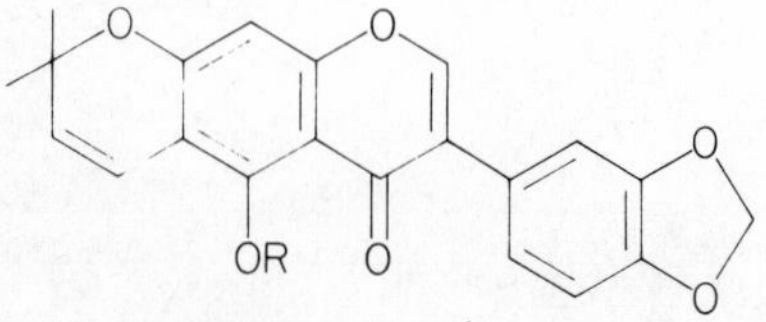

(56, Robustone R = H)
(57, Robustone methyl ether R = Me)

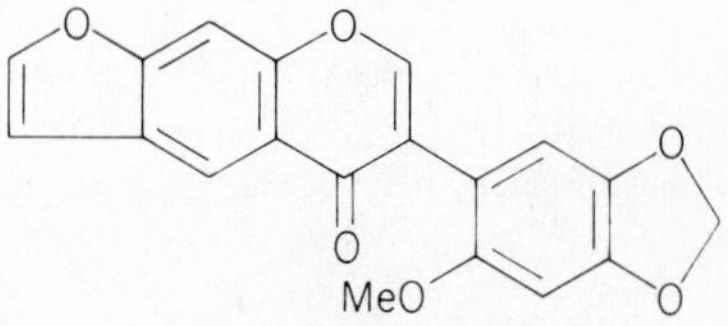

(58, Dehydroneotenone)

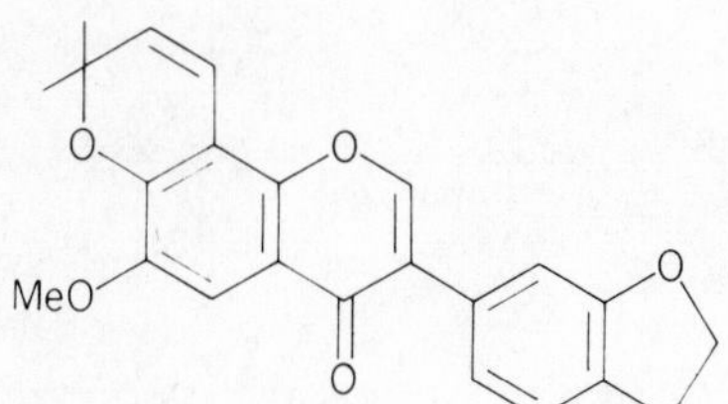

(59, Durmillone)

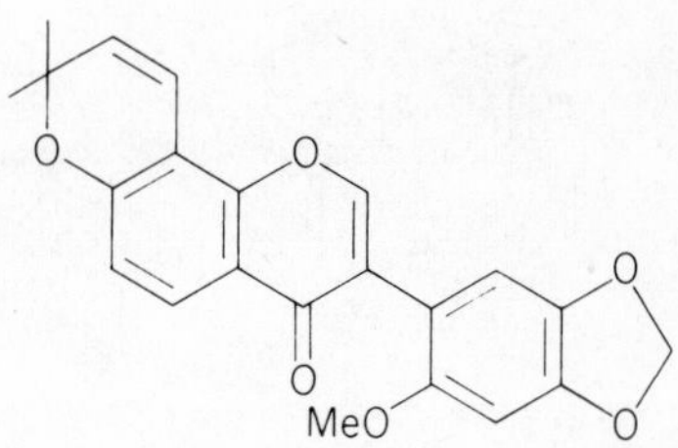

(60, Jamaicin)

Figure 14.2–*Continued*

753

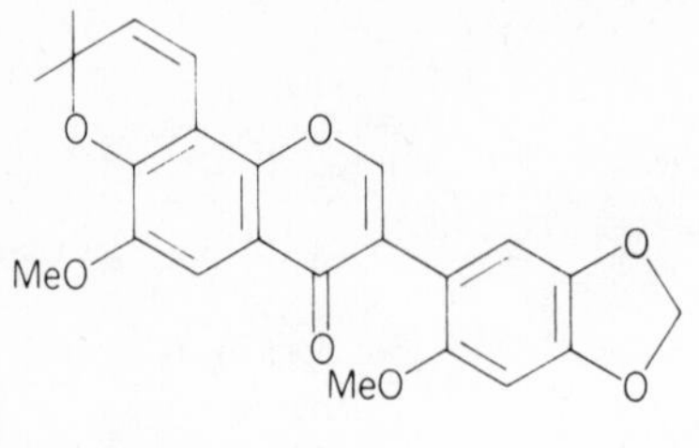

(61, Ichthynone)

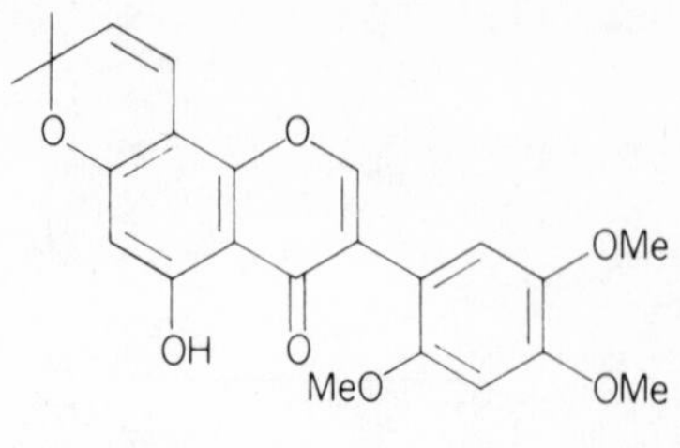

(62, Toxicarol isoflavone)

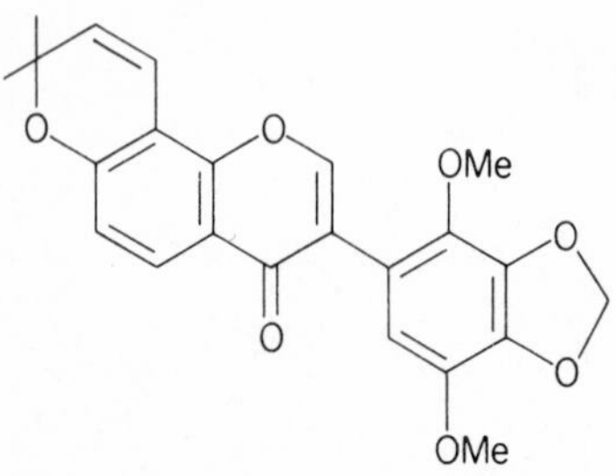

(63, Ferrugone)

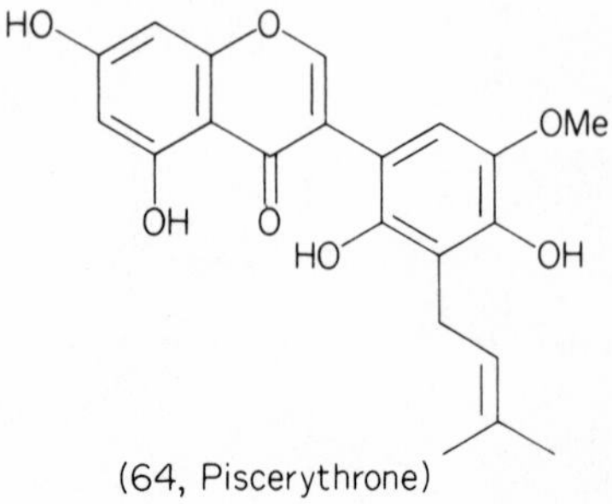

(64, Piscerythrone)

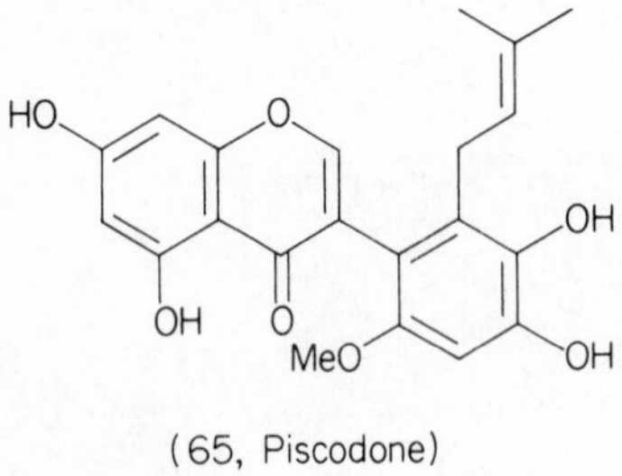

(65, Piscodone)

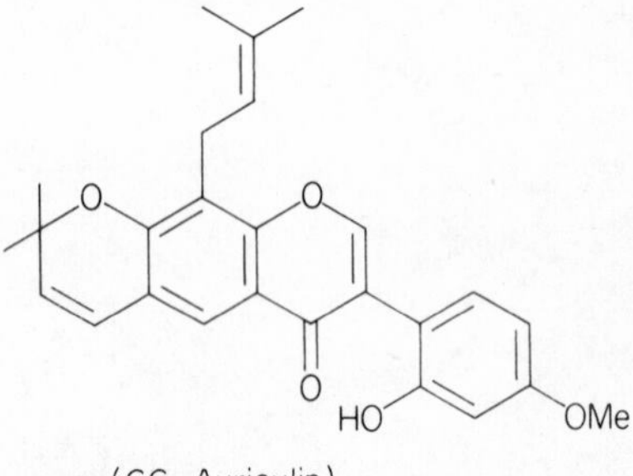

(66, Auriculin)

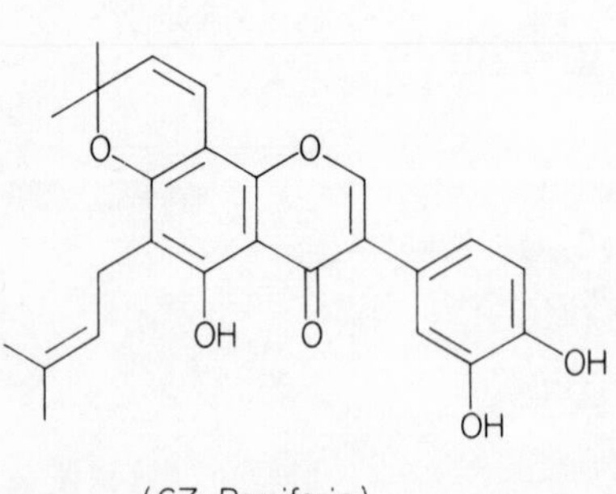

(67, Pomiferin)

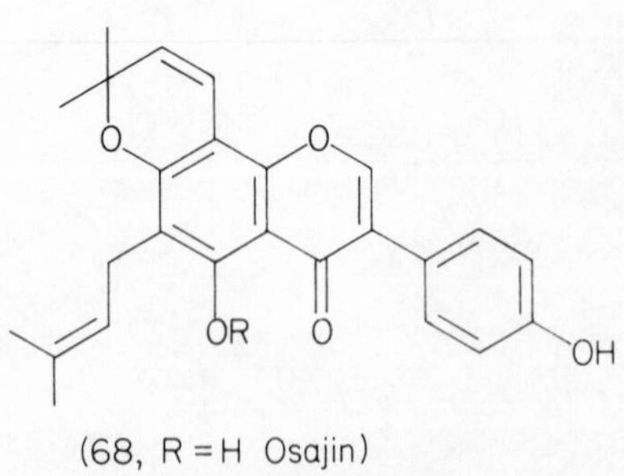

(68, R = H Osajin)
(69, R = Me Scandinone)

Figure 14.2—*Continued*

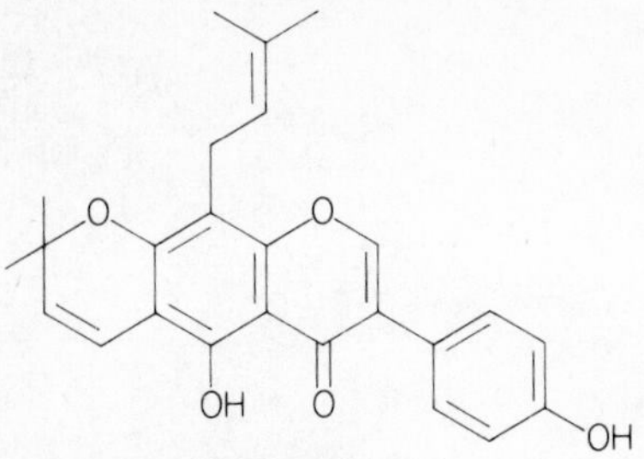

(70, Scandenone)

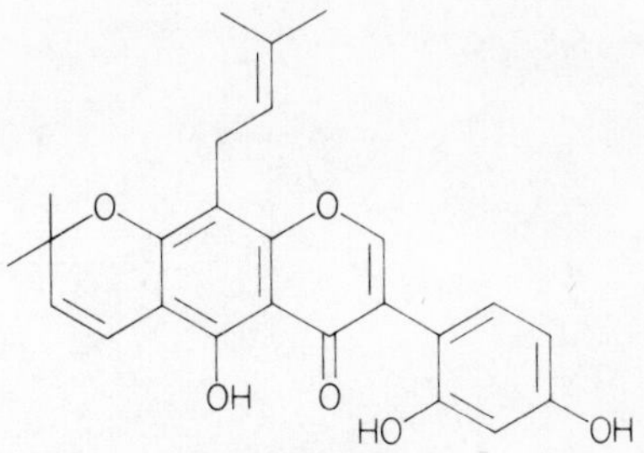

(71, Auriculatin)

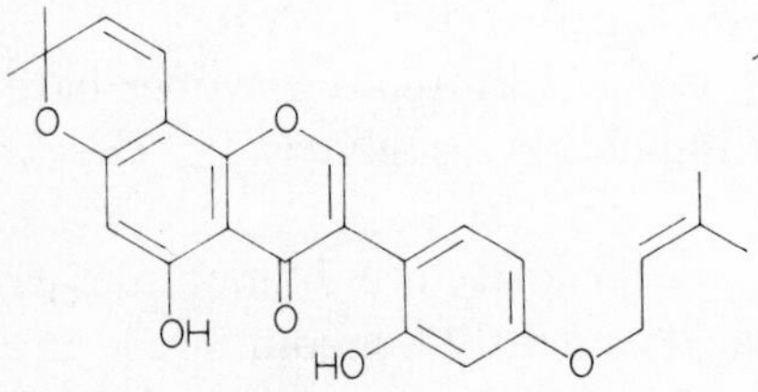

(72, Isoauriculatin)

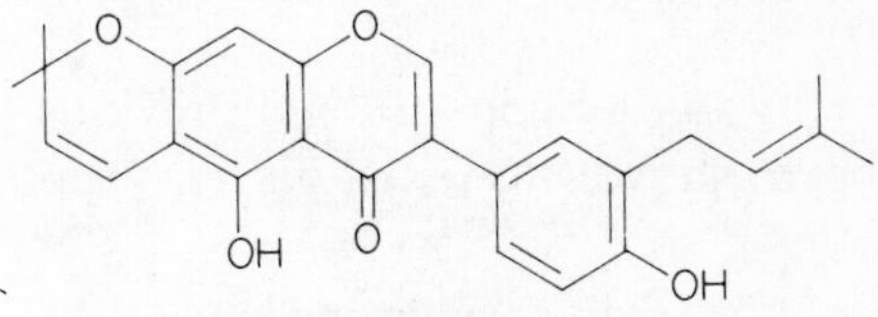

(73, Chandalone)

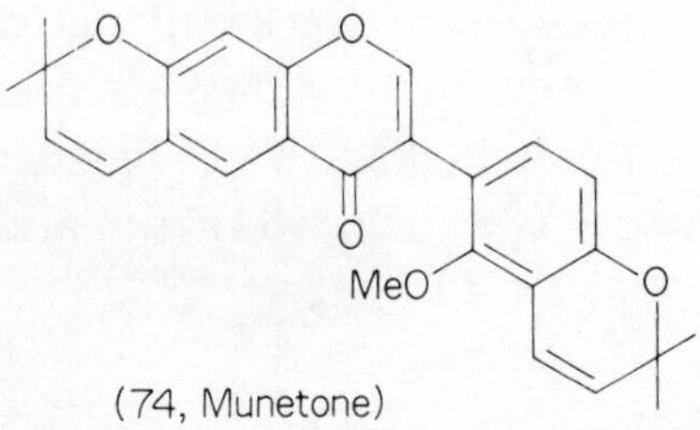

(74, Munetone)

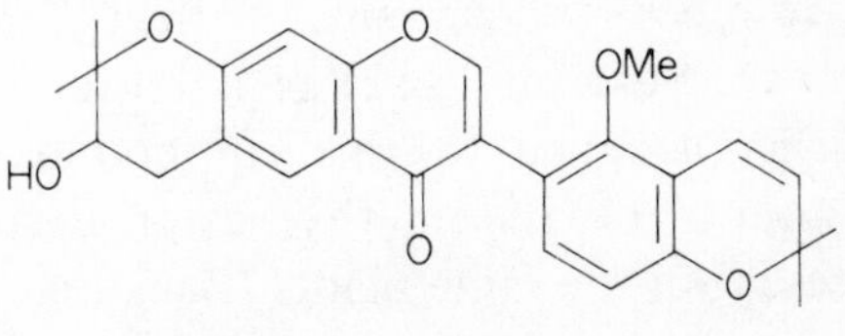

(75, Mundulone)

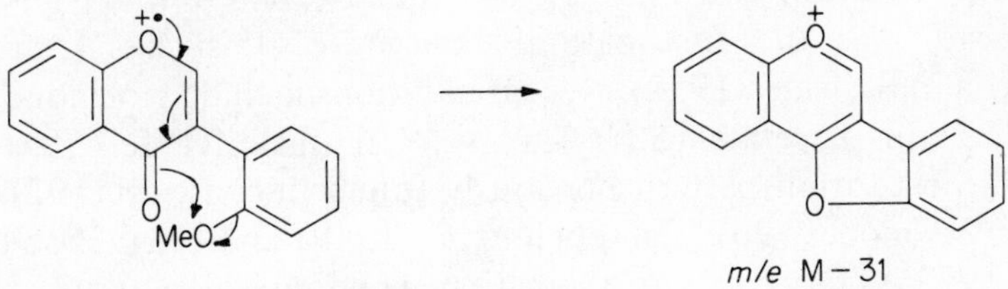

m/e M − 31

Figure 14.3 MS fragmentation of 2′-methoxyisoflavones

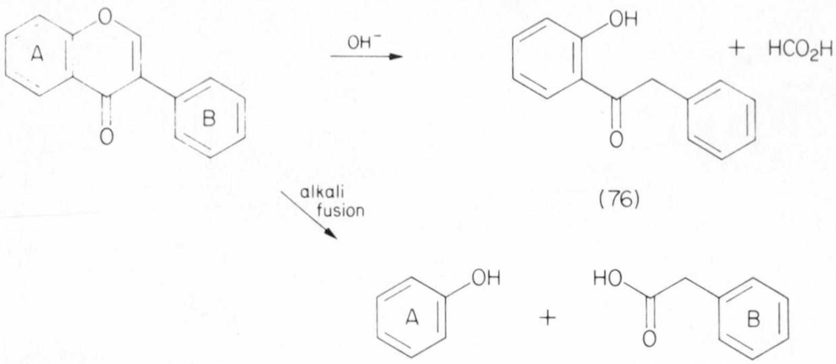

Figure 14.4 Alkali degradation of isoflavones.

many of the constituents previously isolated as free isoflavones have been shown to exist *in vivo* predominantly as glycosides (Beck, 1964).

Of the glycosides in Table 14.2, the majority are 7-glucosides or 7-rhamnosylglucosides. Much less frequently encountered are 4'-glucosides and 4'-rhamnosylglucosides. Among the recent addition to the group are the *C*-glucosides dalpanitin (77), paniculatin (78) and puerarin diacetate (80). The novel 6"-malonyl derivatives of formononetin 7-glucoside and biochanin A 7-glucoside (81), (82), have recently been isolated from *Trifolium* species (Beck and Knox, 1971). Structure (82) is a correction of the 5-malonate structure earlier assigned to this compound (Tamura *et al.*, 1969). Smaller amounts of the methyl ester derivatives (83) and (84) are also present in the same plants (Beck and Knox, 1971).

14.3 Isoflavanones

In contrast to the large number of isoflavones encountered in nature, the number of isoflavanones known is not great (Table 14.3). Padmakastein (85) from the bark of the Indian plant *Prunus puddum* (Rosaceae) was the first natural example of this class (Narasimhachari and Seshadri, 1952). Recent additions include violanone (93) and puerarin diacetate (80). The novel 6"-malonyl derivatives of 2',3',4'-substitution pattern previously found in nepseudin (92). The two isoflavanones (96) (Campbell *et al.*, 1969) and (95) (Braz Filho *et al.*, 1973a) co-occur with their corresponding isoflavones as heartwood constituents. Kievetone (97) is one of four isoflavonoid com-

Table 14.2   *Isoflavone glycosides*

Glycoside	Plant source	References*
Daidzein 7-glucoside (daidzin)	*Glycine max* *Pueraria* spp. *Trifolium pratense* *Baptisia* spp. *Thermopsis* spp.	Tamura *et al.*, 1969. Markham *et al.*, 1970. Dement and Mabry, 1972.
Daidzein 7-rhamnosyl-glucoside	*Baptisia* spp.	Markham *et al.*, 1970.
Daidzein 8-*C*-glucoside (puerarin) (79)	*Pueraria* spp.	
Puerarin xyloside	*Pueraria* spp.	
Puerarin 4′,6″-diacetate (80)	*P. tuberosa*	Bhutani *et al.*, 1969.
Formononetin 7-glucoside (ononin)	*Ononis spinosa* *Baptisia* spp. *Thermopsis* spp. *Trifolium* spp.	Markham *et al.*, 1970. Dement and Mabry, 1972.
Formononetin 7-glucoside 6″-malonate (81)	*Trifolium* spp.	Beck and Knox, 1971.
Formononetin 7-glucoside 6″-malonate, methyl ester (83)	*Trifolium* spp.	
Calycosin 7-glucoside	*Thermopsis* spp. *Baptisia* spp.	Dement and Mabry, 1972. Markham *et al.*, 1970.
Calycosin 7-rhamnosyl-glucoside	*Baptisia* spp.	
Pseudobaptigenin 7-rhamnosylglucoside (pseudobaptisin)	*Baptisia* spp.	Markham *et al.*, 1970.
Baptigenin rhamnoside	*Baptisia tinctoria*	Farkas *et al.*, 1963.
Genistein 7-glucoside (genistin)	*Genista tinctoria* *Glycine max* *Ulex nanus* *Adenocarpus complicatus* *Thermopsis* spp. *Baptisia* spp. *Trifolium* spp.	Paris and Faugeras, 1963. Paris and Faugeras, 1963. Dement and Mabry, 1972. Markham *et al.*, 1970. Beck and Knox, 1971.
Genistein 4′-glucoside (sophoricoside)	*Sophora japonica*	
Genistein 4′-neohesperidoside (sophoribioside)	*Sophora japonica*	Farkas and Nogradi, 1964.
Genistein 7-rhamnosyl-glucoside	*Baptisia* spp.	Markham *et al.*, 1970.
Genistein 7,4′-diglucoside	*Thermopsis* spp.	Dement and Mabry, 1972.
Genistein 6,8-di-*C*-glucoside (78) (paniculatin)	*Dalbergia paniculata*	Narayanan and Seshadri, 1971.
Biochanin A 7-glucoside (sissotrin)	*Dalbergia sisso* *Cicer arietinum* *Baptisia* spp. *Thermopsis* spp.	Banerji *et al.*, 1966. Wong *et al.*, 1965. Markham *et al.*, 1970. Dement and Mabry, 1972.

Continued

Table 14.2—*Continued*

Glycoside	Plant sources	References[*]
Biochanin A 7-glucoside 6″-malonate (82)	*Trifolium* spp. *Trifolium* spp.	Beck and Knox, 1971.
Biochanin A 7-glucoside 6″-malonate, methyl ester (84)	*Trifolium* spp.	
Biochanin A 7-rhamnosyl-glucoside	*Baptisia* spp.	Markham *et al.*, 1970.
Biochanin A 7-apiosyl-glucoside (lanceolarin)	*Dalbergia lanceolaria*	Malhotra *et al.*, 1967.
Prunetin 4′-glucoside (prunetrin)	*Prunus* spp. (Rosaceae)	
Orobol 7-glucoside	*Thermopsis* spp.	Dement and Mabry, 1972.
Orobol 7-rhamnosyl-glucoside	*Baptisia* spp. *Baptisia* spp.	Markham *et al.*, 1970.
Pratensein 7-glucoside 3′-Methylorobol 7-glucoside	*Thermopsis* spp. *Thermopsis* spp.	Dement and Mabry, 1972.
3′-Methylorobol 8-*C*-glucoside (dalpanitin) (77)	*Dalbergia paniculata*	Adinarayana and Rao, 1972a.
6,7,4′-Trihydroxyisoflavone 4′-glucoside	*Glycine max*	Friedlander and Sklarz, 1971.
Glyciteine glucoside	*Glycine max*	Naim *et al.*, 1973.
Texasin 7-glucoside	*Baptisia australis*	Lebreton *et al.*, 1967.
Afrormosin 7-glucoside (wistin)	*Wistaria* spp.	Shibata *et al.*, 1963.
Afrormosin 7-rhamnosyl-glucoside	*Baptisia* spp. *Baptisia* spp.	Markham *et al.*, 1970.
Fujikinetin glucoside (fujikinin)	*Cladrastis platycarpa*	Imamura *et al.*, 1972.
7-Hydroxy-6,2′-dimethoxy-4′,5′-methylenedioxy-isoflavone 7-glucoside (dalpatin)	*Dalbergia paniculata*	Adinarayana and Rao, 1972a.
6-Hydroxygenistein 7-rhamnosylglucoside	*Baptisia hirsuta*	Markham *et al.*, 1968.
Tectorigenin 7-glucoside (tectoridin)	*Iris tectorum* (Iridaceae) *Baptisia* spp. *Dalbergia riparia*	Markham *et al.*, 1970. Braz Filho *et al.*, 1973b.
7-Methyltectorigenin 4′-rhamnosylglucoside	*Dalbergia sisso*	Ahluwalia *et al.*, 1965.
Irigenin 7-glucoside (iridin)	*Iris* spp. (Iridaceae)	Dhar and Kalla, 1972.

[*]Earlier references may be found in the reviews by Ollis (1962) and Dean (1963).

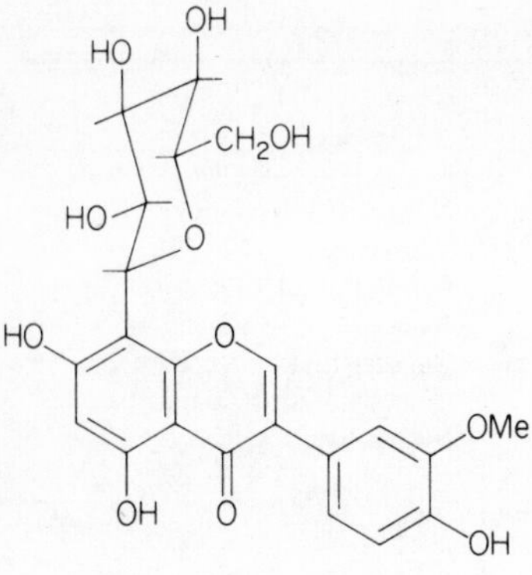

(77, Dalpanitin)

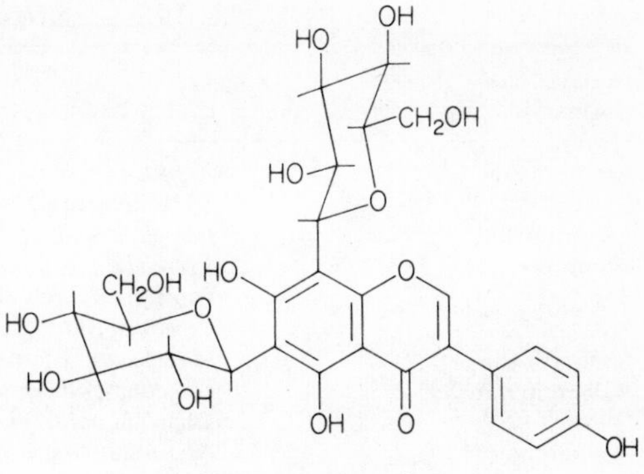

(78, Paniculatin)

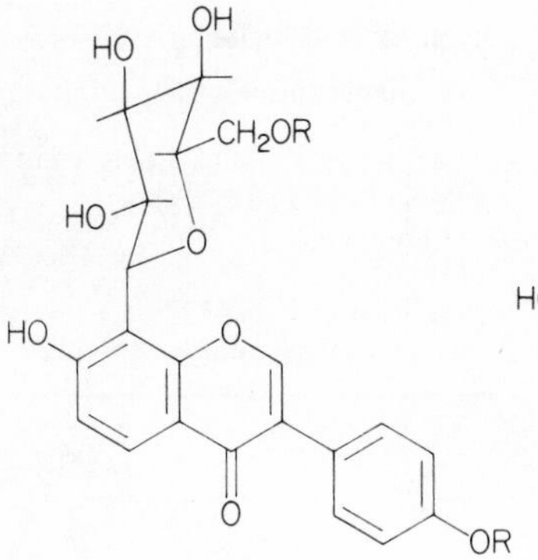

(79, R = H Puerarin)
(80, R = MeCO— Puerarin-4',
                6"-diacetate)

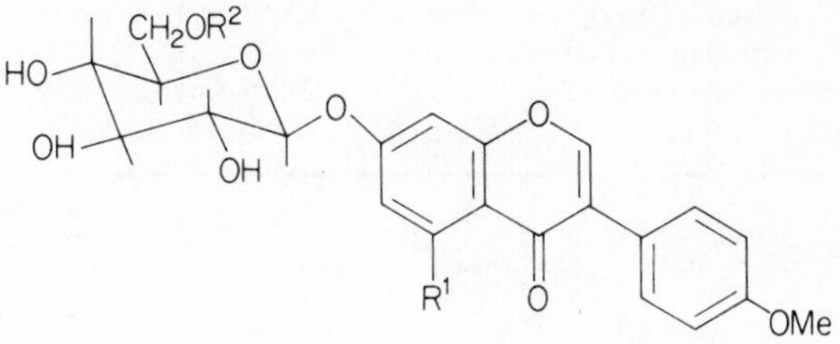

(81, R^1 = H, R^2 = —$COCH_2CO_2H$)
(82, R^1 = OH, R^2 = —$COCH_2CO_2H$)
(83, R^1 = H, R^2 = —$COCH_2CO_2Me$)
(84, R^1 = H, R^2 = —$COCH_2CO_2Me$)

pounds found in fungal- or virus-infected french bean *(Phaseolus vulgaris)* (Burden *et al.*, 1972). The only example of an isoflavanone-C-glycoside, dalpanin (98) was recently isolated from *Dalbergia paniculata* (Adinarayana and Rào, 1972b), from which dalpatin (38, R' = glucosyl-), dalpanitin (77), paniculatin (78) and dalpanol (115) (Section D) have also been isolated.

Isoflavanones show absorption at ~270 nm and ~310 nm which is not very different from that of isoflavones. They also give a positive reaction in the Durham test (i.e. the production of a deep red colour in contact with nitric acid) earlier thought to be characteristic only of rotenoids (Narasimhachari and Seshadri, 1952). Dehydrogenation

Table 14.3 *Isoflavanones*

Isoflavanone (formula)	Plant sources	References
Padmakastein (85)	*Prunus puddum* (Rosaceae)	Narasimhachari and Seshadri, 1952.
Ferreirin (86)	*Ferreirea spectabilis*	King *et al.*, 1952.
Homoferreirin (87)	*Ferreirea spectabilis*	King *et al.*, 1952.
	Ougeinia dalbergioides	Balakrishna *et al.*, 1962.
	Cicer arietinum	Grisebach and Zilg, 1968.
Sophorol (88)	*Maackia amurensis*	Suginome, 1959.
Dalbergioidin (89)	*Ougeinia dalbergioides* ⎫	
Ougenia (90)	*Ougeinia dalbergioides* ⎬	Balakrishna *et al.*, 1962.
Neotenone (91)	*Neorautanenia pseudo-* ⎫ *pachyrrhiza* ⎬ *Pachyrrhizus erosus* ⎭	Crombie and Whiting, 1963.
	Neorautanenia edulis ⎫ *N. amboensis* ⎬	Brink *et al.*, 1965.
Nepseudin (92)	*N. pseudopachyrrhiza*	Crombie and Whiting, 1963.
Violanone (93)	*Dalbergia violacea*	Ollis, 1968.
Parvisoflavanone (94)	*Poecilanthe parviflora*	Assumpcao and Gottlieb, 1973.
5,7-Dihydroxy-4'-methoxy-(95)	*Andira parviflora*	Braz Filho *et al.*, 1973a.
6,7-Dimethoxy-3',4'-methylenedioxy- (96)	*Cordyla africana*	Campbell *et al.*, 1969.
Kievetone (97)	*Phaseolus vulgaris*	Burden *et al.*, 1972.
Dalpanin (98)	*Dalbergia paniculata*	Adinarayana and Rao, 1972b.

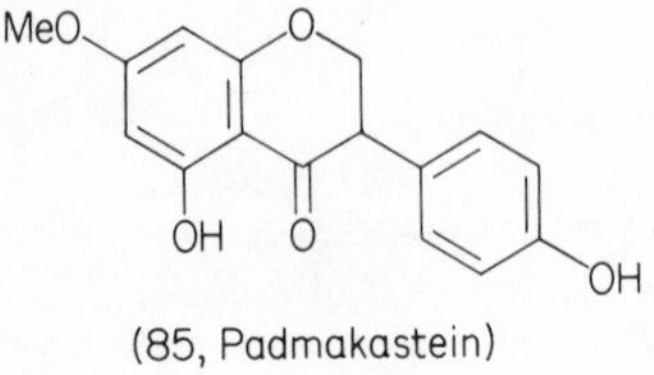

(85, Padmakastein)

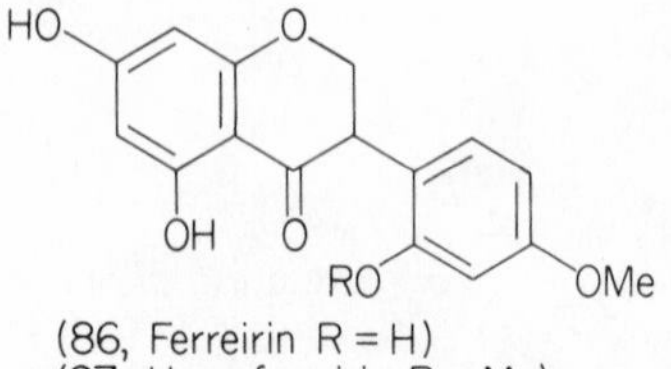

(86, Ferreirin R = H)
(87, Homoferreirin R = Me)

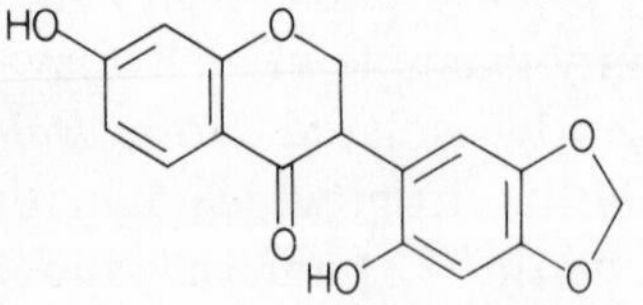

(88, Sophorol)

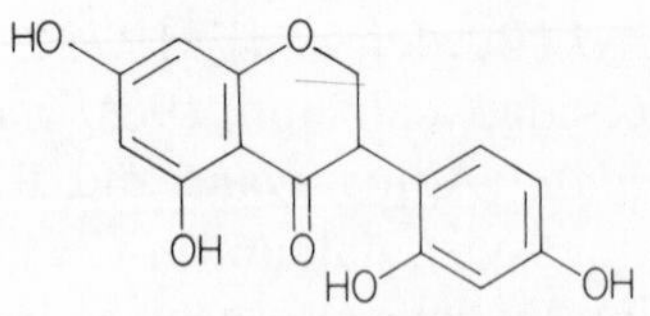

(89, Dalbergioidin)

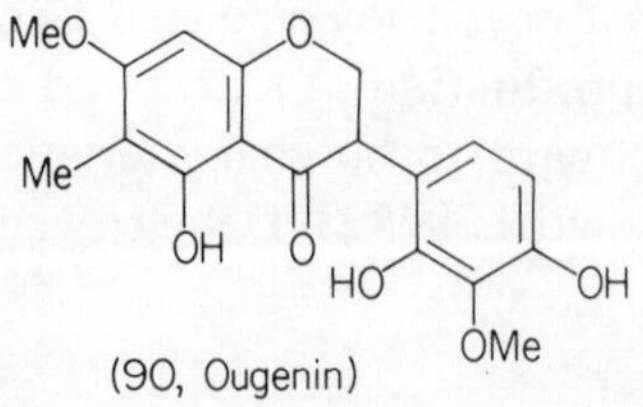

(90, Ougenin)

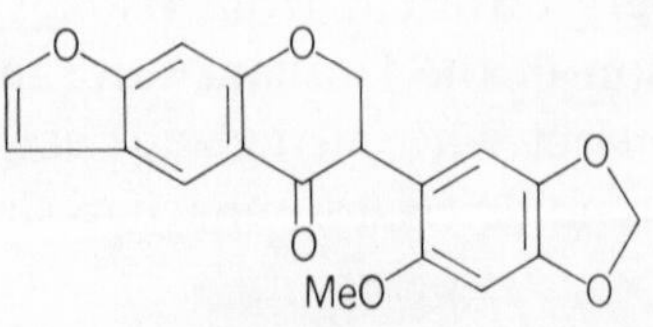

(91, Neotenone)

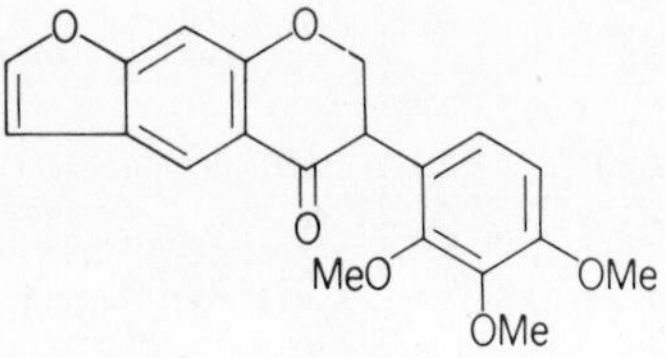

(92, Nepseudin)

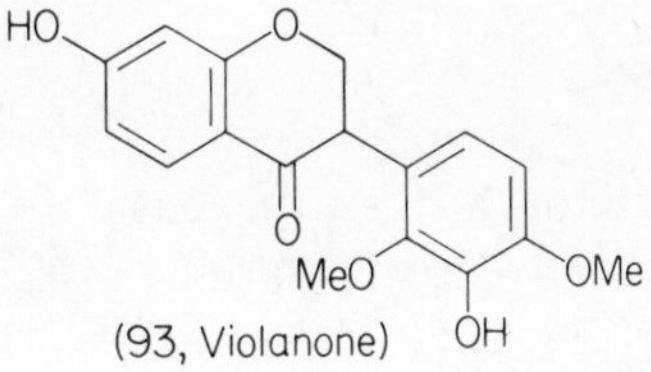

(93, Violanone)

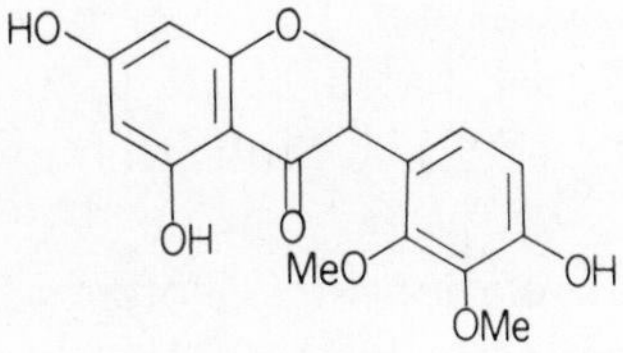

(94, Parvisoflavanone)

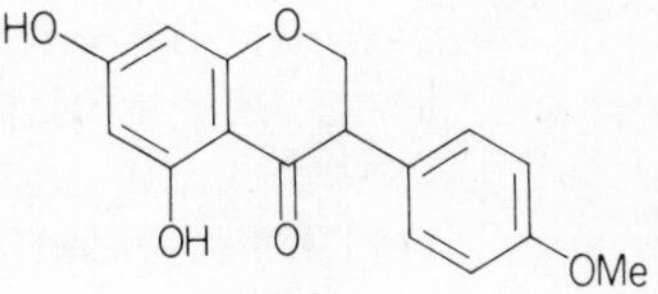

(95)

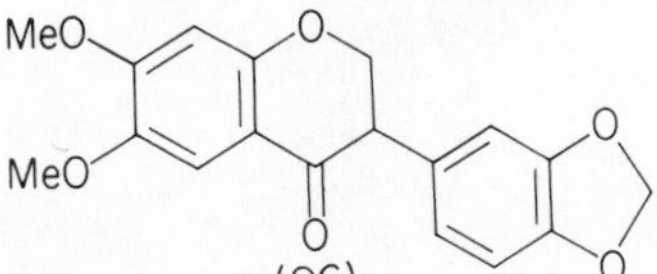

(96)

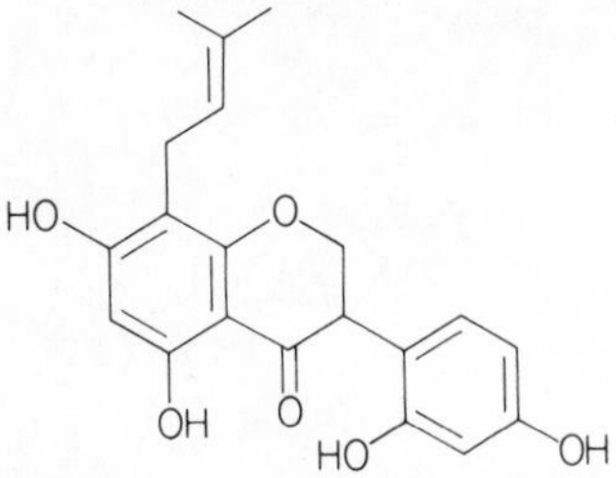

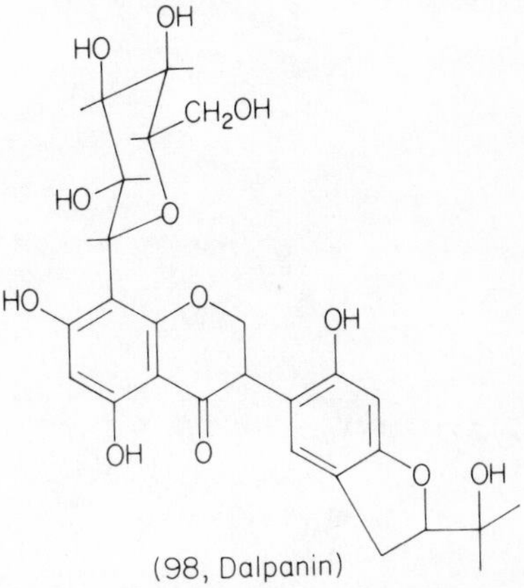

(98, Dalpanin)

of isoflavanones yield the corresponding isoflavones, the structure of which can then be determined in the usual way. This structural approach has been used, for example, on the *Neorautanenia* isoflavanones neotenone (91) and nepseudin (92) (Crombie and Whiting, 1963).

With the exception of sophorol (88) ($[\alpha]_D$ +9.5 acetone, −13.6 ethanol) all the known isoflavanones were isolated only in the optically inactive form. Keto-enol tautomerism offers a possible explanation for the lack of optical activity encountered in these compounds (Clark-Lewis, 1962). It is also noteworthy that an inordinate proportion of the natural isoflavanones possess oxygenation at the 2′-position in ring B. A possible explanation of this characteristic in terms of the biogenetic derivation of isoflavanones via pterocarpans, with which they usually co-occur, will be discussed in Section 14.11.

14.4 Rotenoids

The rotenoids have in common the four-ring chromanochromanone system (99) as the basic structural unit. They may be regarded as formally isoflavanones which have been modified with an 'extra' carbon atom. Such a view is consistent with current knowledge of the biogenetic origin of these compounds (Chapter 16). Rotenone (100), after which the group is named, was isolated at about the same time as the first isoflavone (Geoffroy, 1892). Its full structure was established in 1932, and those of the other early member of the series (101)–(105) revealed close structural relationships. Rotenoids as a group have been reviewed in detail by Crombie (1963), Dean (1963) and more recently by Fukami and Nakajima (1971).

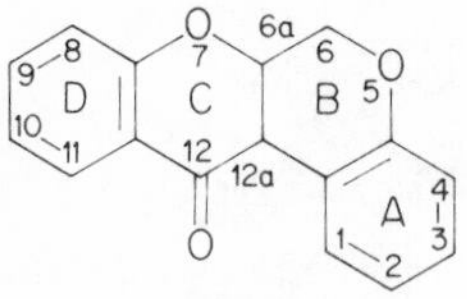

(99)

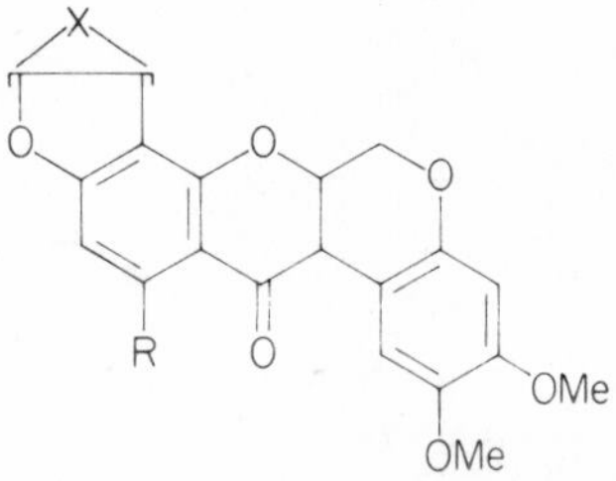

(100, Rotenone R = H)
(101, Sumatrol R = OH) $\quad$ X = CH$_2$=CMe—CH—CH$_2$
(102, Deguelin R = H)
(103, Toxicarol R = OH) $\quad$ X = Me$_2$C—CH=CH
(104, Elliptone R = H)
(105, Malaccol R = OH) $\quad$ X = CH=CH

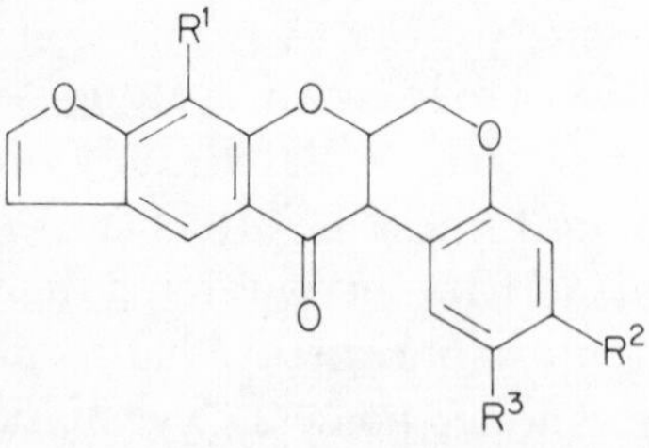

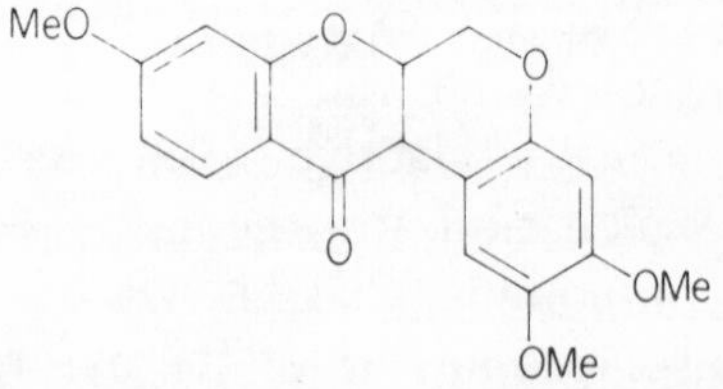

(106, Pachyrrhizone R^1 = OMe, R^2+R^3 = —OCH$_2$O—)
(107, Dolineone R^1 = H, R^2+R^3 = —OCH$_2$O—)
(108, Erosone R^1 = H, R^2 = R^3 = OMe)

(109, Munduserone)

Table 14.4   *Rotenoids*

Rotenoid (formula)	Plant sources	References*
Rotenone (100)	*Derris* spp.	
	Tephrosia spp.	
	Lonchocarpus spp.	
	Piscidia erythrina	
	Neorautanenia ficifolia	
	Pachyrrhizus erosus	
	Millettia dura	Ollis, *et al.*, 1967.
Sumatrol (101)	*Derris malaccensis*	
	Piscidia erythrina	Falshaw *et al.*, 1966.
Deguelin (102)	*Tephrosia* spp.	
	Derris spp.	
	Lonchocarpus nicosa	
Toxicarol (103)	*Tephrosia* spp.	
	Derris spp.	
Elliptone (104)	*Derris elliptica*	
Malaccol (105)	*Derris malaccensis*	
Pachyrrhizone (106) ⎱	*Pachyrrhizus erosus*	
Dolineone (107) ⎬	*Neorautanenia pseudo-*	
Erosone (108) ⎰	*pachyrrhiza*	
Munduserone (109)	*Mundulea sericea*	
Millettone (110)	*Piscidia erythrina*	Falshaw *et al.*, 1966.
	Millettia dura ⎱	Ollis *et al.*, 1967.
Isomillettone (111)	*Millettia dura* ⎰	
Amorphigenin (112)	*Amorpha fruticosa*	Claisse *et al.*, 1964.
Dihydroamorphigenin (114)	*Amorpha fruticosa*	Kasymov *et al.*, 1972.
Dalpanol (115)	*Dalbergia* spp.	Adinarayana *et al.*, 1971.

*Earlier references may be found in the reviews by Crombie (1963) and Dean (1963).

Fifteen rotenoids are currently known (Table 14.4), of which five have been reported since the earlier surveys. Millettone (110) and isomillettone (111) are methylenedioxy analogues of the dimethoxy compounds degulin (102) and rotenone (100), respectively. The other new compounds amorphigenin (112), existing as the viciano-side amorphin (113), dihydroamorphigenin (114), and dalpanol (115), are interesting dihydro and/or hydroxyl derivatives of rotenone.

Because of the presence of the extra ring system (B), chemical and stereochemical reactions undergone in the rotenone series are quite complex. The earlier review of Dean (1963) provides an excellent summary of this specialized field. Some reactions of rotenone, important to the structural elucidation of the fundamental ring system are briefly summarized in Fig. 14.5. A key reaction is the dehydrogenation of rotenone by MnO_2 and other oxidizing agents to

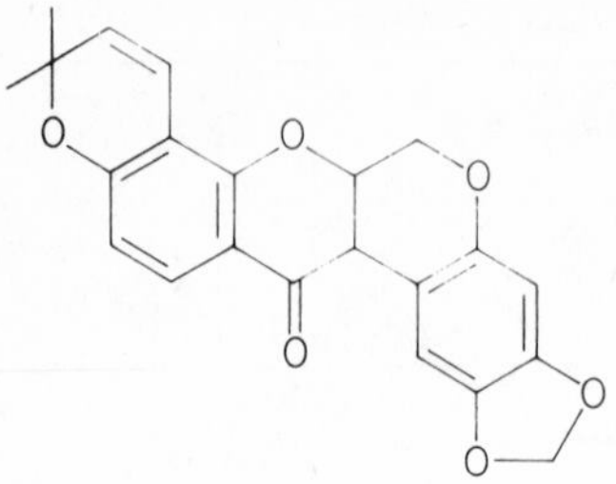

(110, Millettone)

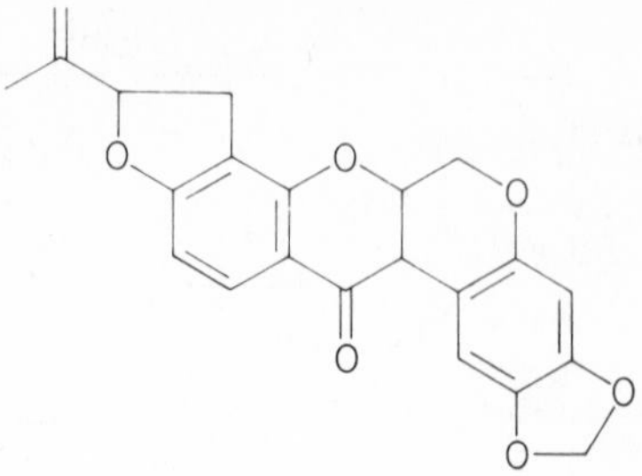

(111, Isomillettone)

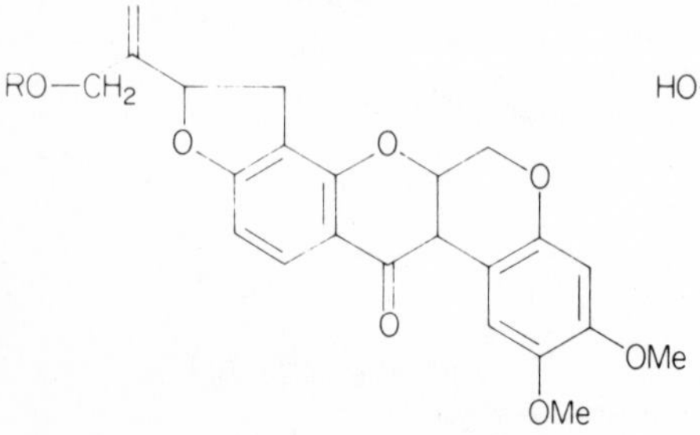

(112, Amorphigenin R = H)
(113, Amorphin R = vicianosyl)

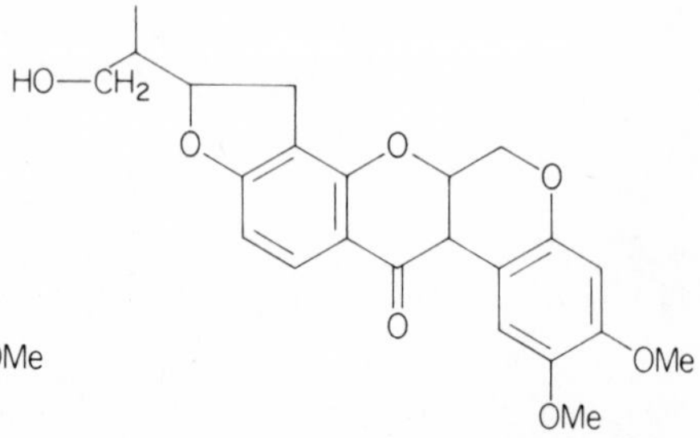

(114, Dihydroamorphigenin)

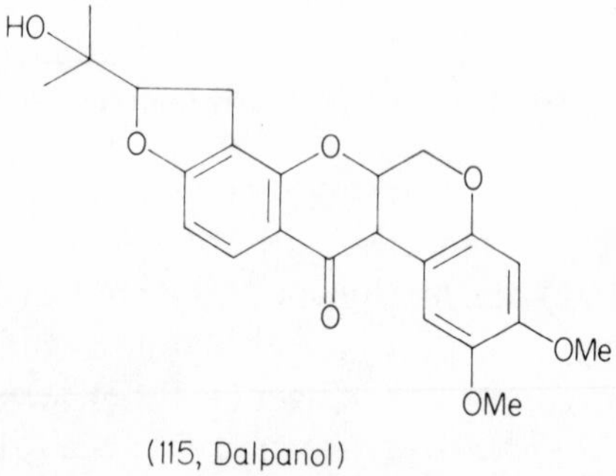

(115, Dalpanol)

6a,12a-dehydrorotenone (116). Dehydrorotenone is easily hydrolysed with alcoholic KOH to derrisic acid (117) without loss of any carbon atoms. Derrisic acid can be recyclized to dehydrorotenone with acetic anhydride/sodium acetate. Dehydrorotenoids such as (116) can be distinguished from isoflavones by oxidation of the methylene group with nitrous acid to yield yellow ketolactones (rotenonones) (118).

The absolute stereochemistry of rotenone is shown in (119)

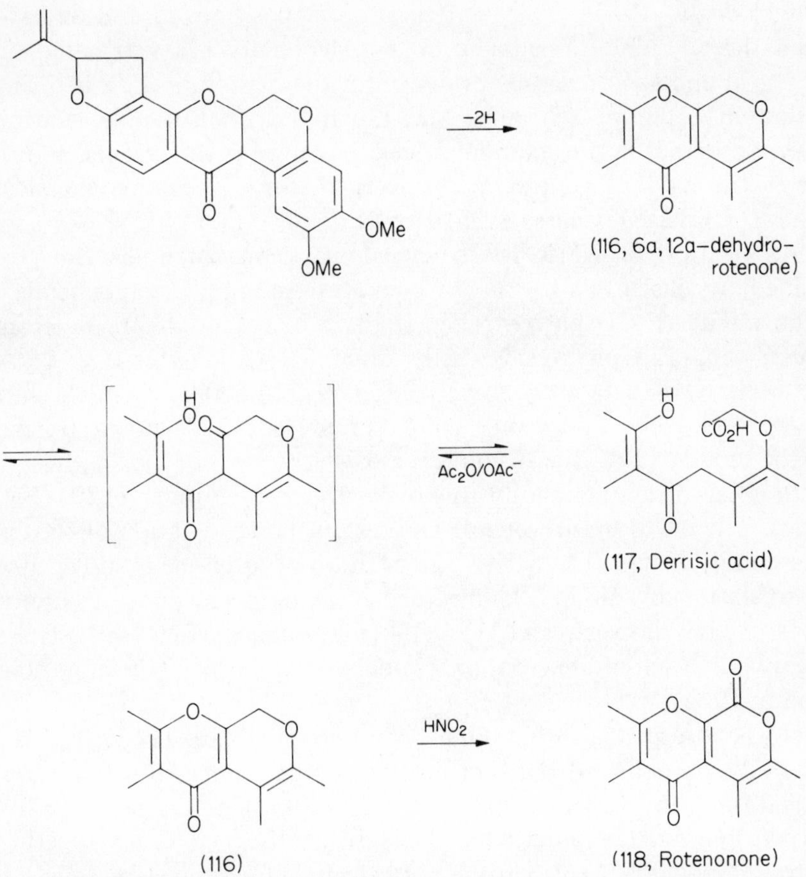
OMe
OMe
−2H
(116, 6a, 12a−dehydro-rotenone)
Ac2O/OAc−
CO2H
(117, Derrisic acid)
HNO2
(116)
(118, Rotenonone)

Figure 14.5 Some degradation products of rotenone

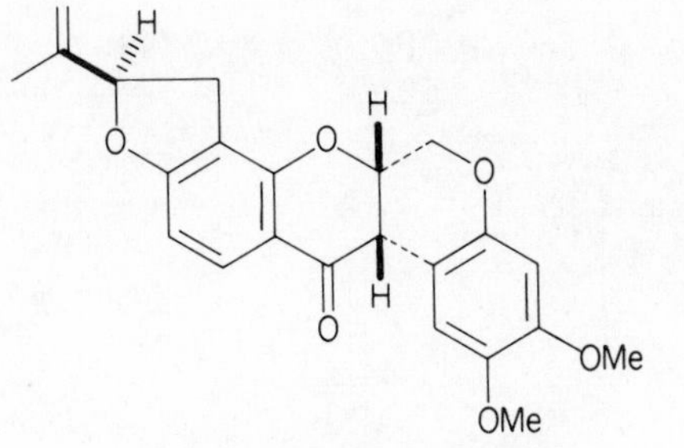
H
H
H
OMe
OMe

(119, 6aS, 12aS, Rotenone)

(Buchi *et al.*, 1961). The configuration at positions 5' and 6a have been directly determined and that assigned to 12a was based on chemical and spectroscopic evidence for the existence of *cis* B/C ring fusion in rotenone (Crombie and Lown, 1962). Rotenone and all other rotenoids so far examined have positive Cotton effects, and so far as the 6a, 12a centres are concerned, form one stereochemical series (6aS, 12aS) (Claisse *et al.*, 1964).

Application of NMR to structural work on rotenoids can give immediate insight into many constitutional and stereochemical details. The ABCD pattern characteristic of rotenoids arising from the 6, 6a, 12a protons is easily discerned (Adams *et al.*, 1966). Relevant NMR data are summarized in Fig. 14.6. The chemical shift of the hydrogen at position 1 (ring A) represents a diagnostic feature of the mode of B/C ring fusion (*cis* ~6.7δ, *trans* ~8.0δ).

In alkali rotenone readily ring opens to a Δ-3-chromen (120) (Fig. 14.7). The reaction is reversible and results in racemization of the 6a, 12a chiral centres. The active 12a position of rotenone is susceptible to oxidation by air in alkaline media, giving a racemic mixture of 12a-hydroxy derivatives (121) (122) (rotenolones) (Fig. 14.7). These are readily dehydrated to 6a,12a-dehydrorotenone (Crombie and Godin, 1961).

(±)-Tephrosin (123) has long been known (Dean, 1963), but has generally been regarded as an artefact of deguelin (102) (via aerial oxidation). The isolation by Ollis *et al.* (1967) of the new 12a-hydroxyrotenoid, (−)millettosin (124) from *Millettia dura*, together with tephrosin (123) also in the optical active form, suggests however that these compounds are indeed natural products.

The isolation of the 6a,12a-dehydrorotenoids, dehydrodeguelin (125) from *M. dura* (Ollis *et al.*, 1967) and dehydromillettone (126) from *Piscidia erythina* (Falshaw *et al.*, 1966) raises the interesting possibilities of these compounds also being natural products. In view

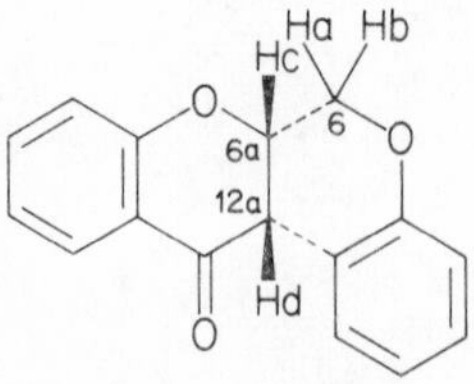

	δ value	Jab	12 Hz
Ha, Hb (multiplet)	4.53	Jad	1–2 Hz
Hc (multiplet)	5.05	Jac	1.5 Hz
Hd (quartet)	4.05	Jbc	3 Hz
		Jcd	4 Hz

Figure 14.6 NMR data for heterocyclic ring protons of rotenoids

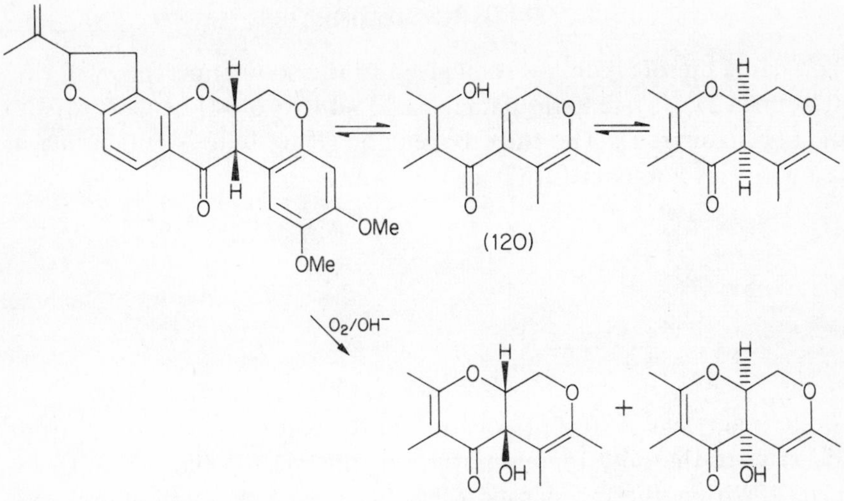

(120)

(121) (122)

Figure 14.7 Some reactions of rotenone in alkali.

of the known ease of dehydration of 12a-hydroxyrotenoids, however, the natural occurrence of dehydrorotenoids cannot be regarded as established.

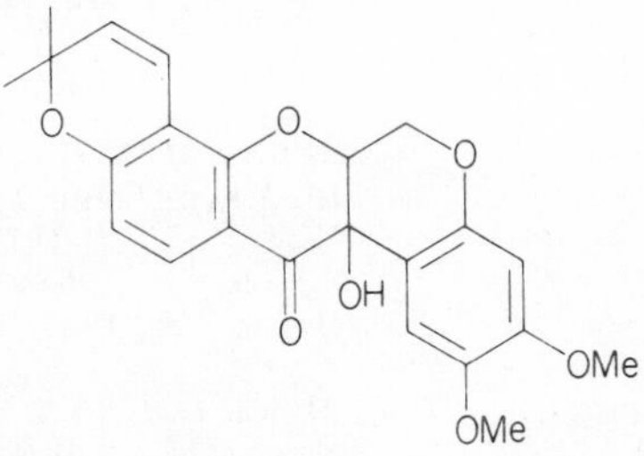

(123, Tephrosin)

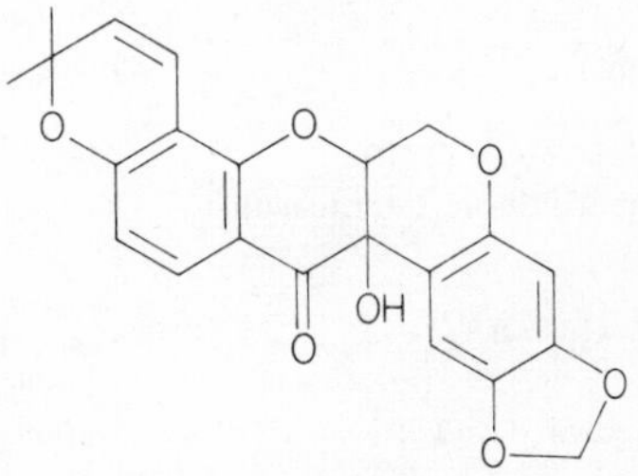

(124, Millettosin)

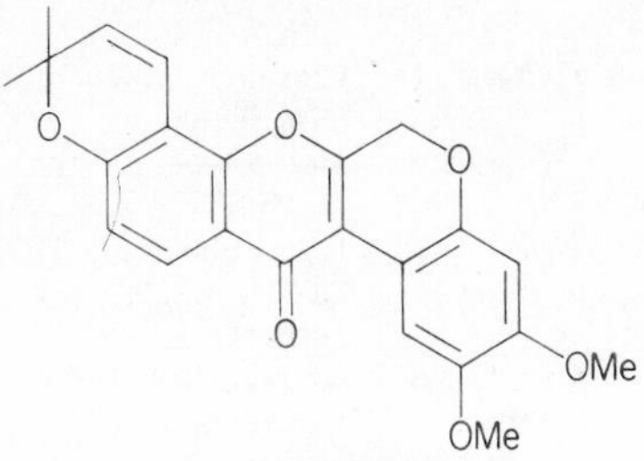

(125, Dehydrodeguelin)

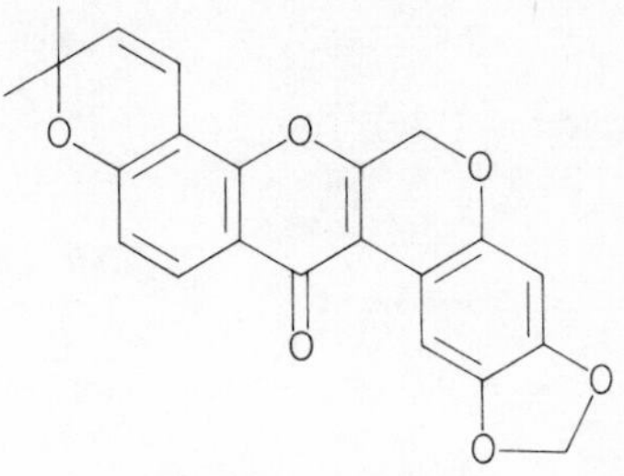

(126, Dehydromillettone)

14.5 Pterocarpans

The name pterocarpan has been given to the coumaranochroman ring system (127) (systematic name: 6a,11a-dihydro-6-H-benzofuro[3,2-c][1]benzopyran). The fully systematic (Ring Index) numbering in use with this nomenclature is also shown.

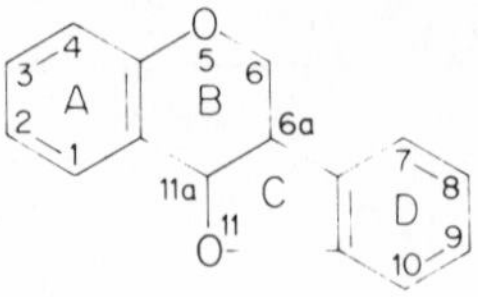

(127)

For many years, (−)-pterocarpin (128) and (−)-homopterocarpin (129) were the only representatives of this class (Ollis, 1962; Dean, 1963). Within the last decade there has been a dramatic number of additions to the pterocarpan group (Table 14.5). These, in the main, occur as heartwood constituents of tropical genera of the Legumi-nosae, including some outside the sub-family Lotoideae. In other leguminous plants, pterocarpans are often found in the roots (e.g. in

Table 14.5   *Natural pterocarpans*

Pterocarpan (formula)	Plant sources	References
(−)-Medicarpin (130) (demethylhomopterocarpin)	*Medicago sativa*	Smith *et al.*, 1971.
	Trifolium pratense	Higgins and Smith, 1972.
	Dalbergia stevensonii	McMurray *et al.*, 1972.
	Andira inermis	Cocker *et al.*, 1965.
	Swartzia madagasca- riensis	Harper *et al.*, 1969.
(+)-Medicarpin (130)	*Dalbergia* spp.	Alencar *et al.*, 1972.
		Matos *et al.*, 1970.
		Braz Filho *et al.*, 1973b.
		Ollis, 1968.
	Machaerium spp.	Ollis, 1968.
	Platymiscium trinitatis	Craveiro and Gottlieb, 1974.
(±)-Medicarpin (130)	*Dalbergia* spp.	Ollis, 1968.
		Matos *et al.*, 1970.
		McMurray *et al.*, 1972.
	Aldina heterophylla	Braz Filho *et al.*, 1973a.
(−)-Homopterocarpin (129)	*Pterocarpus* spp.	Ollis, 1968.
	Barhia nitida	Seshadri, 1972.
	Swartzia madagascariensis	Harper *et al.*, 1969.
	Pericopsis angolensis	Harper *et al.*, 1969.

Continued

Table 14.5—*Continued*

Pterocorpan (formula)	Plant sources	References
(+)-Homopterocarpin (129)	*Machaerium villosum*	Ollis, 1968.
(−)-Maackiain (131)	*Maackia amurensis*	Suginome, 1966a.
	Andira inermis	Cocker *et al.*, 1962.
	Swartzia madagascariensis	Harper *et al.*, 1969.
	Diplotropis purpurea	Braz Filho *et al.*, 1973a.
	Dalbergia stevensonii	McMurray *et al.*, 1972.
(+)-Maackiain (131)	*Aldina heterophylla* ⎫	Braz Filho *et al.*, 1973a.
(±)-Maackiain (131)	*Aldina heterophylla* ⎭	
	Sophora japonica	Shibata and Hishikawa, 1963.
	Pterocarpus dalbergi- *oides*	Parthasarathy *et al.*, 1969.
	Dalbergia stevensonii	McMurray *et al.*, 1972.
(−)-Pterocarpin (128)	*Pterocarpus* spp.	Ollis, 1968.
	Baphia nitida	Seshadri, 1972.
	Sophora subprostrata	Shibata and Hishikawa, 1963.
(−)-3-Hydroxy-4,9-dimethoxy- (134)	*Swartzia madagascariensis*	Harper *et al.*, 1969.
(−)-3,4,9-Trimethoxy (135)	*S. madagascariensis*	Harper *et al.*, 1969.
(−)-3,4-Dihydroxy-8,9-methylenedioxy- (136)	*Dalbergia spruceana*	Ollis, 1968.
(−)-4-Hydroxy-3-methoxy-8,9-methylenedioxy- (137)	*D. spruceana*	Ollis, 1968.
(−)-3-Hydroxy-4-methoxy-8,9-methylenedioxy-(138)	*Dalbergia spruceana*	Ollis, 1968.
	Swartzia madagascariensis	Harper *et al.*, 1969.
(−)-3,4′-Dimethoxy-8,9-methylenedioxy- (139)	*Swartzia madagascariensis*	Harper *et al.*, 1969.
	Neorautanenia ficifolia	Bouwer *et al.*, 1968.
(−)-Philenopteran (140)	*Lonchocarpus laxiflorus*	Pelter and Amenechi, 1969.
(−)-9-Methylphilenopteran (141)	*Lonchocarpus laxiflorus*	Pelter and Amenechi, 1969.
(−)-2-Hydroxypterocarpin (142)	*Neorautanenia edulis*	Rall *et al.*, 1970.
	Swartzia leiocalycina	Donnelly and Fitzgerald, 1971.
(−)-2-Methoxypterocarpin (143)	*Neorautanenia edulis*	Rall *et al.*, 1972.
Neodulin (144a)	*Neorautanenia edulis*	Brink *et al.*, 1965.
Ficinin (144b)	*Neorautanenia ficifolia*	Brink *et al.*, 1966.
(+)-Leiocarpin (145)	*Apuleia leiocarpa*	Braz Filho and Gottlieb, 1971.
(−)-Phaseollin (146)	*Phaseolus vulgaris*	Perrin, 1964.
(−)-Phaseollidin (147)	*Phaseolus vulgaris*	Perrin *et al.*, 1972.
		Burden *et al.*, 1972.
(−)-Neorautane (148)	*Neorautanenia edulis*	Rall *et al.*, 1972.
(−)-2-Isopentenyl-3-hydroxy-8,9-methylene-dioxy- (149)	*Neorautanenia edulis*	Rall *et al.*, 1971.
(−)-Ficifolinol (150)	*Neorautanenia ficifolia* ⎫	
(−)-Folitenol (151)	*Neorautanenia ficifolia* ⎬	Brink *et al.*, 1970.
(−)-Folinin (152)	*Neorautanenia ficifolia* ⎭	
(−)-Gangetin (153)	*Desmodium gangeticum*	Purushothaman *et al.*, 1971.

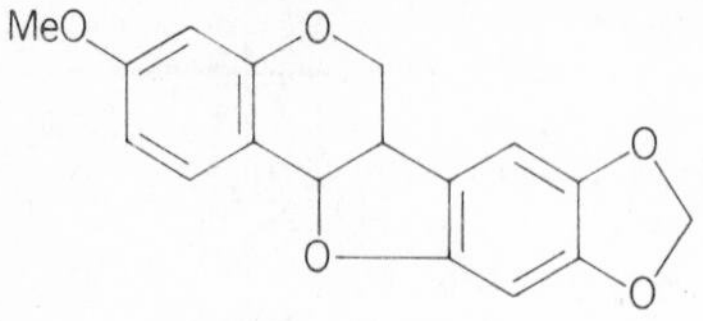

(128, Pterocarpin)

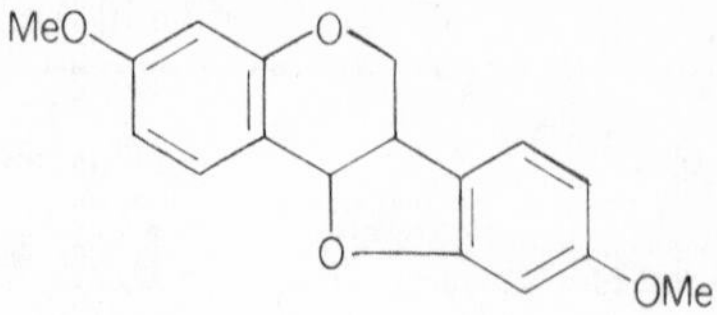

(129, Homopterocarpin)

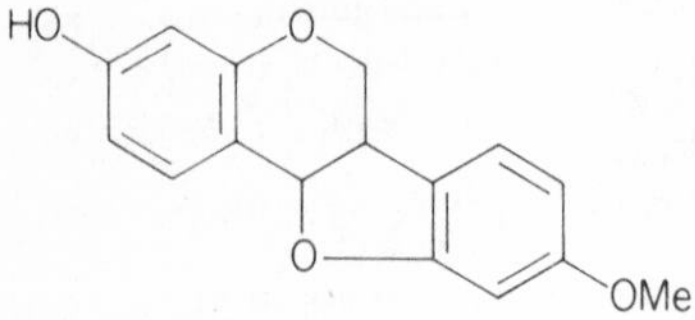

(130, Medicarpin)

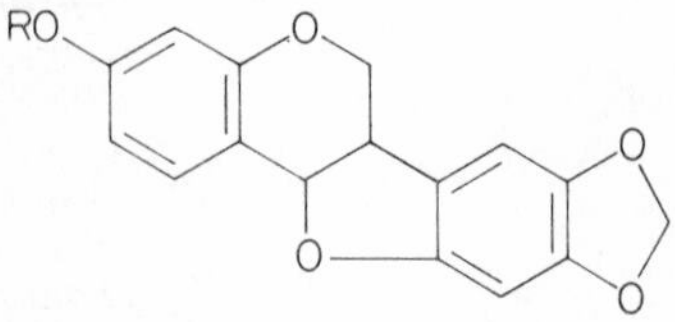

(131, Maackiain R = H)
(132, Trifolirhizin R = glucosyl)
    ((−)-maackiain β-D-glucoside)
(133, Sophorajaponicin R = glucosyl)
    ((+)-maackiain β-D-glucoside)

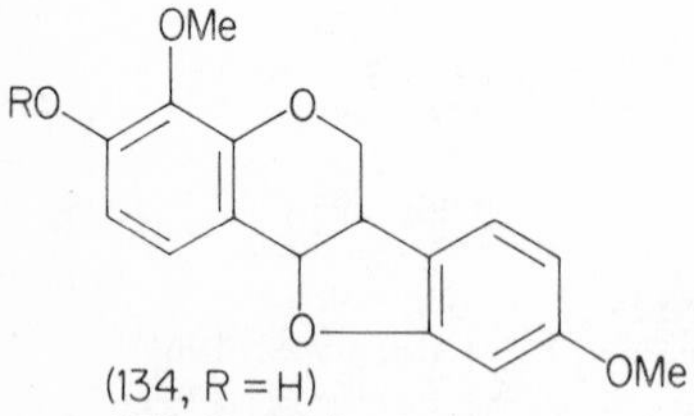

(134, R = H)
(135, R = Me)

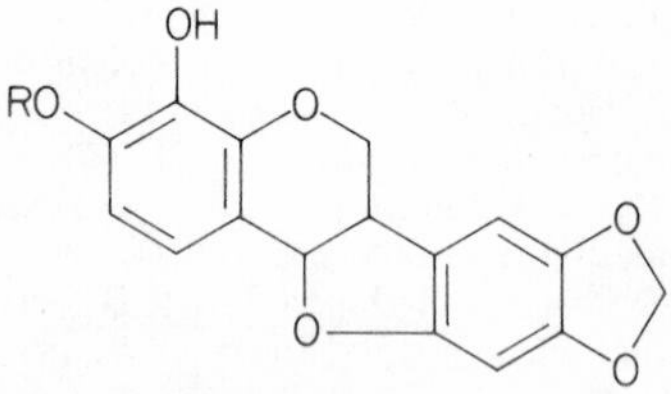

(136, R = H)
(137, R = Me)

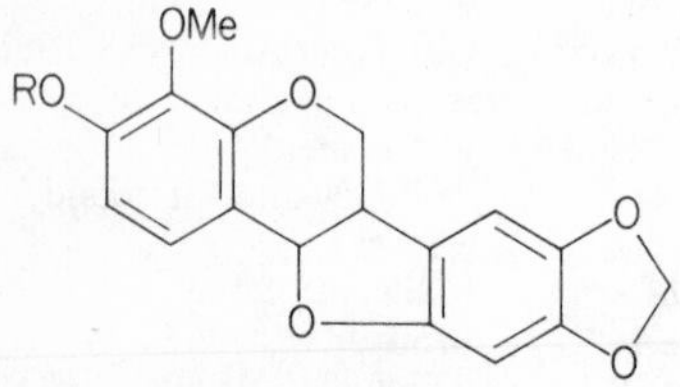

(138, R = H)
(139, R = Me)

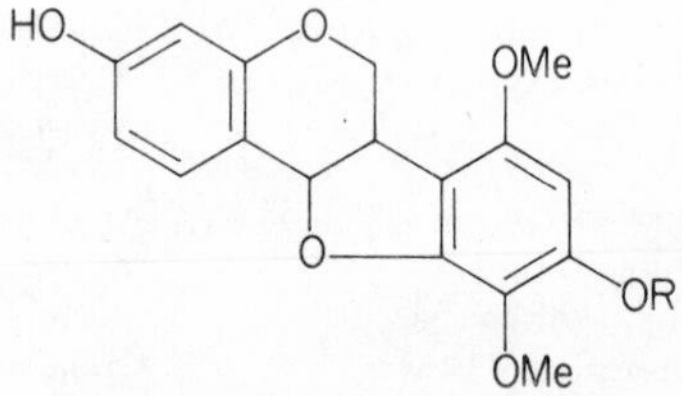

(140, Philenopteran R = H)
(141, 9-O-methylphilenopteran R = Me)

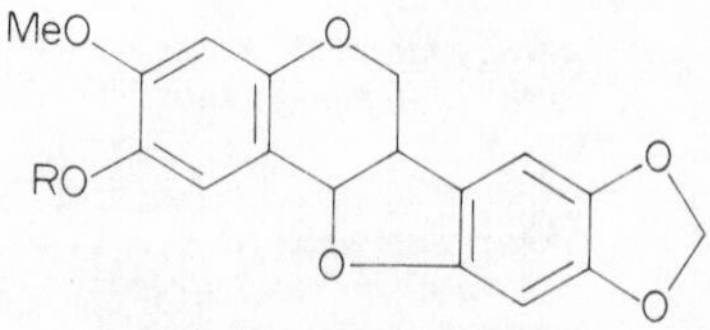

(142, R = H)
(143, R = Me)

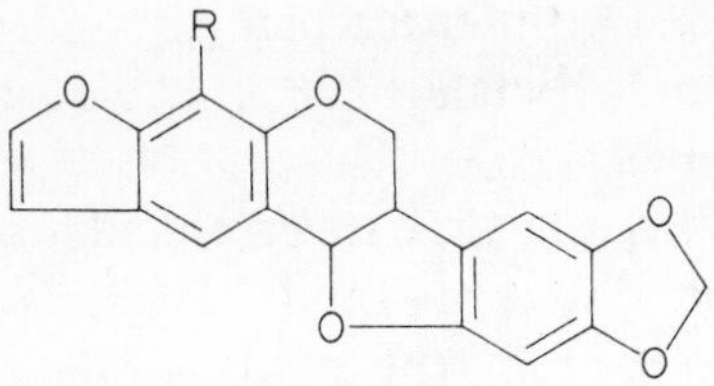

(144a, Neodulin R = H)
(144b, Ficinin R = OMe)

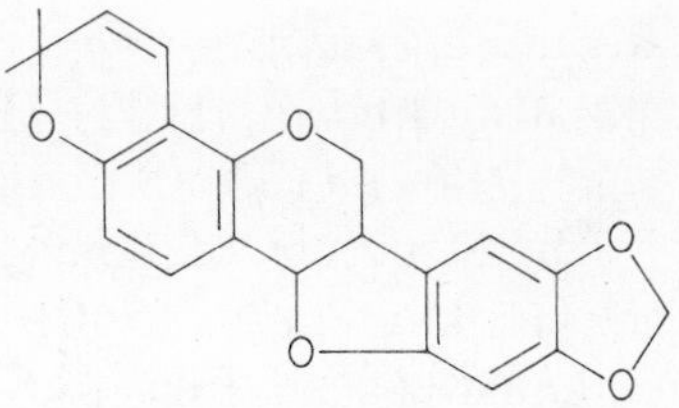

(145, Leiocarpin)

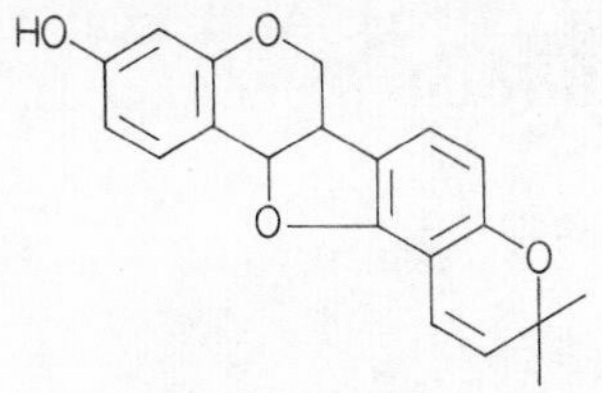

(146, Phaseollin)

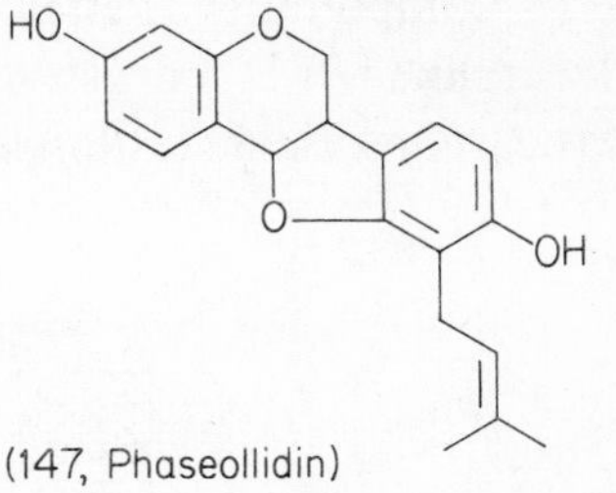

(147, Phaseollidin)

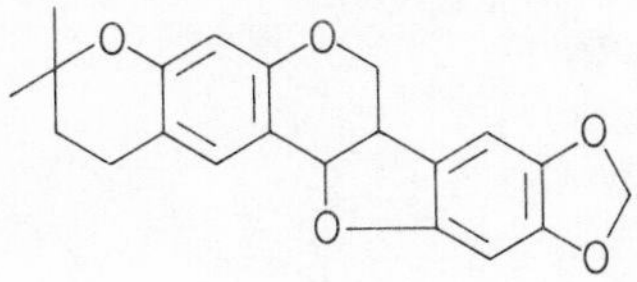

(148, Neorautane)

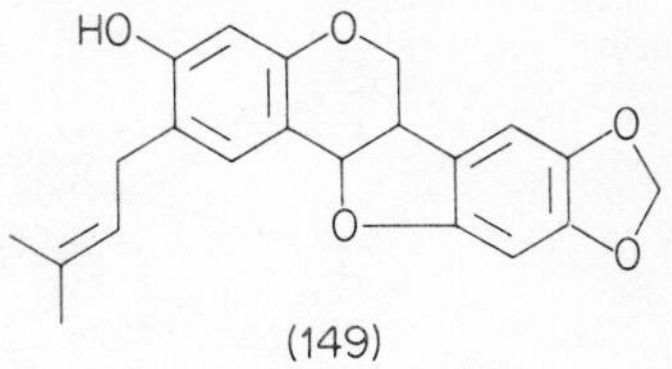

(149)

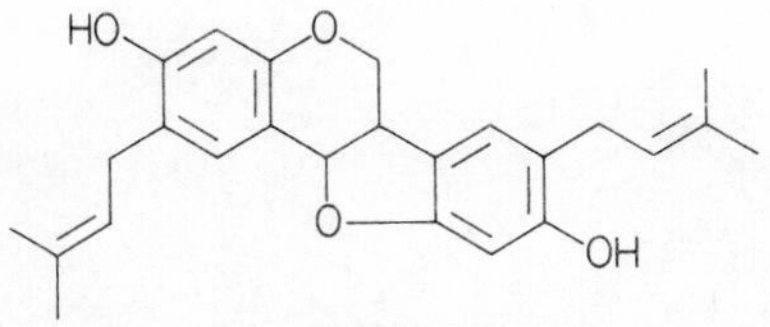

(150, Ficifolinol)

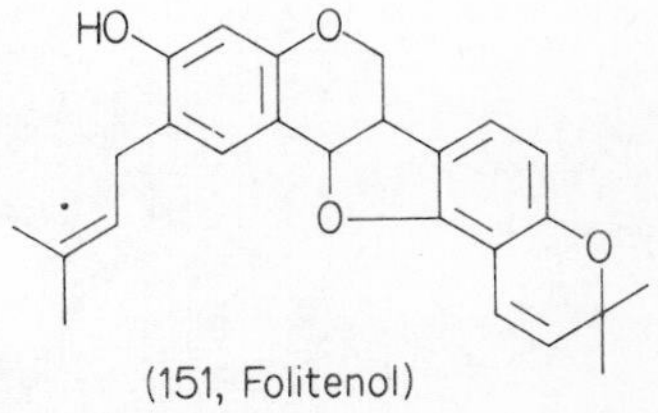

(151, Folitenol)

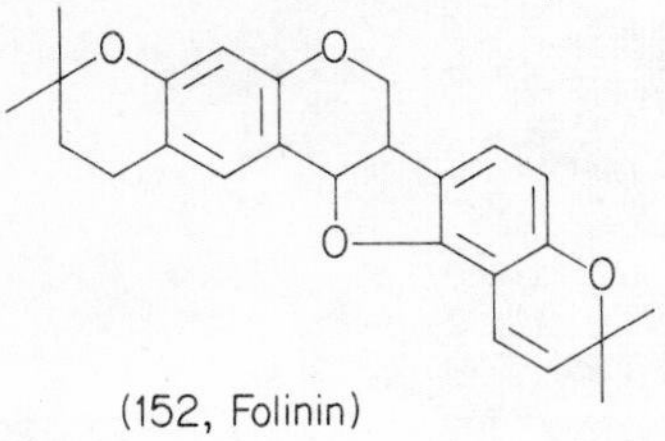

(152, Folinin)

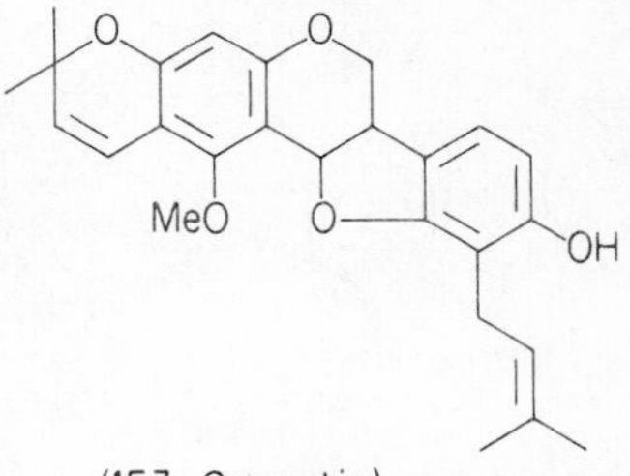

(153, Gangetin)

Neorautanenia, Baptisia, Lonchocarpus). Two pterocarpan glucosides
are also known, (132) (Bredenberg and Hietala, 1961; Lebreton
et al., 1967), (133) (Shibata and Hishikawa, 1963).

The structure of homopterocarpin was the first to be elucidated
(McGookin *et al.*, 1940) and the degradative methods developed
(summarized in Fig. 14.8) have been used in the study of later
members of the series. Reductive ring opening of the pterocarpan
system to a 2′-hydroxyisoflavan constitutes an important reaction of
these compounds (Pelter and Amenechi, 1969; Donnelly and
Fitzgerald, 1971; Purushothaman *et al.*, 1971). Acid treatment of
pterocarpans yields resinous products (Bredenberg and Hietala, 1961;

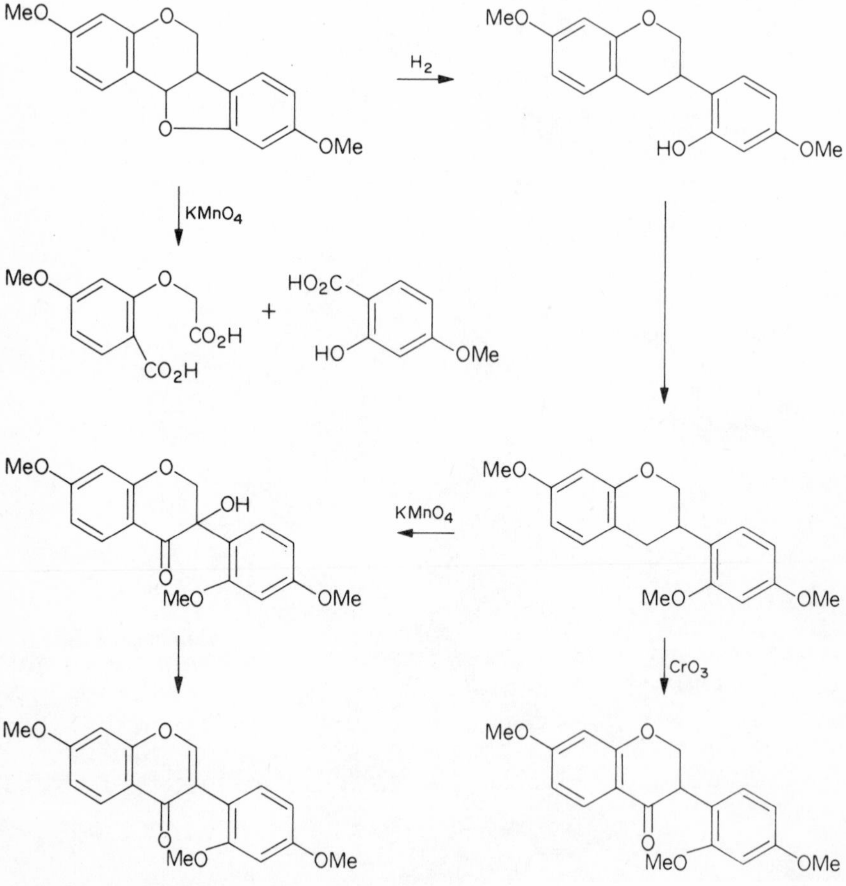

Figure 14.8 Degradation of homopterocarpin.

Shibata and Hishikawa, 1963), presumably from intermediate formation of isoflav-3-ens. These compounds can be obtained from pterocarpans by controlled action of acid (Bevan *et al.*, 1964). Maackiain (131) could not be obtained from its glucoside, trifolirhizin (132) by acid hydrolysis but was isolated pure after enzymic treatment (Bredenberg and Hietala, 1961).

In the NMR spectrum of pterocarpans, the heterocyclic ring protons give rise to a complex four-spin system, with long range coupling preventing first order analysis. A detailed study by Pachler and Underwood (1967) has allowed accurate analysis and assignments of the coupling constants between these protons to be made. Illustrative NMR data for (−)-pterocarpin are reproduced in Fig. 14.9. The UV spectrum of simple pterocarpans generally shows a major peak at ~285–310 nm, together with a lesser band at ~280–287 nm. In the pterocarpan series, the mass spectrum cannot be used to assign the various substituent groups to rings A and B as every fragment can reasonably be formulated as arising from either ring (cf. 154, 155) (Ollis, 1966; Pelter and Amenechi, 1969). Isoflavan derivatives of pterocarpans, however, show typical retro Diels-Alder fragmentation patterns in the mass spectrum (Pelter *et al.*, 1965a).

It is seen in Table 14.5 that pterocarpans can exist in nature in

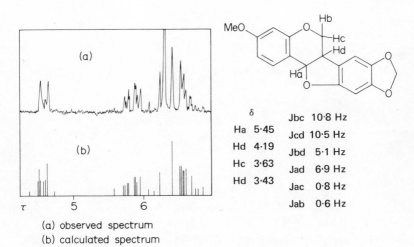

δ		
Ha	5·45	Jbc 10·8 Hz
Hd	4·19	Jcd 10·5 Hz
Hc	3·63	Jbd 5·1 Hz
Hd	3·43	Jad 6·9 Hz
		Jac 0·8 Hz
		Jab 0·6 Hz

(a) observed spectrum
(b) calculated spectrum

Figure 14.9 NMR spectrum and calculated data for the heterocyclic protons of (−)-pterocarpin (Pachler and Underwood, 1967).

both antipodal forms. The majority of known natural pterocarpans have large negative $[\alpha]_D$ values. These compounds have been taken to have the same relative stereochemistry and results from ORD comparisons are consistent with this view (Suginome, 1966b; Pelter and Amenechi, 1969). The absolute configuration of this series of compounds has been established as that shown in type formula (156). The R-configuration of position 6a has been directly established (Ito *et al.*, 1965; Verbit and Clark-Lewis, 1968) and the relative configuration of the 6a,11a chiral centres is *cis* on the basis of NMR evidence for (−)-homopterocarpin (Suginome and Iwadare, 1962). This being accepted, the absolute configuration for the (−)-pterocarpans follows necessarily as 6aR, 11aR (156) and that of the (+)-enantiomers as 6aS, 11aS. The NMR study of Pachler and Underwood (1967) confirmed the *cis* arrangement of the 6a, 11a protons. Furthermore the conclusion was reached by these workers that the normal preferred conformation of the (6aR,11aR)-pterocarpan ring system is (157) rather than the alternative (158).

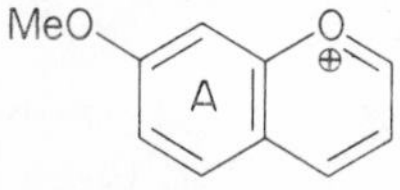

(154, *m/e* 161)

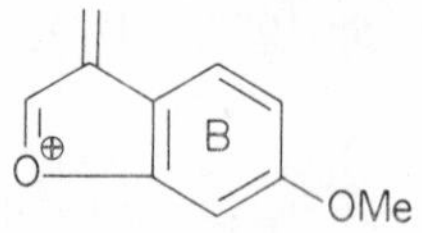

(155, *m/e* 161)

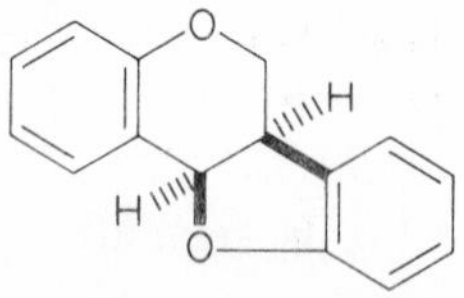

(156, 6aR, 11aR, −(−)−
Pterocarpan)

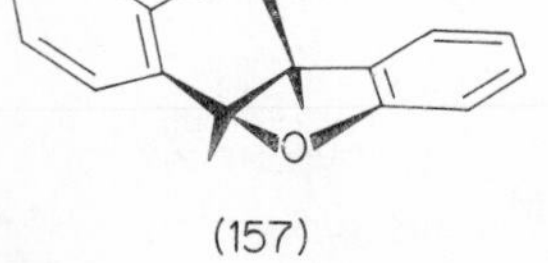

(157)

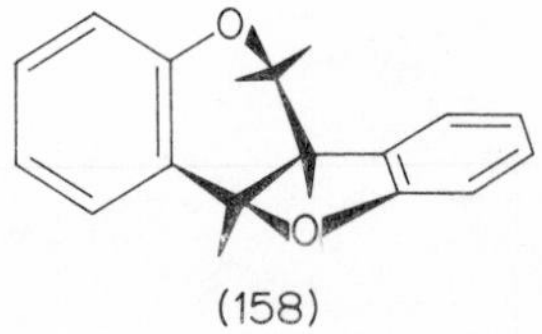

(158)

Recent developments in pterocarpan chemistry have included not only the discovery of new compounds but also the recognition of the existence of variants of the pterocarpanoid theme. The 6a-hydroxy-pterocarpan pisatin (159) accumulates in pods of pea *(Pisum sativum)* after fungal infection (Perrin and Bottomley, 1962). A similar 6a-hydroxy derivative of phaseollin (160) occurs in soya bean under similar conditions (Sims *et al.*, 1972). The isolation of a third compound of this type, variablin (161) from *Dalbergia spruceana* has been reported (Ollis, 1968).

6a,11a-Dehydropterocarpans are also known in nature. Dehydro-homopterocarpin (162) occurs in *Swartzia madagascariensis* together with homopterocarpin (129) (Harper *et al.,* 1969), and in *Dalbergia decipularis* in company with medicarpin (130) (Alencar *et al.,* 1972). The corresponding dehydro analogue of pterocarpin, flemichapparin B (163) was recently isolated from the roots of *Flemingia chappar* (Adityachaudhury and Gupta, 1973). The third compound of this type, leiocalycin (164) occurs in another species of *Swartzia (S. leiocalycina)* (Donnelly and Fitzgerald, 1971).

The dehydropterocarpans are coloured compounds with UV spectral characteristics of stilbenes. They are readily formed by acid dehydration of 2'-hydroxyisoflavanones (Suginome, 1959) or 6a-hydroxypterocarpans (Perrin and Bottomley, 1962). It is note-worthy that the known 6a,11a-dehydropterocarpans all co-occur

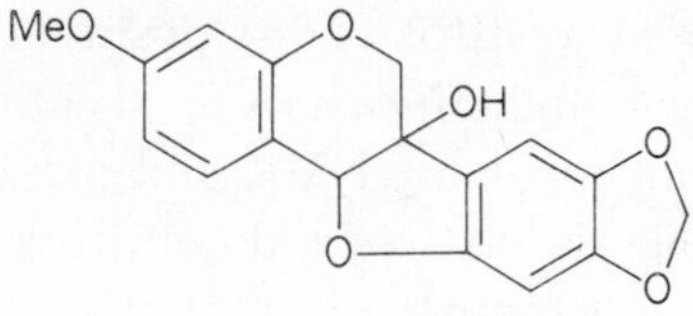

(159, Pisatin)

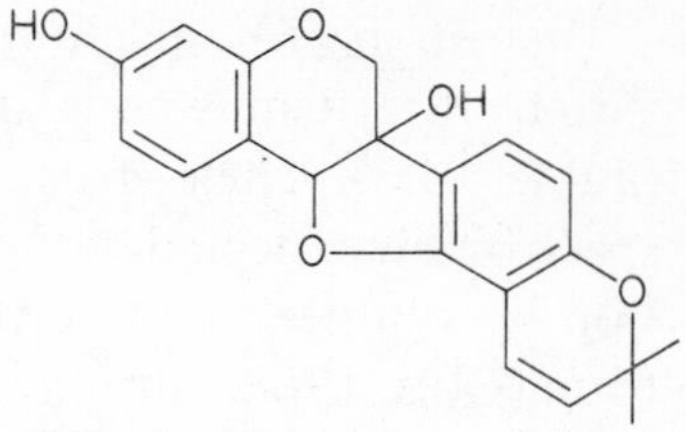

(160, 6a-hydroxyphaseollin)

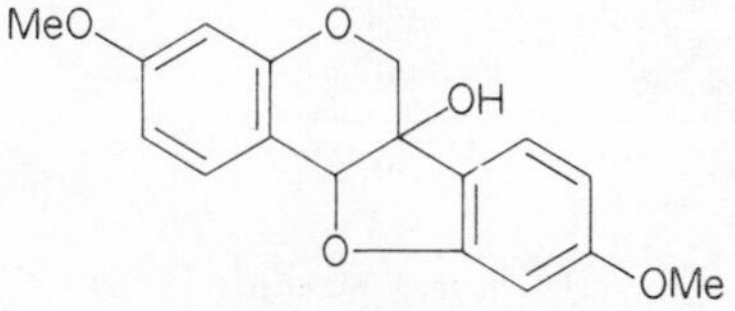

(161, Variablin)

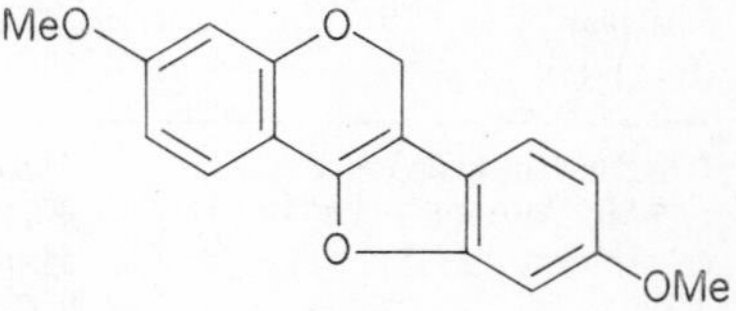

(162)

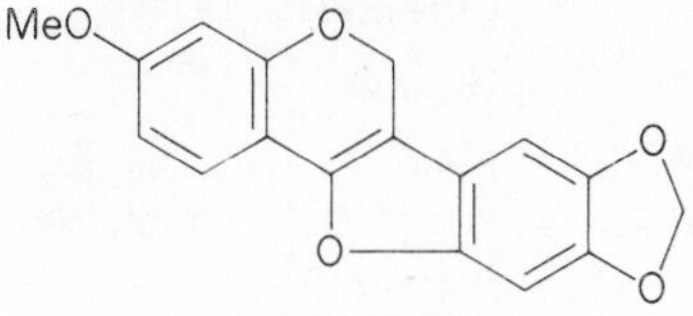

(163, Flemichapparin B)

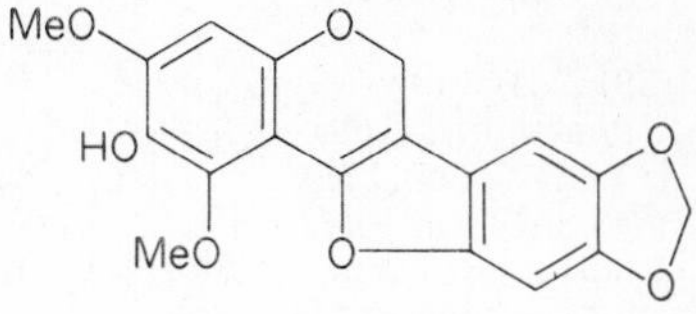

(164, Leiocalycin)

with their corresponding 6-oxo derivatives in the plant. These are compounds of the coumestan class which will be discussed in Section 14.7.

14.6 Isoflavans

Isoflavans represent the most reduced of the isoflavonoid modifications. For many years this class was exemplified in nature by only one compound, the animal metabolite equol (165). The occurrence of isoflavans in plants has been established by several groups of workers in recent years.

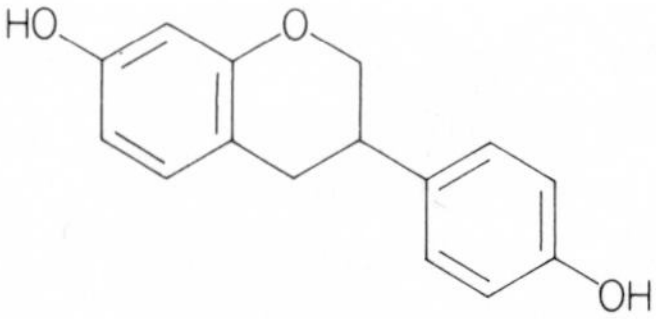

(165, Equol)

Duartin (166), mucronulatol (167) and vestitol (168) occur in various Brazilian woods of the *Dalbergia* and *Machaerium* genera (Table 14.6). Their structural elucidation, involving NMR and mass spectrometry, degradation via isoflavanone derivatives and confirmation by synthesis, has been summarized and reported in preliminary form (Ollis, 1968; Kurosawa *et al.*, 1968a).

Table 14.6   *Isoflavans*

Isoflavan (formula)	Plant sources	References
(−)-Duartin (166)	*Machaerium* spp.	
(−)-Mucronulatol (167)	*Machaerium* spp.	
(±)-Mucronulatol (167)	*M. mucronulatum, Dalbergia variablis*	Kurosawa *et al.*, 1968a.
(+)-Vestitol (168)	*M. vestitum, D. variablis*	
	D. ecastophyllum	Matos *et al.*, 1970.
(−)-Vestitol (168)	*Cyclolobium claussenii C. vecchi*	Oliveira *et al.*, 1971a.
(3S)-2′-Hydroxy-7,4′-dimethoxy- (169)	*Dalbergia ecastophyllum*	Matos *et al.*, 1970.
(+)-Laxifloran (170)	*Lonchocarpus laxiflorus*	Pelter and Amenechi, 1969.
(+)-Lonchocarpan (171)	*Lonchocarpus laxiflorus*	Pelter and Amenechi, 1969.
Phaseollinisoflavan (172)	*Phaseolus vulgaris*	Burden *et al.*, 1972.
(+)-Licoricidin (173)	*Glycyrrhiza glabra*	Shibata and Saitoh, 1968.

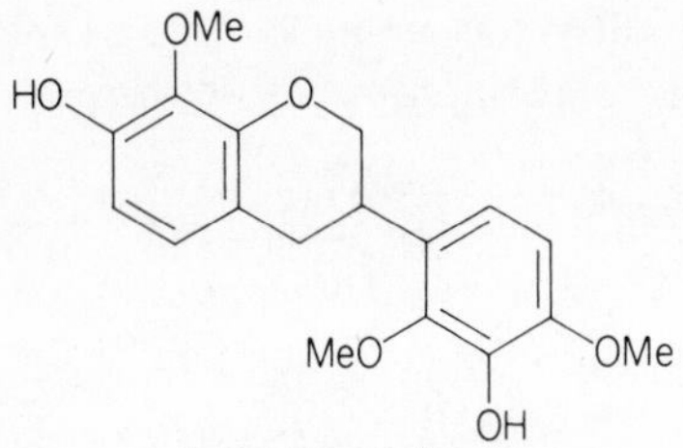

(166, Duartin)

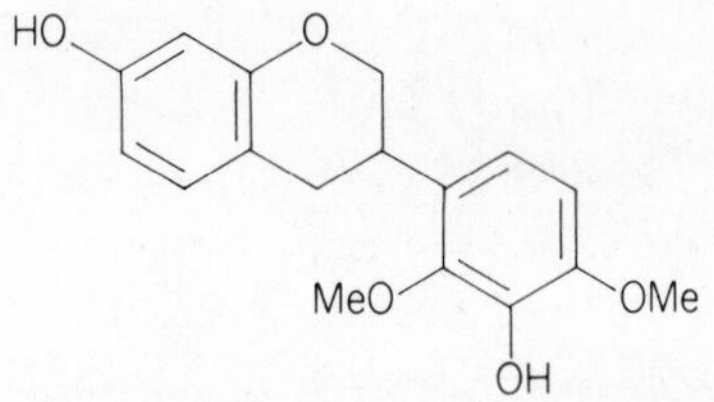

(167, Mucronulatol)

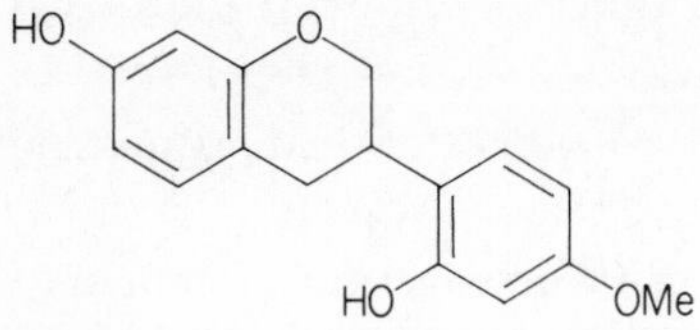

(168, Vestitol)

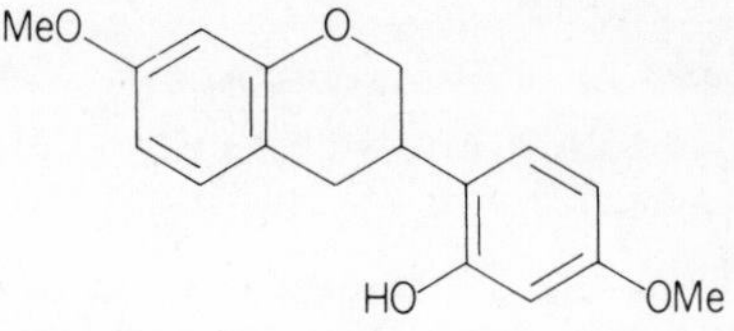

(169)

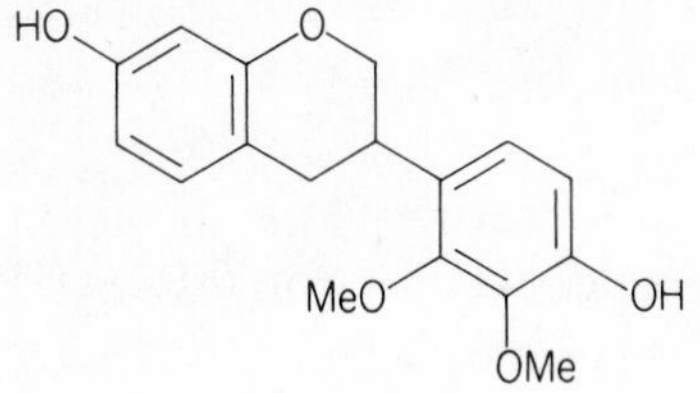

(170, Laxifloran)

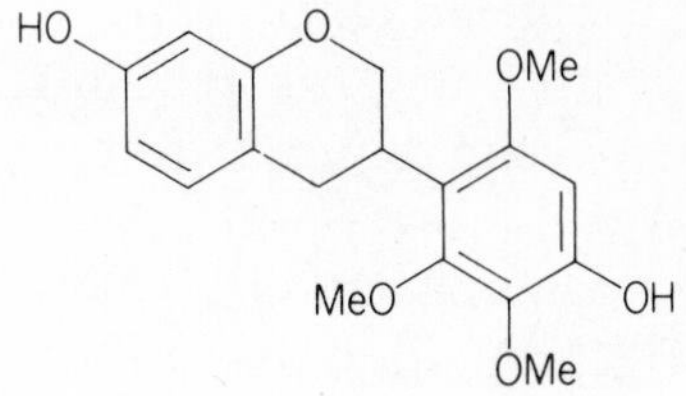

(171, Lonchocarpan)

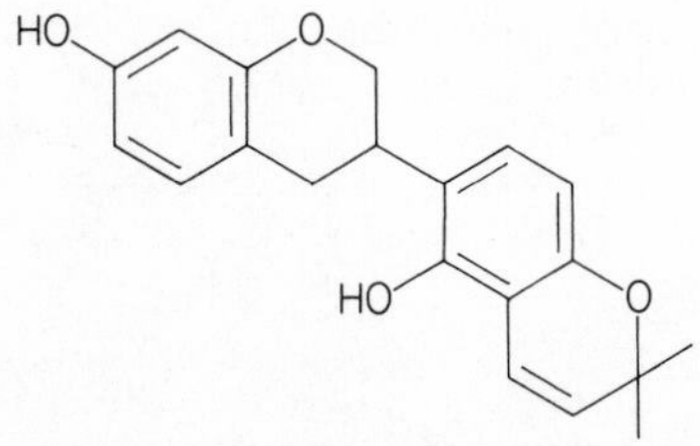

(172, Phaseollin isoflavan)

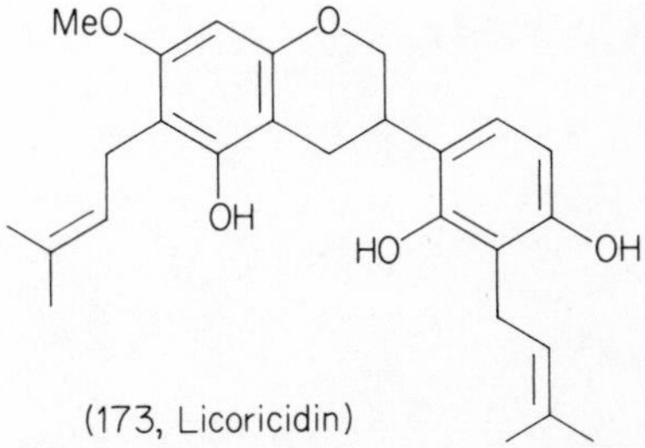

(173, Licoricidin)

These isoflavans were isolated in optically active forms and their absolute configurations have been studied by comparison of their ORD curves in the region 200–300 nm with that of (3S)-(−)-5,7,3′,4′-tetramethoxyisoflavan (175). Thus (+)-vestitol, (−)-duartin and (−)-mucronalatol were all shown to have the 3S configuration (174) (Kurosawa et al., 1968b). ORD comparison confined to the long wavelength region only has been shown to be unsuitable for deciding configurational relationships of isoflavans (Verbit and Clark-Lewis, 1968).

Laxifloran (170) and lonchocarpan (171) occur together with the pterocarpans (140) and (141) in the African plant *Lonchocarpus laxiflorus.* The assignment of the substituent groups in the highly substituted B-ring in these compounds presented problems which were solved by application of a combination of instrumental techniques, particular use being made of benzene-induced methoxyl shifts in NMR spectroscopy.

(+)-Laxifloran dimethyl ether and (+)-lonchocarpan dimethyl ether have been shown to have the R configuration at C-3 (176) (177), in contrast to the enantiomeric 3S configuration assigned to the isoflavans (166)–(168) from *Machaerium* spp. described earlier. (+)-Lonchocarpan dimethyl ether (177), the 3R configuration of which can be proved by its preparation from 6aR,11aR-(–)-philenopteran (140), gave an atypical ORD curve (Fig. 14.10) which was attributed to the different conformation adopted by the heterocyclic ring, due to steric effects of the 2′,6′- substituted ring B. This conformation effect was also reflected in the NMR spectrum (Pelter and Amenechi, 1969). Other anomalous effects on ORD and NMR properties due to steric factors in these conformationally mobile systems have also been reported by Kurosawa *et al.* (1968b).

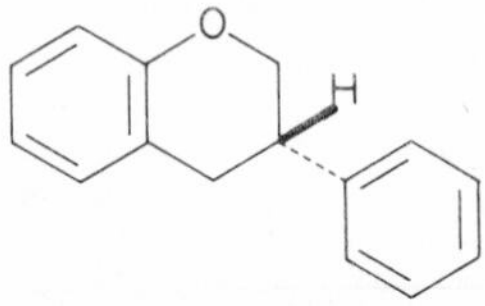

(174, (3S)-isoflavans)

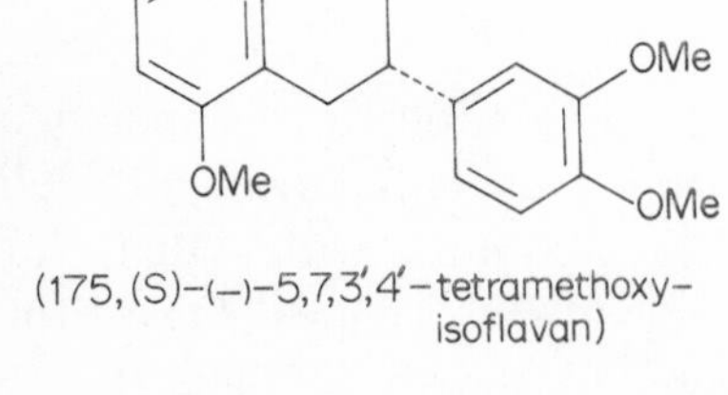

(175, (S)-(–)-5,7,3′,4′-tetramethoxy-
isoflavan)

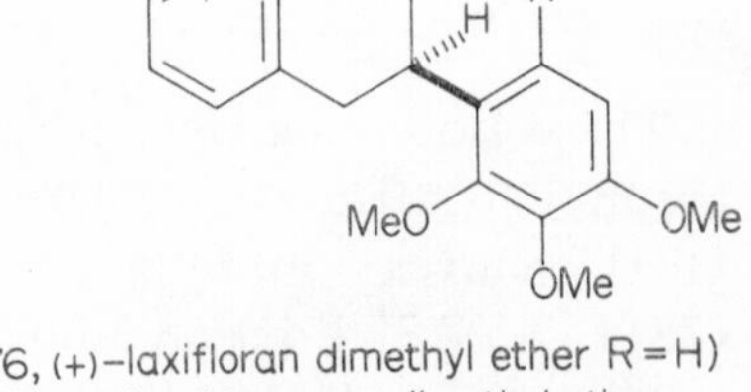

(176, (+)-laxifloran dimethyl ether R = H)
(177, (+)-lonchocarpan dimethyl ether
R = OMe)

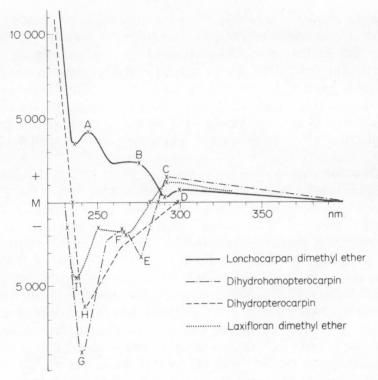

Figure 14.10 ORD curves for some (3R)-isoflavans (Pelter and Amenichi, 1969).

Recent additions to the isoflavan group include (3R)-(−)-vestitol (Oliveira *et al.*, 1971a) and (3S)-2'-hydroxy-7,4'-dimethoxyisoflavan (169) (Matos *et al.*, 1970). Licoricidin (173), and the isoflavan (172), occurring in fungal-infected bean *(Phaseolus vulgaris)* with its pterocarpan analogue phaseollin (146), are the first examples of complex isoflavans.

The existence in nature of antipodal forms of isoflavans resembles the situation in pterocarpans. Furthermore, in cases where both classes of compounds are found in the same plant, the chirality with respect to the equivalent assymmetric carbon centre (C-3 or C-6a) in the two classes is identical. Thus in *L. laxiflorus* a common R configurational relationship is found, whilst in *Macherium* and *Dalbergia* species, both common S or common R relationships are found (Oliveira *et al.*, 1971b). These findings provide strong circumstancial evidence that pterocarpans and isoflavans are biogenetically

directly related. Chemically 2'-hydroxyisoflavans are readily obtained by hydrogenation of pterocarpans and it is likely that this is
also the natural route to isoflavans. In this connection it is
noteworthy that all the known natural isoflavans show oxygenation
at the 2'-position. In terms of the biogenetic relationship indicated,
vestitol (168), 2'-hydroxy-7,4'-dimethoxyisoflavan (169) and phaseollinisoflavan (172) can be regarded as the isoflavan analogues of
medicarpin (130), homopterocarpin (129) and phaseollin (146)
respectively.

Two coloured 2',5'-isoflavanquinones have been isolated from
Brazilian woods bearing isoflavans. (3S)-(−)-Mucroquinone (178)
from *Machaerium mucronulatum* was assigned the quinonoid structure on the basis of combined spectral characteristics and its full
structure was proved by synthesis via oxidation of 7,3'-dihydroxy-
8,4'-dimethoxyisoflavan with Fremy's salt (Kurosawa *et al.*, 1968a).
The enantiomeric (3R)-mucroquinone is also known. It has been
isolated from *Cyclolobium claussenii* heartwood together with a
second isoflavanquinone, (3R)-claussequinone (179) (Oliveira *et al.*,
1971a). This latter compound has also been obtained from *C. vecchi*
(Oliveira *et al.*, 1971a). The possibility that the 2',5'-iso-
flavanquinones are artifacts of corresponding 2',5'-dihydroxyiso-
flavans (Farkas, 1973) is unlikely, since freshly cut *Cyclobium* wood
is brightly coloured (Gottlieb, 1973).

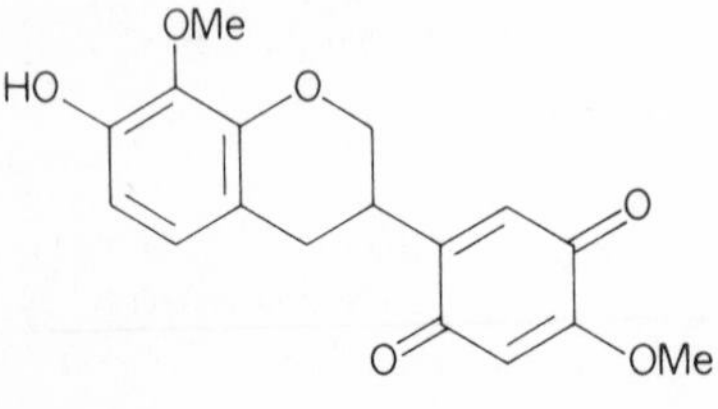

(178, Mucroquinone) 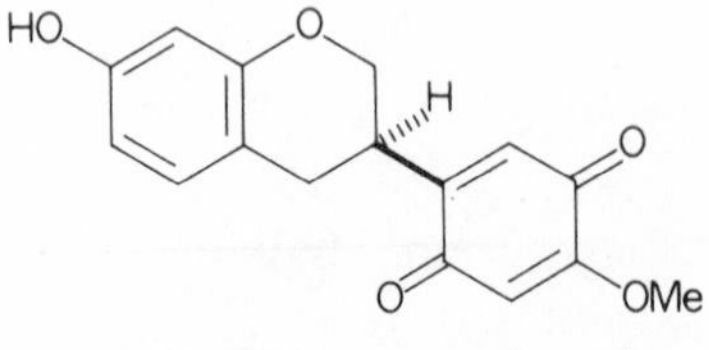   (179, (3R)−Claussequinone)

14.7 Coumestans

The highest oxidation level possible for the isoflavonoid skeleton is
represented by the coumaranocoumarin structure (180), for which
the trivial name coumestan has come into general use. The
numbering system in use with this nomenclature (180a) is arbitrary
and it is proposed that the systematic numbering shown in (180) be

used instead. This will bring, with advantage, the numbering system into line with that already in use with the pterocarpan series.

The first example of a natural coumestan, wedelolactone (181) was reported in 1957 (Govindachari *et al.*, 1957). Since then twenty additional compounds bearing this ring system have been isolated from various leguminous plants (Table 14.7). Study of the important forage legumes alfalfa *(Medicago sativa)* and ladino clover *(Trifolium repens)* has led to the discovery of coumestrol (184), trifoliol (185), medicagol (186), lucernol (187), sativol (188) and the three compounds represented by formulae (189)–(191). A review of this work by Bickoff *et al.* (1969) contains much chemical and biological information on these coumestans. Notably, as in the pterocarpans series, many of the coumestans occur in roots of leguminous plants. These include the complex coumestans sojagol (199), isoglycyrol (200), and glycyrol (201) and 1-methylglycyrol (202). Demethyl-wedelolactone 7-glucoside (183) from *Eclipta alba* leaves, is the only example of a coumestan glycoside known (Bhargava *et al.*, 1972).

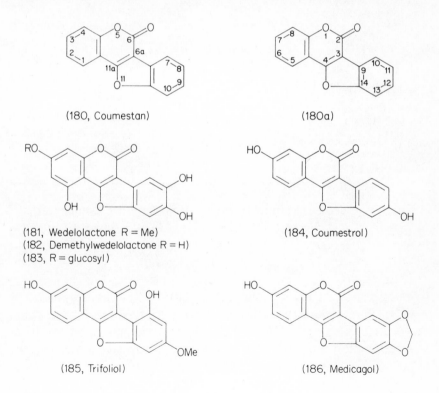

(180, Coumestan)

(180a)

(181, Wedelolactone R = Me)
(182, Demethylwedelolactone R = H)
(183, R = glucosyl)

(184, Coumestrol)

(185, Trifoliol)

(186, Medicagol)

Table 14.7   *Natural coumestans*

Coumestan (formula)	Plant sources	References
Coumestrol (184)	*Trifolium* spp. *Medicago sativa*	Bickoff *et al.*, 1969.
Trifoliol (185)	*Trifolium* spp.	Livingston *et al.*, 1964. Wong and Latch, 1971b.
Medicagol (186)	*Medicago sativa*	Livingtson *et al.*, 1965.
Lucernol (187)	*Medicago sativa*	Spencer *et al.*, 1966a.
Sativol (188)	*Medicago sativa*	Spencer *et al.*, 1966a.
Repensol (193)	*Trifolium repens*	Wong and Latch, 1971b.
3,9-Dihydroxy-8-methoxy- (189)	*Medicago sativa*	Bickoff *et al.*, 1966.
3-Hydroxy-8,9-dimethoxy- (190)	*Medicago sativa*	Spencer *et al.*, 1966b.
3-Hydroxy-9-methoxy- (191)	*Trifolium repens* *Medicago sativa*	Wong and Latch, 1971b. Bickoff *et al.*, 1965.
3,9-Dimethoxy- (192)	*Swartzia madagascariensis* *Dalbergia decipularis*	Harper *et al.*, 1969. Alencar *et al.*, 1972.
Flemichapparin C (194)	*Flemingia chappar*	Adityachaudhury and Gupta, 1973.
2-Hydroxy-3-methoxy-8,9-methylenedioxy- (195)	*Swartzia leiocalycina*	Donnelly and Fitzgerald, 1971.
2-Hydroxy-1,3-dimethoxy 8,9-methylenedioxy- (196)	*Swartzia leiocalycina*	Donnelly and Fitzgerald, 1971.
Wedelolactone (181)	*Wedelia calendulaceae* *Eclipta alba* (Compositae)	Seshadri, 1964.
Demethylwedelolactone (182)	*Eclipta alba*	Krishnaswamy and Prasanna, 1970.
Erosnin (197)	*Pachyrrhizus erosus*	Seshadri, 1964.
Psoralidin (198)	*Psoralea corylifolia*	Seshadri, 1964.
Sojagol (199)	*Phaseolus aureus*	Zilg and Grisebach, 1968.
Isoglycyrol (200)	*Glycyrrhiza* spp.	Saitoh and Shibata, 1969.
Glycyrol (201)	*Glycyrrhiza* spp.	Saitoh and Shibata, 1969.
1-Methylglycyrol (202)	*Glycyrrhiza* spp.	Saitoh and Shibata, 1969.

The UV spectral characteristics of coumestans are similar to those of flavones and flavonols (Bickoff *et al.*, 1969). Their IR spectra, however, differ from those of the γ-chromones in having a band ~ 1710 cm^{-1} due to the lactone carbonyl group.

Coumestans in general can be systematically degraded (Fig. 14.11) by methylation, ring opening (203, R=CO$_2$Me), hydrolysis (203, R=CO$_2$H), and decarboxylation to their more soluble benzofuran derivatives (203, R=H). Further degradation via ozonolysis results in the production of acidic and aldehydic fragments (Govindachari *et al.*, 1957; Bickoff *et al.*, 1969).

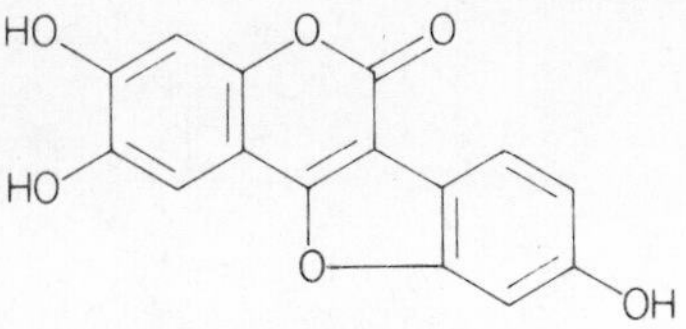

(187, Lucernol)

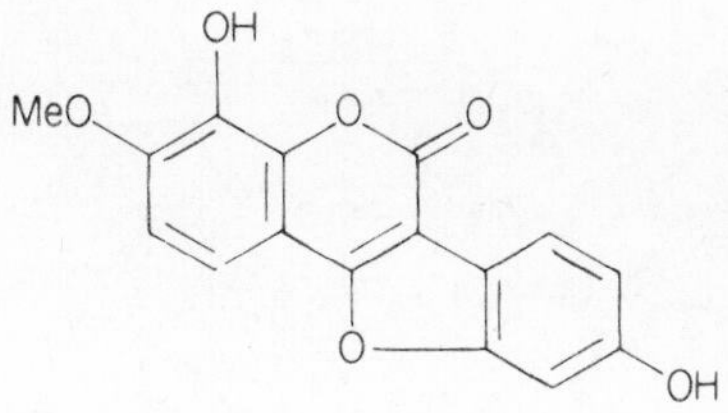

(188, Sativol)

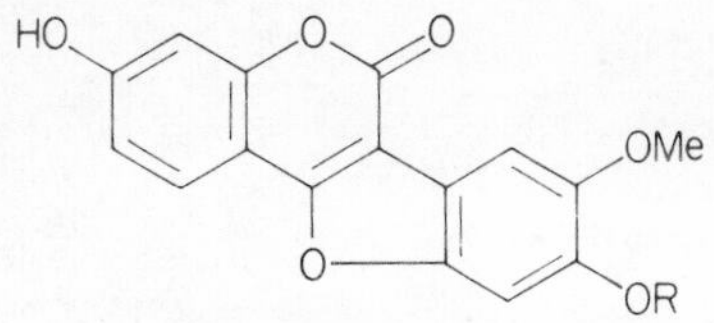

(189, R = H)
(190, R = Me)

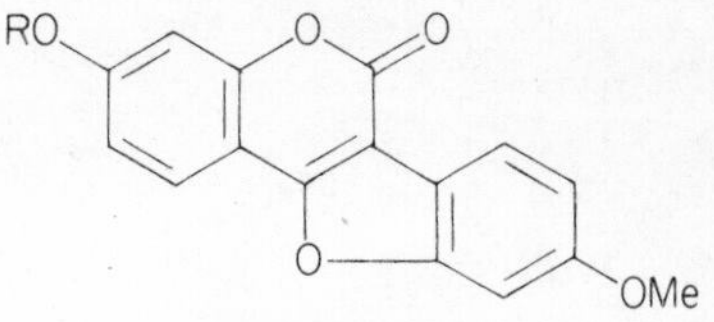

(191, R = H)
(192, R = Me)

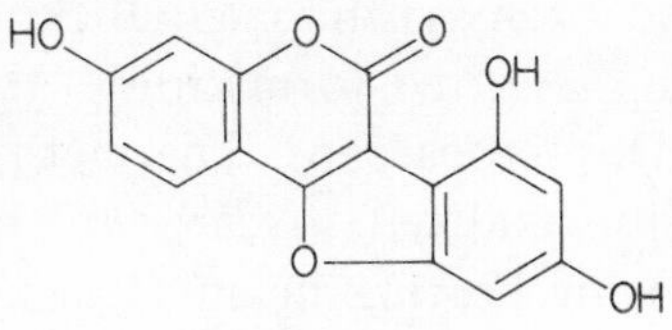

(193, Repensol)

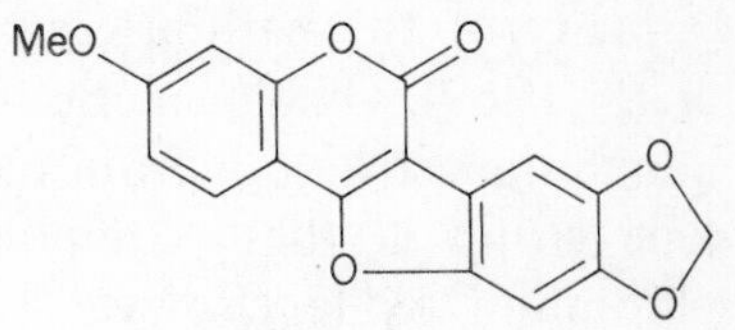

(194, Flemichapparin C)

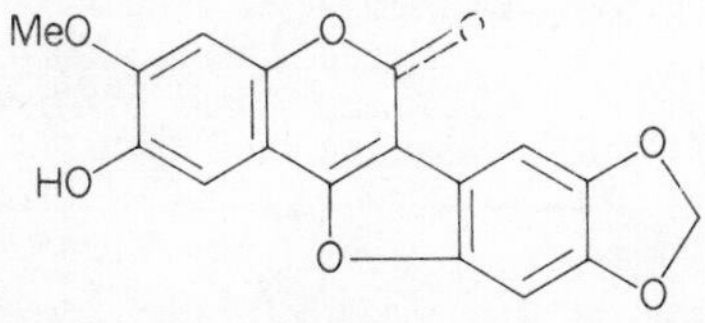

(195)

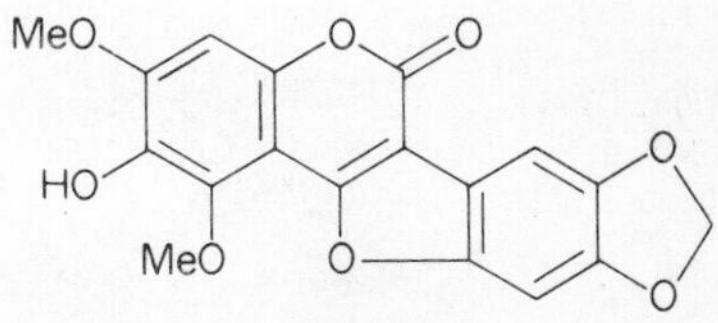

(196)

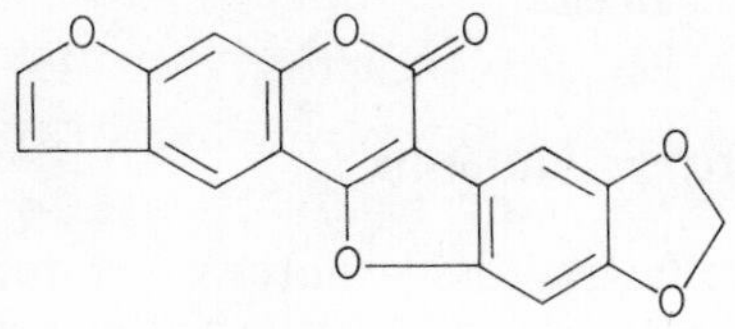

(197, Erosnin)

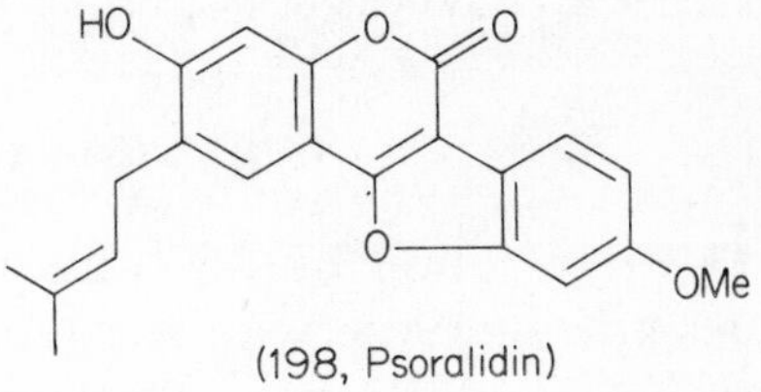

(198, Psoralidin)

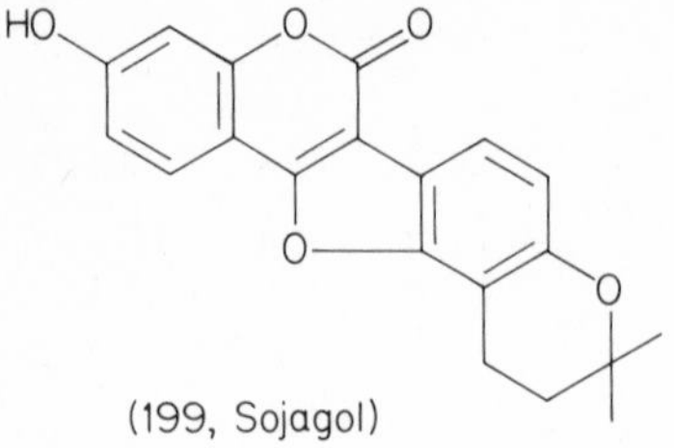

(199, Sojagol)

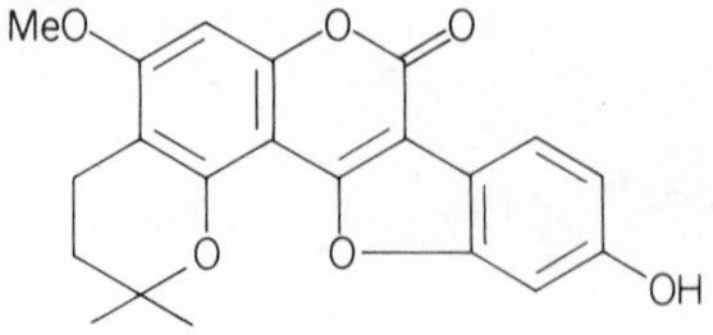

(200, Isoglycyrol)

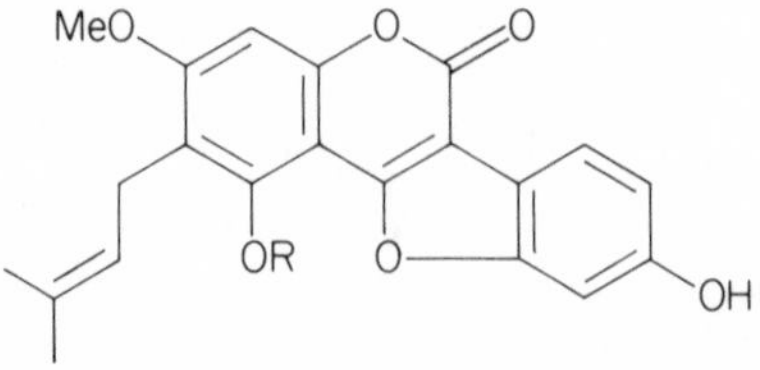

(201, Glycyrol R = H)
(202, 1-methylglycyrol R = Me)

Many of the natural coumestans have been synthesized (Bickoff *et al.,* 1969). Of biogenetic interest is the reported formation of the coumestan structure from 6a,11a-dehydropterocarpans. The methylene group in these compounds is easily oxidized to carbonyl by chromium oxide (Bowyer *et al.,* 1964), or heating in air (Dewick *et al.,* 1969). The natural co-occurrence of 3,9-dimethoxycoumestan (192), (196) and flemichapparin C (194) with their respective dehydropterocarpan analogues has already been alluded to (Section 14.5).

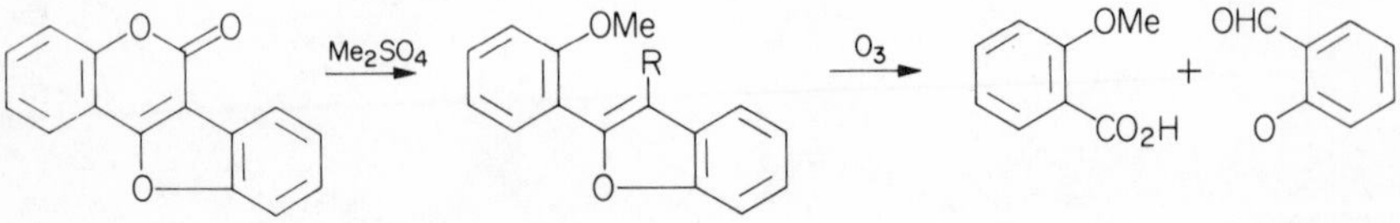

(203)

Figure 14.11 Systematic degradation of coumestans.

14.8 3-Aryl-4-hydroxycoumarins

The recognition of this class of isoflavonoids as natural products dates from 1966, with the report by Johnson *et al.* (1966) of the structural elucidation of lonchocarpic acid (204) and scandenin

(205) from *Derris scandens.* These workers later characterized robustic acid (206) and its methyl ester (207) from the related *D. robusta* species (Johnson and Pelter, 1966). Ollis and his associates, working independently, isolated in addition lonchocarpin (208) (from *D. scandens*) (Falshaw *et al.,* 1969) and robustin (209), robustin methyl ether (210) and derrusnin (211) (from *D. robusta*) (East *et al.,* 1969). These compounds thus form a closely related series of new plant constituents.

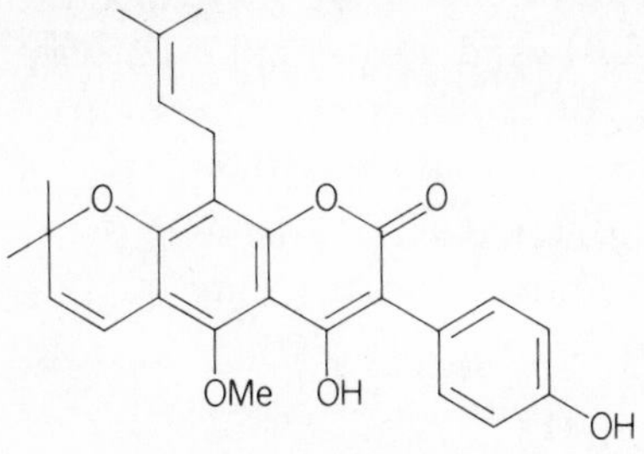

(204, Lonchocarpic acid)

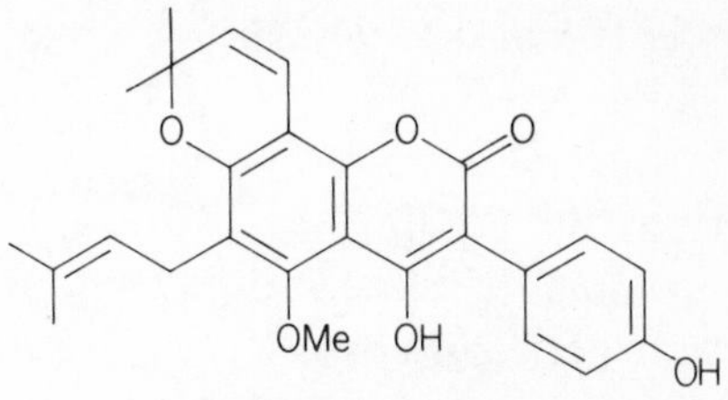

(205, Scandenin)

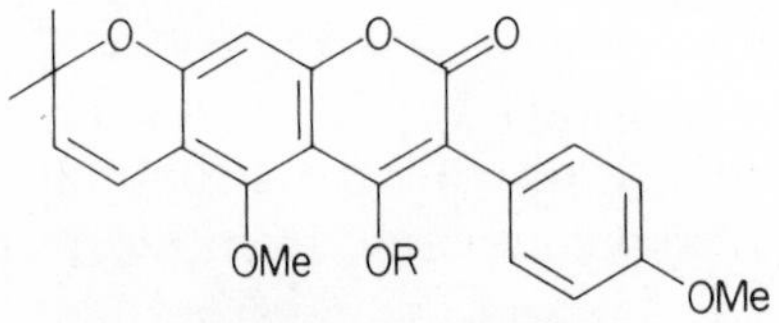

(206, Robustic acid R = H)
(207, Robustic acid methyl ether
R = Me)

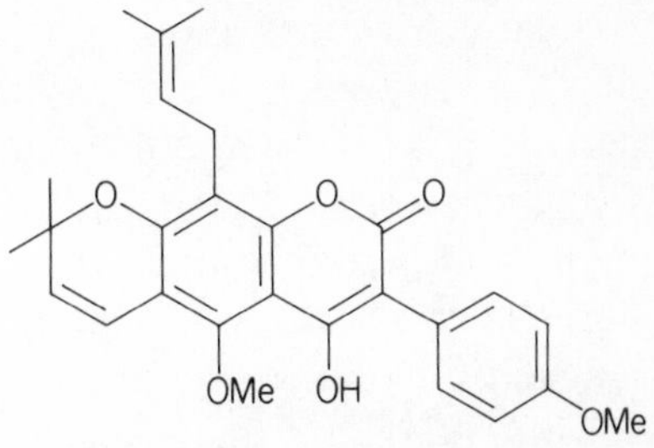

(208, Lonchocarpin)

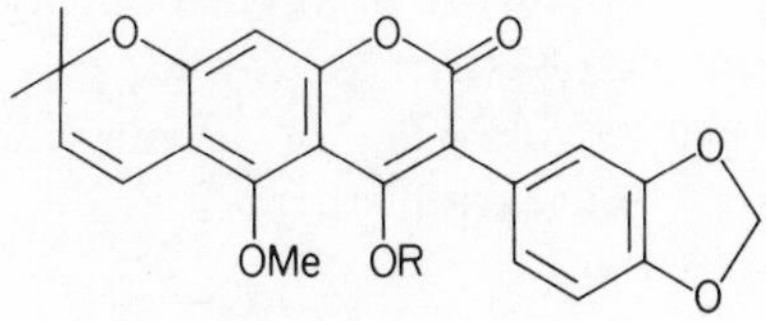

(209, Robustin R = H)
(210, Robustin methyl ether R = Me)

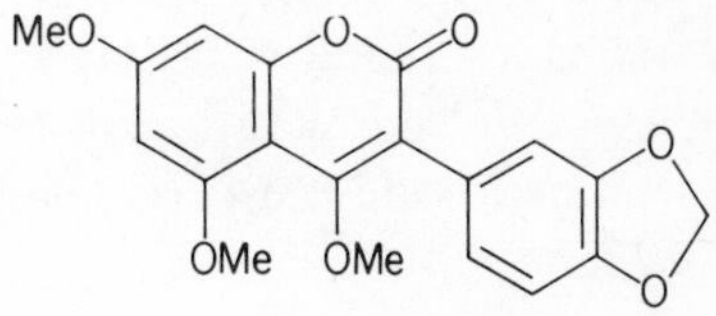

(211, Derrusnin)

The 3-aryl-4-hydroxycoumarin structure is tautomeric with 2-hydroxyisoflavone (Fig. 14.12). All such compounds from the *Derris* species have a 5-methoxyl substituent in ring A and exist in

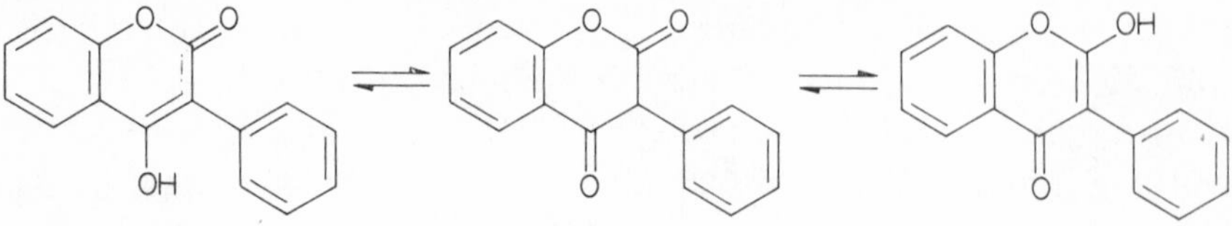

(212)

Figure 14.12 Tautomerism of 3-aryl-4-hydroxycoumarins.

the coumarin form. This is best indicated, *inter alia,* by their spectral characteristics (UV λ_{max} ~230, ~270 and ~350 nm; IR ν_{co} ~1710 cm^{-1}) which resemble those of coumarins rather than isoflavones. When a free 5-hydroxyl group is present in the molecule however, the 2-hydroxyisoflavone tautomer exists in large proportion in equilibrium (Johnson and Pelter, 1966). This is due presumably to the stabilising effect of hydrogen bonding between the peri 5-hydroxyl and 4-carbonyl groups. 3-Aryl-4-hydroxycoumarins can therefore be regarded as isoflavonoid counterparts of flavonols (isoflavonols).

Like isoflavones and isoflavanones, 3-aryl-4-hydroxycoumarins are degraded by alkali to deoxybenzoins and phenylacetic acids. Condensation of the deoxybenzoin with methyl chloroformate followed by hydrolysis regenerates the 3-aryl-4-hydroxycoumarin. Alkaline hydrogen peroxide oxidation also yields substituted benzoic acids deriving from rings A and B (East *et al.,* 1969). In the structural elucidation of the complex 3-aryl-4-hydroxycoumarins from *Derris* species, NMR and mass spectrometric techniques were indispensible. These compounds apparently fragment in the diketone form (212) on mass spectral analysis (Pelter *et al.,* 1965b).

Two novel routes to 3-aryl-4-hydroxycoumarins involving boron trifluoride-etherate catalysed rearrangements of either a 2-α-hydroxy-benzyl-2-methoxyaurone (213) (Jain *et al.,* 1967), or a related aurone epoxide (214) (Geoghegan *et al.,* 1966), are illustrated in Fig. 14.13.

14.9 Other types of isoflavonoids

Within this heading are included three compounds each of which is the sole natural example thus far of a new isoflavonoid type. The α-methyldeoxybenzoin (−)-angolensin (215) has been known since

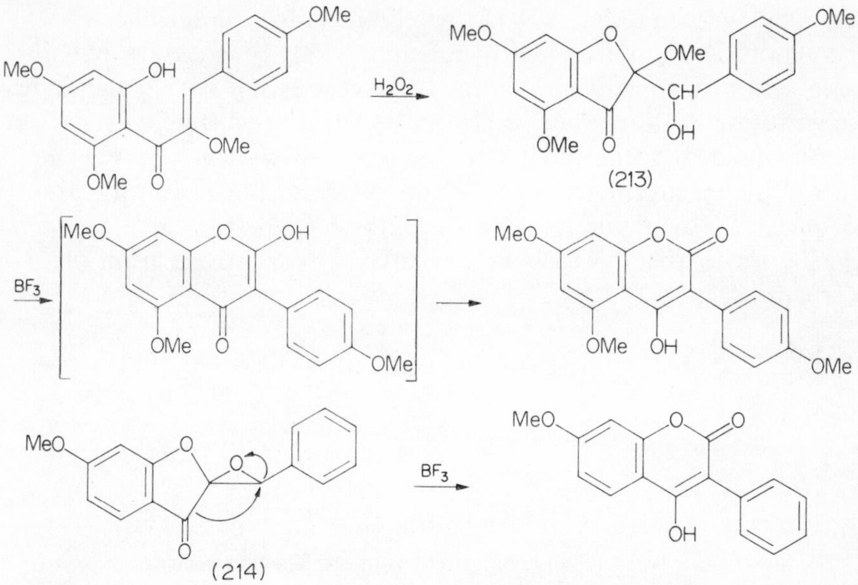

Figure 14.13 Novel synthetic routes to 3-aryl-4-hydroxycoumarins.

1952. It occurs in the heartwood of *Pterocarpus* species *(P. angolensis, P. indicus, P. erinaceous)* and in teak wood (*Pericopsis* sp.) (Ollis, 1962, Imamura *et al.*, 1968). The absolute configuration of (−)-angolensin has been determined as R as shown in (215) (Ollis *et al.*, 1965).

Pachyrrhizin from *Pachyrrhizus erosus*, was first characterised as the 3-arylcoumarin (216a) by Simonitsch *et al.* (1957). Its occurrence in the related species *Neorautanenia pseudopachyrrhiza, N. edulus* and *N. amboensis* has since been reported (Brink *et al.*, 1965; Crombie and Whiting, 1963) as also its methoxy derivative neofolin (216b) in *N. ficifolia* (Brink *et al.*, 1966). Pachyrrhizin (216a)

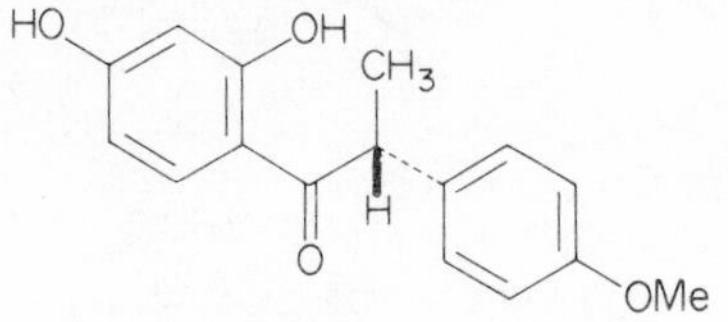

(215, (−)−angolensin)

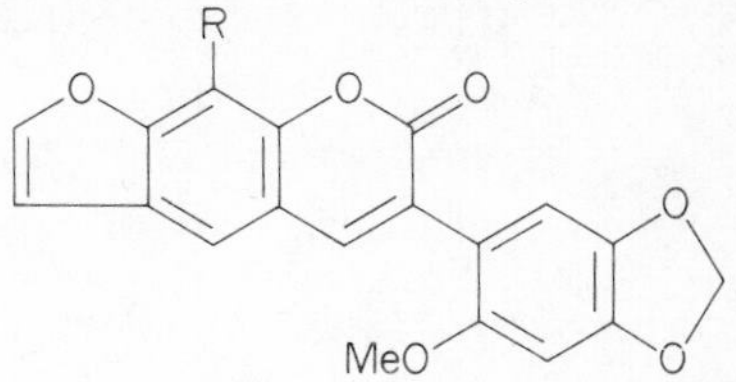

(216a, Pachyrrhizin R = H)
(216b, Neofolin R = OMe)

together with erosnin (197), neodulin (144), dolineone (107),
neotenone (91), and dehydroneotenone (58), constitute a remarkable series of constiuents within these species, all having the same
substitution pattern but each representing a different class of
isoflavonoid. It seems likely that 2'-hydroxy-3-arylcoumarins exemplified by pachyrrhizin (216a) are derived biogenetically from
coumestans by reductive ring opening (Fig. 14.14), in a manner
analogous to that postulated for isoflavan biosynthesis from pterocarpans (Section 14.6).

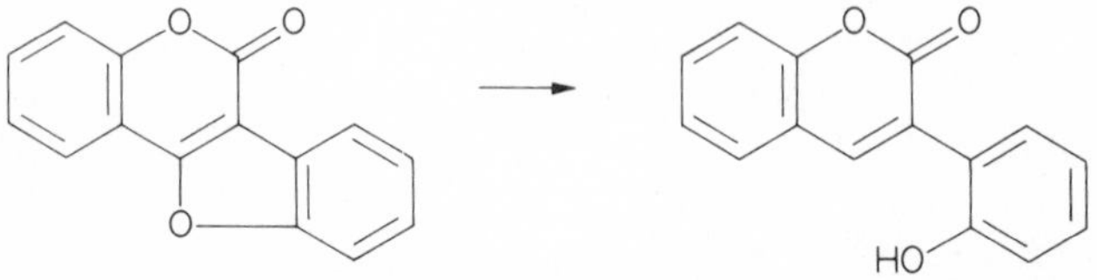

Figure 14.14 Possible biogenetic route to 3-arylcoumarins

Among the isoflavonoid constituents of *Piscidia erythrina,* lisetin
has recently been found to be a coumaronochromone, with the
constitution shown in (217) (Falshaw *et al.,* 1966). The oxidation
level of lisetin, $C_{15}H_3O_3(OH)_3(OCH_3)(C_5H_9)$, excluded its formulation as an isoflavone but indicated a coumestan or a coumaronochromone type structure. The IR data for lisetin ($\gamma_{co} = 1693$ cm^{-1})
indicated, however, that lisetin was not a compound of the former type
($\gamma_{co} \sim 1700-1740$ cm^{-1}). Lisetin trimethyl ether (218) on hydrolysis by dilute alkali yielded the 3-aryl-4-hydroxyxoumarin
(219). This provided confirmatory evidence that lisetin is a coumaronochromone and not a coumestan, since coumestans give
hydroxy-acids under these conditions. The alkaline hydrolysis of

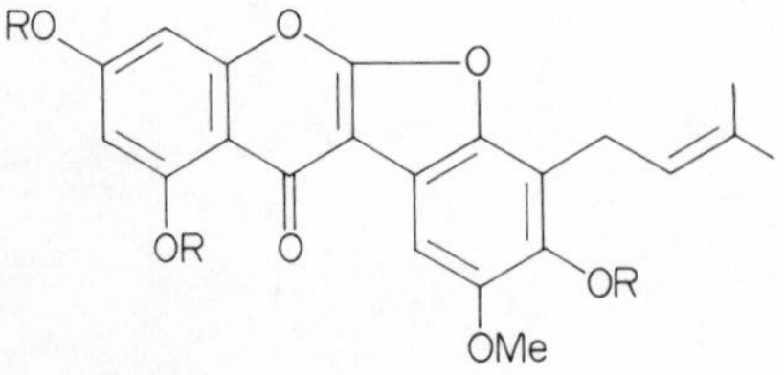

(217, Lisetin R = H)
(218, R = Me)

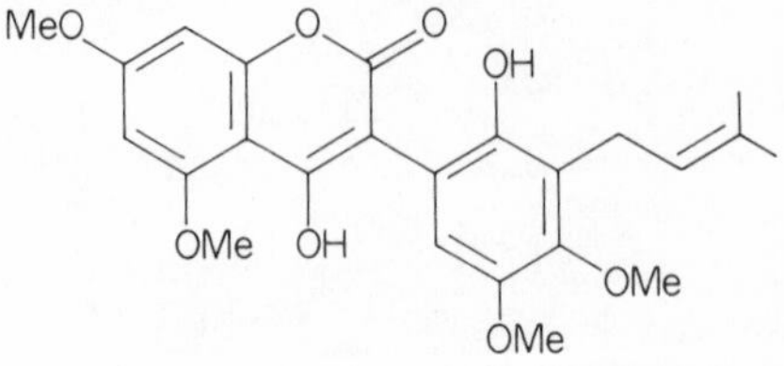

(219)

coumaronochromones to 3-aryl-4-hydroxycoumarins (with 2'-hy-droxylation) has been suggested as a useful diagnostic reaction for this new type of natural product (Falshaw *et al.*, 1966).

Final proof of the structure of lisetin was provided by its partial synthesis from the isoflavone piscerythrone (64), by means of oxidation with potassium ferricyanide.

14.10 Biogenetic relationships among the isoflavonoids

Undoubtedly the different classes of isoflavonoids encountered in nature are biogenetically closely related. The possession of a common 1,2-diphenylpropane skeleton by these compounds implies that the rearrangement step, involving the 1,2-shift of the aromatic B ring (Chapter 16), is the distinguishing feature of their derivation from intermediates of the flavonoid series. Very probably only one such rearrangement step is needed, giving rise to a primary class of isoflavonoids which is the common progenitor of all the other classes.

Direct experimental evidence for the existence of biogenetic relationships among classes of isoflavonoids is available in a few cases (Chapter 16), but the precise nature of these relationships cannot be said to have been established. A comparative study of the structures, distribution and co-occurrence patterns of all the known natural isoflavonoids has shown up some striking features and has led to deductions of plausible biogenetic relationships among these com-pounds (Wong, 1970). A slightly modified version of the relation-ships proposed is given in Fig. 14.15.

The scheme in Fig. 14.15 provides only a broad outline for interrelating isoflavonoid biosynthesis; details of the various bio-genetic pathways postulated must await elucidation by biochemical and enzymic studies. One detail, however, in the problem of isoflavonoid biosynthesis is worthy of special comment. This concerns the origin of 2'-oxygenation.

The preponderance of 2'-oxygenation is the most striking second-ary structural feature of the isoflavonoids as compared to other flavonoids. The presence of 2'-oxygen function is explicit in the majority of isoflavanones, isoflavans and complex isoflavones and implicit in the ring systems of pterocarpans, coumestans and rotenoids. Two types of mechanisms have been proposed to account

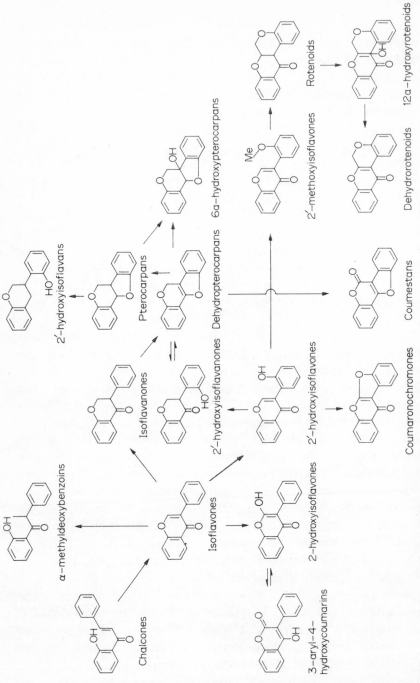

Figure 14.15 Possible biogenetic relationships among the different classes of isoflavonoids.

for the apparent facile oxygenation of the 2'-position of the isoflavonoid B-ring. The first (Ollis, 1968) envisages hydroxylation taking place via attack of an electrophilic oxygenating species (formally OH$^+$) for which the 2'-position is activated by the heterocyclic ring oxygen in isoflavones but not in flavones (Fig. 14.16).

An alternative scheme explaining in particular the presence of 2'-oxygenation in pterocarpanoids, isoflavans and isoflavanones, postulates an oxidative ring cyclization involving the C-4 oxygen in isoflavanones (Fig. 14.16) (Wong, 1970). This scheme in effect explains the preferential 2'-oxygenation in ring B of isoflavonoids essentially as a consequence of the proximity of ring B to the C-4 oxygen function. Both mechanisms presented in Fig. 14.16 are invoked as alternative routes to 2'-hydroxyisoflavanones in the biogenetic scheme given in Fig. 14.15.

14.11 Biological properties of isoflavonoids

A distinctive feature of the isoflavonoids is the possession of biological activity within the group. In contrast to other flavonoids which on the whole are innocuous substances (Whalley, 1959), isoflavonoids have oestrogenic, insecticidal, piscicidal and anti-fungal properties.

Isoflavones are weak oestrogens (Biggers, 1959) and their presence in forage legumes such as subterranean clover (*Trifolium subterraneum*) and red clover (*T. pratense*) has been recognized as the cause of infertility problems sometimes occurring in animals grazing these pasture species (Bickoff, 1968; Braden and McDonald, 1970).

Formononetin (220), genistein (224) and biochanin A (223) are the main oestrogenic isoflavones of subterranean clover. Recent studies of their metabolism in sheep (Batterham *et al.*, 1971; Braden *et al.*, 1967) have demonstrated an interesting and important difference in the degradation of the 5-oxy and 5-deoxyisoflavones. The major degradation products of genistein and biochanin A are simple phenols such as *p*-ethylphenol (225) whose production results in a loss of oestrogenic activity, whilst formononetin degradation produces the oestrogenically active compounds equol (221) and *O*-demethylangolensein (222) (Fig. 14.17). This inactivation pathway for genistein and biochanin A, however, is developed only in the

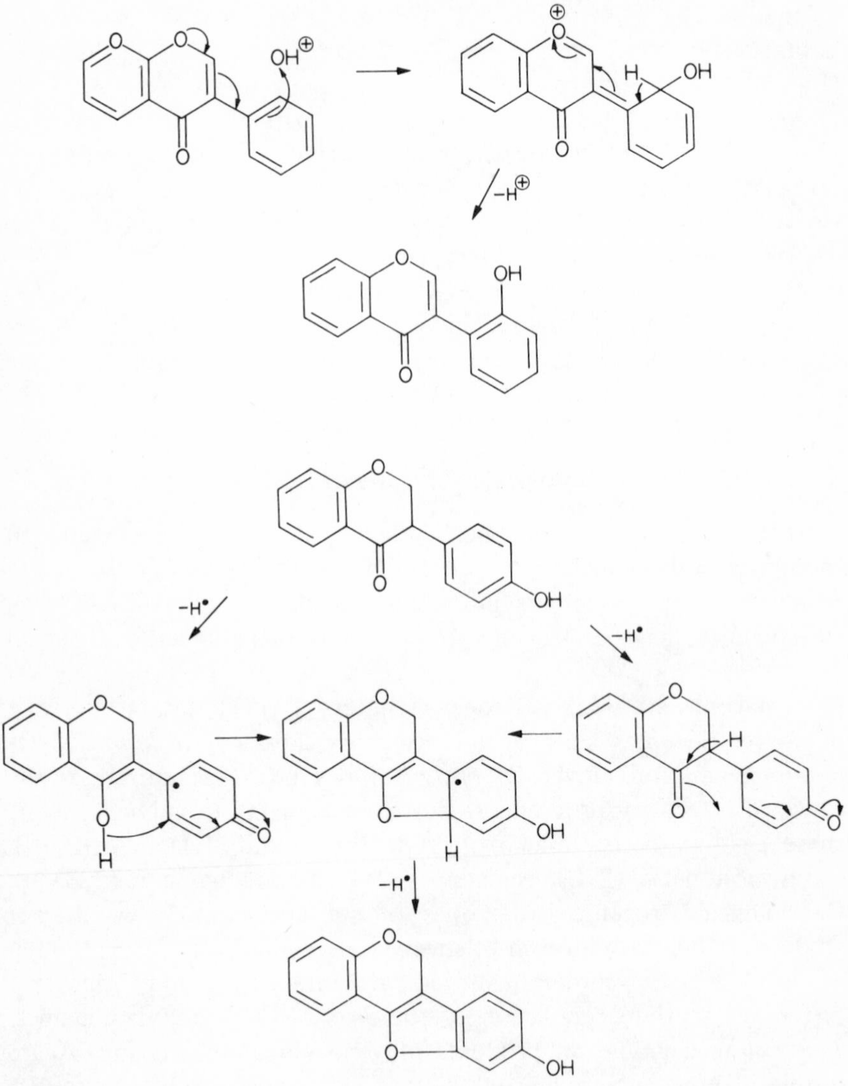

Figure 14.16 Possible mechanisms for 2'-oxygenation of isoflavonoids.

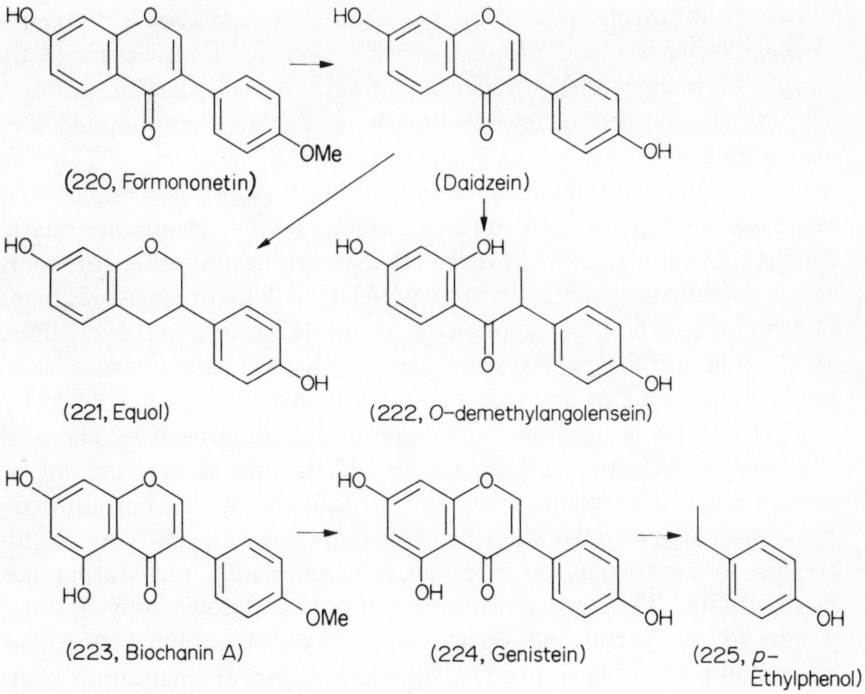

Figure 14.17 Metabolism of isoflavones in sheep.

animal several days after ingestion of pasture rich in these iso-flavones. The oestrogenicity of genistein and biochanin A in sheep is thus a function of the animal's pre-conditioning (Lindsay and Kelly, 1970). Formononetin on the other hand is apparently oestrogenic in sheep at all times.

The weak oestrogenic activity occasionally found in alfalfa (lucerne) (*Medicago sativa*) and ladino clover (*Trifolium repens*) provided the impetus for the studies by Bickoff and co-workers which resulted in the discovery of coumestrol and other coumestans as the oestrogenic principle in these forage plants (review, Bickoff *et al.*, 1969). These compounds are somewhat more active than the isoflavones in mice and sheep (Braden *et al.*, 1967) but because of their much lower concentrations in the plant, coumestans are not important in general as forage oestrogens. Extrinsic factors however can change this situation markedly (Bickoff *et al.*, 1969; Wong and Latch, 1971a).

Rotenoid-bearing plants of the related genera *Derris, Loncho-carpus, Tephrosia* and *Mundulea* have been used for centuries by people of tropical Asia, Africa and South America as fish poisons. The insecticidal properties of rotenone and related compounds have also been known for a long time (Fukami and Nakajima, 1971). The major effect of rotenoids on insects as well as fish is a remarkable decrease in the oxygen uptake which finally results in death. Biochemical studies have revealed that rotenone and other rotenoids act as inhibitors specifically at the $NADH_2$ –dehydrogenase segment of the mitochondrial respiratory chain (Fukami and Namajima, 1971). These findings, however, leave unanswered the question as to why rotenoids are of low toxicity to mammals.

Phytoalexins are anti-microbial compounds produced by plants in response to infection by pathogens. Their role as determinant of disease reaction in plants is currently receiving much attention from phytopathologists (Wood, 1972; Deverall, 1972). Of the phyto-alexins so far recognized from leguminous plants, pterocarpanoids predominate. Thus in addition to the first known phytoalexin, pisatin (159) (Perrin and Bottomley, 1962), phaseollin (146), and phaseollidin (147) have been isolated as anti-fungal compounds from infected-French bean (Perrin *et al.*, 1972; Burden *et al.*, 1972), and 6a-hydroxyphaseollin (160) has been found to be the anti-fungal principle in soya bean (Keen *et al.*, 1971). Medicarpin (130) is the phytoalexin in alfalfa (Smith *et al.*, 1971). This compound together with maackiain (131) have recently been recognized as the anti-fungal agents in red clover and the same compounds appear to play a similar role in white clover and subterranean clover (Higgins and Smith, 1972).

Pisatin and phaseollin are inhibitory to a wide range of phyto-pathogenic fungi. In a recent study (Perrin and Cruickshank, 1969) the anti-fungal activity of some naturally occurring pterocarpans towards *Monilinia fructicola* was compared. Compounds found to be as active as (−)-phaseollin and (+)-pisatin were: (±)-pisatin, (−)-maackiain, (±)-maackiain, (−)-homopterocarpin and (±)-variab-lin. In the preferred conformation (157) of natural pterocarpans already mentioned (Section E), rings A, C and D of the two enantiomeric forms are superimposable. This circumstance has been used by Perrin and Cruickshank (1969) to explain the above findings that both (+) and (−) forms of the pterocarpans tested possess comparable biological activity.

Many of the phytoalexins mentioned above are also formed after treatment of plant tissues with agents other than the pathogenic fungi. Another significant point is that other related flavonoids are sometimes found to be accumulated in infected tissues of these plants. Coumestan, flavone and flavonol concentrations, for example, are greatly increased in alfalfa and white clover after fungal infection (Bickoff *et al.*, 1967; Wong and Latch, 1971a). Understanding of the mechanisms of phytoalexin induction (Cruickshank and Perrin, 1971) in these legumes must therefore await elucidation of the metabolic pathways and associated enzymes involved in flavonoid and isoflavonoid biosynthesis.

References

Adams, D. L., Crombie, L. and Whiting, D. A. (1966), *J. chem. Soc.* (C) 542.

Adinarayana, D., Radhakrishniah, M., Rao, J. R., Campbell, R. and Crombie, L. (1971), *J. chem. Soc.* (C) 29.

Adinarayana, D. and Rao, J. R. (1972a), *Tetrahedron* 28, 5377.

Adinarayana, D. and Rao, J. R. (1972b), 8th Internatl. Symposium on the Chemistry of Natural Products (IUPAC) Abstracts, p. 96.

Adityachaudhury, N. and Gupta, P. K. (1973), *Phytochemistry* 12, 425.

Ahluwalia, V. K., Sachdev, G. P. and Seshadri, T. R. (1965), *Indian J. Chem.* 3, 474.

Akisanya, A., Bevan, C. E. L. and Hirst, J. (1959), *J. chem. Soc.* 2679.

Alencar, R. D., Braz Filho, R. and Gottlieb, O. R. (1972), *Phytochemistry* 11, 1517.

Almeida, M. E. L. and Gottlieb, O. R. (1974), *Phytochemistry* 13, 751.

Assumpcao, R. M. V. and Gottlieb, O. R. (1973), *Phytochemistry* 12, 1184.

Balakrishna, S., Ramanathan, J. D., Seshadri, T. R. and Venkataramani, B. (1962), *Proc. Roy. Soc. (London)* A268, 1.

Banerji, A., Murti, V. V. S. and Seshadri, T. R. (1965), *Curr. Sci.* 34, 431.

Banerji, A., Murti, V. V. S. and Seshadri, T. R. (1966), *Indian J. Chem.* 4, 70.

Banerji, A., Murti, V. V. S., Seshadri, T. R. and Thakur, R. S. (1963) *Indian J. Chem.* 1, 25.

Batterham, T. J., Shutt, D. A., Hart, N. K., Braden, A. W. H. and Tweeddale, H. J. (1971), *Aust J. agric. Res.* 22, 131.

Beck, A. B. (1964), *Aust. J. Chem.* 15, 223.

Beck, A. B. and Knox, J. R. (1971), *Aust. J. Chem.* 24, 1509.

Bevan, C. W. L., Birch, A. J., Moore, B. and Mukerjee, S. K. (1964), *J. chem. Soc.* 5991.

Bevan, C. W. L., Ekong, D. E. U., Obasi, M. E. and Powell, J. W. (1966), *J. chem. Soc.* (C) 509.

Bhargava, K. K., Krishnaswamy, N. R. and Seshadri, T. R. (1972), *Indian J. Chem.* 10, 810.

Bhutani, S. P., Chibber, S. S. and Seshadri, T. R. (1969), *Indian J. Chem.* **7**, 210.

Bickoff, E. M. (1968), Rev. Ser. 1/1968, Commonwealth Bur. of Pastures and Field Crops, pp. 1–39. Hurley, Berkshire, England.

Bickoff, E. M., Livingston, A. L., Witts, S. C., Lundin, R. E. and Spencer, R. R. (1965), *J. agric. Fd. Chem.* **13**, 597.

Bickoff, E. M., Loper, G. M., Hanson, C. H., Graham, J. H., Witts, S. C. and Spencer, R. R. (1967), *Crop Sci.* **7**, 259.

Bickoff, E. M., Spencer, R. R., Knuckles, B. E. and Lundin, R. E. (1966), *J. agric. Fd. Chem.* **14**, 444.

Bickoff, E. M., Spencer, R. R., Witt, S. C. and Knuckles, B. E. (1969), 'Studies on the Chemical and Biological Properties of Coumestrol and Related Compounds'. Tech. Bull. No. 1408, U.S.D.A.

Biggers, J. D. (1959), In *The Pharmacology of Plant Phenolics* (J. W. Fairbairn, ed.), pp. 51–69. Academic Press, London.

Bouwer, D., Brink, C. v. d. M., Engelbrecht, J. P. and Rall, G. J. H. (1968), *J. S. African chem. Inst.* **21**, 159.

Bowyer, W. J., Chatterjea, J. N., Dhoubhadel, S. P., Handford, B. O. and Whalley, W. B. (1964), *J. chem. Soc.* 4212.

Braden, A. H., Hart, N. K. and Lamberton, J. A. (1967), *Aust. J. agric. Res.* **18**, 335.

Braden, A. H. and McDonald, I. W. (1970). In *Australian Grasslands* (R. M. Moore, ed.), pp. 38–392. Australian National University Press, Canberra.

Braz Filho, R., Almeida, M. E. L. and Gottlieb, O. R. (1973b), *Phytochemistry* **12**, 1187.

Braz Filho, R. and Gottlieb, O. R. (1971), *Phytochemistry* **10**, 2433.

Braz Filho, R., Gottlieb, O. R. and Assumpcao, R. M. V. (1971), *Phytochemistry* **10**, 2835.

Braz Filho, R., Gottlieb, O. R., Pinho, S. L. V., Monte, F. J. Q. and Rocha, A. I. D. (1973a), *Phytochemistry* **12**, 1184.

Bredenberg, J. B-Son. and Hietala, P. K. (1961), *Acta. chem. scand.* **15**, 696.

Brink, C. v. d. M., Dekker, J. J., Hanekom, E. C., Meiring, D. H. and Rall, G. J. H. (1965), *J. S. African chem. Inst.* **18**, 21.

Brink, C. v. d. M., Nel, W., Rall, G. J. H., Weitz, J. C., Pachler, K. G. R. (1966), *J. S. African chem. Inst.* **19**, 24.

Brink, C. v. d. M., Engelbrecht, J. P. and Graham, D. Z. (1970), *J. S. African chem. Inst.* **23**, 24.

Buchi, G., Crombie, L., Godin, P. J., Kaltenbronn, J. S., Siddalingaiah, K. S. and Whiting, D. A. (1961), *J. chem. Soc.* 2843.

Burden, R. S., Bailey, J. A. and Dawson, G. W. (1972), *Tetrahedron Letters* 4175.

Campbell, R. V. M., Harper, S. H. and Kemp, A. D. (1969), *J. chem. Soc.* (C) 1787.

Claisse, J., Crombie, L. and Pearce, R. (1964), *J. chem. Soc.* 6023.

Clark-Lewis, J. W. (1962), *Rev. pure appl. Chem.* **12**, 96.

Cocker, W., Dahl, T., Dempsey, C. and McMurray, T. B. H. (1962), *J. chem. Soc.* 4906.

Cocker, W., McMurray, T. B. H. and Staniland, P. A. (1965), *J. chem. Soc.* 1034.

Craveiro, A. A. and Gottlieb, O. R. (1974), *Phytochemistry* 13, 1629.

Crombie, L. (1963), *Fortschr. Chem. organ. Naturst.* 21, 275.

Crombie, L. and Godin, P. J. (1961), *J. chem. Soc.* 2861.

Crombie, L. and Lown, J. W. (1962), *J. chem. Soc.* 775.

Crombie, L. and Whiting, D. A. (1963), *J. chem. Soc.* 1569.

Cruickshank, I. A. M. and Perrin, D. R. (1971), *J. Indian bot. Soc. Golden Jubilee, Vol.* 50A, 1.

Dean, F. M. (1963), *Naturally Occurring Oxygen Ring Compounds*, Butterworths, London.

De Laire, G. and Tiemann, F. (1893), *Chem. Ber.* 26, 2010.

Dement, W. A. and Mabry, T. J. (1972), *Phytochemistry* 11, 1089.

Deverall, B. J. (1972), *Proc. Roy. Soc.* B. 181, 233.

Dewick, P. M., Barz, W. and Grisebach, H. (1969), *Chem. Commun.* 466.

Dhar, K. L. and Kalla, K. (1972), *Phytochemistry* 11, 3097.

Donnelly, D. M. X. and Fitzgerald, M. A. (1971), *Phytochemistry* 10, 3147.

Dyke, S. F., Ollis, W. D. and Sainsbury, M. (1963), *Proc. chem. Soc.* 179.

Eade, R. A., Hinterberger, H. and Simes, J. J. H. (1963), *Aust. J. Chem.* 16, 188.

East, A. J., Ollis, W. D. and Wheeler, R. E. (1969), *J. chem. Soc.* (C) 365.

Elgamal, M. H. A. and Fayez, M. B. E. (1972), *Indian J. Chem.* 10, 128.

Falshaw, C. P., Harmer, R. A., Ollis, W. D., Wheeler, R. E., Lalitha, V. R. and Subba Rao, N. V. (1969), *J. chem. Soc.* (C) 374.

Falshaw, C. P., Ollis, W. D., Moore, J. A. and Magnus, K. (1966), *Tetrahedron, Suppl. No. 7* 333.

Farkas, L. (1973), personal communication.

Farkas, L. and Nogradi, M. (1964), *Tetrahedron Letters* 3919.

Farkas, L., Varady, J. and Gottsegen, A. (1963), *Chem. Ber.* 96, 1865.

Francis, C. M., Millington, A. J. and Bailey, E. T. (1967), *Aust. J. argic. Res.* 18, 47.

Friedlander, A. and Sklarz, B. (1971), *Experientia* 27, 762.

Fukami, H. and Nakajima, M. (1971). In *Naturally Occurring Insecticides* (M. Jacobsen and D. G. Crosby, eds.), pp. 71–97. Marcel Dekker Inc, New York.

Geoffroy, J. (1892), *J. pharm. Chim.* 26, 454.

Geoghegan, M., O'Sullivan, W. I. and Philbin, E. M. (1966), *Tetrahedron* 22, 3209.

Gopinath, K. W., Kidwai, A. R. and Prakash, L. (1961), *Tetrahedron* 16, 201.

Gottlieb, O. R. (1973), personal communication.

Gottlieb, O. R. and Rocha, A. I. (1972), *Phytochemistry* 11, 1183.

Govindachari, T. R., Nagarajan, K., Pai, B. R. and Parthasarthy, P. C. (1957), *J. chem. Soc.* 545.

Grisebach, H. and Zilg, H. (1968), *Ztsch. Naturf.* 23b, 494.

Guggolz, J., Livingston, A. L. and Bickoff, E. M. (1961), *J. agric Fd. Chem.* 9, 330.

Gyorgy, P., Murata, K. and Ikehata, H. (1964), *Nature* 203, 870.

Harborne, J. B. (1967), *Comparative Biochemistry of the Flavonoids*, Academic Press, London.

Harborne, J. B. (1969), *Phytochemistry* **8**, 1449.

Harborne, J. B. (1971). In *Chemotaxonomy of the Leguminosae* (J. B. Harborne, D. Boulter and B. L. Turner), pp. 31–71. Academic Press, London.

Harborne, J. B., Gottlieb, O. R. and Magalhaes, M. (1963), *J. org. Chem.* **28**, 881.

Harper, S. H., Kemp, A. D., Underwood, W. G. E. and Campbell, R. V. M. (1969), *J. chem. Soc.* (C) 1109.

Harper, S. H. and Underwood, W. G. E. (1965), *J. chem. Soc.* 4203.

Higgins, V. T. and Smith, D. G. (1972), *Phytopathol.* **62**, 235.

Highet, R. J. and Highet, P. F. (1967), *J. org. Chem.* **32**, 1055.

Imamura, H., Hibino, Y. and Ohashi, H. (1972), *Mokuzai Gakkaishi* **18**, 6; (*Chem. Abstr.* **77**, 85677).

Imamura, H., Tanno, Y. and Takahashi, T. (1968), *Mokuzai Gakkaishi* **14**, 295; (*Chem. Abstr.* **70**, 44835).

Ito, S., Fujise, Y. and Mori, A. (1965), *Chem. Commun.* 595.

Jackson, B., Owen, P. J. and Scheinmann, F. (1971), *J. chem. Soc.* (C) 3389.

Jain, A. C., Rohatgi, V. K. and Seshadri, T. R. (1967), *Tetrahedron* **23**, 2499.

Jay, M., Lebreton, P. and Letoublon, R. (1971), *Boissiera* **19**, 219.

Johnson, A. P. and Pelter, A. (1966), *J. chem. Soc.* 606.

Johnson, A. P., Pelter, A. and Stainton, P. (1966), *J. chem. Soc.* 192.

Jurd, L., Stevens, K. and Manners, G. (1972), *Phytochemistry* **11**, 2535.

Kasymov, A. U., Kondratenko, E. S. and Abubakirov, N. K. (1972), *Khim Prir. Soedin.* **8**, 115; [*Chem. Abstr.* **77**, 85584].

Keen, N. T., Sims, J. J., Erwin, D. C., Rice, E. and Partridge, J. E. (1971), *Phytopathology* **61**, 1084.

King, F. E., Grundon, M. F. and Neil, K. Q. (1952), *J. chem. Soc.* 4580.

Krishnaswamy, N. R. and Prasanna, S. (1970), *Indian J. Chem.* **8**, 761.

Kukla, A. S. and Seshadri, T. R. (1962), *Tetrahedron* **18**, 1443.

Kurosawa, K., Ollis, W. D., Redman, B. T., Sutherland, I. O., Gottlieb, O. R. and Alves, H. M. (1968b), *Chem. Commun.* 1265.

Kurosawa, K., Ollis, W. D., Redman, B. T., Sutherland, I. O., Oliveira, A. B., Gottlieb, O. R. and Alves, H. M. (1968a), *Chem. Commun.* 1263.

Lebreton, P., Markham, K. R., Swift, W. T., Oung-Boran and Mabry, T. J. (1967), *Phytochemistry* **6**, 1675.

Lindsay, R. and Kelly, R. W. (1970), *Aust. Veter. J.* **46**, 219.

Livingston, A. L., Bickoff, E. M., Lundin, R. E. and Jurd, L. (1964), *Tetrahedron* **20**, 1963.

Livingston, A. L., Witt, S. C., Lundin, R. E. and Bickoff, E. M. (1965), *J. org. Chem.* **30**, 2353.

Malhotra, A., Murti, V. V. S. and Seshadri, T. R. (1967), *Tetrahedron* **23**, 405.

Markham, K. R., Mabry, T. J. and Swift, W. T. (1968), *Phytochemistry* **7**, 803.

Markham, K. R., Mabry, T. J. and Swift, W. T. (1970), *Phytochemistry* **9**, 2359.

Matos, F. J. D. B., Gottlieb, O. R., Ollis, W. D. and Andrade, C. H. S. (1970), *An. Acad. Brasil. Cienc.* **42**, 61.

McGookin, A., Robertson, A. and Whalley, W. B. (1940), *J. chem. Soc.* 787.

McMurray, T. B. H., Martin, E., Donnelly, D. M. X. and Thompson, J. C. (1972), *Phytochemistry* **11**, 3283.

Morgan, J. W. W. and Orsler, R. J. (1967), *Chem. Ind.* 1173.

Murakami, T., Nisikawa, Y. and Ando, T. (1960). *Chem. pharm. Bull. (Tokyo)* **8**, 688.

Naim, M., Gestetner, B., Kirson, I., Birk. Y. and Bondi, A. (1973), *Phytochemistry* **12**, 169.

Narasimhachari, N. and Seshadri, T. R. (1952), *Proc. Indian Acad. Sci.* **35A**, 202.

Narayanan, V. and Seshadri, T. R. (1971), *Indian J. Chem.* **9**, 14.

Oliveira, A. B., Gottlieb, O. R., Goncalves, T. M. M. and Ollis, W. D. (1971a), *An. Acad. Brasil. Cienc.* **43**, 129.

Oliveira, A. B., Gottlieb, O. R. and Ollis, W. D. (1968), *An. Acad. Brasil Cienc.* **40**, 147.

Oliveira, A. B., Gottlieb, O. R., Ollis, W. D. and Rizzini, C. T. (1971b), *Phytochemistry* **10**, 1863.

Oliveira, A. B., Silva, L. G. F. and Gottlieb, O. R. (1972), *Phytochemistry* **11**, 3515.

Ollis, W. D. (1962). In *The Chemistry of Flavonoid Compounds* (T. A. Geissman, ed.), p. 353—405. Pergamon Press, Oxford.

Ollis, W. D. (1966), *Experientia* **22**, 777.

Ollis, W. D. (1968). In *Recent Advances in Phytochemistry* (T. J. Mabry, R. E. Alston and V. C. Runeckles, eds.), Vol. 1, pp. 329—378. Appleton-Century-Crofts, New York.

Ollis, W. D., Ramsay, M. V. J. and Sutherland, I. O. (1965), *Aust. J. Chem.* **18**, 1787.

Ollis, W. D., Rhodes, C. A. and Sutherland, I. O. (1967), *Tetrahedron* **23**, 4741.

Pachler, K. G. R. and Underwood, W. G. E. (1967), *Tetrahedron* **23**, 1817.

Paris, R. R. and Faugeras, G. (1963), *Séances Acad. Sci.* **257**, 1728.

Parthasarathy, M. R., Puri, R. N. and Seshadri, T. R. (1969), *Indian J. Chem.* **7**, 118.

Pelter, A. and Amenechi, P. I. (1969), *J. chem. Soc.* (C) 887.

Pelter, A. and Stainton, P. (1966), *J. chem. Soc.* (C) 701.

Pelter, A., Stainton, P. and Barber, M. (1965a), *J. heterocycl. Chem.* **2**, 262.

Pelter, A., Stainton, P., Johnson, A. P. and Barber, M. (1965b), *J. heterocycl. Chem.* **2**, 256.

Perrin, D. R. (1964), *Tetrahedron Letters* 29.

Perrin, D. R. and Bottomley, W. (1962), *J. Am. chem. Soc.* **84**, 1919.

Perrin, D. R. and Cruickshank, I. A. M. (1969), *Phytochemistry* **8**, 971.

Perrin, D. R., Whittle, C. P. and Batterham, T. J. (1972), *Tetrahedron Letters* 1673.

Prakash, L., Zaman, A. and Kidwai, A. R. (1965), *J. org. Chem.* **30**, 3561.

Purushothaman, K. K., Kishore, V. M., Narayanaswami, V. and Connolly, J. D. (1971), *J. chem. Soc.* (C) 2420.

Rall, G. J. H., Bring, A. J. and Engelbrecht, J. P. (1972), *J. S. African chem. Inst.* **25**, 25.

Rall, G. J. H., Engelbrecht, J. P. and Brink, A. J. (1970), *Tetrahedron* 26, 5007.

Rall, G. J. H., Engelbrecht, J. P. and Brink, A. J. (1971), *J. S. African chem. Inst.* 24, 56.
24, 56.

Saitoh, T. and Shibata, S. (1969), *Chem. pharm. Bull. (Tokyo)* 17, 729.

Schwarz, J. S. P., Cohen, A. I., Ollis, W. D., Kalzka, E. A. and Jackman, L. M. (1964), *Tetrahedron* 20, 1317.

Seshadri, T. R. (1964). In *Symposium on Phytochemistry, Hong Kong 1961* (H. R. Arthur, ed.), pp. 145–163. Hong Kong University Press.

Seshadri, T. R. (1972), *Phytochemistry* 11, 881.

Shabbir, M. and Zaman, A. (1970), *Tetrahedron* 26, 5041.

Shabbir, M., Zaman, A., Crombie, L., Tuck, B. and Whiting, D. A. (1968), *J. chem. Soc.* 1899.

Shamma, M. and Stiver, L. D. (1969), *Tetrahedron* 25, 3887.

Shibata, S. and Hishikawa, Y. (1963), *Chem. pharm. Bull. (Tokyo)* 11, 167.

Shibata, S., Murata, T. and Fujita, M. (1963), *Chem. pharm. Bull. (Tokyo)* 11, 382.

Shibata, S. and Saitoh, T. (1968), *Chem. pharm. Bull. (Tokyo)* 16, 1932.

Simonitsch, E., Frei, H. and Schmid, H. (1957), *Monatsh. Chem.* 88, 541.

Sims, J. J., Keen, N. T. and Honwad, V. K. (1972), *Phytochemistry* 11, 827.

Smith, D. G., McInnes, A. G., Higgin, V. J. and Millar, K. L. (1971), *Physiol. Pl. Pathol.* 1, 41.

Spencer, R. R., Bickoff, E. M., Lundin, R. E. and Knuckles, B. E. (1966a) *J. agric. Fd. Chem.* 14, 162.

Spencer, R. B., Knuckles, B. E. and Bickoff, E. M. (1966b), *J. org. Chem.* 31, 988.

Suginome, H. (1959), *J. org. Chem.* 24, 1655.

Suginome, H. (1966a), *Bull. chem. Soc. Japan* 39, 1529.

Suginome, H. (1966b), *Bull. chem. Soc. Japan* 39, 1544.

Suginome, H. and Iwadare, T. (1962), *Experientia* 18, 163.

Suginome, H. and Kio, T. (1966), *Bull. chem. Soc. Japan* 39, 1541.

Tamura, S., Chang, C. F., Suzuki, A. and Kumai, S. (1969), *Agric. biol. Chem.* 33, 391.

Verbit, L. and Clark-Lewis, J. W. (1968), *Tetrahedron* 24, 5519.

Whalley, W. B. (1959). In *The Pharmacology of Plant Phenolics* (J. W. Fairbairn, ed.), pp. 27–37. Academic Press, London and New York.

Wong, E. (1963), *J. org. Chem.* 28, 2336.

Wong, E. (1970), *Fortschr. Chem. organ. Naturst.* 28, 1.

Wong, E. and Latch, G. C. M. (1971a), *New Zeal. J. agric. Res.* 14, 633.

Wong, E. and Latch, G. C. M. (1971b), *Phytochemistry* 10, 465.

Wong, E., Mortimer, P. I. and Geissman, T. A. (1965), *Phytochemistry* 4, 89.

Wood, R. K. S. (1972), *Proc. Roy. Soc.* B 181, 213.

Zilg, H. and Grisebach, H. (1968), *Phytochemistry* 7, 1765.

Chapter 15

Neoflavanoids

DERVILLA M. X. DONNELLY

15.1 Introduction

15.2 Summary of structural types

15.3 Structural elucidation

15.3.1 4-Phenylcoumarins of the Guttiferae
15.3.2 Exostemin (Rubiaceae)
15.3.3 4-Arylcoumarins of the Leguminosae
15.3.4 Dalbergiones
15.3.5 Neoflavenes
15.3.6 Dalbergiquinols
15.3.7 4-Arylchromans
15.3.8 Coumarinic acids

15.4 Biosynthesis

15.5 Interconversions

15.1 Introduction

The term neoflavanoid, suggested by Dr T. Swain, was first used by Ollis in the 1965a Eyton publication to describe the group of natural products with a 4-arylchroman skeleton (1). These C_{15} compounds relate closely to the flavanoids (2) and the isoflavanoids (3). The open-chain compounds, the dalbergiones (4) (Gonçalves da Lima and Dalia Maia, 1961) and the 3,3-diarylpropenes (5) have been included in the neoflavanoid class, in line with the assignment of 2'-hydroxychalcone (6) and of angolensin (7) to the flavanoid and isoflavanoid classes respectively.

In 1955, Robinson remarked that brazilin and haemotoxylin had the 4-phenylchroman skeleton but that no simple member of this class had as yet been isolated. In fact the first neoflavanoid, calophyllolide, was isolated from *Calophyllum inophyllum* L. (Guttiferae) (Ormancey-Potier *et al.*, 1951), but its structure at the

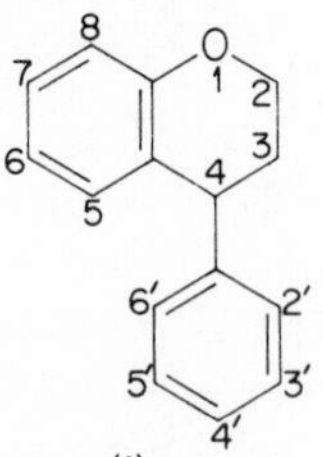

(1) 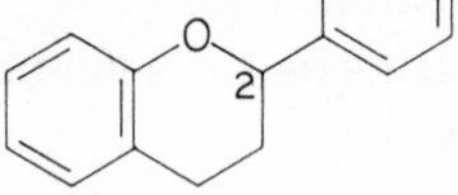　　　　(2) 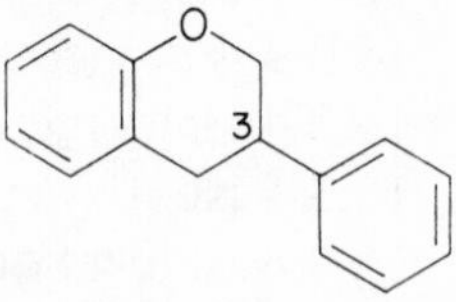　　　　(3)

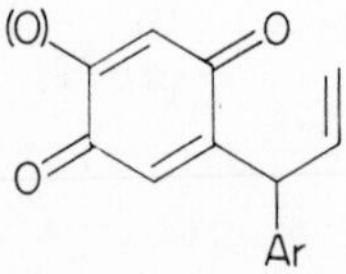

(4) 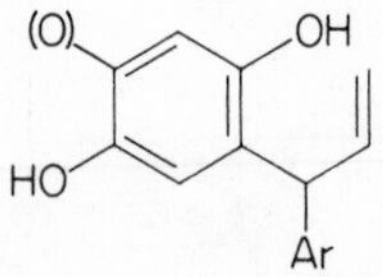　　　　(5) 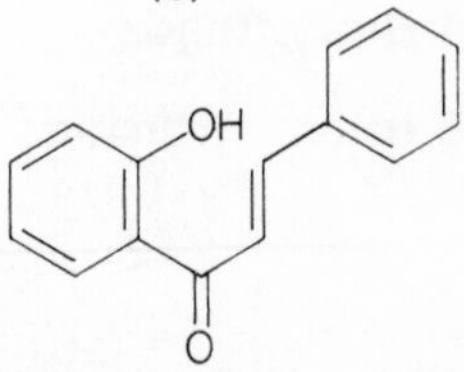　　　　(6)

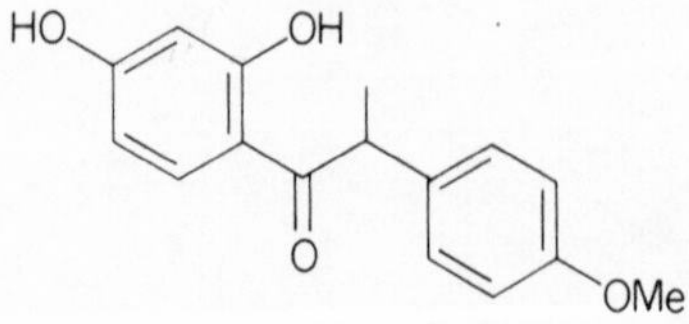

(7)

time was not realized. It was then described as 'une lactone insaturée possédant un groupe methoxyle'. Two years later, Kathpalia and Dutt (1953) obtained a compound from the heartwood extract of *Dalbergia sissoo* (Leguminosae) which they named dalbergin and to which they assigned incorrectly a 3-phenylcoumarin structure. Dalbergin was subsequently shown to have a 4-phenylcoumarin structure and was the first neoflavanoid from the Leguminosae family.

The elucidation of the 4-phenylcoumarin structure for a natural product was first achieved by Polonsky (1955, 1956, 1957) when she established the structures of calophyllolide (8) and inophyllolide (9).

15.2 Summary of structural types

The naturally occurring neoflavanoids have been arranged conveniently (although arbitrarily) in groups according to their structural type and source (Table 15.1).

15.3 Structural elucidation

The procedure for structural elucidation of natural neoflavanoids has been particularly modified in cases where the compound under analysis can be related to a number of substances of known structure. In the descriptions which follow on the structural elucidation of the neoflavanoids, the discussion emphasizes mainly the chemical reactions involved and only refers to data from physical measurements when necessary to derive the correct formula. Compounds whose structures are similar or which have been elucidated by similar procedures have been grouped together. The nomenclature used is that employed by the respective authors in their publications.

15.3.1 4-Phenylcoumarins of the Guttiferae

Twenty-six 4-phenylcoumarins have been encountered in species of the genera *Calophyllum, Mesua* and *Mammea*. These neoflavanoids co-occur with 4-*n*-alkylcoumarins which have similar substitution patterns. Interest in the phytochemical study of the genera mentioned arose from a report by Grosourdy (1864) on the insecticidal activity of some species. For example, extracts from seeds of *Mammea americana* L. gave neoflavanoids and 4-*n*-alkyl-

Table 15.1   *The structures and natural occurrence of neoflavanoids*

Compound	Organ examined	Family, Genus, Species	References
4-ARYLCOUMARINS		Guttiferae	
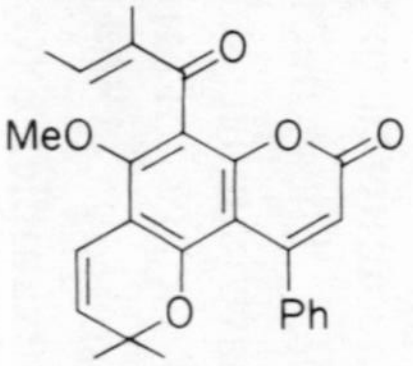	nuts	*Calophyllum inophyllum*	Polonsky, 1957.
(8) Calophyllolide	nuts	*C. bracteatum*	Somanathan and Sultanbawa, 1972.
(9) (±) and (+)-*trans*-inophyllolide	leaves	*C. inophyllum*	Polonsky, 1957.
(10) (+)-*cis*-inophyllolide			Kawazu et al, 1968, 1972
	leaves	*C. inophyllum*	*Ibid.*
	leaves	*C. inophyllum*	*Ibid.*

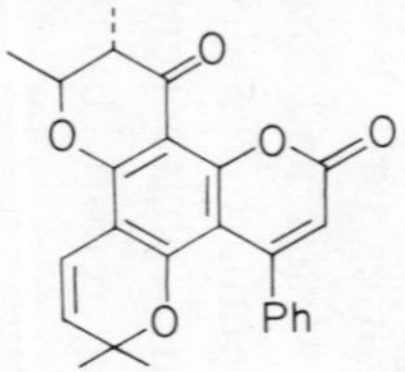

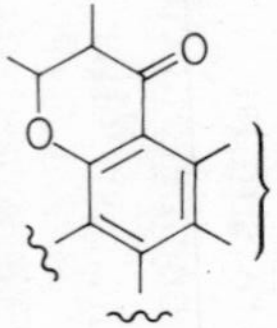

(8) Calophyllolide

(9)

(10)

(9) (±) and (+)-*trans*-inophyllolide
(10) (+)-*cis*-inophyllolide

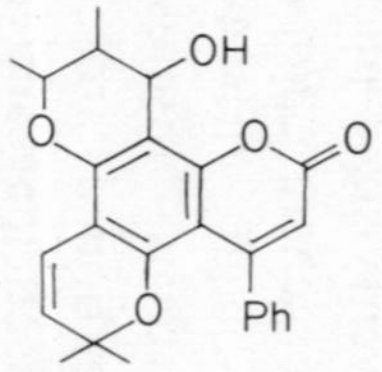

(11a)

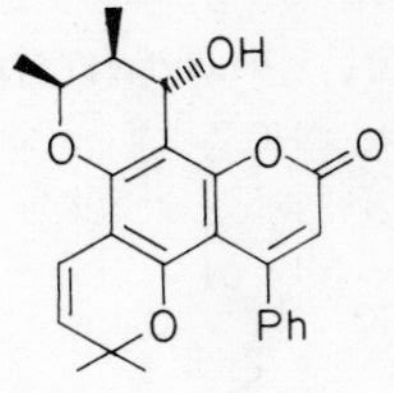

(11b)

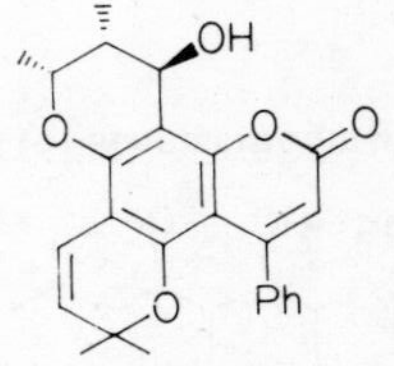

(11c)

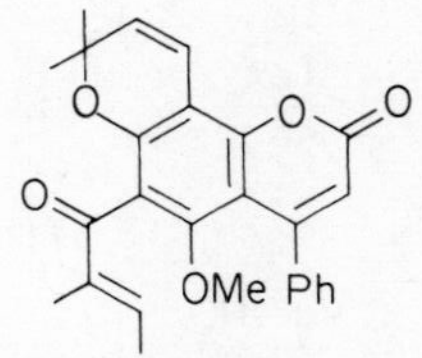

(12) apetalolide

(13) tomentolide A

OH Ph

OH Ph

OMe Ph

Ph

nuts   *C. apetalum*   Nigam *et al.*, 1967.

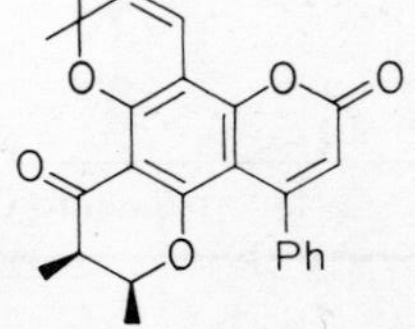

nuts   *C. tomentosum*   *Ibid.*

Continued

Table 15.1–Continued

Compound	Organ examined	Family, Genus, Species	References
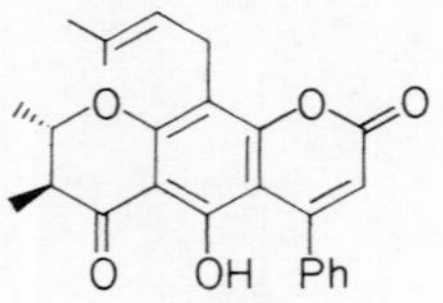 (14)	resins	C. australianum	Breck and Stout, 1969.
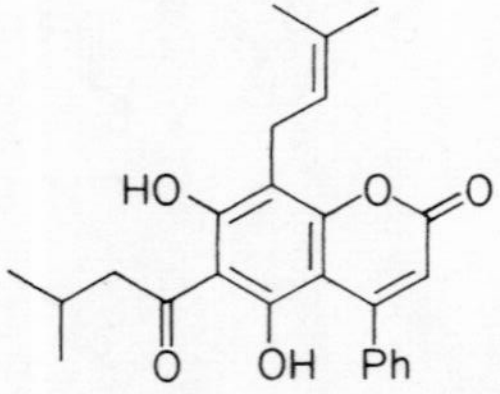 (15) Mammeisin (mammea A/AA)	seeds/bark	Mammea africana	Carpenter et al., 1970
	fruit peels	Mammea americana	Finnegan et al., 1961.
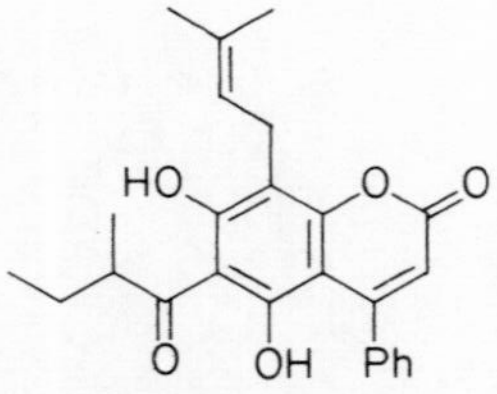 (16) (Mammea A/AB)	seeds	M. americana	Crombie et al., 1966, 1967.
	bark	M. africana	Carpenter et al., 1970.

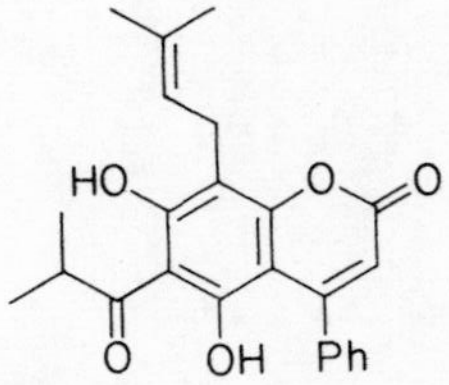

(17) Mesuol

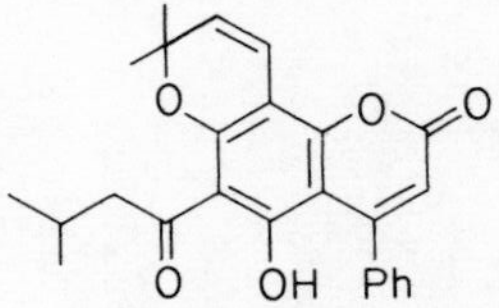

(18) Mammeigin
(Mammea A/AcycloD)

(19) (M.A.B.5)

(20) mesuagin

seed oil	*Mesua ferrea*	Chakraborty and Das, 1966.
seeds	*Mammea americana*	Games, 1972.
seed oil	*M. americana*	Finnegan and Mueller, 1964, 1965
	M. africana	Crombie *et al.*, 1967.
seeds		Games, 1972.
bark	*M. africana*	Carpenter *et al.*, 1970.
	M. africana	Games, 1972.
seed oil	*Mesua ferrea*	Chakraborty and Chatterji, 1969.
seeds	*Mammea americana*	Games, 1972.

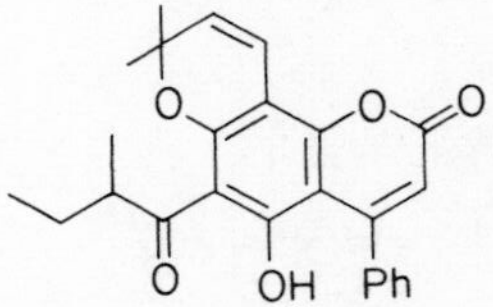

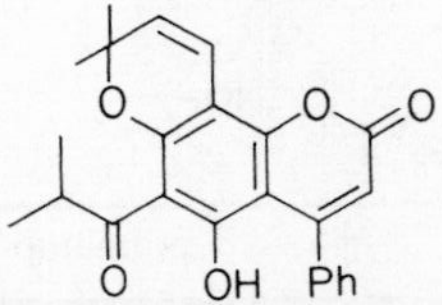

Continued

807

Table 15.1—Continued

Compound	Organ examined	Family, Genus, Species	References
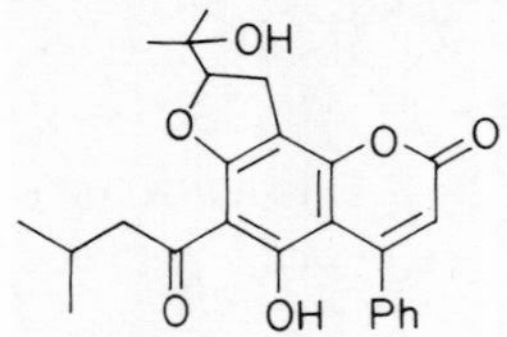 (21)	seeds	M. americana	Crombie et al., 1970.
(22)	seeds	M. americana	Crombie et al., 1972a.
	bark	M. africana	Carpenter et al., 1970.
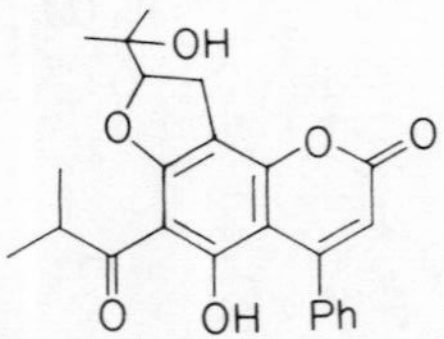 (23)	seeds	M. americana	Crombie et al., 1972a.

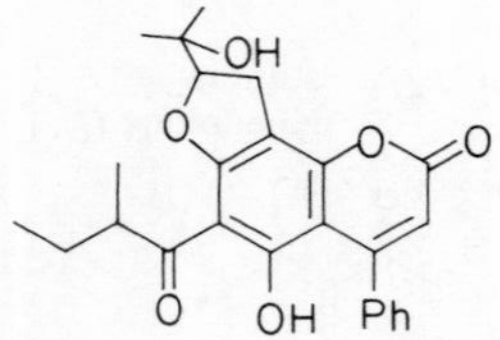

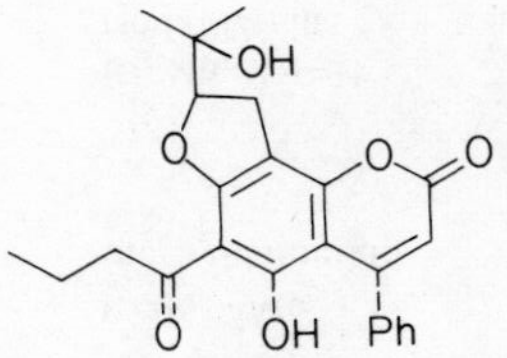

(24)

(25) (Mammea A/AB)

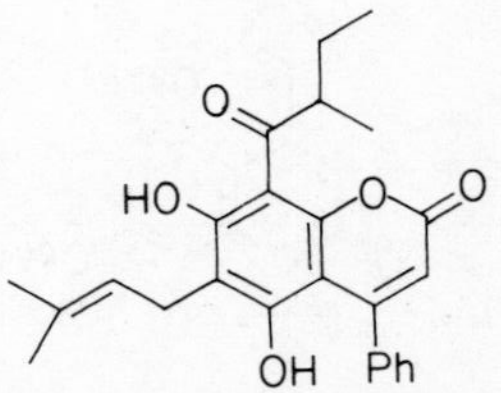

(26) (Mammea A/BB)

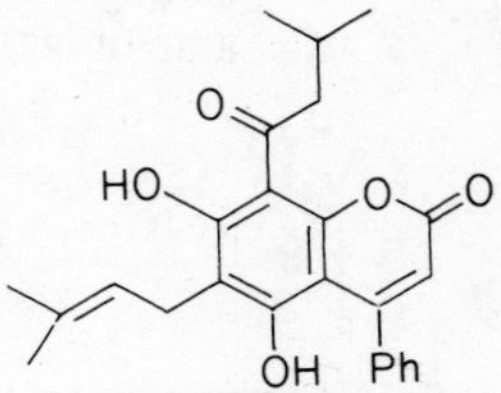

seeds	*M. americana*	*Ibid.*
seeds	*M. americana*	Crombie *et al.*, 1967.
	M. africana	
seeds	*M. americana*	Crombie *et al.*, 1967.

Continued

Table 15.1—*Continued*

Compound 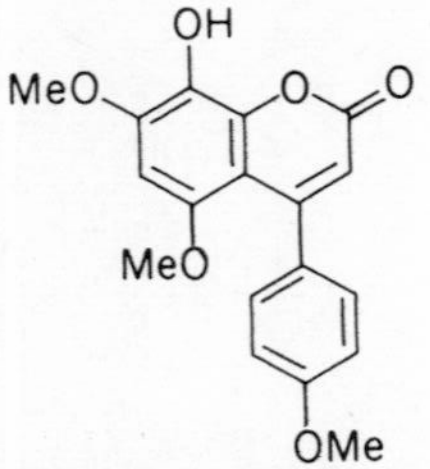	Organ examined	Family, Genus, Species	References
		Rubiaceae	
	wood	*Exostemma caribaerum*	Sánchez-Viesca *et al.*, 1967.
(27) exostemin			
		Leguminosae	
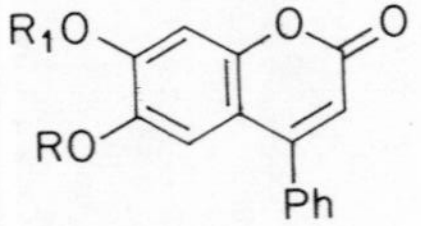	stembark	*Dalbergia sissoo*	Ahluwalia and Seshadri, 1957.
	bark	*D. latifolia*	Dhingra *et al.*, 1971.
	sapwood	*D. melanoxylon*	Donnelly *et al.*, 1966.
	heartwood	*D. nigra*	
		D. spruceana	Ollis, 1966.
		D. violacea (Vog) Malme	*Idem*, 1967; Gregson *et al.*, 1968.
		D. cearensis	*Ibid.*
		D. baroni	Donnelly, B. J. *et al.*, 1968.
		D. cultrata	Donnelly *et al.*, 1972.
		Macherium scleroxylon	Eyton *et al.*, 1965b.
		M. pedicellatum	Ogiyama and Yasue, 1973.
(28) R=H; R$_1$ =Me Dalbergin	stembark	*D. sissoo*	Mukerjee *et al.*, 1971.
	bark	*D. latifolia*	Dhingra *et al.*, 1971.
R=R$_1$ =Me methyldalbergin	sapwood	*M. scleroxylon*	Eyton *et al.*, 1965b.
	heartwood	*M. pedicellatum*	Ogiyama and Yasue, 1973.
R=Me; R$_1$ =H isodalbergin	heartwood	*D. sissoo*	Mukerjee *et al.*, 1971.

R=R₁=H
nordalbergin

heartwood	*D. sissoo*	*Ibid.*	

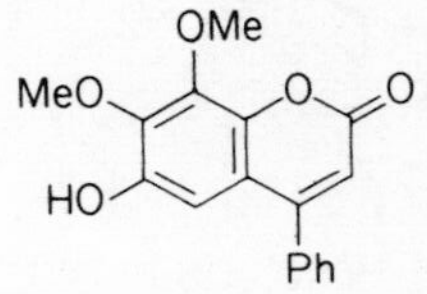

(29) kuhlmannin

wood | *M. nictitans* *M. kuhlmannii* *M. pedicellatum* | Ollis *et al.*, 1968. *Ibid.* Ogiyama and Yasue, 1973.

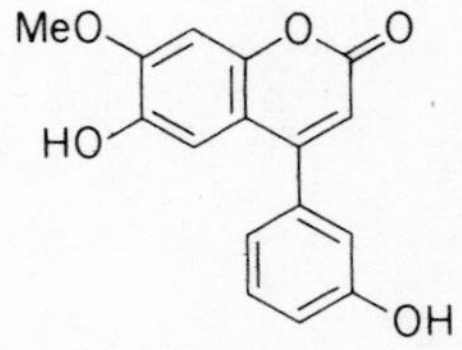

(30) stevenin

heartwood | *D. stevensonii* *D. cultrata* | Donnelly *et al.*, 1973b. Donnelly *et al.*, 1972.

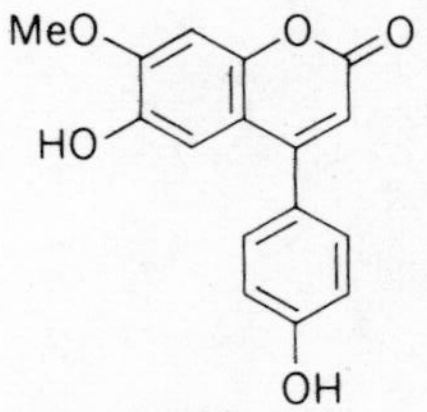

(31) melannin

heartwood | *D. melanoxylon* | Donnelly, 1971.

Continued

Table 15.1–*Continued*

Compound	Organ examined	Family, Genus, Species	References
(32) melannein	heartwood	*D. melanoxylon* *D. baroni*	Donnelly *et al.*, 1966. Donnelly *et al.*, 1966.
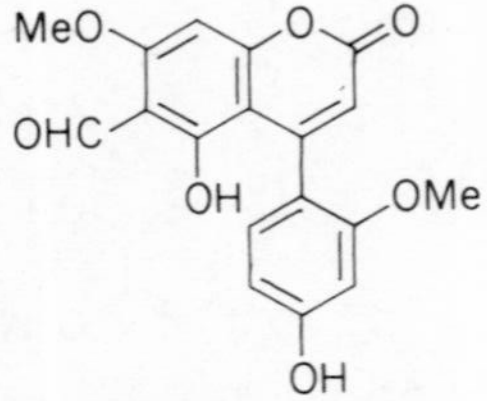 (33) sisafolin	seeds	*D. latifolia*	Saxena *et al.*, 1970.
DALBERGIONES (34)	stembark sapwood heartwood	*Dalbergia baroni* *D. sissoo* *D. cultrata* *D. violacea* *D. barretoana* *D. villosa* *D. stevensonii* *D. melanoxylon*	Donnelly *et al.*, 1965. Ahluwalia and Seshadri, 1963. Mukerjee *et al.*, 1971. Donnelly *et al.*, 1972. Eyton *et al.*, 1965a. Braga, da Silva *et al.*, 1967. *Ibid.* Donnelly, B. J. *et al.*, 1973b. Donnelly *et al.*, 1968, 1969.

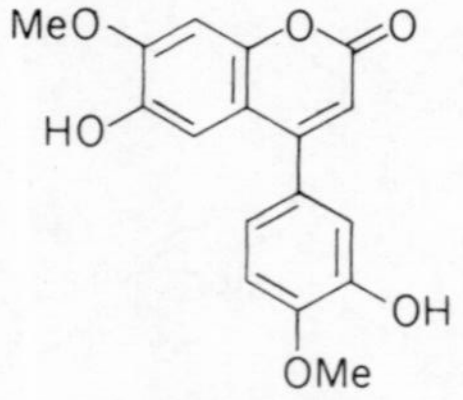

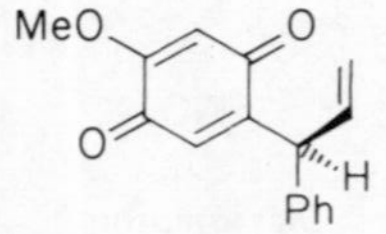

D. spruceana	bark	Ollis, 1966.
D. inundata	sapwood	Almeida and Gottlieb, 1974.
D. riparia		Braz Filho et al., 1973
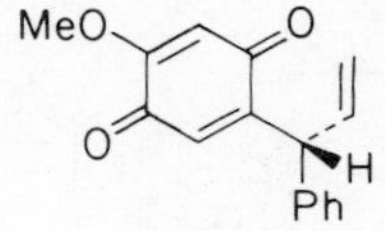 *D. nigra*		Marini-Bettolo et al., 1962. Eyton et al., 1962, 1965a; Braga de Oliveira et al., 1971.
D. latifolia	heartwood	Rao and Seshadri, 1963. Donnelly et al., 1965. Dhingra et al., 1971.
D. obtusa		Gregson et al., 1968.
D. cochinchinensis		Donnelly, B. J. et al., 1968.
D. retusa	heartwood	Jurd et al., 1972.
D. nigra	heartwood	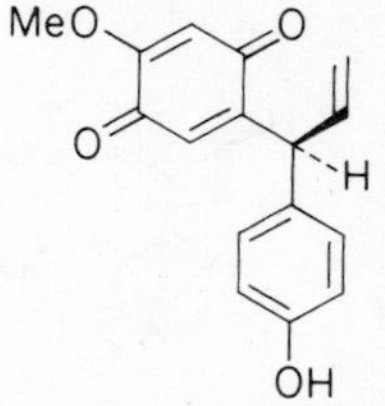 Eyton et al., 1962, 1965a; Marini-Bettolo et al., 1962.
D. melanoxylon		Donnelly, B. J. et al., 1968, 1969.
D. stevensonii		Donnelly et al., 1973b.
D. violacea		Eyton et al., 1965a; Gregson et al., 1968.
D. nigra	heartwood	Eyton et al., 1962, 1965a.
D. violacea		*Ibid.*: Gregson et al., 1968.

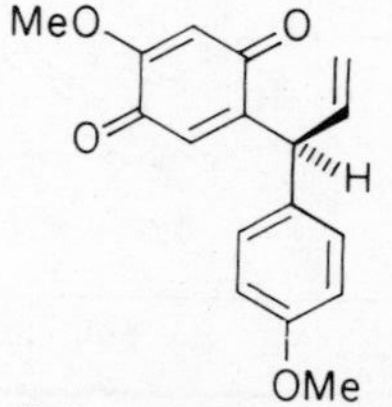

RS-4-methoxydalbergione

(35)

(36)

Continued

Table 15.1—*Continued*

Compound	Organ examined	Family, Genus, Species	References
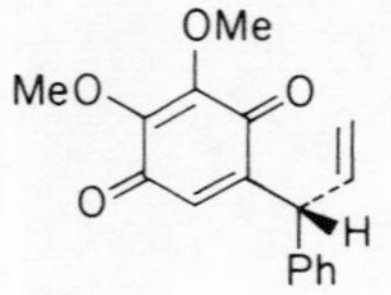 (37)	heartwood	*Macherium nictitans* *M. scleroxylon* *M. kuhlmannii* *M. pedicellatum*	Ollis et al., 1968. Eyton et al., 1965b. Ollis et al., 1968. Ogiyama and Yasue, 1973.
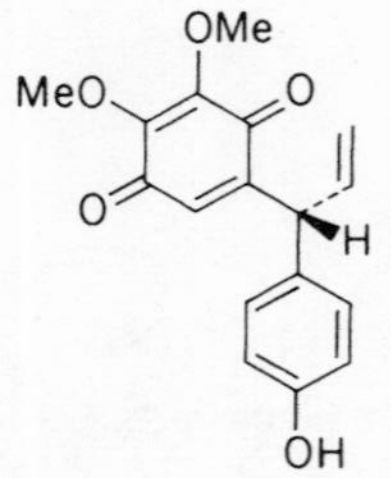 (38)	heartwood	*M. nictitans*	*Ibid.*
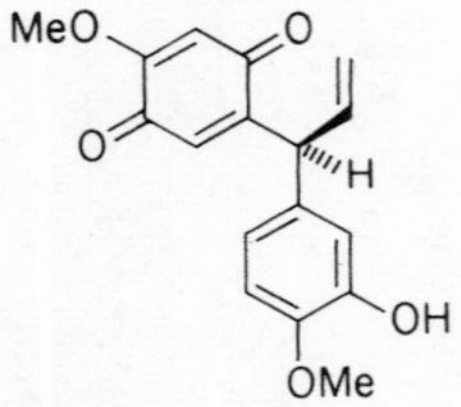 (39)	heartwood	*D. melanoxylon*	Donnelly et al., 1973.

(37)

(38)

(39)

NEOFLAVENES

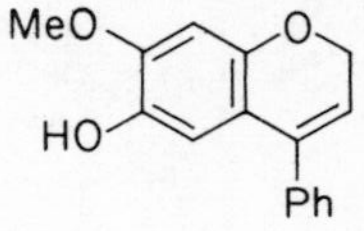

(40)

stembark	*D. sissoo*	Mukerjee *et al.*, 1971.
bark	*D. latifolia*	Dhingra *et al.*, 1971.

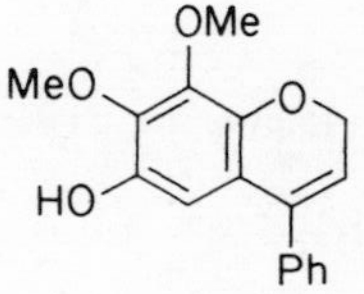

(41)

heartwood	*M. nictitans*	Ollis *et al*, 1968.
	M. kuhlmannii	*Ibid.*
	M. pedicellatum	Ogiyama and Yasue, 1973.

DALBERGIQUINOLS

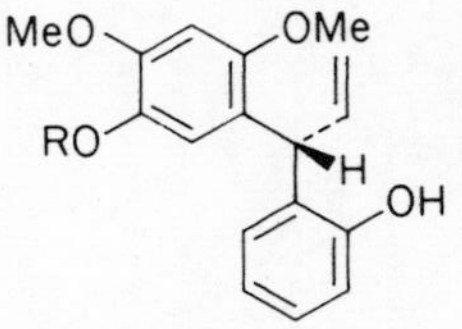

(42)

bark	*D. latifolia*	Balakrishna *et al.*, 1962;
sapwood		Dempsey *et al.*, 1963;
heartwood		Dhingra *et al.*, 1971;
		Rao and Seshadri, 1963;
		Kumari *et al.*, 1965.

R=H latifolin

	D. cochinchinensis	Donnelly, D. M. X. *et al.*, 1968.

R=Me
5-*O*-methyllatifolin

heartwood	*D. cochinchinensis*	*Ibid.*

Continued

815

Table 15.1—*Continued*

Compound	Organ examined	Family, Genus, Species	References
(43)	heartwood	*D. cultrata*	Donnelly *et al.*, 1972.
(44) obtusaquinol	heartwood	*D. retusa* *D. obtusa*	Jurd *et al.*, 1972. Gregson *et al.*, 1968.
(45)	heartwood	*M. scleroxylon* *M. pedicellatum*	Eyton *et al.*, 1965b. Ogiyama and Yasue, 1973.

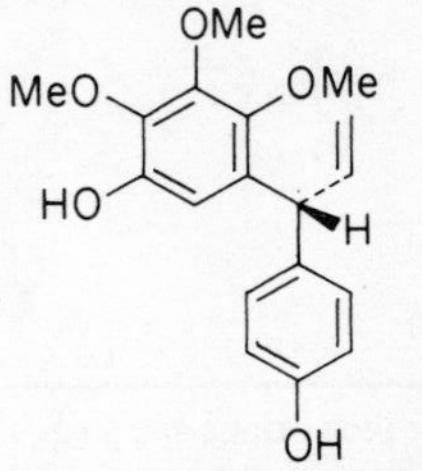

(46) kuhlmanniquinol

wood   *M. kuhlmannii*
M. nicitans   Ollis *et al.*, 1968.
Ibid.

4-ARYLCHROMANS

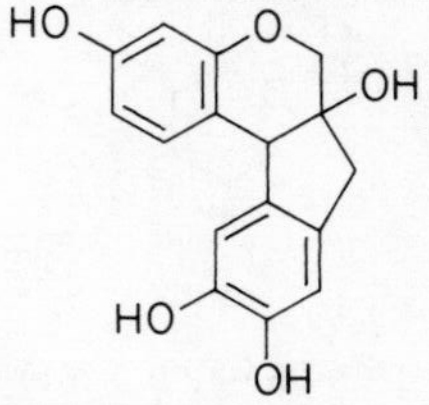

(47) brazilin

wood   *Caesalpinia crista*
C. braziliensis   Chevreul, 1808.
Perkin *et al.*, 1928.

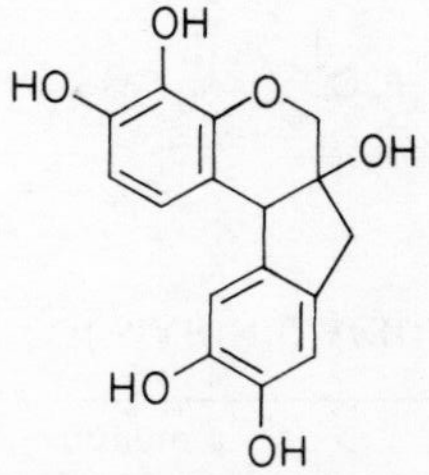

(48) haematoxylin

wood   *Haematoxylon*
campechianum   Perkin and Robinson,
1908.

Continued

Table 15.1—*Continued*

Compound	Organ examined	Family, Genus, Species	References
COUMARINIC ACIDS			
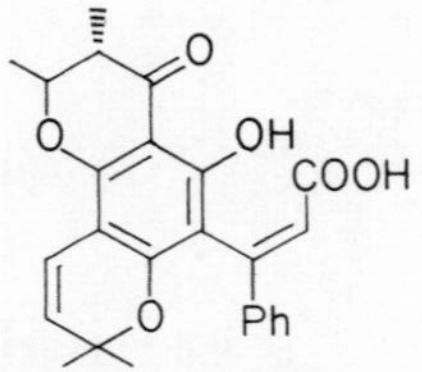 (49) Calophyllic acid	nuts	*C. inophyllum*	Polonsky, 1957.
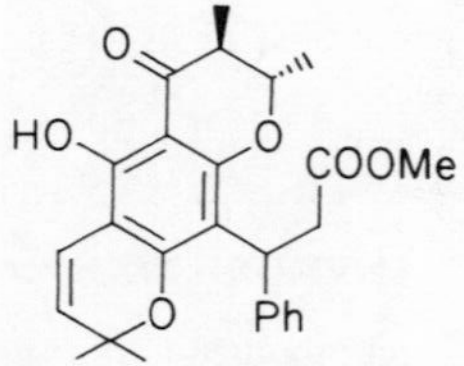 (50) Chapelieric acid methyl ester	nuts	*C. chapelieri*	Guerreiro *et al.*, 1971.

coumarins; however, none possessed insecticidal properties relative to those of the crude extract. More recent studies by Crombie and his co-workers (1972b) described the isolation of an active constituent that has a coumarin structure with a 4-(1-acetoxypropyl) residue.

The examination of a number of species belonging to the Guttiferae has led to the isolation of a closely related series of neo-flavanoids (8)–(26).

15.3.1.1 4-Phenylcoumarins from *Calophyllum* species

Calophyllolide (8), $C_{26}H_{24}O_5$, the first compound to be assigned a 4-phenylcoumarin structure, was isolated by Ormancey-Potier and co-workers (1951) and the structure later eludicated (Polonsky, 1955, 1956). Calophyllolide is optically inactive and its hydrogenation with Adam's catalyst in ethanol gives a mixture of tetrahydro- and hexahydrocalophyllolide and the ethanolysis product of the latter. A comparison of the IR spectra of the partially hydrogenated derivatives (including products of reduction by $NaBH_4$ and $LiAlH_4$) intimated that calophyllolide has an $\alpha\beta$-unsaturated lactone, a double bond and an $\alpha\beta$-unsaturated carbonyl. A systematic degradative study followed involving ozonolysis, oxidation with $Pb(OAc)_4$, hydrolysis with KOH, and hydrolysis with acid. The volatiles from reaction with ozone afforded acetaldehyde and acetone, the latter arose from the dimethylchromene ring. The isolation of ethyl cinnamate from a lead tetra-acetate oxidation showed that the unsubstituted phenyl was β to a carbonyl function. However, the most meaningful degradation was that resulting from hydrolysis with base.

The five compounds isolated from the degradation were acetaldehyde, acetone, propionic acid, acetophenone and 5-hydroxy-7-methoxy-4-phenylcoumarin (51). The coumarin (51) has characteristic UV (λ_{max} 260 nm (log ϵ 4.13), 330 (4.13)) and IR (ν_{CO} 1689 cm^{-1}; ν_{OH} 3154 cm^{-1}) spectra. The assigned structure (51) was substantiated by the chemical reactions (Scheme 15.1) and confirmed by synthesis. The degradative product (51) was the key to the skeletal structure of calophyllolide and determined the relative positions of (i) the phenyl group, (ii) the methoxyl group, (iii) the presence of a phloroglucinol-type A ring and corroborated the results observed in the oxidation with $Pb(OAc)_4$.

The identity of the 6- and 8-substituents was revealed in the

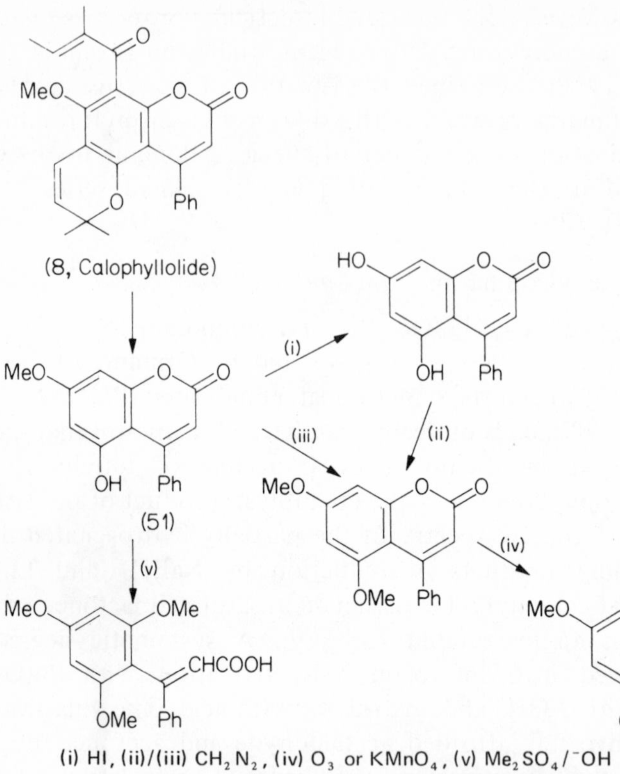

(i) HI, (ii)/(iii) CH_2N_2, (iv) O_3 or $KMnO_4$, (v) $Me_2SO_4/^-OH$

Scheme 15.1   Hydrolysis of calophyllolide with KOH

degradative products from acid hydrolysis of (8). Five compounds were identified: tiglic acid (52), 5-hydroxy-2,2-dimethyl-7,8-(4'-phenyl-α-pyrano)chroman (53, R=H), the methyl ether (53, R=Me), 5,7-dihydroxy-2,3-dimethylchromanone (54) and 5-hydroxy-2,3-dimethyl-7,8-(2',2'-dimethyldihydropyrano)chromanone (55) (Scheme 15.2).

The tiglic acid (52) arises from the acyl substituent. The coumarin (53, R=H), on fusion with base gave 5,7-dihydroxy-2,2-dimethyl-chroman (56), benzoic acid and oxalic acid. The structure of the methyl ether (53, R=Me) was confirmed by synthesis involving esterification of 5-hydroxy-7-methoxy-4-phenylcoumarin (51) with 3-methylcrotonyl chloride followed by a Fries rearrangement and subsequent reduction. The chromanone (54) was synthesized from phloroglucinol and tiglic acid (52) and the structure of compound

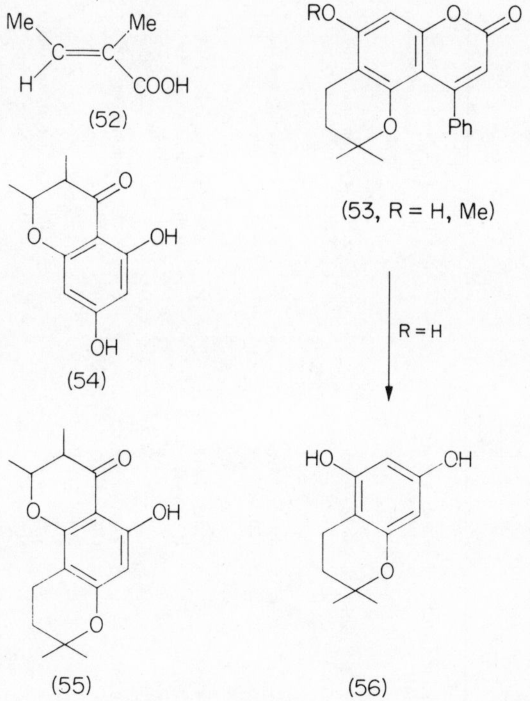

Scheme 15.2   Products of acid hydrolysis of calophyllolide

(55) was assigned on the basis of its IR and UV spectra. From the results of these degradations, structure (8) was assigned to calo-phyllolide. The NMR spectrum of calophyllolide was obtained later (Nigam *et al.*, 1967) and its assessment was in agreement with structure (8).

The chemical shift values and coupling constants for the naturally occurring 4-arylcoumarins have been tabulated (Table 15.2).

(±)- *and* (+)-*trans-Inophyllolide* (9), $C_{25}H_{22}O_5$; (+)-cis-*inophyllolide* (10) *and their corresponding alcohols* (11a,b,c), $C_{25}H_{24}O_5$. The similarity of (±)-inophyllolide (9) to calophyllolide (8), with which it co-occurred in the nuts of *C. inophyllum,* was remarked on by Polonsky (1957). The structure (9) was established by demethylation of the co-occurring calophyllolide (8) to give calophyllic acid (49) and reformation of the lactone ring gave inophyllolide (9).

In two papers Kawazu *et al.* (1968, 1972) described five optically

TABLE 15.2 The NMR spectra of coumarins from Guttiferae. τ-values (no of H, J. in Hz)

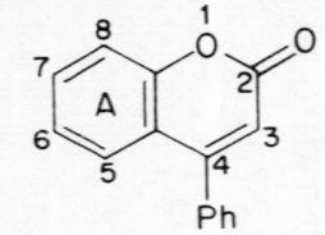

3-H	4-C₆H₅	5-OH	C-6 substituent					7-OH (tert-OH)	C-8 substituent					References
			1'-H	2'-H	2'-Me	3'-H	3'-Me		1'-H	2'-H	2'-Me	3'-H	3'-Me	
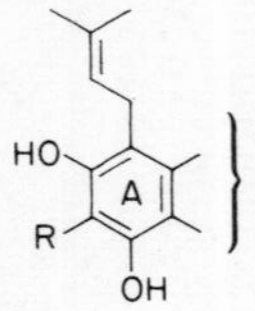 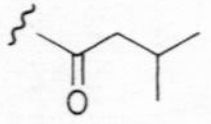 4.21	2.6m	−1.07	—	7.18d (2H,7)	—	7.80m	9.11d (6H,7)	0.20	6.56d (2H,7)	4.80m	—	—	8.15 8.30	Crombie et al.,1967; Finnegan and Mueller, 1965.
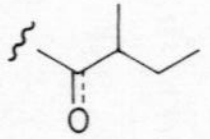 4.18	2.55	−0.93	—	6.35m	8.93d (7)	8.40m (2H)	9.18 (7)	0.10	6.51d (2H,7)	4.78m	—	—	8.12 8.28	Crombie et al., 1967.
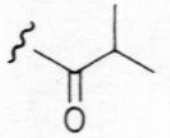 4.08	2.52	−0.94	—	6.25m	8.90d (6H,7)	—	—	0.36	6.45 (2H,7)	4.72m	—	—	8.12 8.25	Chakraborty and Das, 1966.
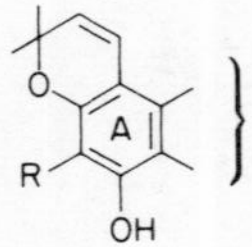 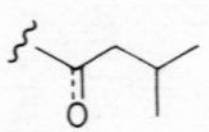 4.02	2.64	−4.75	—	7.07d (2H,7)	—	8.00m	9.07d (6H,7)	—	3.40d (12)	4.41d (12)	—	—	8.46 (6H)	Crombie et al., 1967; Finnegan and Mueller, 1965.
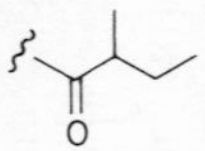 4.00	2.60	−4.72 −4.83	—	6.31m	8.82d (7)	8.40m (2H)	9.09t (7)	—	3.08d (10)	4.36d (10)	—	—	8.42 (6H)	Carpenter et al., 1971.
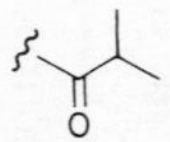 4.03	2.67	−4.63	—	6.27m	8.75 (6H,10)	—	—	—	3.12 (10)	4.38 (10)	—	—	8.42 (6H)	Chakraborty and Chatterji, 1969.
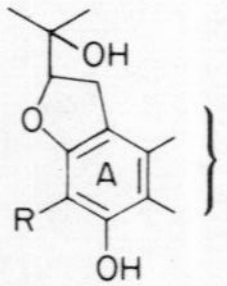 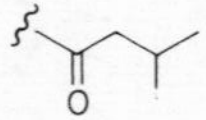 4.09	2.64	−4.46	—	7.02 (2H)	—	7.8m	9.04d (6H,7)	(8.03)	6.69 (2H,9)	5.09t (9)	—	—	8.57 8.69	Crombie et al., 1970, 1972a.
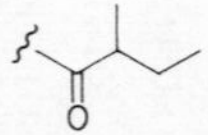 4.09	2.64	−4.60	—	6.35m	8.85dd (7,2)	8.2–8.4m (2H)	9.11t (7)	(8.03)	6.69d (2H,9)	5.09t	—	—	8.57 8.69	Ibid.; Carpenter et al., 1970.

Crombie et al., 1967.

Ibid.

Nigam et al., 1967.

Kawazu et al., 1968.
Nigam et al., 1967.

Kawazu et al., 1968.

Kawazu et al., 1972.

Ibid.

Ibid.

Structure													Reference
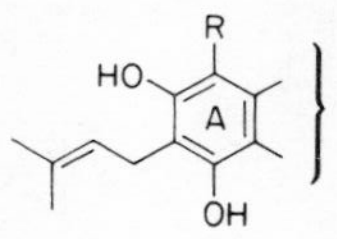 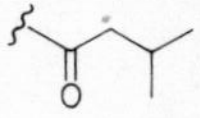 (isopentyl ketone)	4.21	2.57	—	6.81d (2H,7)	4.97m	—	8.37 / 8.32	−4.53	—	6.92 (2H,7)	—	7.8m	8.96d (6H,7)
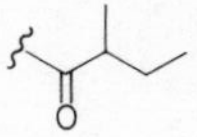 (methylbutyl ketone)	4.18	2.54	—	6.80d (2H,7)	4.90m	—	8.36 / 8.30	−4.51	—	6.13m	8.74d (7)	8.40m (2H)	8.98t (7)
R = OMe R' =	4.11	—	3.58d (10)	4.62d (10)	—	9.06 (6H)	—	—	—	—	8.05	3.50m	8.08d (7)
R,R' 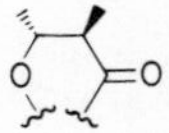 =	4.22	2.7	3.44d (10)	4.58d (10)	—	9.10 / 9.04	—	7.41 (7.8; 11.5)	8.76 (7.2)	—	5.68 (6.6; 11.5)	8.44 (6.6)
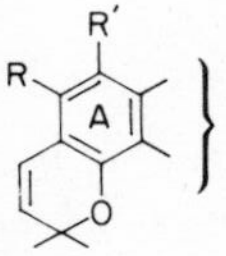 R,R' 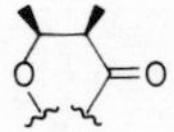 =	3.95	2.73	3.44d (10)	4.58d (10)	—	9.03	—	7.33 (7.2; 3.7)	8.82 (7.2)	—	5.27 (6.8; 3.7)	8.56 (6.8)
R,R' 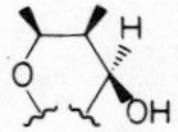 =	4.04	2.73	3.45d (10)	4.64d (10)	—	9.06	—	4.83d (5.4)	7.73m	8.83 (7.2)	5.57m (7.0; 3.3)	8.57d (7.0)
R,R' 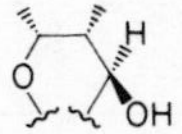 =	4.02	2.7	3.41d (10.2)	4.64d (10.2)	—	9.05	—	5.05d (2.0)	8.01m (7.2; 2.0 2.0)	9.17d (7.2)	5.41 (6.7; 2.0)	8.55d (6.7)
R,R' 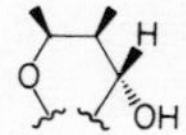 =	4.04	2.7	3.47d (10)	4.63d (10)	—	9.09 / 9.03	—	5.21d (7.4)	7.97m (7.0, 8.9, 7.4)	8.83d (7.0)	6.03 (6.8, 8.9)	8.53d (6.8)

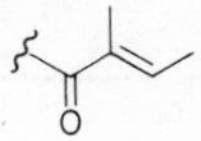

R = OMe

Continued

823

TABLE 15.2—Continued

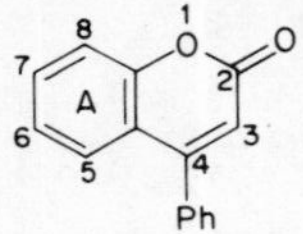

3-H	4-C₆H₅	5-OH	C-6 substituent					7-OH (tert-OH)	C-8 substituent					References
			1'-H	2'-H	2'-Me	3'-H	3'-Me		1'-H	2'-H	2'-Me	3'-H	3'-Me	
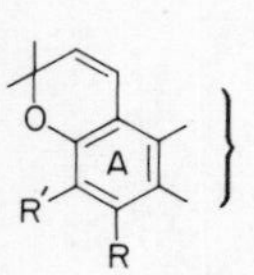 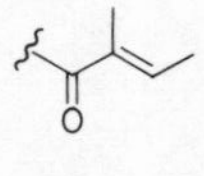 3.95					8.05	3.40m	8.08d (7)		3.50d (10)	4.33d (10)			8.62 (6H)	Nigam et al., 1967.
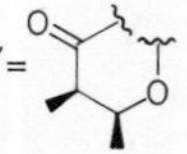 4.15				7.82m	9.00d (6)	6.22m	9.29d (6)		3.22d (10)	4.42d (10)			8.51 8.48	Ibid.
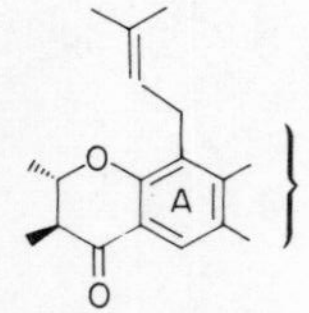 4.03	2.67	−2.8		7.74	8.80d (6)	5.75m (6, 11)	8.45d (6)	—	6.64d (7)	4.75t (7)			8.15 8.32	Breck and Stout, 1969.

$R' =$

$R, R' =$

824

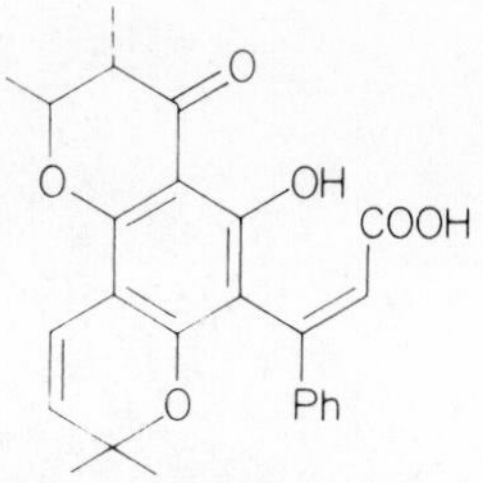

(49, Calophyllic acid) 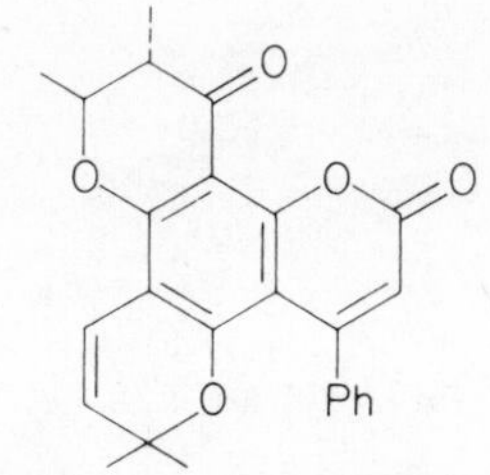 (9, *trans*—inophyllolide)

active piscicidal constituents (+)-(9), (10), (+)-(11a,b,c) which they isolated from the leaves of *C. inophyllum* The compounds (+)-(9) and (10) are isomeric and have m.p.s and rotations of 188–191°; $[\alpha]_D + 13°$ (CHCl$_3$), and 149–151; $[\alpha]_D + 70$ (CHCl$_3$) respectively. Their UV (Table 15.3) and IR spectra agree closely with spectral data published (Polonsky, 1957) for (±)-inophyllolide. The NMR spectra for (+)-(9) and (10) differ only in the chemical shifts of the methyl protons and of the protons in the chromanone ring, and in the splitting pattern of the latter. These observations indicated a *trans-cis* relationship for the chromanone methyl groups. A comparison of coupling constants (J 11.5 Hz for (9) and 3.7 Hz for (10)) reveals the relative stereochemistry (Table 15.2).

The three chromanols (11a,b,c), also present in the leaves of *C. inophyllum,* are isomeric and optically active $[\alpha]_D + 43°$ (acetone) (11a); and $[\alpha]_D + 35°$ (CHCl$_3$) (11b); $[\alpha]_D + 36$ (CHCl$_3$) (11c). The chromanol structures were assigned primarily on spectral evidence. The mass spectrum for (11a) has intense peaks due to the ions M−H$_2$O (m/e 386) and M−CH$_3$−H$_2$O (m/e 371), in addition to those observed in the spectra of compounds (9) and (10). Structure (11a) was assigned and confirmed by an oxidation with CrO$_3$-pyridine to yield *cis*-inophyllolide (10). Comparison of the NMR spectra of (11b) and (11c) with those of (11a), (9) and (10) and analysis of their oxidation products suggested that compound (11b) is the hydroxy analogue of (10) and the epimer of (11a), whilst compound (11c) is the hydroxy analogue of (9).

Dehydration of the compounds (11a) and (11c) with *p*-toluenesulphonic acid in benzene afforded the one anhydro product indicating a similar configuration for 3′-Me$_3$ of the 8-substituent. The chromanol ring has been encountered previously attached to the 4-*n*-propylcoumarin, costatolide (57) (Stout and Stevens, 1964).

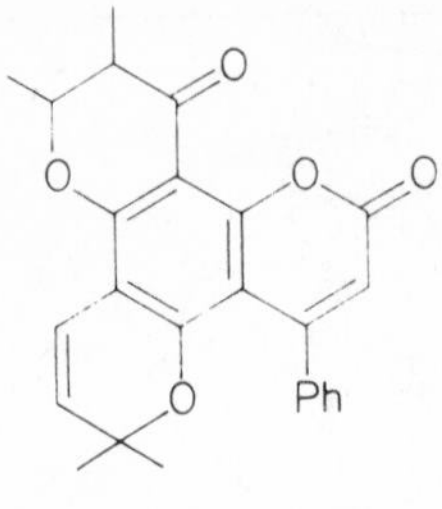

(10) 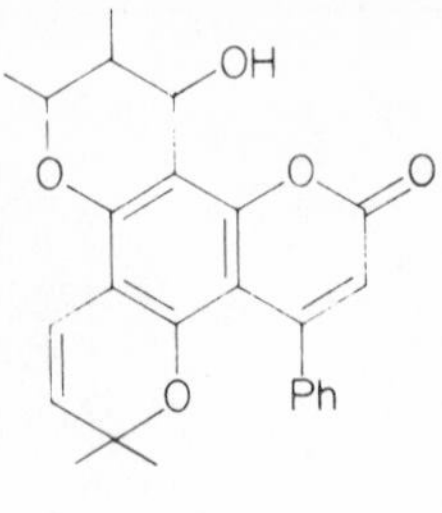   (11a) 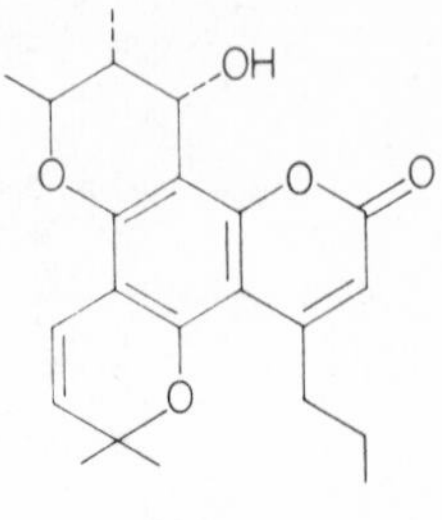   (57)

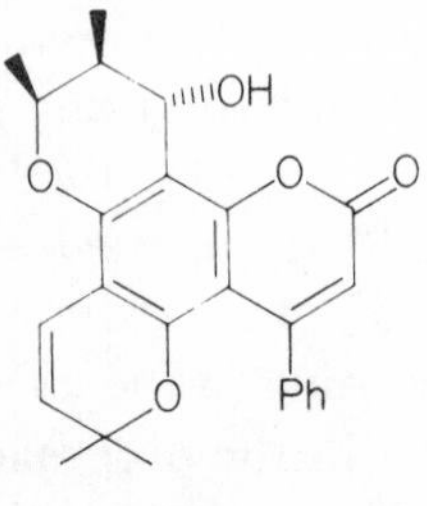

(11b) 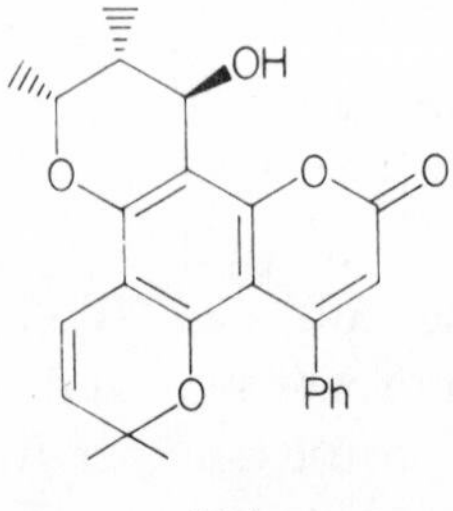   (11c)

Apetalolide (12), $C_{26}H_{24}O_5$ is an isomer of calophyllolide (8) and this was shown in a study of UV (Table 15.3), IR and mass spectral data (Nigam *et al.*, 1967). The methoxyl substituent (τ 6.97) is upfield from that registered for the corresponding group (τ 6.28) in the NMR spectrum of calophyllolide; whilst the signals due to the *gem*-dimethyl groups (τ 9.06) in calophyllolide (8) and in apetalolide (12) (τ 8.62) are shifted downfield. This emphasizes the relative positions of the methyls to the shielding effect of the 4-phenyl group (Table 15.2). On this evidence apetalolide was assigned structure (12).

Tomentolide A (13), $C_{25}H_{22}O_5$ is optically inactive and isomeric with inophyllolide (9). Both coumarins have a similar fragmentation pattern in their mass spectra. Tomentolide A has an intense peak at *m/e* 387 (M-15) which is considered due to the ready formation of a stable benzopyrylium ion from the 2,2-dimethylchromene. The NMR data (Table 15.2) provide the evidence for assignment of structure (13) for the compound. The signals for the quaternary methyl groups (τ 8.51 and 8.48) show no shielding by the 4-phenyl substituent as in inophyllolide (9) (τ 9.10 and 9.04) whilst the methyl groups of the pyranone ring show higher field signals (τ 9.0 and 9.29) than the corresponding pair in inophyllolide (τ 8.42 and 8.79) (Nigam *et al.*, 1957).

5-Hydroxy-10(3-methylbut-2-enyl)-7,8-trans-dimethyl-4-phenyl-2H, 6H-benzo-3,4-dehydrodipyran-2,6-dione (14), $C_{25}H_{24}O_5$ is present in the bark of *C. australianum*. The structure (14) for the coumarin is based on a comparison of the UV, IR and NMR spectra with those of the known coumarins (9), (13), (15) and (16) (see Tables 15.2, 15.3). The NMR spectrum of the compound (14) has signals due to a *trans*-2,3-dimethylchromanone, an isopentenyl chain, a chelated hydroxyl and a phenyl ring. The carbonyl absorptions at 1718 cm^{-1} and 1645 cm^{-1} are due to the stretching frequency of the group in a coumarin structure and in a chromanone with an adjacent hydroxyl group respectively. The UV spectrum is similar to that of a number of 6-acyl-5,7-dihydroxy-4-phenylcoumarins. The presence of the chelated hydroxyl rules out an 8-acyl substituent. The normal values (τ 8.45 and 8.80) recorded for the methyl groups of the pyrone ring in the NMR spectrum may be accounted for in structure (14) only (Breck and Stout, 1969).

15.3.1.2 4-Phenylcoumarins from *Mesua* and *Mammea* species

The 4-phenylcoumarins isolated from species of the genera *Mesua* and *Mammea* may be divided into two groups (Tables 15.4 and 15.5). This arbitrary division is based on the position (6- or 8-) of the acyl substituents (Crombie *et al.*, 1967). The 6-acylcoumarins (15)–(24) are yellow compounds, give an olive green colour with ferric chloride, contain a hydrogen-bonded hydroxyl (ν3480 cm^{-1} (C < 0.001M in CCl$_4$)) and display characteristic UV shifts to longer wavelengths in alkaline solution (Table 15.3). The two 8-acyl-coumarins (25) and (26) are colourless compounds which give a brown to purple colour with ferric chloride, contain a hydrogen-bonded hydroxyl (ν3500 cm^{-1} (C < 0.001M in CCl$_4$)) and have different UV shifts in alkaline solution to the coumarins (15)–(24).

Crombie and co-workers (1967) have shown that the 8-acyl-4-phenylcoumarins (25) and (26) when treated with methanolic potassium hydroxide give mammeisin (15) and 5,7-dihydroxy-8-(3-methyl-but-2-enyl)-6-(2-methylbutyryl)-4-phenylcoumarin (16) respectively. These rearrangements were incomplete but the equilibrium favours a

Table 15.3 UV data for natural 4-arylcoumarins

Compound	Solvent	λmax (nm) (log ε)					References
(15)	EtOH	234 (4.20)	283 (4.39)	330 (3.95)	394 (3.94)	428 (4.07)	Crombie et al., 1967.
	0.1N HCl						
	0.1N KOH	238 (4.27)	302 (4.21)				
(16)	EtOH	233 (4.14)	283 (4.47)	333 (4.01)	394 (3.96)	428 (4.11)	Ibid.
	0.1N HCl						
	0.1N KOH	238 (4.35)	293 (4.31)				
(17)	EtOH	234 (4.17)	284 (4.37)	337 (3.97)	394 (3.84)	426 (4.03)	Ibid.
	0.1N HCl						
	0.1N KOH	234 (4.33)	303 (4.21)				
(18)	EtOH	231 (4.56)	252 (4.40)	285 (4.56)	361 (3.83)		Ibid.
	0.1N HCl						
	0.1N KOH	225 (4.45)	312 (4.36)			417 (3.76)	
(19)	0.1N HCl	233 (4.36)	286 (4.46)	335 (3.76)			Carpenter et al., 1970.
	0.1N KOH	250 (4.30)	310 (4.31)			410 (3.74)	
(20)	EtOH	235 (4.31)	285 (4.40)	362 (3.79)			Chakraborty and Chatterji, 1969.
(21)	EtOH	232 (4.10)	249 (4.21)	282 (4.39)	348 (3.99)		Crombie et al., 1972a
	0.1N HCl						
	0.1N KOH			316 (4.04)		430 (3.98)	
	EtOH						

No.	Solvent						Reference
(22)	0.1N HCl 0.1N KOH EtOH	228 (4.10)	247 (4.20)	280 (4.38) 312 (4.05)	347 (3.97)	425 (3.96)	*Ibid.*; Carpenter *et al.*, 1971.
(25)	0.1N HCl 0.1N KOH	225 (4.46) 233 (4.40)	261 (4.19)	294 (4.36)	332 (4.25) 337 (4.60)		Crombie *et al.*, 1967
(26)	EtOH 0.1N HCl 0.1N KOH	227 (4.45) 234 (4.39)	263 (4.19)	294 (4.38)	333 (4.24) 337 (4.58)		*Ibid.*
(8)	Alcohol	235 (4.47)	270 (4.27)	295 (4.23)	325 sh.		Ormancey-Potier *et al.*, 1951.
(10)	EtOH	240 sh(4.32)	257 sh(4.46) 266 (4.49)	302 (4.39)			Kawazu *et al.*, 1968.
(11a)	EtOH	235 (4.18)	280 (4.15)	286 (4.18)	337 (3.95)		Kawazu *et al.*, 1972.
(11b)	EtOH	232 (4.23)	277 sh(4.15)	286 (4.2)	334 (3.98)		*ibid.*
(11c)	EtOH	233 (4.25)	277 sh(4.18)	287 (4.23)	335 (4.0)		*ibid.*
(12)		236 (4.55)		275 (4.40)			Nigam *et al.*, 1967.
(13)		237 (4.63)		275−280 (4.59)	350 (3.9)		*Ibid.*
(14)	EtOH			286 (4.2)	337 (3.95)		Breck and Stout, 1969.
(28)	MeOH	237 (4.23)	260 (4.05)	301 (3.85)	355 (4.0)		Donnelly *et al.*, 1973a.
(30)	MeOH	222 sh(4.74)		284 (4.14)	340 (4.05)		
(32)	NaOMe EtOH	234 (4.38)	252 (4.49) 256 (4.17)	312 (3.97) 308 (4.00)	344 (4.09)	397 (4.08)	Donnelly, B. J. *et al.*, 1968.

6-acyl isomer. Interestingly, Chakraborty and Das have reported (1966) otherwise for mesuol (17).

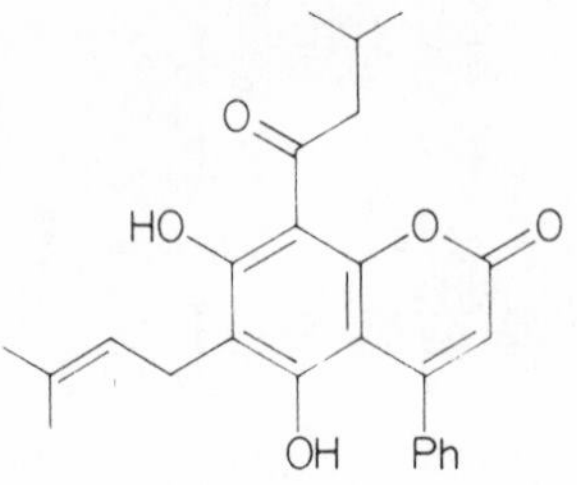

(25) 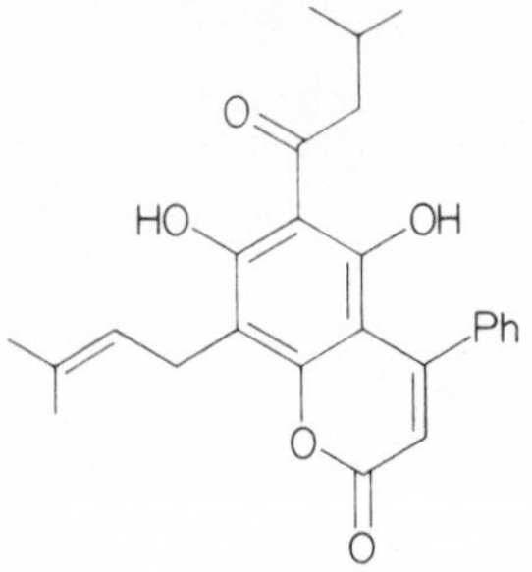   (15)

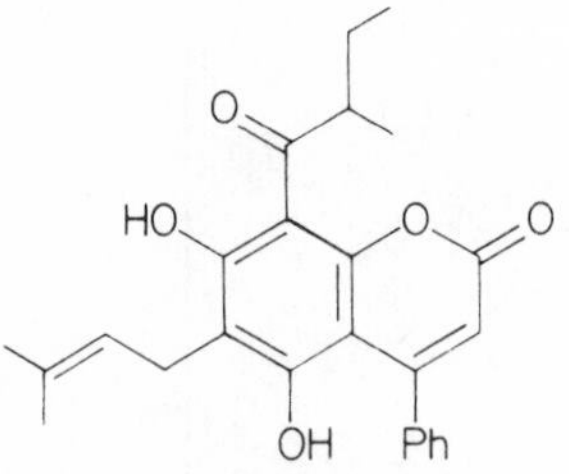

(26) 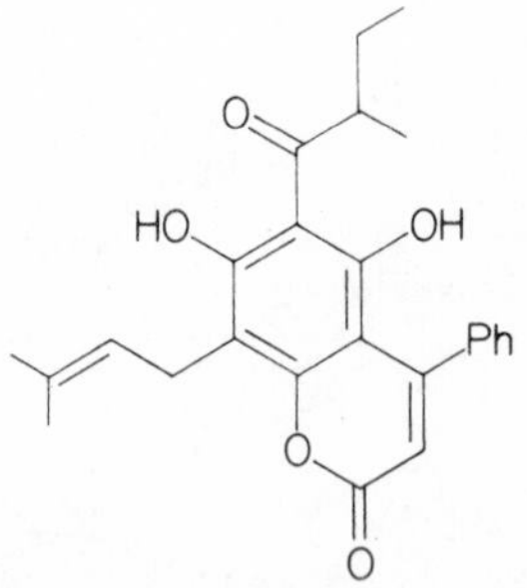   (16)

The assignment of the orientation of the 6- and 8-substituents in the coumarins (15)–(20) was based on some chemical degradative work but mainly on a combination of spectroscopic data. Model compounds were synthesized (Crombie *et al.*, 1967) and the orientation of the acyl substituent was confirmed by application of the Gibbs reaction by a spectroscopic method. 5,7-Dihydroxy-6-(3-methylbutyryl)- and 5,7-dihydroxy-6-(2-methylbutyryl)-4-phenyl-coumarins show clear maxima at about 614 nm whilst the corresponding 8-substituted isomers have no maxima in this region. The second valuable criterion for the assignment of orientation of the 6- and 8-acyl isomers is the difference in the hydroxy resonance patterns in their NMR spectra.

Mammeisin (15), 5,7-dihydroxy-8-(3-methylbut-2-enyl)-6-(2-methyl-butyryl)-4- phenylcoumarin (16), and Mesuol (17).
The 4-phenylcoumarins (15), (16) and (17) possess a 6-(3-methyl-

butyryl), 6-(2-methylbutyryl) and 6-(isobutyryl) substituent respectively attached to a common 5,7-dihydroxy-8-(3-methylbut-2-enyl)-4-phenylcoumarin system.

Initially mammeisin (15) was considered to be the toxic principle in the fruit peelings of *M. americana* L. (Finnegan *et al.*, 1961) from which it was first isolated. Later, its presence was observed in seeds of the same plant (Crombie *et al.*, 1966, 1967) and in the bark of *M. africana* (Carpenter *et al.*, 1970, 1971). Crombie and co-workers (1972b) reported mammeisin as insecticidally inactive. Degradative studies (Scheme 15.3) were used in the structural proof of mammeisin (15) (Finnegan *et al.*, 1961). Facile hydrogenation of the coumarin with an uptake of 1 mole gave a yellow dihydride (58), with a UV spectrum similar to the parent compound; therefore an unconjugated olefinic link is present. Acetone was obtained on ozonolysis, and alkaline degradation afforded acetophenone, isovaleric acid, isopentenylphloroglucinol and 5,7-dihydroxy-8-isopentenyl-4-phenylcoumarin (59). Methylation and subsequent hydrogenation of (59) gave 5,7-dimethoxy-8-isopentyl-4-phenylcoumarin (60, R=Me) identical with the product of a Pechmann condensation of 3,5-dimethoxy-2-isopentylphenol with benzoyl acetic ester.

These results are consistent with the structure (15) or its isomer (25). Structure (15) was considered more likely from the analysis of the alkaline shift in the UV spectrum (Crombie *et al.*, 1967) (Table 15.3). Unambiguous proof for the assigned structure (15) was obtained from a deacylation reaction with 75% H_2SO_4. The dihydride (58) and H_2SO_4 gave the coumarin (60, R=H). Mammeisin on acid treatment underwent deacylation and cyclization and gave the coumarin (61).

The assignment of the structure 5,7-dihydroxy-2-(3-methylbut-2-enyl)-6- (2-methylbutyryl)-4-phenylcoumarin (16) is based solely on spectroscopic evidence (Crombie *et al.*, 1966, 1967). It gives an alkaline shift in the UV expected of a 6-acylcoumarin (Table 15.3). A comparative analysis of the NMR spectrum with that of mammeisin (Table 15.2) shows differences only in the nature of the acyl substituent. Both hydroxyl groups (τ −1.0 and 0.1) are bonded to the acyl function and disappear on deuteration. The 5-hydroxyl group in the 6-acylcoumarin series (Table 15.4) causes the 4-phenyl substituent to be twisted out of plane, and this has been envisaged (Crombie *et al.*, 1972a) as bringing the energies of the two forms (62a,b) closer together.

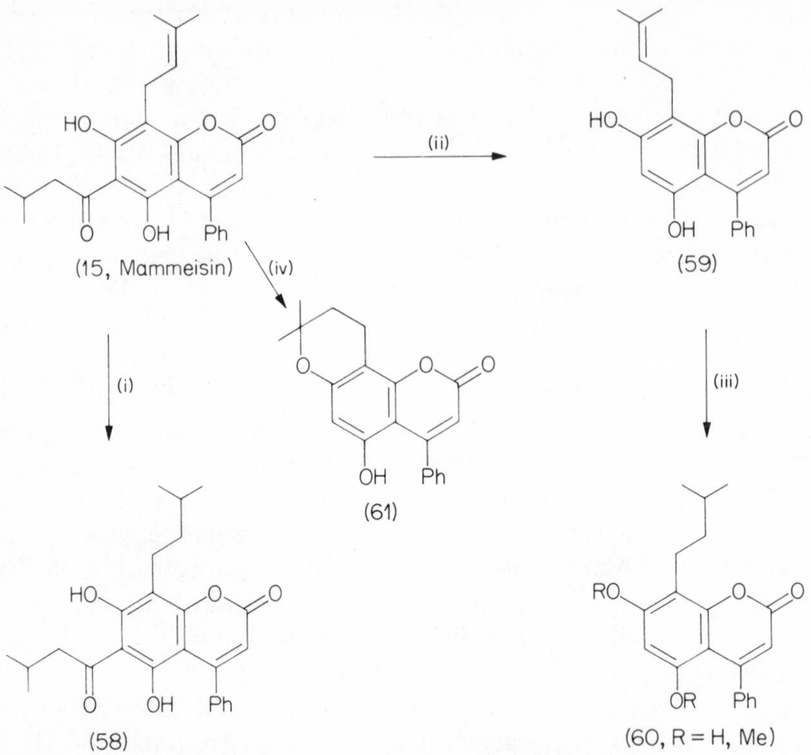

(i) H$_2$ Pd/C, (ii) KOH, (iii) Me$_2$SO$_4$/$^-$OH, H$_2$, Pd/C, (iv) 75%H$_2$SO$_4$

Scheme 15.3   Degradation of mammeisin.

Mammeisin (15) and the coumarin (16) are chromatographically similar but show differences in their IR spectra in the range 1100–1250 cm^{-1} (Carpenter et al., 1971). A similar electron impact fragmentation pattern was obtained for both compounds. The main fragmentation path is loss of a butyl radical (M-57) from the acyl side chain and the subsequent elimination of a butene molecule from the 3-methylbut-2-enyl substituent. The butenyl substituent loses a propyl radical (M-43) and β-fission gives rise to the ion (M-55). Low abundance of this ion is used as diagnostic marker for the presence of a 2-methylbutyryl substituent. The observation of this consistent reduction in abundance of this ion for coumarins with 2-methyl-butyryl group was based on a study of model compounds and analogous 4-*n*-propylcoumarins (Crombie et al., 1967).

The third coumarin in this sub-group, mesuol (17), was assigned a

Table 15.4   *6-Acyl-4-phenylcoumarins from species of* Mesua *and* Mammea

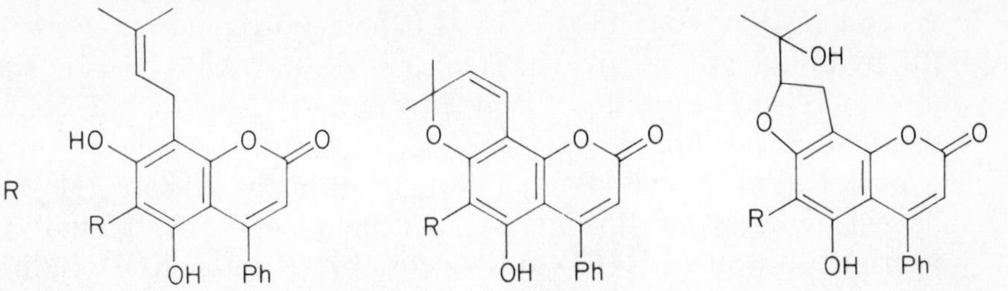

R			
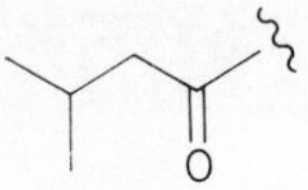 ![isovaleryl]	$C_{25}H_{26}O_5$ (15) 83–84°; 98–109°[a]	$C_{25}H_{24}O_5$ (18) 144–146°[e]	$C_{25}H_{26}O_6$ (21) 115–117°[g]; 148–150°[h]
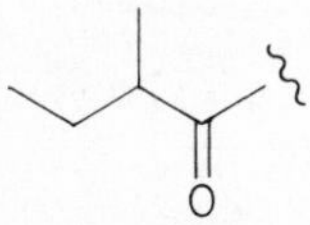 ![2-methylbutyryl]	$C_{25}H_{26}O_5$ (16) 107–108°[b,c]	$C_{25}H_{24}O_5$ (19) 78–80°[c]	$C_{25}H_{26}O_6$ (22) 115–117°[g]
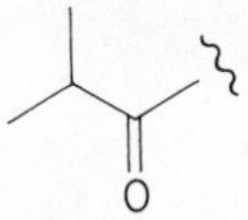 ![isobutyryl]	$C_{24}H_{24}O_5$ (17) 154°[d]	$C_{24}H_{22}O_5$ (20) 152–153°[f]	$C_{24}H_{24}O_6$ (23) M⁺ 408
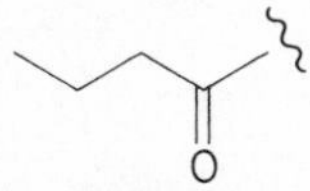 ![butyryl]	–	–	$C_{24}H_{24}O_6$ (24) M⁺ 408

[a]Finnegan *et al.,* 1961; [b]Crombie *et al.,* 1967; [c]Carpenter *et al.,* 1970, 1971; [d]Chakraborty and Das, 1966; [e]Finnegan and Mueller, 1965; [f]Chakraborty and Chatterji, 1969; [g]Crombie *et al.,* 1972a; [h]Merkel, 1970.

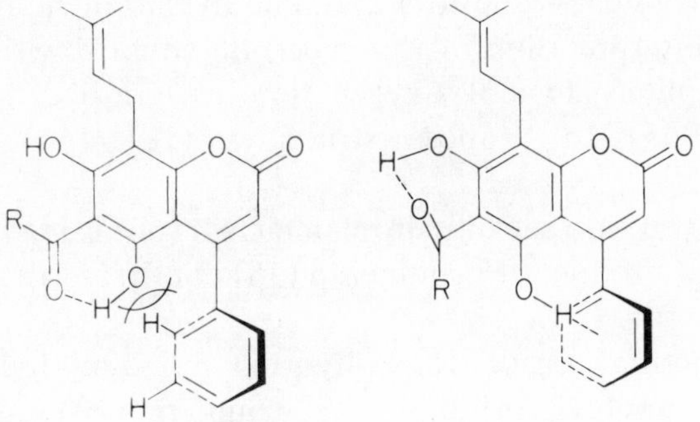

(62)

partial structure (Chakraborty and Bose, 1960) on results of a degradation. The correct structure was established by analysis of its mass and NMR spectra (Table 15.2) (Chakraborty and Das, 1966). Both hydroxyl groups are chelated (τ-0.94 and 0.36) placing the isobutyryl chain at position 6 and therefore the 3-methylbut-2-enyl substituent at position-8. The mass spectrum confirms the structural assignment of (17). The loss of a propyl radical gives the ion (M-43), followed by loss of a butene molecule (M-43−56). Mesuol is isomerized to isomesuol (63) by aqueous or methanolic KOH. In this rearranged coumarin (63) only one of the two hydroxyls (τ-4.7 and 4.01) is chelated to the acyl substituent.

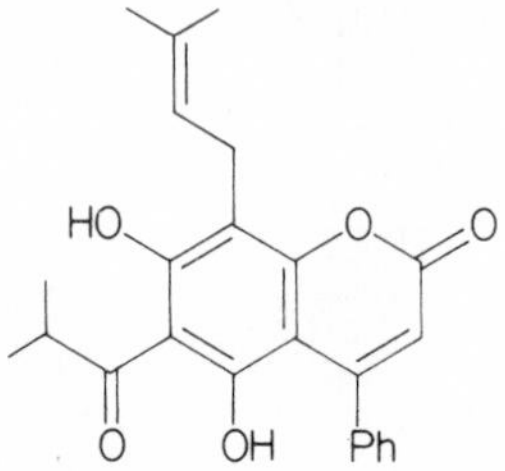

(17, Mesuol) 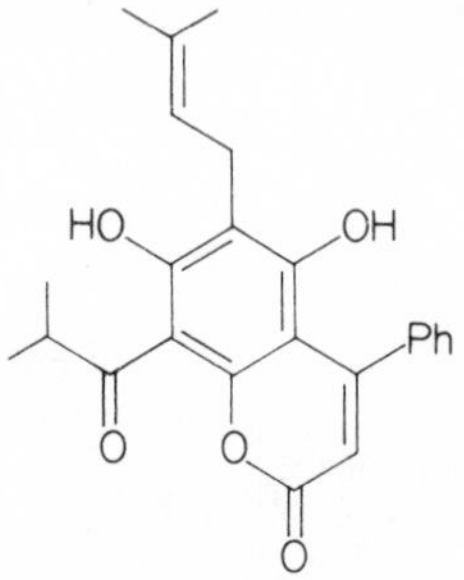   (63, Isomesuol)

Mammeigin   (18),   5-hydroxy-6,6'-dimethyl-6-(2-methylbutyryl)-4-phenylpyrano [2',3' : 7,8] coumarin (19) and mesuagin (20).

The IR and UV spectra of mammeigin (18) show that it is structurally related to mammeisin (15). However, unlike that for mammeisin, the NMR spectrum of mammeigin shows no isopentyl group but has signals due to a 2,2-dimethylchromene moiety (Table 15.2). The interpretation of this spectrum coupled with the assumption of a phloroglucinol oxygenation pattern led Finnegan and Mueller (1965) to propose structures (18), (64) or (65) for mammeigin.

The observation that dihydromammeigin (66) is identical with the acid cyclized product of mammeisin (15) confirmed that mammeigin had the structure (18).

The structural proof for 5-hydroxy-6,6'-dimethyl-6-(2-methylbutyryl)-4-phenylpyrano [2',3' : 7,8] coumarin (19) was based solely on a comparison of its spectroscopic data with those of known 4-phenylcoumarins (Carpenter *et al.*, 1971). The NMR spectrum has

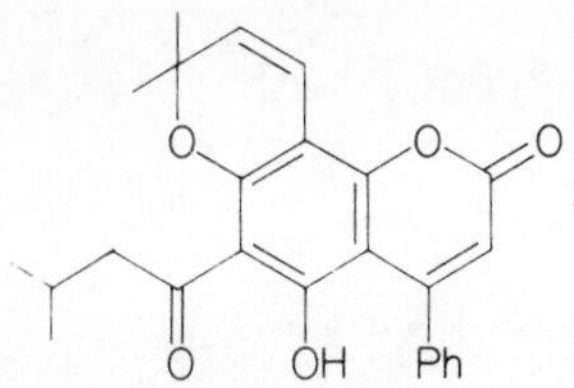

(18, Mammeigin)

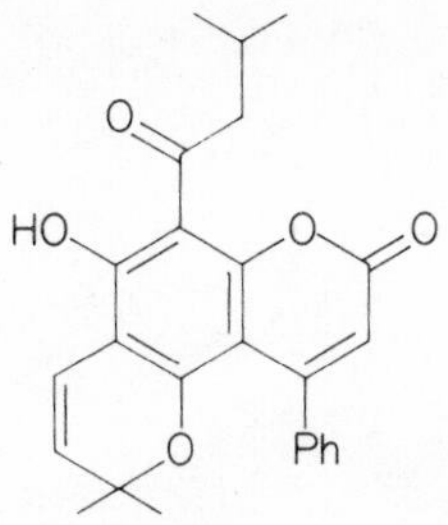

(64)

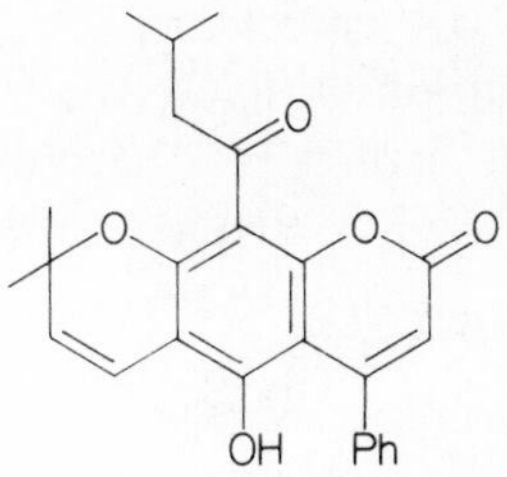

(65)

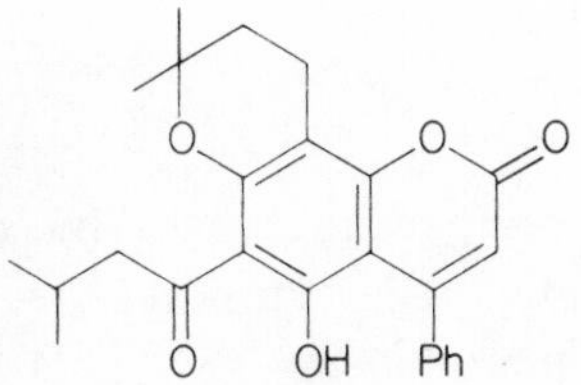

(66)

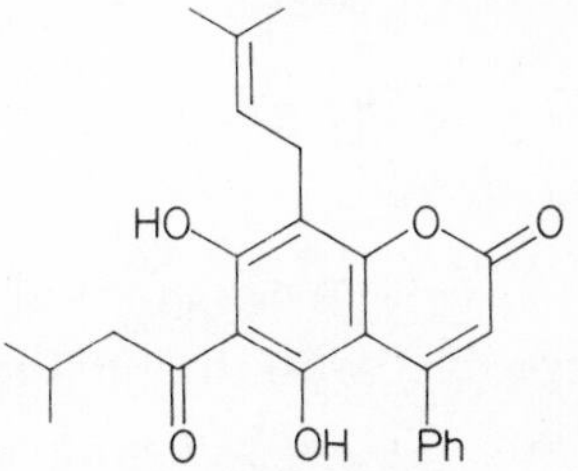

(15, Mammeisin)

signals due to a phenyl ring, a chelated hydroxyl, a 2-methylbutyryl substituent and two doublets at τ 3.08 and 4.36 consistent with the presence of a 2,2-dimethylpyran ring (Table 15.2). On the basis of the UV spectrum, the 2-methylbutyryl group is placed at position 6 and a base peak of M-15 in the mass spectrum characterises the 2,2-dimethylpyran system. Games (1972) has shown by GC—MS that this extractive contains a trace of the isomeric 6-(3-methylbutyryl)-pyranocoumarin.

Mesuagin (20) co-occurs with mammeigin (18) in the extractives of *Mesua ferrea* L. Its structural elucidation was described by Chakraborty and Chatterji (1969). Alkaline degradation of the compound furnished acetone, acetophenone, isobutyric acid and 5,7-dihydroxy-4-phenylcoumarin. The acetone and isobutyric acid arise from the 2,2-dimethyl-Δ^3-pyran ring and an isobutyryl residue respectively. The Δ^3-pyran ring was characterized by the NMR spectrum [τ 8.42 s, 6H; 4.38, d (1H) and 3.12 d (1H] and by the high intensity (M-15) peak in the mass spectrum. The ring fusion was considered to be at position -7,8- rather than position -5,6- as the two methyl groups register as a sharp non-shielded singlet in the

NMR spectrum. A comparison was made with the spectra of calophyllolide (8) and inophyllolide (9). To confirm the assignment of structure (20) for mesuagin, it was deacylated with acid and shown to give the coumarin (67).

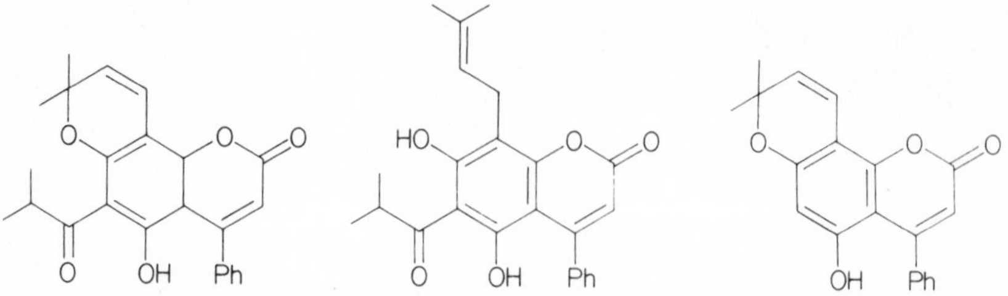

(20)   (17)   (67)

Recently Bala and Seshadri (1971) obtained mesuagin by oxidative cyclisation of mesuol with DDQ. This facile reaction was used also in obtaining mammeigin (18) from mammeisin (15) (Bala and Seshadri, 1971; Carpenter *et al.*, 1971; Finnegan and Merkel, 1972).

4-Phenylcoumarins of Mammea americana
A fraction from the seed oil of *M. americana* L. extract contained a mixture of 1,2-dihydro-5-hydroxy-2-(1-hydroxy-1-methylethyl)-4-(3-methylbutyryl)-6-phenylfuro {2,3-h} {1} benzopyran-8-one (21), 1,2-dihydro-5-hydroxy-2-(1-hydroxy-1-methylethyl)-4-(2-methylbutyryl)-6-phenylfuro {2,3-h} {1} benzopyran-8-one (22), 1,2-dihydro-5-hydroxy-2-(1-hydroxy-1-methylethyl)-4-(isobutyryl)-6-phenylfuro {2,3-h} {1} benzopyran-8-one (23) and 1,2-dihydro-5-hydroxy-2-(1-hydroxy-1-methylethyl)-4-(butyryl)-6-phenylfuro {2,3-h} {1} - benzopyran-8-one (24). (Crombie *et al.*, 1970, 1972a). A separation resulted in predominance of either (21) or (22) in each fraction. The IR and UV spectra of these fractions established that they had the 6-acylhydroxy-4-phenylcoumarin skeleton. The NMR spectra (Table 15.2) had the proton relationships confirmed by double resonance whilst the mass spectral fragmentations for both compounds (21) and (22) were consistent with an α-(hydroxyisopropyl)dihydrofuran system (Scheme 15.4).
    These proposed structures (21) and (22) were proved correct

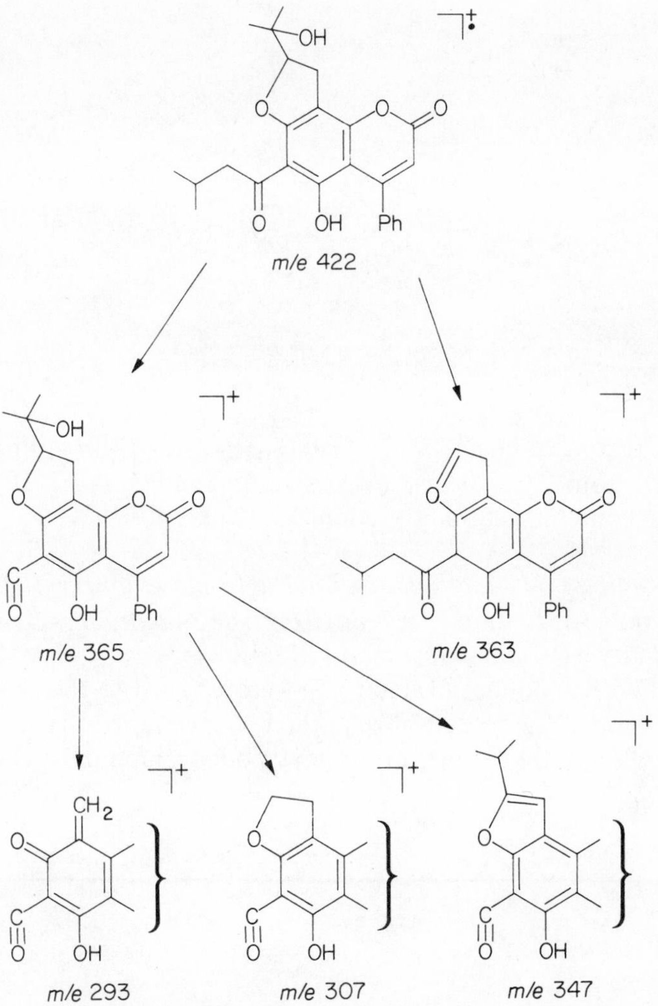

Scheme 15.4   Mass spectral fragmentation of coumarin (21)

by their similarity with the products from reaction of mammeisin (15) and 5,7-dihydroxy-6-(2-methylbutyryl)-8-(3-methylbut-2-enyl)-4-phenylcoumarin (16) with *m*-chloroperbenzoic acid respectively. Acid catalyzed dehydration of the 6-(3-methylbutyryl)coumarin (21) gave the known benzofuran (68).

The two remaining coumarins (23) and (24) were identified only by mass spectrometry (Crombie *et al.*, 1972a). It has been shown by

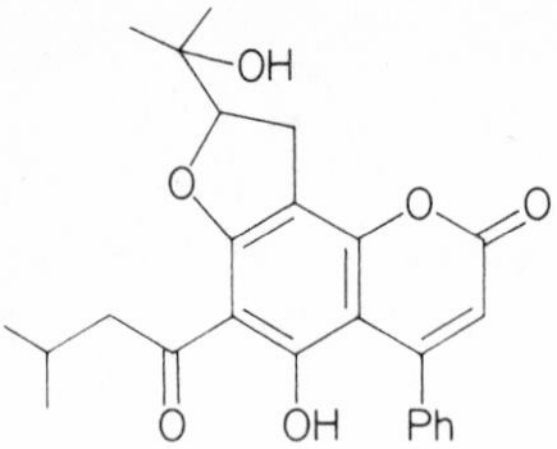

GC-MS (Crombie *et al.*, 1972a; Games, 1972) that the coumarin (22) isolated from *M. africana* (MAB-5) (Carpenter *et al.*, 1971) is a mixture containing a trace of coumarin (21).

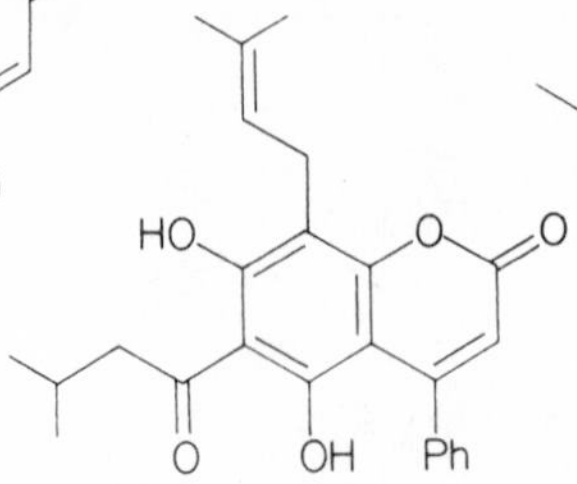

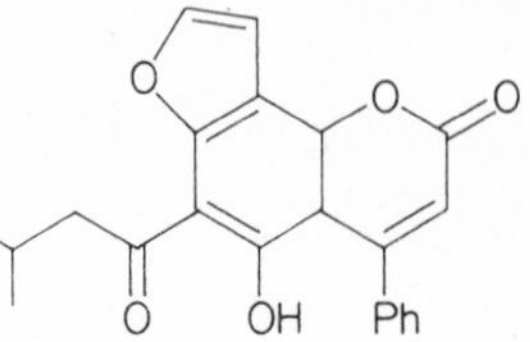

5,7-Dihydroxy-6-(3-methylbut-2-enyl)-8-(3-methylbutyryl)-4-phenyl-coumarin (25) and 5,7-dihydroxy-6-(3-methylbut-2-enyl)-8-(2-methylbutyryl)-4-phenylcoumarin (26).
These two compounds (Table 15.5) were assigned to the 8-acyl series on the basis of the resemblance in the UV spectra in ethanol and base to the UV spectra of authentic 8-acyl model compounds (Crombie *et al.*, 1967).

Table 15.5   *8-Acyl-4-phenylcoumarins from* M. americana

R	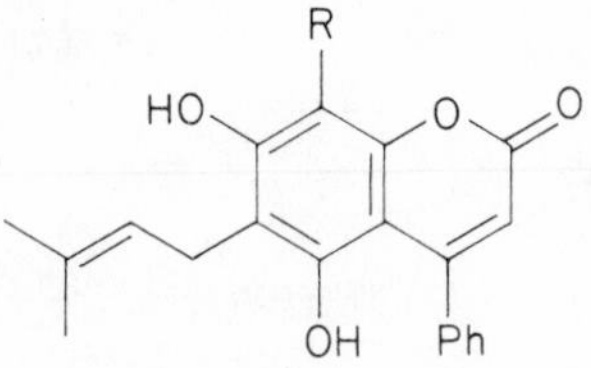	
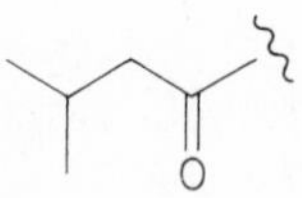	$C_{25}H_{26}O_5$ 125–126°	(25)
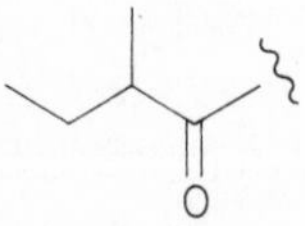	$C_{25}H_{26}O_5$ 124–125°	(26)

The 7-hydroxyl group in each compound (25) and (26) shows a resonance in the spectra due to chelation (τ −4.5) and a signal due to the unchelated 5-hydroxyl at τ 4.1. The mass spectra of these compounds were similar to mammeisin (15) and 5,7-dihydroxy-6-(2-methylbutyryl)-8-(3-methylbut-2-enyl)-4-phenylcoumarin (16). Base catalyzed rearrangement of the coumarins (25) and (26) gave the 6-acyl isomers (15) and (16).

15.3.2 Exostemin (Rubiaceae)

Studies on the structural assignment of the only 4-arylcoumarin, present in the Rubiaceae, namely exostemin, $C_{18}H_{16}O_6$ from *Exostemma caribaeum*, have had a chequered history. In a preliminary report, Sánchez-Viesca *et al.* (1967) proposed 8-hydroxy-5,7-dimethoxy-4-(4-methoxyphenyl)-coumarin (27) as the tentative structure. The proposal was based on the analytical and spectral data of the phenol, its derivatives and on a product (69) of CrO_3 oxidation.

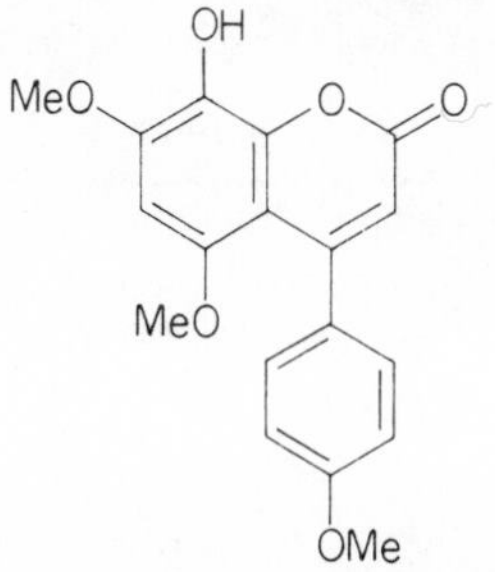

(27, Exostemin) 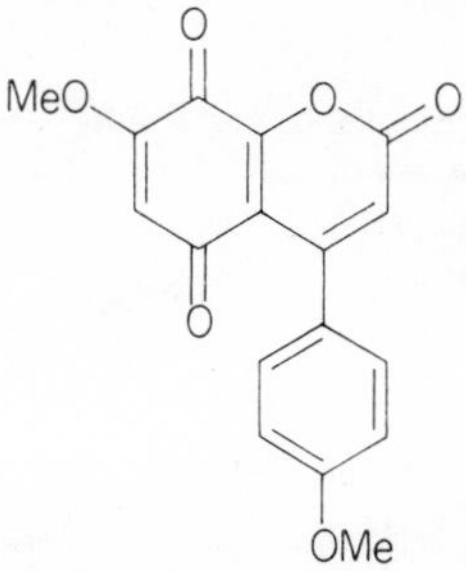   (69)

The alternative *o*-quinone structure for compound (69) was discounted on the basis of a negative reaction with guanidine carbonate. Later the validity of the structure (27) was questioned (Mukerjee *et al.*, 1968) as the melting points of the synthesized phenol and its acetate were at variance with those of the natural product and its acetate. The synthesis of the phenol was achieved by Pechmann condensation of 1,2-dihydroxy-4,6-dimethoxybenzene and *p*-methoxybenzoyl acetic ester. A second paper (Sánchez-Viesca, 1969) claims confirmation of structure (27) for exostemin and corrects the anomaly in melting point of the coumarin. Similarity in IR and NMR spectra are quoted in support of the assignment. However, the possibility of a 5,6,7-arrangement of substituents has not been conclusively excluded.

15.3.3 4-Arylcoumarins of the Leguminosae

To date, nine 4-arylcoumarins have been isolated from species of the genera *Dalbergia* and *Macherium* (Table 15.6). The 4-arylcoumarins with the exception of sisafolin, isolated from seeds of *D. latifolia* (Saxena *et al.*, 1970), have a distinctive oxygenation pattern in the 6,7-positions. Nordalbergin (28, $R = R_1 = H$), known synthetically for many years, was isolated recently from the wood bark of *D. sissoo* (Mukerjee *et al.*, 1971). It was suggested that this 6,7-dihydroxy-4-arylcoumarin is the likely precursor of the co-occurring isodalbergin (28, R=Me, R_1=H), dalbergin (28, R=H, R_1=Me) and methyl-dalbergin (28, $R = R_1$ =Me).

Methyldalbergin was the first of this group of natural products to be degraded chemically. Oxidation with neutral $KMnO_4$ gave benzoic acid, oxalic acid and 2-hydroxy-4,5-dimethoxybenzophenone. Confirmation of the suggested structure for dalbergin (28, R=H, R_1 =Me) was obtained by partial methylation of 4-phenylaesculetin (Ahluwalia and Seshadri, 1957).

Table 15.6　*4-Arylcoumarins of the Leguminosae*

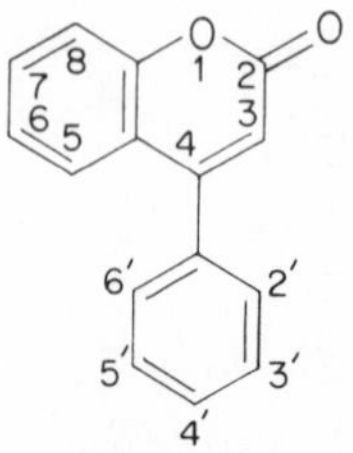

Compound	m.p. (°C)	Mol. formula	Substituents								
			5	6	7	8	2'	3'	4'	5'	6'
dalbergin (28)	212–214	$C_{16}H_{12}O_4$	H	OH	OH	H	H	H	H	H	H
methyldalbergin (28)	142–143	$C_{17}H_{14}O_4$	H	OMe	OMe	H	H	H	H	H	H
isodalbergin (28)	195–196	$C_{16}H_{12}O_4$	H	OMe	OH	H	H	H	H	H	H
nordalbergin (28)	268–269	$C_{15}H_{10}O_4$	H	OH	OMe	H	H	H	H	H	H
kuhlmannin (29)	211	$C_{17}H_{14}O_5$	H	OH	OMe	OMe	H	H	H	H	H
stevenin (30)	250–254 (dec.)	$C_{16}H_{12}O_5$	H	OH	OMe	H	H	OH	H	H	H
melannin (31)		$C_{16}H_{12}O_5$	H	OH	OMe	H	H	H	OH	H	H
melannein (32)	221–223	$C_{17}H_{14}O_6$	H	OH	OMe	H	H	OH	OMe	H	H
sisafolin (33)	259–260	$C_{18}H_{14}O_7$	OH	CHO	OMe	H	OMe	H	OH	H	H

An interesting transformation of methyldalbergin (28, R= R$_1$ =Me) to the corresponding coumarilic acid (71) was achieved with mercuric oxide in alkali. The formation of β-phenylcoumaric acid (7) as an intermediate in the reaction was considered (Ahluwalia *et al.*, 1958). The coumarilic acid on decarboxylation gave 3-phenylcoumarone (72). The latter compound was obtained by alkali treatment of 3-bromomethyldalbergin and subsequent decarboxylation of the intermediate acid.

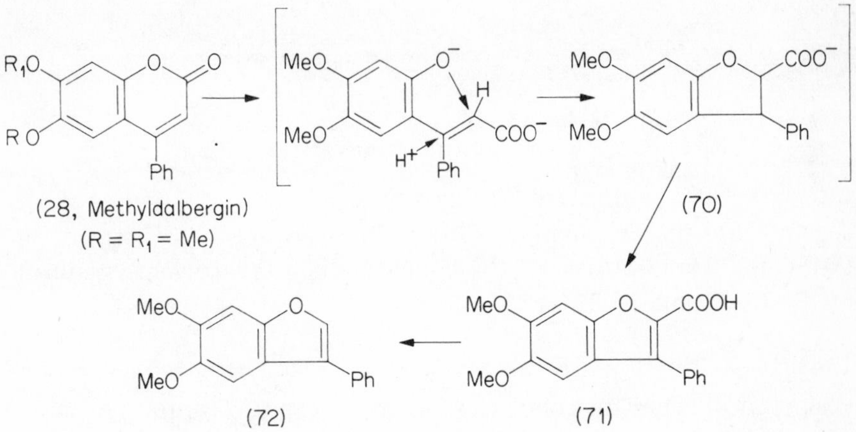

(28, Methyldalbergin)
(R = R$_1$ = Me)

(70)

(72)

(71)

Oxidative degradation was also used in the structural proof of melannein and stevenin (Donnelly, B. J. *et al.*, 1968; Donnelly, D. M. X. *et al.*, 1973b).

15.3.3.1 Spectroscopic measurements

Spectroscopic analysis is used extensively in structural assignments of the 4-arylcoumarins of the Leguminosae. In the IR spectra, a low ν_{CO} value (1664 cm^{-1}) is recorded in the solid state spectra of the coumarins (28), (30)–(32) (Donnelly *et al.*, 1973b). The methylated derivatives in the series exhibit the expected lactone frequency (1710 cm^{-1}). The magnitude of the difference ν_{CO}^{KBr} methyl ether $-\nu_{CO}^{KBr}$ phenolic coumarin is independent of the position of the hydroxyl group. The mass spectra of the 4-arylcoumarins show the loss of CO from the molecular ion resulting in a benzofuran cation; however the lower mass fragments are not intense. The fragmentation pathway is demonstrated for mellanein (32) (Scheme 15.5). Because of the low solubility in carbon tetrachloride and deuteriochloroform, the NMR

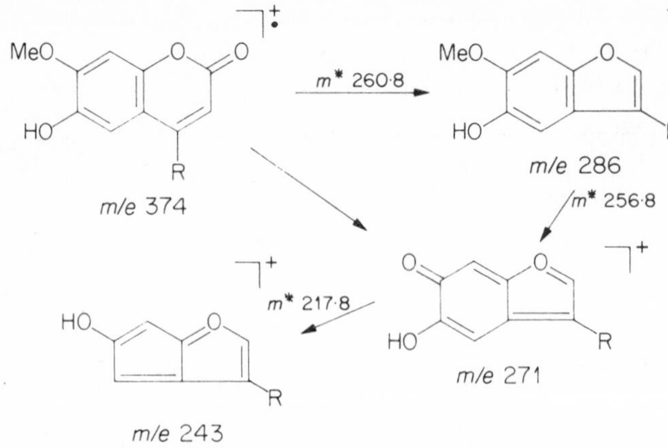

Scheme 15.5   Mass spectrum fragmentation pattern for melannein (32)

spectra of the coumarins (28)–(32) have been examined in DMSO-d$_6$. The 3-proton of the compounds in this solvent occurs at τ 3.5–3.9 (Table 15.2).

15.3.3.2 Synthesis

Perkin and Pechmann condensations have been used in the syntheses of the 4-arylcoumarins. Choosing the appropriate benzophenones, the syntheses of the methyl ethers of dalbergin (Ahluwalia and Seshadri, 1957), melannein (32) (Donnelly, B. J. et al., 1968) and stevenin (30) (Donnelly et al., 1973b) were achieved via the Perkin condensation. The Pechmann condensation of the appropriate phenols with ethyl benzoyl acetate (Ollis et al., 1968) or with substituted benzoyl acetic esters (Mukerjee et al., 1969a) afforded kuhlmannin and a series of 4-arylcoumarins respectively.

A partial methylation technique was perfected for the synthesis of the natural products (Mukerjee et al., 1969b), advantage being taken of the acidity of those hydroxyl groups in conjugation with the lactone carbonyl. The course of the partial methylation is solvent dependent and improved yields were obtained by use of acetate derivatives and DMF as solvent. The coumarins (28, R=H; R'=Me), (29) and (32) were obtained in high yields on the oxidation of the corresponding neoflavenes with CrO$_3$ in pyridine (Donnelly et al., 1973a). Some of the more recent general syntheses for 4-phenyl-coumarins include the use of BF$_3$ dietherate complex (Chatterji and Chakraborty, 1972); the hydrolysis of 2-amino-4-phenyl-2H-1-benzopyran (Shighiro *et al.*, 1970); the condensation of resorcinol and PhCH=CHCN in $ZnCl_2$, followed by dehydrogenation (Gupta and Paul, 1970).

15.3.3.3 Sisafolin

Sisafolin was the first of the 5,7-oxygenated coumarins reported in the Leguminosae. The assignment of the tentative structure (33) was based on physical measurements (IR, UV, NMR) (Tables 15.2, 15.3) and oxidation of *O*-dimethylsisafolin with neutral potassium permanganate. This oxidation gave β-resorcyclic acid dimethyl ether, oxalic acid and a benzophenone which has not been assigned a structure (Saxena *et al.*, 1970). The published evidence for the structure (33) for sisafolin is inconclusive.

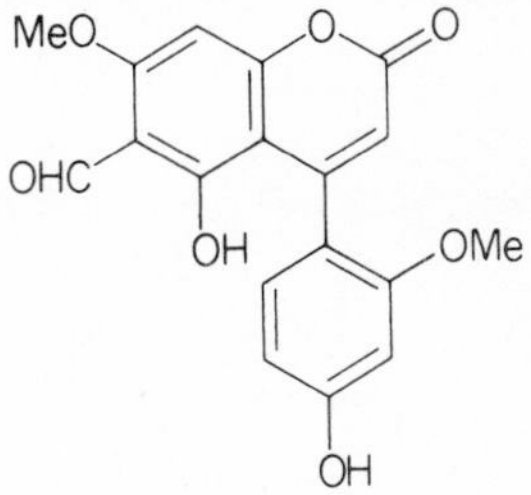

(33, Sisafolin)

15.3.4 Dalbergiones

The name dalbergione was introduced by Gonçalves da Lima and Dalia Maia (1961) to describe a type of benzoquinone belonging to the neoflavonoid group. The term dalbergenone was later used by Rao and Seshadri (1963) to describe these dalbergiquinones. The introduction of this second term is unfortunate and, on the basis of 'precedence and general acceptance', these benzoquinones are called dalbergiones (Eyton *et al.*, 1965a). There are eight known naturally occurring examples (Table 15.7). A review on the dalbergiones was published recently (Thomson, 1971).

R-, S-, RS- 4-methoxydalbergiones (34), S-4'-hydroxy-4-methoxydalbergione (35), S-4,4'-dimethoxydalbergione (36), R-3,4-dimethoxydalbergione (37), R-4'-hydroxy-3,4-dimethoxydalbergione (38), and S-3'-hydroxy-4,4'-dimethoxydalbergione (39).

The dalbergiones show normal redox properties and form dihydrodalbergiquinols which can be reoxidized to dihydrodalbergiones

Table 15.7 Dalbergiones in the Leguminosae

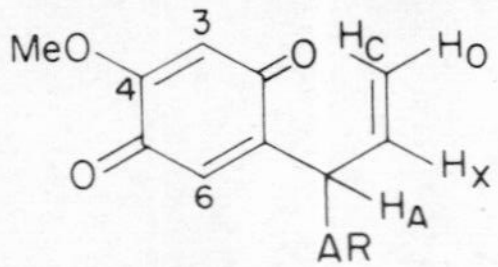

Structure (column header): MeO–substituted cyclohexadienedione ring with positions 3, 4, 6 indicated; pendant vinyl group labelled H$_C$, H$_O$, H$_X$, H$_A$, and AR (O at 1 and 5 positions).

	m.p. (°C)	[α]$_D$ (solvent)	λ (solvent)nm(log ε)	ν(CHCl$_3$)	3-H	3-OMe	4-OMe	6-H	H$_B$ + H$_C$	H$_A$	H$_X$	References
(S)-4-Methoxy-dalbergione (34) di-Ac	118–120 65	−13°(CHCl$_3$) +99°(C$_6$H$_6$) −20.9°(CH$_3$)CO	207 (4.13) 262 (4.09) (EtOH)	1646, 1667	4.05	—	6.2	3.47 (.09)	4.75 + 5.02	5.07	3.86	Eyton et al., 1965. Donnelly et al., 1965; Barnes et al., 1965.
(R)-4-Methoxy-dalbergione di-Ac	114–116 98–99	+13°(CHCl$_3$) −51°(dioxan) +21°(CH$_3$)$_2$CO										Eyton et al., 1962, 1965a; Donnelly et al., 1965.
(RS)-4-Methoxy-dalbergione di-Ac	125–126	0.00										
(S)-4'-Hydroxy-4-methoxy-dalbergione (35) tri-Ac	104–105 172–178 dec. 126.5–127.5	−51° (dioxan) −176° (C$_6$H$_6$) −69.9° (CH$_3$)$_2$CO +5 (CHCl$_3$)	228 (4.10) 262 (4.12) (EtOH) 330 (3.20) (EtOH)	1645, 1672 1664, 1633 (Nujol)	4.15							Jurd et al., 1972. Eyton et al., 1962, 1965a.
(S)-4,4'-Dimethoxy-dalbergione (36) di-Ac	109–111.5	−139°(CHCl$_3$) −32°(dioxan) −176° (C$_6$H$_6$)	228 (4.16) 258 (4.14) (EtOH) 333 (3.23) (EtOH)	1645, 1672 1667 (Nujol)	4.15		6.25	3.6 (0.9)	4.82 + 5.08	5.18	3.94	Ibid. Barnes et al., 1965.
(R)-3,4-Dimethoxy-dalbergione (37) di-Ac	113–115 41–42	+5° (CHCl$_3$) ?60° (CHCl$_3$)	260 (4.06) 405 (3.00) (MeOH)	1660	—	6.0	6.05	3.65 (1.0)	4.78 + 5.05	5.18	3.9	Ibid., Eyton et al., 1965b; Ollis et al., 1968.
(R)-4'-Hydroxy-3,4-dimethoxy-dalbergione (38)	an oil	—	—	1650	—	6.0	6.07	3.64 (1.0)				Ibid.
(S)-3'-Hydroxy-4,4'-dimethoxy-dalbergione (39) tri-Ac	101–102 99–100	−143° (CHCl$_3$) −14.3° (CHCl$_3$)	260h (3.09) 260 (3.86) 290 (1.08) (MeOH)	1662, 1640	4.0	—	6.1	3.45 (1.09)	4.7 + 5.05	5.12	3.9	Donnelly, 1971.

(Eyton *et al.*, 1965a); these derivatives retain any existing optical activity. The UV absorption of the dalbergiones is similar to that of 5-methoxy-2-methyl-1,4-benzoquinone (λ_{max} 262 nm ϵ 17 000). A further absorption at 228 nm was observed for those dalbergiones substituted in the aromatic ring. Confirmation of this benzenoid moiety was obtained by permanganate oxidation of 4-methoxy-dalbergione (34) and 4,4'-dimethoxydalbergione (36) to yield. benzoic acid and *p*-anisic acid respectively. With the exception of RS-4-methoxydalbergione obtained from the heartwood of *D. retusa* (Jurd *et al.*, 1972), the dalbergiones isolated are optically active. The optically active centre must be contained in the $C_3 H_4$ fragment, a residue remaining after subtraction of benzoquinone and benzenoid moieties. The spectra (IR, NMR) of the dalbergiones are in accord with a 3,3-disubstituted propene. The IR absorption bands at 910 and 990 cm^{-1} are assignable to a vinyl group, whilst their disappearance in the IR spectra of the dihydrodalbergiones indicates the reduction of the vinyl group. The UV spectra of the reduced quinones (73) were unchanged from the respective parent quinone (34), thus confirming the absence of direct conjugation of the vinyl group with either the quinonoid or aryl chromophores. An ABCX system for the $C_3 H_4$ propene residue was evident in the NMR spectra. The ring A protons in the dalbergiones were clearly identifiable. A singlet in the range τ 4.05—4.15 and doublet at τ 3.47—3.6 (J 0.9—1.1 Hz) are assignable to 3-H (hydrogen adjacent to the methoxyl) and 6-H (coupled to the benzylic proton) respectively. Dalbergiones with a *p*-substituent in ring B showed the signals of the $A_2 B_2$ system.

15.3.4.1 Absolute stereochemistry

The absolute stereochemistry of 4-methoxydalbergione, isolated from *D. nigra* (Eyton *et al.*, 1965a) was established chemically by ozonolysis of the dihydro derivative (73) to yield (−)-α-ethylphenyl-acetic acid (74) a compound of established absolute configuration. Dihydro-4,4'-dimethoxydalbergione (also obtained from the quinone from *D. nigra*) on ozonolysis gave (+)-α-ethyl-*p*-methoxyphenylacetic acid (75). These acids have negative and positive plain ORD curves respectively.

It can be concluded that the 4-methoxydalbergione in *D. nigra* has a R-configuration and the 4,4'-dimethoxydalbergione has an

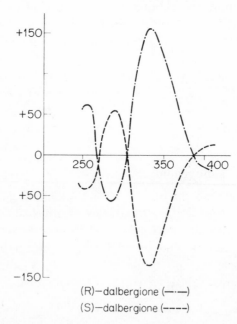

(34)   (73)   (74)   (75)

S-configuration. The antipodal relationship was fully supported by
ORD and CD curves (Eyton *et al.*, 1965a). These two antipodes were
present in *D. nigra* as a quasiracemate. Closely following this first
report of the occurrence of quasiracemates in nature was the finding
of a quasiracemic alkaloid narcissamine (Laiho and Fales, 1964).

ORD curves of the quinones (34) (Fig. 15.1) show three apparent
Cotton effects. The effect about 300–350 nm is positive and
negative for compounds with S- and R-configurations and has been
used for the assignment of absolute stereochemistry. The existence of

(R)–dalbergione (—·—)
(S)–dalbergione (----)

Fig. 15.1 Optical rotatory dispersion curves.

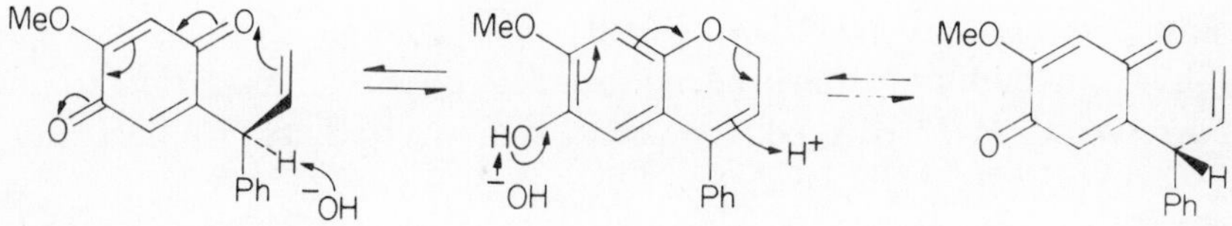

Scheme 15.6

two antipodal forms may arise from a transformation (Scheme 15.6) analogous to that of the ubiquinone-ubichromenol (Links, 1960).

15.3.4.2 Synthesis

Synthesis of the dalbergiones were devised for the confirmation of structure assignment. A Claisen rearrangement of a cinnamyl ether (76), achieved by heating with dimethylaniline, gave two products (77) and (78). The o-allylphenol (77) on oxidation with Fremy's salt gave 4-methoxydalbergione (34). These rearrangements were equated as laboratory equivalents of biogenetic C-alkylation. However, the search for less extreme experimental conditions and support for

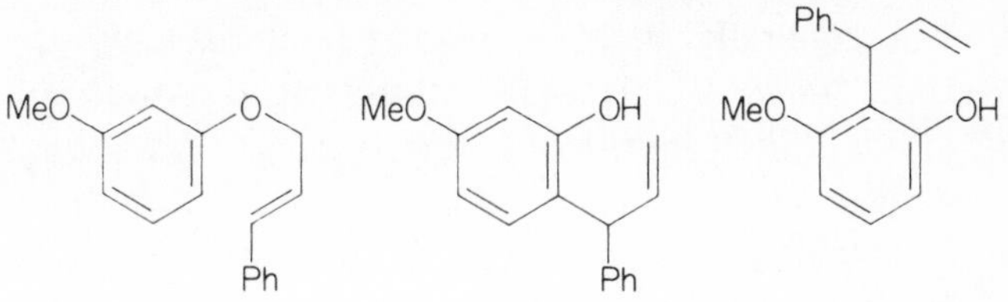

(76)   (77)   (78)

biogenetic proposals (Eyton et al., 1965b, Ollis et al., 1968) led to acetic acid catalyzed alkylations of phenols with protonated cinnamyl alcohol (Jurd, 1968, 1969a,b) cinnamyl acetates (Mageswaran et al., 1969) or cinnamyl pyrophosphate (Larkin et al., 1970). Citric acid (with ascorbic acid) has given improved yields of neoflavanoids over the cinnamyl phenols and minimized undesirable oxidation. The dalbergiquinols are converted on TLC with silver nitrate impregnated silica gel (Mageswaran et al., 1969) or by O_2/K_2CO_3 (Eyton et al., 1965a) into the corresponding dalbergiones.

The reaction of phenols with (±)-1-phenylallyl alcohol in aqueous propionic acid give mainly the neoflavanoid and only traces of the cinnamyl phenols (Mageswaran et al., 1969). The composition of the condensation products varied with the organic acid used, for example aqueous formic acid suppressed neoflavanoid formation. The dihydrodalbergiones have been obtained by CrO_3 oxidation of the corresponding dihydrodalbergiquinol dimethyl ethers (Kumari *et al.*, 1966).

15.3.4.3 Physiological and pathological function

The presence of R-3,4-dimethoxydalbergione (37) in *Macherium schleroxylon* (Eyton *et al.*, 1965b) is considered to be responsible for the dermatitis in workers handling it (Morgan *et al.*, 1968). Tests have shown that the quinone (37) and its quinol are the sensitizing constituents. Interestingly, it has been observed in the author's laboratory that the dalbergiones are less causitive of dermatitis than their isomeric cinnamyl phenols.

The dalbergiones have been shown to inhibit the drying of oil and polyester lacquers (Dietrichs and Hausen, 1971) and antibiotic activity is claimed for crude extracts of *D. nigra*. The active component R-4-methoxydalbergione inhibits the growth of *Bacillus anthracis* and *Candida albicans*. Further reports have shown that this compound is generally inactive against bacteria and fungi (except *Reticulitermes flavipes*). It will inhibit the growth of *Candida tropicalis* and a limited number of other yeasts (Jurd *et al.*, 1971).

15.3.5 Neoflavenes

The isomerization (Ollis, 1967; Ollis *et al.*, 1968) of dalbergiones to give neoflavenes (Scheme 15.6) is mechanistically similar to the ubiquinone-ubichromenol transformation (Links, 1960). The isolation of kuhlmannene (40) from *Macherium nictitans* and *M. kuhlmannii* (Ollis *et al.*, 1968) strongly supports the proposal of its biosynthesis from the congeneric (R)-3,4-dimethoxydalbergione (37). The second example of this class of compound is dalbergichromene (41) isolated from the stembark of *Dalbergia sissoo* (Mukerjee *et al.*, 1971), and *D. latifolia* (Dhingra *et al.*, 1971). The NMR spectra of kuhlmannene and dalbergichromene are characterized by an AX_2 system.

The 4-phenylchromenes have been synthesized from phenylpropargyl ethers (Iwai and Ide, 1963); from chromen-4-ones by Grignard reaction and subsequent dehydration (Baranton *et al.*, 1968); by cyclization of diphenyl allyl alcohols in benzene with a cation exchange resin (Mukerjee *et al.*, 1971); by thermal cyclization of

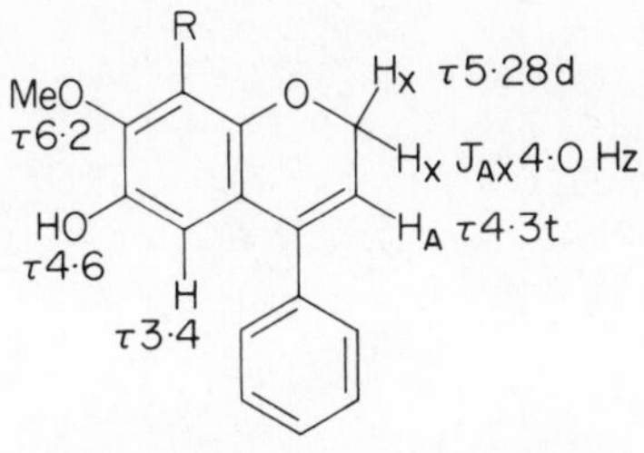

MeO
τ6·2
HO
τ4·6
R
Hₓ τ5·28d
Hₓ Jₐₓ4·0 Hz
Hₐ τ4·3t
H
τ3·4

(40, Dalbergichromene R = H) (3·5τ)
(41, Kuhlmannene R = OMe) (6·07τ)

α-vinylsalicyl alcohols (Hug *et al.*, 1972); and by facile isomerization of the dalbergiones on neutral alumina (Ollis *et al.*, 1968), or hot pyridine (Mukerjee *et al.*, 1971) or *N*,*N*-dimethylpyridine in chloroform at room temperature (Donnelly *et al.*, 1973a).

15.3.6 Dalbergiquinols

The known naturally occurring dalbergiquinols are listed in Table 15.8. Of the six 3,3-diarylpropenes isolated, five are optically active and the stereochemistry of the sixth, obtusaquinol, is unspecified (Gregson *et al.*, 1968), for the sample isolated from *Dalbergia obtusa*. Obtusaquinol is also present in *D. retusa* but as a racemic component of an easily dissociated molecular complex (quinhydrone) (Jurd *et al.*, 1972).

Latifolin (42) was the first of the six known 3,3-diarylpropenes isolated (Balakrishna *et al.*, 1962; Dempsey *et al.*, 1963). It is optically active and the assigned structure (42) is based on spectroscopic and chemical evidence. Degradative experiments were carried out on dimethyllatifolin (79) and its dihydro derivative. The dimethyllatifolin, on ozonolysis gave formaldehyde, while oxidation with permanganate afforded *o*-methoxybenzoic acid and 2′,2,4,5-tetramethoxybenzophenone (80). The positions of the hydroxyl groups were determined by similar degradation of the diethyllatifolin (Dempsey *et al.*, 1963; Kumari *et al.*, 1965).

The structures of the dalbergiquinols (43)–(46) were determined by conversion to the corresponding dalbergiquinones. Kuhlmanni-quinol (46) was oxidized with sodium periodate in acetic acid to give 3,4-dimethoxydalbergione (Ollis *et al.*, 1968).

Table 15.8 *Dalbergiquinols in the Leguminosae*

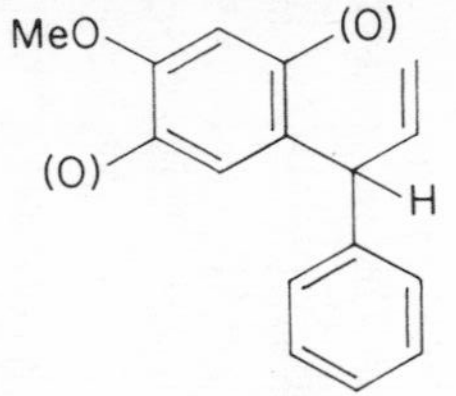

Compound	Mol. formula	m.p. (°C) (Acetate deriv.)	φ	ORD* (nm)	$a \times 10^{-2}$	UV λ (nm)	log ε
R-Latifolin (42)	$C_{17}H_{18}O_4$	124	+1020tr	278	-55	285	3.62
			+6400i	254	+589		
			+12700pk	242			
			-46200tr	218			
R-5-O-Methylatifolin (42)	$C_{18}H_{20}O_4$	91–92	-1420tr	287	-30	230i	3.98
			+1530sh	282	+401	285	3.7
			+6780pk	250		293	3.72
			-33300tr	224		232i	3.93
						214	4.27
S-3'-Hydroxy-4-methoxydalbergiquinol (43)	$C_{17}H_{18}O_4$	93–95	-4460	296	-646	233	4.1
			-2240sh	279			
			-15600tr	248			
			+49000pk	225			
Obtusaquinol (44)	$C_{16}H_{16}O_3$	104–105 b.p. 100°/ 10^{-3} mm					
R-3,4-Dimethoxydalbergiquinol (45)	$C_{17}H_{18}O_4$		-420tr	278	-7	270	3.00
			+340sh	263	+55	278	3.31
			+715pk	242			
			+4800tr	226			
R-Kuhlmanniquinol (46)	$C_{18}H_{20}O_5$						

pk = Peak; tr = trough; sh = shoulder; i = inflection.
*Reference Donnelly et al., 1972 and references therein.

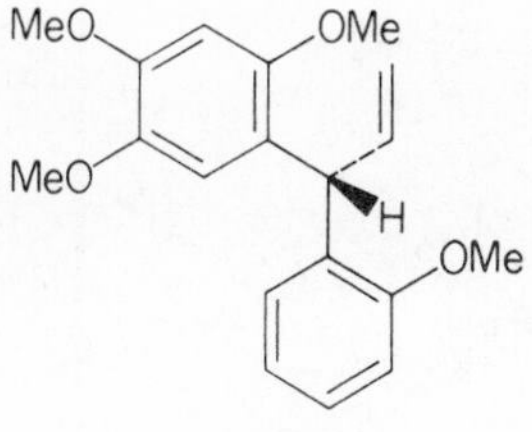

(79) 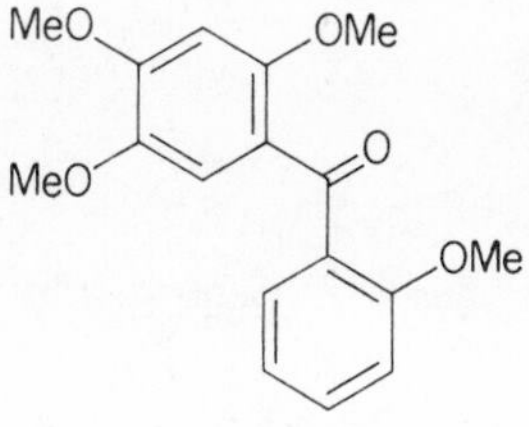   (80) 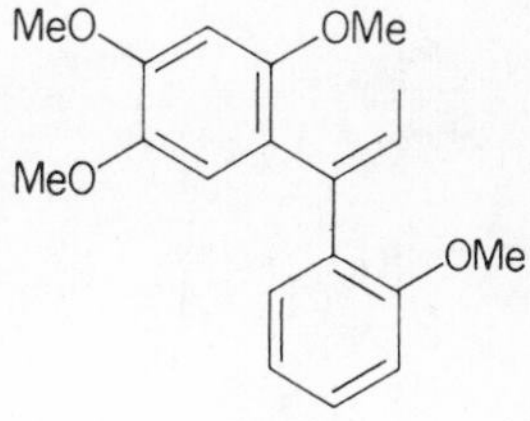   (81)

15.3.6.1 Absolute stereochemistry

An absolute configuration (R) was assigned to latifolin by analysis of the Cotton effects in the ORD curve (Donnelly *et al.*, 1967) without prior oxidation of the dalbergiquinol to the quinonoid structure (dalbergiones) (Kumari *et al.*, 1966). Latifolin and all its derivatives show a positive Cotton effect at short wavelength (Table 15.8). Alteration of substituents in the 5- and 2'-positions does not affect the sign of the Cotton effect but affects its amplitude and wavelength. Saturation of the vinyl group also causes an increase in the amplitude of the Cotton effect. The quinol diacetates derived from the dalbergiones (34) and (36) of established configuration provided the closely related compounds for comparison with the 3,3-diarylpropenes. (Fig. 15.2)

15.3.6.2 Rearrangements

Isomerization of dimethyllatifolin (79) with aqueous base gives isolatifolin dimethyl ether (81), identical with the synthetic compound obtained from a Grignard reaction (ethyl-magnesium bromide) on 2',2,4,5-tetramethoxybenzophenone (80). The facile formation of a cyclopropane ring by photo-rearrangement of methyllatifolin in benzene was observed (Kumari and Mukerjee, 1967). Confirmation of structure (82) for the photo-isomer was provided by hydrogenation of the product to yield 1-(2,4,5-trimethoxyphenyl)-3-(2-methoxyphenyl)propane (83). The intramolecular nature of this transformation was demonstrated by irradiation of a mixture of the dimethyl and diethyl ethers of latifolin.

 Thermal rearrangement of (±)-obtusaquinol (44) affords the dihydrobenzofuran, obtusafuran (84) (Jurd *et al.*, 1973). This result is in agreement with the work published (Schmid *et al.*, 1972) on thermal rearrangements of a series of 3,3-diarylpropenes, which give dihydrobenzofurans via a cyclic intramolecular process.

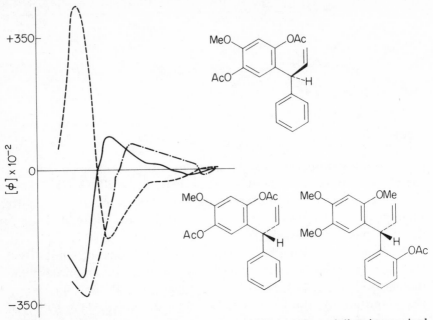

Fig. 15.2 Optical rotatory dispersion curves of (S)-4-methoxydalbergione quinol diacetate (———), (R)-4-methoxydalbergione quinol diacetate (····), (R)-2-O-acetyl-5-O-methyllatifolin (—·—·—).

The acid catalysed transformations on 3,3-diarylpropenes have been studied by three groups of workers (Scheme 15.7). The Indian school (Kumari *et al.*, 1967) treated dimethyllatifolin (79) with BF$_3$ and isolated the isomeric benzylstyrene. They observed that HCl, AlCl$_3$, and other Lewis acids were without effect in bringing about

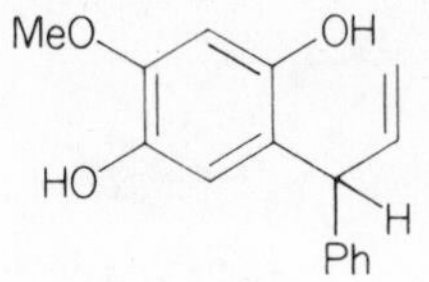

(44, Obtusaquinol) 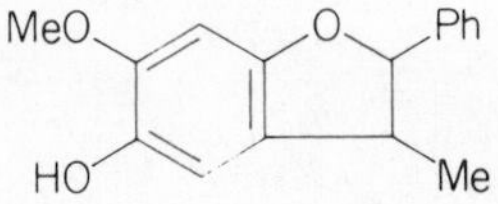   (84, Obtusafuran)

this rearrangement. Schmid and co-workers (1972) using HBr in HOAc effected a rearrangement of the diarylpropenes to mixtures of *cis* and *trans* 2-methyl-3-phenyl- and 3-phenyl-2-methyl- dihydro-benzofurans. They argued that the initial step involved protonation of the ethylenic bond and subsequent cyclization. Jurd (1973) in his studies of the highly oxygenated 2,2-diarylpropenes proposes preferential protonation of ring A rather than the ethylenic bond. Elimination of the cinnamyl cation from the intermediate protonated species would yield methoxyquinol which, in turn, would recombine with the cinnamyl cation to give the stable 2-cinnamyl-5-methoxyquinol. This reaction resembles the direct alkylation of methoxyquinol with protonated cinnamyl alcohol. Confirmation for the proposed mechanism, i.e. fission of ring A, was sought and found when obtusaquinol was treated with aqueous formic acid in the presence of excess pyrogallol. This type of rearrangement depends on hydrogen ion concentration, the nucleophilicity of the ring A and the stability of the potential cinnamyl cation.

15.3.6.3 Synthesis

The dalbergiquinols result from sodium dithionite reduction of the dalbergiones (Eyton *et al.*, 1965a) or from the acid catalyzed condensation of methoxyquinol with cinnamyl alcohol (Jurd, 1968, 1969a,b) cinnamyl acetates (Mageswaran *et al.*, 1969, cinnamyl pyrophosphate (Larkin *et al.*, 1970) or with (±)-1-phenyl-allyl alcohol. The conversion of 4-phenylcoumarins into dalbergiquinols occurs by two-stage reduction involving initial formation of a cinnamyl alcohol and subsequent reductive elimination of the alcoholic hydroxyl group (Mukerjee *et al.*, 1970).

15.3.7 Arylchromans

Brazilin was first isolated by Chevreul in 1808 from brazil wood. Later it was assigned structure (47) on the basis of the results of degradative studies (Perkin and Robinson, 1908). This work and

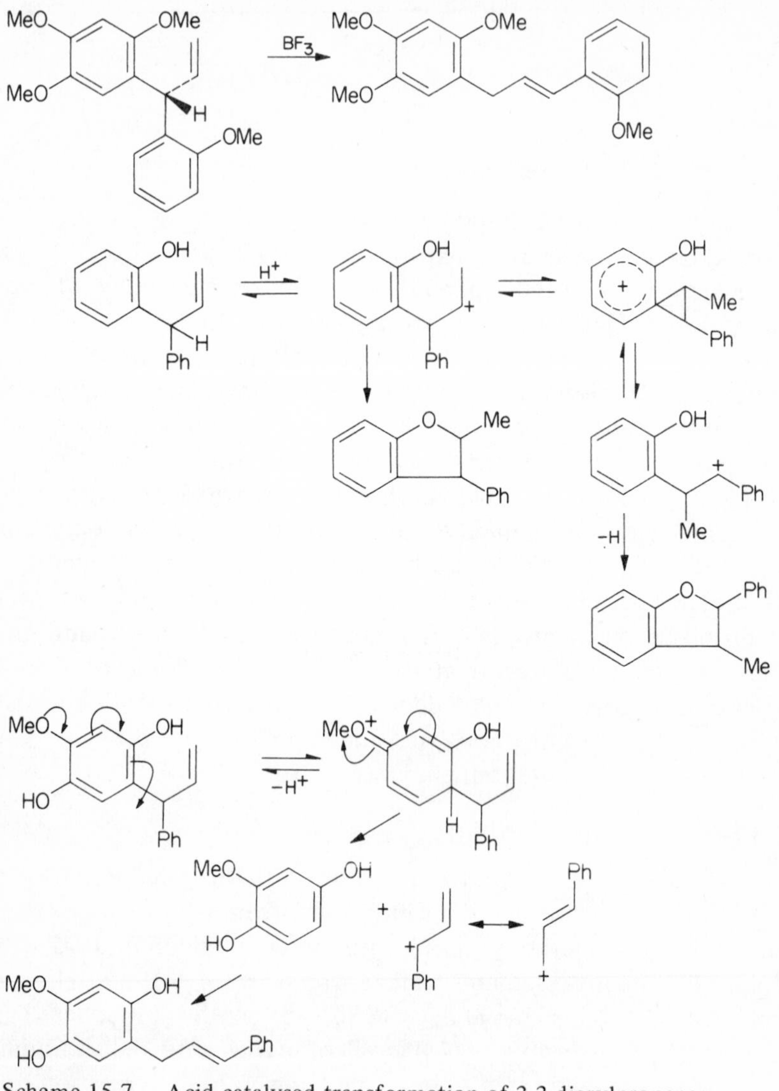

Scheme 15.7   Acid catalysed transformation of 3,3-diarylpropenes

subsequent synthetic work on brazilin and haematoxylin (45), a close relative, were reviewed by Robinson (1962). It was only recently (Morsingh and Robinson, 1970) that details of some of the synthetic work were published. In this synthesis of brazilin and haematoxylin, the fundamental problem was to retain the tertiary hydroxyl group in the position-3 of the chromanone ring. The synthesis involved dehydration of a *p*-hydroxydiphenylcarbinol to a quinonemethide (85) which was then reduced.

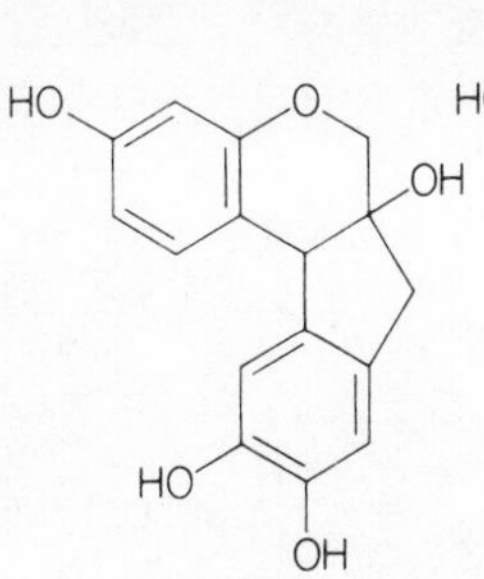

(47, Brazilin) 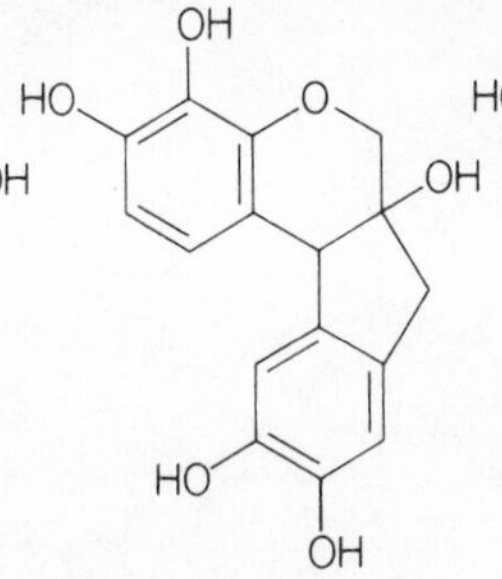   (48, Haematoxylin) 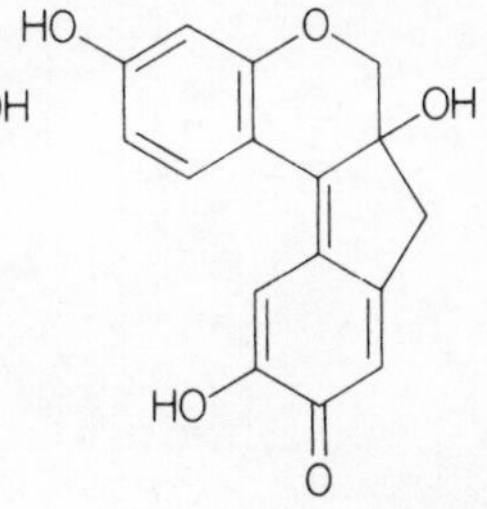   (85)

Other syntheses of brazilin and haematoxylin were published by Dann and Hofmann (1963) and more recently by Kirkiacharian and Billet (1972).

In a discussion on the possible biosynthesis of (47) and (48), Whalley (1956) proposed the isoflavanone as a precursor that would undergo a 1,2-aryl migration. This type of pathway was also proposed for the biosynthesis of 4-phenylcoumarins but was discounted on results obtained from feeding experiments in *Calophyllum* species (Kunesch and Polonsky, 1967). The isolation recently of a group of homoisoflavones (86)–(88) from *Eucomis bicolor* and *E. autumnalis* (Liliaceae) (Böhler and Tamm, 1967; Sidwell and Tamm, 1970) leads one to consider the chromodogenide as a precursor of the 'brazilin' group. Interestingly, one of the synthetic routes to anhydrobrazilin (89) was from chromandogenide (90) by treatment with P_2O_5 (Robinson, 1962).

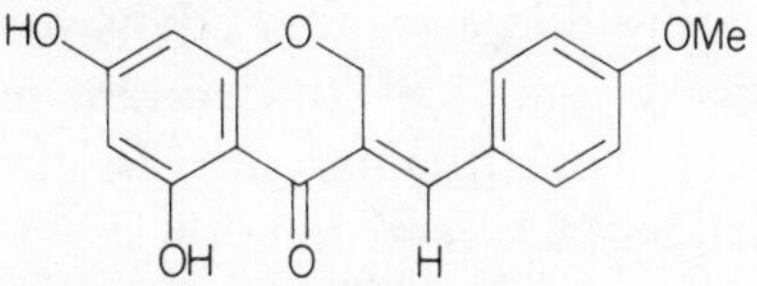

(86, Eucomin) 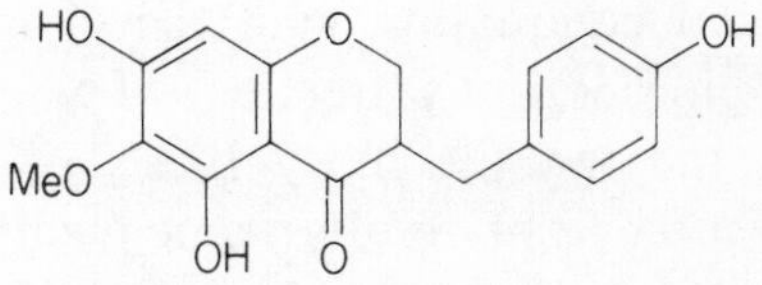   (87)

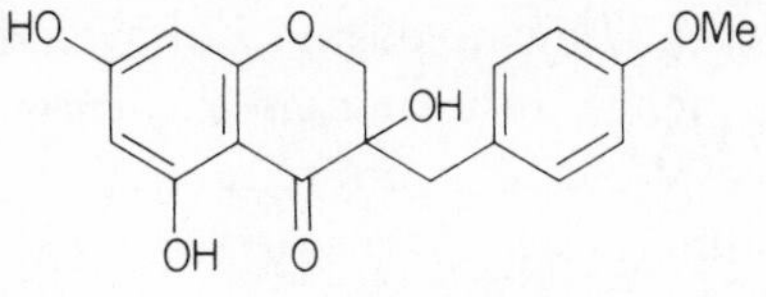

(88, (−)−eucomol)

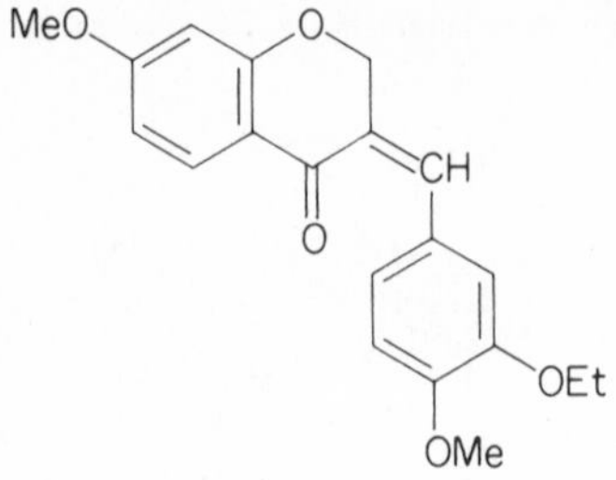

(89) 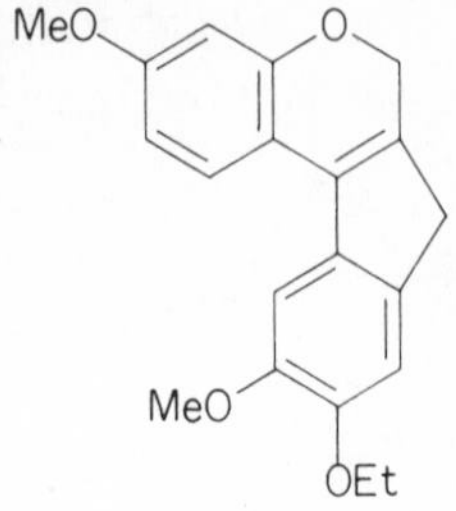   (90)

15.3.8 Coumarinic acids

Calophyllic acid (49) $C_{25}H_{24}O_6$ and chapelieric acid methyl ester, (5), $C_{26}H_{28}O_6$, were isolated from extracts of *Calophyllum inophyllum* and *C. chapelieri* Drake (Guerreiro *et al.*, 1971). Calophyllic acid co-occurs with calophyllolide (8) and inophyllolide (9) and may be obtained from the former by demethylation with $AlCl_3$ and then cyclization to the latter.

Chapelieric acid methyl ester has a NMR spectrum that supports the formula (50) assigned to this compound. The signal at τ 2.8 in the NMR and peaks at 760 cm^{-1} and 695 cm^{-1} in IR confirm the presence of a phenyl group. The sequence $Ph-\overset{|}{C}H-CH_2 \cdot CO_2 Me$ is represented by an AB_2 NMR pattern and is supported by the presence of a peak at m/e 361 ($M^+ - CH_2 CO_2 Me$) in the mass spectrum.

15.4 Biosynthesis

The similarity in the oxygenation pattern in the dalbergiones, the dalbergiquinols and the 4-arylcoumarins suggests that they are biogenetically related. There are many practical difficulties in studying the biosynthesis of heartwood constituents, and, in the absence of results from feeding experiments, several hypothesis for their genesis have been published (Gottlieb *et al.*, 1970). Among the theories proposed are: a biosynthetic pathway parallel to the established route to the flavanoids and isoflavanoids (Whalley, 1961); a bio-oxidative sequence resulting in the formation of the 4-aryl-coumarin (Ollis, 1966); and a bio-reductive sequence commencing with the 4-arylcoumarin (Mukerjee *et al.*, 1970) which arises by cyclization of the corresponding cinnamic ester.

The bio-oxidative sequence envisages *C*-alkylation of polyphenols with cinnamyl pyrophosphate under mild acid catalysis to give the dalbergiquinols, *cis*- and *trans*- cinnamylphenols or dihydrobenzofurans (Eyton *et al.*, 1965a; Ollis and Gottlieb, 1968). The mechanistic feasibility of this proposal was demonstrated by the observation (Jurd, 1969a,b) that protonated cinnamyl alcohol condenses with phenols in aqueous acetic or citric acid solutions, to yield both cinnamylphenols and 3,3-diarylpropenes. The bio-oxidation proposal by Ollis and his co-workers gains further support by the evident structural correlations among compounds isolated from some species (Scheme 15.8).

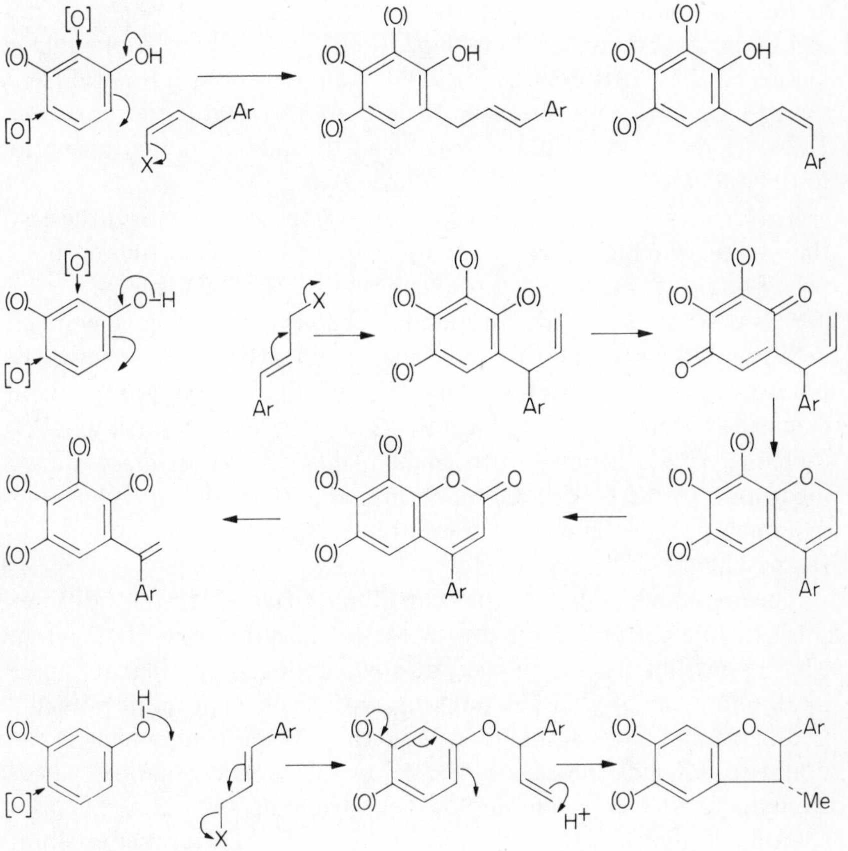

Scheme 15.8 Biogenetic relations among the neoflavanoids and their congeners (Ollis and Gottlieb, 1968).

The 4-phenylcoumarins of the Guttiferae are more amenable to tracer studies. The first experiments involving feeding (+)-3-^{14}C-phenylalanine to young shoots of *Calophyllum inophyllum* showed an incorporation into calophyllolide (8) in which 92% of the label was located at C-4 (Kunesch and Polonsky, 1967). In an extension of these studies (Gautier *et al.*, 1972), the order of biosynthesis of the minor congeners, calophyllic acid (49) and inophyllolide (9) was sought. A comparison of specific activities in incorporation of 3-^{14}C-phenylalanine showed an eightfold specific activity for the acid over inophyllolide. U-^{14}C-isoleucine, an efficient precursor of the 2,3-dimethylchromanone ring, was used as the control.

On the assumption that the acid (49) is the immediate precursor of the neoflavanoid (8), then the earlier proposal by Seshadri (1959) would be favoured for the biosynthetic pathway to 4-phenylcoumarins in the Guttiferae. Additional support for this proposal lies in the isolation of chapelieric acid from *C. inophyllum* (Guerreiro *et al.*, 1971). These acids could be envisaged to undergo dehydrogenation to the α,β-unsaturated acids.

Tracer experiments were used to investigate the biosynthesis of the phloroglucinol ring and its 6- and 8- substituents in *C. inophyllum.*   L-(U-^{14}C)-isoleucine,   L-(U-^{14}C)-leucine   and 1-^{14}C-acetate were incorporated. The isoleucine is specifically incorporated into the tigloyl side chain, the acetate into the phloroglucinol ring and C_3 and C_4 of the tigloyl group, whilst leucine is metabolized to acetate (Kunesch and Polonsky, 1969) (Scheme 15.9). Studies with small plants of *Mesua ferrea* showed incorporation of L-(U-^{14}C)-valine into the 3-methylbutyryl chain in mammeisin (15) and into the isobutyryl chain of mesuol (17) (Kunesch and Polonsky, 1971).

The neoflavanoids of the Guttiferae have distinctly different substitution patterns from those of the Leguminoae. The acyl and isoprenoid substituents in the phloroglucinol-type A ring are present in all Guttiferae 4-phenylcoumarins, with the exception of ponnalide (Adinarayama and Seshadri, 1965) the structure of which is now doubtful (Games and Haskins, 1971). A 6,7-oxygenation pattern is characteristic of Leguminosae neoflavanoids with one exception, sisafolin, found in the seeds of *D. latifolia.* Therefore, the possibility of different biogenetic pathways for the neoflavanoids in the two families cannot be excluded (Ollis, 1970; Gautier *et al.*, 1972).

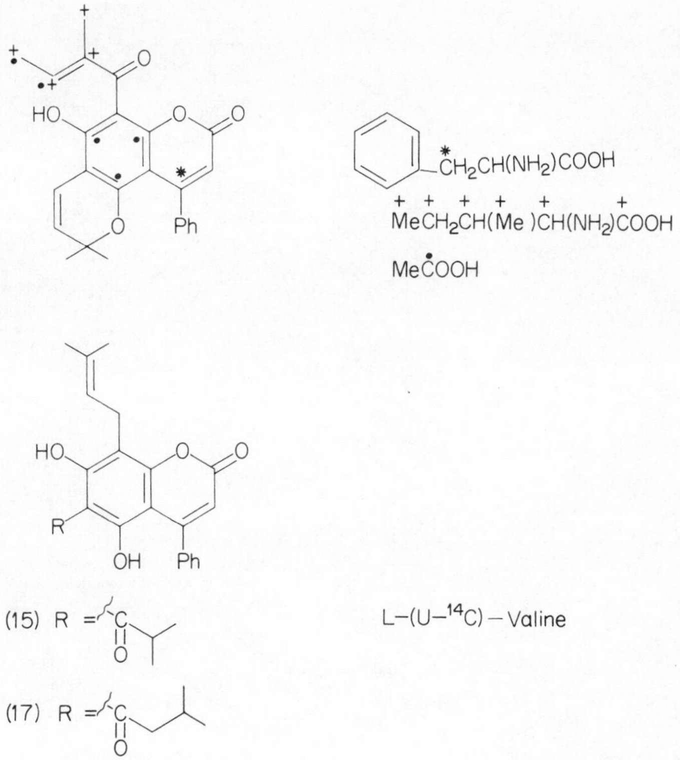

$$\cdot \overset{*}{C}H_2CH(NH_2)COOH$$

$$\overset{+}{Me}\overset{+}{C}H_2\overset{+}{C}H(\overset{+}{Me})\overset{+}{C}H(NH_2)\overset{+}{C}OOH$$

$$Me\overset{\bullet}{C}OOH$$

(15) R = $\overset{\diagup}{\underset{\overset{\|}{O}}{C}}$⟨⟩

L–(U–^{14}C) – Valine

(17) R = $\overset{\diagup}{\underset{\overset{\|}{O}}{C}}$⟨⟩

Scheme 15.9   *Incorporation of radioactivity into neoflavanoids*

15.5 Interconversions

The isolation of closely related neoflavanoids of similar configuration (Ollis *et al.*, 1968; Donnelly *et al.*, 1969) from the genera *Dalbergia* and *Macherium* lend support to the proposals for their biosynthesis. High yield biological-type conversions of the neoflavanoids have been obtained *in vitro* (Scheme 15.10) (Donnelly *et al.*, 1973a). In this cyclical interconversion, cognisance was taken of the absence of a 6-hydroxy-substituent in all the presently known 4-phenylcoumarins isolated from the Guttiferae (Games, 1972), and in exostemin (Sánchez-Viesca, 1969) from the family Rubiaceae.

Chemical inter-relations within the series of the phloroglucinol-4-phenylcoumarins have been used for structural elucidation. Examples are: the cyclization of a 3-methylbut-2-enyl group with DDQ as in mammeigin (18), the oxidation product from mammeisin (15),

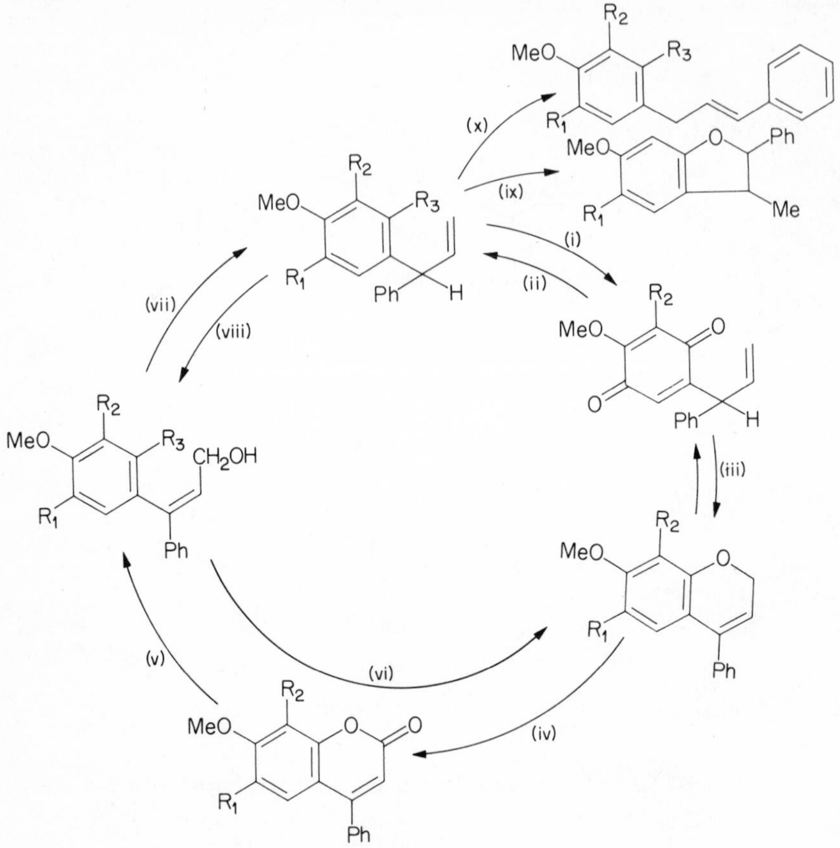

a, $R_1 = R_3 = OH (OAc)$, $R_2 = H$

b, $R_1 = R_3 = OH (OAc)$, $R_2 = OMe$

c, $R_1 = R_3 = OH (OAc)$, $R_2 = H$, $C_6H_3-3-OH-4-OMe$

d, $R_1 = R_2 = H$, $R_3 = OAc$

Scheme Reagents	References
(i) $O_2 - O.1N - K_2CO_3$ If $R' = H$	Eyton *et al.*, 1965a.
m-Chloroperbenzoic acid — toluenesulphonic acid — $CHCl_3$	Donnelly *et al.*, 1973a.
$DDQ - C_6H_6$	*Ibid.*
(ii) Sodium dithionite (13%)	Eyton *et al.*, 1965a.

Continued

(iii) Neutral alumina   Ollis *et al.*, 1968. Mukerjee
                                                         et al., 1971.
     Reflux in pyridine   *Ibid.*
     NN-Dimethylaminopyridine − $CHCl_3$ Donnelly *et al.*, 1973a.
(iv) CrO_3 − pyridine   *Ibid.*
     SeO_2 − dioxan   *Ibid.*
(v) LAH − Et_2O Mukerjee *et al.*, 1970.
(vi) IR 120 H^+ form − C_6H_6 Mukerjee *et al.*, 1971.
     HCl − EtOH   Donnelly *et al.*, 1973a.
(vii) LAH − $AlCl_3$ − Et_2O Mukerjee *et al.*, 1970.
     $SnCl_2$ − HOAc
     Zn − HCl − HOAc
     DMF − Zn − HCl
(viii) $Hg(OAc)_2$ − HOAc   Donnelly *et al.*, 1973a.
     SeO_2 − HOAc   *Ibid.*
(ix) BF_3 Kumari *et al.*, 1967.
     Formic acid   Jurd *et al.*, 1973.
(x) 220°, 30 min.   *Ibid.*

Scheme 15.10   Interconversions of the neoflavanoids

which compound may be obtained by base catalysed rearrangement of 5,7-dihydroxy-6-(3-methylbut-2-enyl)-8-(3-methylbutyryl)-4-phenylcoumarin (25); and the application of the epoxidation reaction on mammeisin to give the isomer (21) (Crombie *et al.*, 1972a; Finnegan and Merkel, 1972).

References

Adinarayana, D. and Seshadri, T. R. (1965), *Bull. natl. Inst. Sci. India* 31, 90.

Ahluwalia, V. K., Mehta, A. C. and Seshadri, T. R. (1958), *Tetrahedron* 4, 271.

Ahluwalia, V. K. and Seshadri, T. R. (1957), *J. chem. Soc.* 970.

Ahluwalia, V. K. and Seshadri, T. R. (1963), *Curr. Sci.* 32, 455.

Almeida, M. E. L. and Gottlieb, O. R. (1974), *Phytochemistry* 13, 751.

Bala, K. R. and Seshadri, T. R. (1971), *Phytochemistry* 10, 1131.

Balakrishna, S., Rao, M. M. and Seshadri, T. R. (1962), *Tetrahedron* 18, 1503.

Baranton, F., Fontaine, G. and Maitte, P. (1968), *Bull. Soc. chim. France* 10, 4203.

Barnes, M. F., Ollis, W. D., Sutherland, I. O., Gottlieb, O. R. and Taveira Magalhães, M. and Ollis, W. D. (1965), *Tetrahedron* 21, 2707.

Böhler, P. and Tamm, Ch. (1967), *Tetrahedron Letters* 3479.

Braga, A. da Silva, Arndt, V. H., Magalhães Alves, H., Gottlieb, O. R., Taveira Magalhães, M. and Ollis, W. D. (1967), *An. Acad. Brasil. Cienc.* 39, 249.

Braga de Oliveira, A., Gottlieb, O. R., Ollis, W. D. and Rizzini, C. T. (1971), *Phytochemistry* 10, 1863.

Braz Filho, R., Almeida, M. E. L. and Gottlieb, O. R. (1973), *Phytochemistry* 12, 1187.

Breck, G. D. and Stout, G. H. (1969), *J. org. Chem.* **34**, 4203.

Carpenter, I., McGarry, E. J. and Scheinmann, F. (1970), *Tetrahedron Letters* 3983.

Carpenter, I., McGarry, E. J. and Scheinmann, F. (1971), *J. chem. Soc.* (C) 3783.

Chakraborty, D. P. and Bose, P. K. (1960), *Proc. natl. Inst. Sci. India* **26A**, 1.

Chakraborty, D. P. and Chatterji, D. (1969), *J. org. Chem.* **34**, 3784.

Chakraborty, D. P. and Das, B. C. (1966), *Tetrahedron Letters* 5727.

Chatterji, D. and Chakraborty, D. P. (1972), *J. Indian chem. Soc.* **49**, 1045.

Chevreul, P. M. (1808), *Ann. chim. Phys.* **66**, 225.

Crombie, L., Games, D. E., Haskins, N. J. and Reed, G. F. (1972a), *J. chem. Soc. Perkin I* 2248.

Crombie, L., Games, D. E., Haskins, N. J. and Reed, G. F. (1972b), *J. chem. Soc. Perkin I* 2255.

Crombie, L., Games, D. E., Haskins, N. J., Reed, G. F., Finnegan, R. A. and Merkel, K. E. (1970), *Tetrahedron Letters* 3979.

Crombie, L., Games, D. E. and McCormick, A. (1966), *Tetrahedron Letters* 145.

Crombie, L., Games, D. E. and McCormick, A. (1967), *J. chem. Soc.* (C) 2553.

Dann, O. and Hofmann, H. (1963), *Liebigs Ann. Chem.* **667**, 116.

Dempsey, C. B., Donnelly, D. M. X. and Laidlaw, R. A. (1963), *Chem. Ind.* 491.

Dhingra, V. K., Mukerjee, S. K., Saroja, T. and Seshadri, T. R. (1971), *Phytochemistry* **10**, 2551.

Dietrichs, H. H. and Hausen, B. M. (1971), *Holzforschung* **25**, 183.

Donnelly, B. J., Donnelly, D. M. X. and O'Sullivan, A. M. (1966), *Chem. Ind.* 1498.

Donnelly, B. J., Donnelly, D. M. X. and O'Sullivan, A. M. (1968), *Tetrahedron* **24**, 2617.

Donnelly, B. J., Donnelly, D. M. X., O'Sullivan, A. M. and Prendergast, J. P. (1969), *Tetrahedron* **29**, 4409.

Donnelly, B. J., Donnelly, D. M. X. and Sharkey, C. B. (1965), *Phytochemistry* **4**, 337.

Donnelly, D. M. X. (1971), Lecture at a Symposium on recent Chemistry and Biochemistry of Phenolic Compounds, Dublin.

Donnelly, D. M. X., Kavanagh, P., Polonsky, J. and Kunesch, G. (1973a), *J. chem. Soc. Perkin I*, 965.

Donnelly, D. M. X., Nangle, B. J., Hulbert, P. B., Klyne, W. and Swan, R. J. (1967), *J. chem. Soc.* (C) 2450.

Donnelly, D. M. X., Nangle, B. J., Prendergast, J. P. and O'Sullivan, A. M. (1968), *Phytochemistry* **7**, 647.

Donnelly, D. M. X., O'Reilly, J. and Thompson, J. C. (1972), *Phytochemistry* **11**, 823.

Donnelly, D. M. X., Thompson, J. C., Whalley, W. B. and Ahmad, S. (1973b), *J. chem. Soc. Perkin I*, 1737.

Eyton, W. B., Ollis, W. D., Fineberg, M., Gottlieb, O. R., Salignac de Souza Guimarães and Taveira Magalhães, M. (1965b), *Tetrahedron* **21**, 2697.

Eyton, W. B., Ollis, W. D., Sutherland, I. O., Gottlieb, O. R., Taveira Magalhães, M. and Jackman, L. M. (1965a), *Tetrahedron* 21, 2683.

Eyton, W. B., Ollis, W. D., Sutherland, I. O., Jackman, L. M., Gottlieb, O. R. and Magalhães, M. T. (1962), *Proc. chem. Soc.* 301.

Finnegan, R. A. and Merkel, K. E. (1972), *J. pharm. Sci.* 61, 1603.

Finnegan, R. A., Morris, N. P. and Djerassi, C. (1961), *J. org. Chem.* 26, 1180.

Finnegan, R. A. and Mueller, W. H. (1964), *Chem. Ind.* 1065.

Finnegan, R. A. and Mueller, W. H. (1965), *J. org. Chem.* 30, 2342.

Games, D. E. (1972), *Tetrahedron Letters* 3187.

Games, D. E. and Haskins, N. J. (1971), *Chem. Commun.* 1005.

Gautier, J., Cave, A., Kunesch, G. and Polonsky, J. (1972), *Experientia* 28, 759.

Gonçalves da Lima, O. and Dalia Maia, M. H. (1961), *Lecture at The Third Annual Meeting Associacãa Brasileira de Quimca,* Rio de Janeiro.

Gottlieb, O. R., Mageswaran, S., Ollis, W. D., Roberts, R. J. and Sutherland, I. O. (1970), *An. Acad. Brasil. Cienc.* 42, (Supple.) 412.

Gregson, M., Kurosawa, K., Ollis, W. D., Redman, B. T., Roberts, R. J., Sutherland, I. O., Braga de Oliveira, A., Eyton, W. B. and Gottlieb, O. R. (1968), *Chem. Commun.* 1390.

Grosourdy, D. de (1864), *El médico Botanical Criollo,* Vol. II(2), p. 511. F. Brachet, Paris.

Guerreiro, E., Kunesch, G. and Polonsky, J. (1971), *Phytochemistry* 10, 2139.

Gupta, A. K. Das and Paul, M. S. (1970), *J. Indian chem. Soc.* 47, 1017.

Hug, R., Hansen, H. J. and Schmid, H. (1972), *Helv. chim. Acta* 55, 1675.

Iwai, I. and Ide, J. (1963), *Chem. pharm. Bull.* 11, 1042.

Jurd, L. (1968), *Experientia* 24, 858.

Jurd, L. (1969a), *Tetrahedron Letters* 2863.

Jurd, L. (1969b), *Tetrahedron* 25, 1407.

Jurd, L., King, A. D., Mihara, K. and Stanley, W. L. (1971), *Appl. Microbiol.* 21, 507.

Jurd, L., Stevens, K. and Manners, G. (1972), *Phytochemistry* 11, 3287.

Jurd, L., Stevens, K. and Manners, G. (1973), *Tetrahedron* 29, 2347.

Kathpalia, Y. P. and Dutt, S. (1953), *Indian Soap J.* 18, 213.

Kawazu, K., Ohigashi, H. and Mitsui, T. (1968), *Tetrahedron Letters* 2383.

Kawazu, K., Ohigashi, H., Takahashi, N. and Mitsui, T. (1972), *Bull. Inst. Chem. Res. Kyoto Univ.* 50, 160.

Kirkiacharian, B. S. and Billet, D. (1972), *Bull. Soc. chim. France* 3292.

Kumari, D. and Mukerjee, S. K. (1967), *Tetrahedron Letters* 4196.

Kumari, D., Mukerjee, S. K. and Seshadri, T. R. (1965), *Tetrahedron* 21, 1495.

Kumari, D., Mukerjee, S. K. and Seshadri, T. R. (1966), *Tetrahedron Letters* 3767.

Kumari, D., Mukerjee, S. K. and Seshadri, T. R. (1967), *Tetrahedron Letters* 1153.

Kunesch, G. and Polonsky, J. (1967), *Chem. Commun.* 317.

Kunesch, G. and Polonsky, J. (1969), *Phytochemistry* 8, 1221.

Kunesch, G. and Polonsky, J. (1971), *Biochemie* 53, 431.

Laiho, S. M. and Fales, H. M. (1964), *J. Am. chem. Soc.* 86, 4434.

Larkin, J., Nonhebel, D. C. and Wood, H. C. S. (1970), *Chem. Commun.* 455.

Links, J. (1960), *Biochim. Biophys. Acta* **38**, 193.

Mageswaran, S., Ollis, W. D., Roberts, R. J. and Sutherland, I. O. (1969), *Tetrahedron Letters* 2897.

Marini-Bettolo, G. B., Casinovi, C. G., Conçalves da Lima, O., Dalia Maia, M. H. and D'Albuquerque, I. L. (1962), *Ann. Chim.* 119.

Merkel, K. E. (1970). Ph.D. thesis, University of New York, Buffalo.

Morgan, J. W. W., Orsler, R. J. and Wilkinson, D. S. (1968), *Brit. J. ind. Med.* **25**, 119.

Morsingh, F. and Robinson, R. (1970), *Tetrahedron* **26**, 281.

Mukerjee, S. K., Saroja, T. and Seshadri, T. R. (1968), *Tetrahedron* **24**, 6527.

Mukerjee, S. K., Saroja, T. and Seshadri, T. R. (1969a), *Indian J. Chem.* **7**, 671.

Mukerjee, S. K., Saroja, T. and Seshadri, T. R. (1969b), *Indian J. Chem.* **7**, 844.

Mukerjee, S. K., Saroja, T. and Seshadri, T. R. (1970), *Indian J. Chem.* **8**, 21.

Mukerjee, S. K., Saroja, T. and Seshadri, T. R. (1971), *Tetrahedron* **27**, 799.

Nigam, S. K., Mitra, G. R., Kunesch, G., Das, B. C. and Polonsky, J. (1967), *Tetrahedron Letters* 2633.

Ogiyama, K. and Yasue, M. (1973), *Phytochemistry* **12**, 2544.

Ollis, W. D. (1966), *Experientia* 777.

Ollis, W. D. (1967). In *Recent Advances in Phytochemistry* (T. J. Mabry, R. E. Alston and V. C. Runeckles, eds.), Vol. I, pp. 361–373. Appleton-Century-Crofts, New York.

Ollis, W. D. (1970), *An. Acad. Brasil. Cienc.* **42**, 9.

Ollis, W. D. and Gottlieb, O. R. (1968),*Chem. Commun.* 1396.

Ollis, W. D., Redman, B. T., Roberts, R. J., Sutherland, I. O. and Gottlieb, O. R. (1968), *Chem. Commun.* 1392.

Ormancey-Potier, A., Buzas, A. and Lederer, E. (1951), *Bull. Soc. chim. France* 577.

Perkin, W. H. Jr., Ray, J. N. and Robinson, R. (1928),*J. chem. Soc.* 1504.

Perkin, W. H., Jr. and Robinson, R. (1908),*J. chem. Soc.* **93**, 489.

Polonsky, J. (1955), *Bull. Soc. chim. France* 541.

Polonsky, J. (1956), *Bull. Soc. chim. France* 914.

Polonsky, J. (1957), *Bull. Soc. chim. France* 1079.

Rao, M. M. and Seshadri, T. R. (1963), *Tetrahedron Letters* 211.

Robinson, R. (1955), *The Structural Relations of Natural Products*, p. 42. Clarendon Press, Oxford.

Robinson, R. (1962). In *Chemistry of Natural and Synthetic Colouring Matters* (T. S. Gore, B. S. Joshi, S. V. Sunthankar and B. D. Tilak, eds.). Academic Press, New York and London.

Sánchez-Viesca, F. (1969), *Phytochemistry* **8**, 1821.

Sánchez-Viesca, F., Diaz, E. and Chávez, G. (1967), *Ciencia Mexico* **25**, 135.

Saxena, V. K., Tiwari, K. P. and Tandon, S. P. (1970), *Proc. natl. Acad. Sci. India* **40A**, 165.

Schmid, E., Fráter, Gy., Hansen, H.-J. and Schmid, H. (1972), *Helv. chim. Acta* **55**, 1625.

Seshadri, T. R. (1959), *Tetrahedron* **6**, 173.

Shighiro, A., Sato, K., Asami, T., Amakasu, K., Itutura, T. and Nishio, N. (1970), *Chem. Abstr.* **72**, 78881u.

Sidwell, W. T. L. and Tamm, Ch. (1970), *Tetrahedron Letters* 475.

Somanathan, R. and Sultanbawa, M. U. S. (1972), *J. chem. Soc. Perkin I* 1935.

Stout, G. H. and Stevens, K. L. (1964), *J. org. Chem.* **29**, 3604.

Thomson, R. H. (1971), *Naturally Occurring Quinones*, pp. 117–123. Academic Press, London and New York.

Whalley, W. B. (1956), *Chem. Ind.* 1049.

Whalley, W. B. (1961), In *Recent Developments in the Chemistry of Natural Phenolic Compounds* (ed. W. D. Ollis), p. 31. Pergamon, London.

Chapter 16

Biosynthesis of Flavonoids

KLAUS HAHLBROCK AND HANS GRISEBACH

16.1 Introduction

Interest in the biosynthesis of flavonoids was first stimulated by studies on genetic aspects of flower colour (cf. Harborne, 1967; Hess, 1968) and by chemical speculations on the mode of formation of the carbon skeleton of this class of compounds (Birch and Donovan, 1953; Robinson, 1955); tracer studies were first applied to the problem around 1957. Investigations with intact plants or plant tissues led to a basic knowledge of the precursors required and to an understanding of some details of the biosynthesis of flavonoids. From this, a general picture of the interrelationships between various classes of these compounds has emerged. In the course of these tracer experiments, however, it became apparent that a more detailed knowledge of the nature and sequence of the individual biosynthetic steps and their regulation could only be gained by investigations of the enzymes involved.

The beginning of enzymic studies on flavonoids was marked by the discovery of the first enzyme of the phenylpropanoid pathway, phenylalanine ammonia-lyase, by Koukol and Conn (1961). Further progress in this field was limited for some years, probably due to difficulties involved in the isolation of enzymes from higher plants and to the relatively low concentration and variation in the activities of enzymes of secondary metabolism during most stages of plant growth (see Sections 16.3.3 and 16.3.4). Such difficulties have been circumvented more recently by the use of plant cell suspension cultures and this has led to the discovery and isolation of a number of enzymes of flavonoid biosynthesis. These investigations now form the basis of our knowledge of the enzymology of flavonoid biosynthesis.

Since a number of reviews covering the tracer work have recently appeared (Neish, 1964; Grisebach, 1965, 1967, 1968; Grisebach and Barz, 1969; Pachéco, 1969), the main emphasis of this chapter is placed on enzymic studies.

16.2 Origin of flavonoids from primary metabolism

16.2.1 Tracer experiments

It was deduced from feeding experiments with radioactively labelled compounds that flavonoids originate from 'acetate units' and from a phenylpropanoid intermediate derived from the shikimic acid pathway. Results of this work have been reviewed extensively (see above). Basically, ring A is formed by head-to-tail condensation of three 'acetate units' while ring B as well as C-atoms 2, 3 and 4 arise from a phenylpropanoid precursor (Fig. 16.1). A more detailed knowledge of this reaction and of the chemical nature of the immediate precursors was recently obtained from enzymic studies. This confirmed the proposal that CoA esters of malonic and cinnamic acids are the substrates of an enzyme-mediated condensation reaction (see Section 16.3.3.2). Such a mechanism was originally proposed by analogy with similar condensations which take place in fatty acid biosynthesis (Grisebach, 1962).

In a few cases, it has been reported that shikimic acid is a better precursor for flavonoid biosynthesis than cinnamic acid or L-phenylalanine as judged from incorporation rates and dilution values. For instance, the dilution value determined for the incorporation of $(3\text{-}^{14}C)$-cinnamate into delphinidin in *Viola cornuta* is 7700 as opposed to a dilution value of 55 for the incorporation of $(1,2\text{-}^{14}C)$-shikimate (Pla *et al.*, 1967). Similarly, shikimic acid is a more efficient precursor of catechins and flavonols in tea shoots than L-phenylalanine and cinnamic acid (Zaprometov and Bukhleava, 1968, 1971). Meier and Zenk (1965) reported the incorporation of labelled 3,4,5-trihydroxycinnamic acid into delphinidin in *Campanula medium* with a higher efficiency (1.4%) than *p*-coumaric

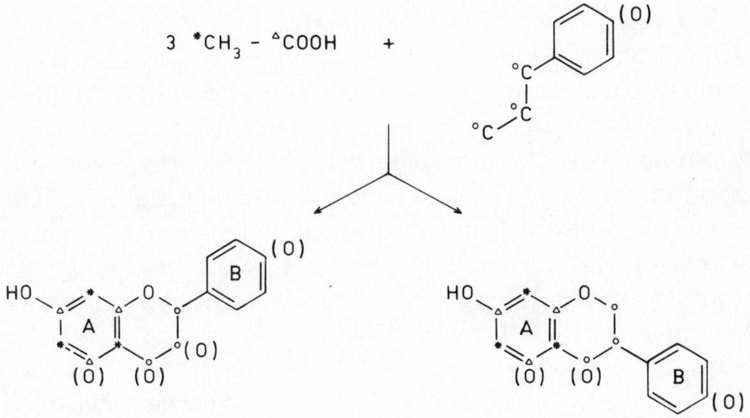

Figure 16.1 Origin of the carbon atoms in flavonoid and isoflavonoid compounds.

(0.35%) or caffeic acid (0.7%). On the basis of the results obtained with *Viola cornuta*, Pla *et al.* (1967) postulated that 3,4,5-tri-hydroxycinnamic acid could be formed from shikimic acid by a more direct route which bypassed cinnamic acid. However, enzymic evidence for a route from shikimic acid to highly substituted cinnamic acids which does not pass through phenylalanine is completely lacking. The possibility must therefore be considered that the above results were due to differences in permeability or pool size, to compartmentation, or to the trapping of intermediates by other pathways.

Different incorporation rates of various substituted cinnamic acids into flavonoids have led to the question whether the substitution pattern of ring B of flavonoid compounds is determined at the cinnamic acid stage or whether further substitution of this ring occurs after formation of the chalcone/flavanone intermediate. This has been discussed in detail in earlier reviews (see Section 16.1) and by Hess (1968) in connection with his 'cinnamic acid starter hypothesis'. According to this hypothesis, the enzyme catalysing the condensation reaction between an activated cinnamic acid and malonyl CoA (see Section 16.3.3) exhibits a high degree of specificity with regard to the substitution pattern of a certain cinnamic acid from a given pool of precursor molecules. Hence, caffeic acid would be the precursor of 3',4'-dihydroxyflavonoids, ferulic acid of 4'-hydroxy,3'-methoxyflavonoids and so on.

Hess (1964) has shown that ferulic and sinapic acids labelled in the O-methyl groups were best incorporated into the correspondingly substituted anthocyanins in *Petunia hybrida*. However, only 12 to 65% of the radioactivity associated with the anthocyanins were found in the O-methyl groups indicating extensive demethylation of these compounds. Results reported by Steiner (1970) on the incorporation of $(2\text{-}^{14}C)$-sinapic acid into the flower anthocyanins of the same plant also indicated that demethylation and 'demethoxyl-ation' reactions occurred, even though this acid was preferentially incorporated into anthocyanins with a substitution pattern of ring B corresponding to that of the labelled precursor.

When the incorporation of p-(Me-^{14}C, Me-^{3}H)-methoxy-(β-^{14}C)-cinnamic acid into acacetin (5,7-dihydroxy-4'-methoxyflavone) was studied in older leaves of *Robinia pseudoacacia* (64 hr. feeding time), a decrease in the ^{3}H/^{14}C ratio to 23% in the isolated product relative to that in the precursor was observed. This result again suggested that demethylation had taken place. However, when p-(Me-^{3}H)-methoxy-(β-^{14}C)-cinnamic acid was administered to younger leaves of the same plant for shorter periods of time (8, 20 and 30 h, respectively) the ^{3}H/^{14}C ratio determined for acacetin remained the same, within the limits of experimental error, as that of the precursor (Ebel *et al.*, 1970b). The latter observation proves the incorporation of p-methoxycinnamic acid intact into acacetin and is therefore in agreement with Hess's hypothesis mentioned above.

On the other hand, much evidence has been accumulated from tracer experiments (Section 16.3.1) and enzymic studies (Sections 16.3.3 and 16.3.8) which show that modifications in the substitution pattern of ring B also occur at the chalcone/flavanone stage or at a later stage of flavonoid biosynthesis. Obviously both possibilities, the intact incorporation of substituted cinnamic acids and the further modification of ring B at, or after, the chalcone/flavanone stage have to be considered.

16.2.2 Enzymology of general phenylpropanoid metabolism

16.2.2.1 Definition

The term 'general phenylpropanoid metabolism' is used in this context to describe the sequence of reactions involved in the conversion of phenylalanine and/or tyrosine to activated cinnamic acids. The known enzymes related to this pathway are phenylalanine ammonia-lyase, cinnamic acid 4-hydroxylase, p-coumarate:CoA ligase, and possibly phenolases and methyltransferases (Fig. 16.2). The products of this biosynthetic sequence are not only intermediates in flavonoid biosynthesis, but also precursors of a number of other phenolic compounds in higher plants. The enzymes of this general pathway will therefore be treated separately from those exclusively involved in the formation of flavonoids (see Section 16.3). That a very close metabolic relationship exists between the enzymes of general phenylpropanoid metabolism is apparent from their interdependent regulation in a number of plant tissues (Amrhein and Zenk, 1970a; Hahlbrock *et al.* 1971a, b; Hahlbrock and Wellmann, 1973).

16.2.2.2 Phenylalanine ammonia-lyase

Phenylalanine ammonia-lyase (PAL, E.C. 4.3.1.5) catalyses the formation of *trans*-cinnamic acid from L-phenylalanine (Fig. 16.2)

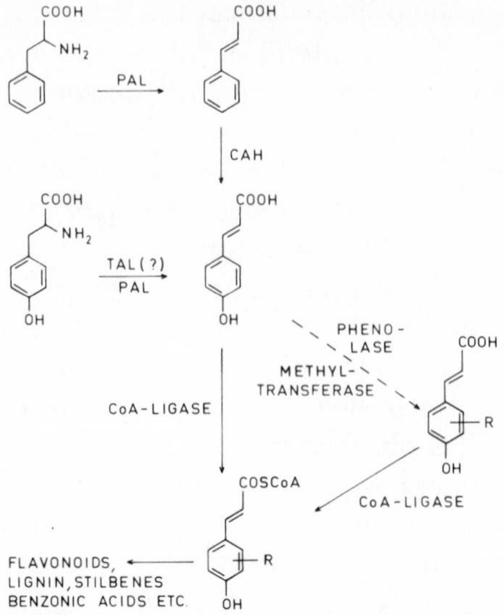

Figure 16.2 Reactions catalysed by the enzymes of general phenylpropanoid metabolism. PAL = Phenylalanine ammonia-lyase; TAL = tyrosine ammonia-lyase; CAH = cinnamic acid 4-hydroxylase; R = hydroxyl and/or methoxyl groups in various positions.

and thus provides the metabolic link between primary metabolism and the phenylpropanoid pathway. The isolation and partial purification of this enzyme was first described by Koukol and Conn (1961). Numerous reports on various aspects of this enzyme have since accumulated, including its purification from different plant species to apparent homogeneity, the determination of its molecular weight and studies on the substrate specificity and on the reaction mechanism at the catalytic centre. Most of this work has recently been reviewed by Hanson and Havir (1972a, b) and by Camm and Towers (1973). The enzyme from maize seems to be composed of four, probably identical, subunits each bearing one active site (Havir and Hanson, 1972). Dehydroalanine was identified as an essential constituent of the catalytic centre (Hanson and Havir, 1970). No co-factor requirements could be demonstrated, except that thiol reagents were found to be essential in some cases.

The isolation of ammonia-lyases active towards L-tyrosine and

β(3,4-dihydroxyphenyl)-L-alanine from higher plants has also been reported (Neish, 1961; Macleod and Pridham, 1963). However, no tyrosine ammonia-lyase has so far been found which does not possess PAL activity. There is much evidence to support the assumption that the same enzyme acts on phenylalanine and on its hydroxylated derivatives (Havir et al., 1971) albeit at different relative rates with enzyme preparations from different sources (Hanson and Havir, 1972a).

Much effort has been devoted to the demonstration of PAL isoenzymes, especially in those cases when a certain level of enzyme activity could be considerably increased by treatment of the plant material with light (Havir and Hanson, 1968; Ahmed and Swain, 1970; Amrhein and Zenk, 1971; Schopfer, 1971; Hahlbrock et al., 1971b). However, convincing evidence for the separation of two distinct isoenzymes of PAL has only been reported by Boudet and his associates (Boudet et al., 1971; Alibert et al., 1972) working with leaves or roots of Quercus pedunculata. On the basis of differences in the inhibitory effects of various cinnamic and benzoic acids on PAL activity, the two isoenzymes were assigned to phenylpropanoid and benzoic acid metabolism respectively.

Another interesting aspect of this enzyme is the correlation between flavonoid production and the expression of its activity in developing plant tissues or after treatment of plants, plant tissues, or plant cell cultures with light, micro-organisms, or hormones and other chemicals (see Sections 16.3.3 and 16.3.4 and Chapter 18). Regulatory effects on PAL activity in vivo may be exerted through end-product inhibition by flavonoids, as Attridge et al. (1971) concluded from experiments with crude enzyme preparations in vitro.

16.2.2.3 Cinnamic acid 4-hydroxylase

The enzymic conversion of cinnamic acid to 4-hydroxycinnamic acid (p-coumaric acid) was first reported by Nair and Vining (1965) with extracts from spinach leaves. However, several attempts to reproduce these observations were unsuccessful (cf. Russell, 1971). Two years later, Russell and Conn (1967) described the isolation of cinnamic acid 4-hydroxylase from Pisum sativum. The enzyme was associated with the microsomal fraction isolated from apical buds and catalyzed the reaction depicted in Fig. 16.3. More detailed

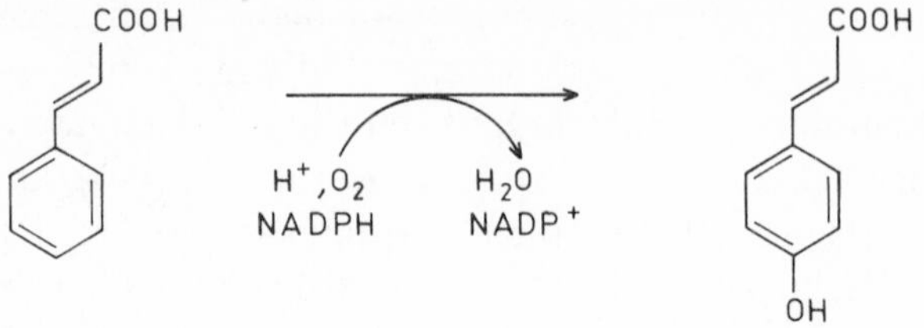

Figure 16.3 Conversion of cinnamic acid to 4-hydroxy-cinnamic (p-coumaric) acid by cinnamic acid 4-hydroxylase.

studies (Russell, 1971) revealed that this enzyme is a mixed function oxidase which requires molecular oxygen, NADPH, and mercapto-ethanol for the hydroxylation of *trans*-cinnamic acid. The reaction product, *p*-coumaric acid, was shown to exert a strong regulatory effect at concentrations of 3×10^{-7} to 10^{-6} M, that is at concentrations significantly lower than those reported for the inhibition of PAL from *Sorghum vulgare* by cinnamic or *p*-coumaric acid (Koukol and Conn, 1961). Studies on the effects of various inhibitors on the activity of cinnamic acid 4-hydroxylase from etiolated sorghum seedlings, in particular on the light reversal of the inhibition by carbon monoxide as a function of the wavelength of the light, indicated that cytochrome P-450 is involved in the enzyme-mediated reaction (Potts *et al.*, 1974).

With regard to the mechanism of the reaction, Russel *et al.* (1968) have shown that when (4-^{3}H)-cinnamic acid was used as a substrate of the enzyme from pea seedlings, the label was almost exclusively retained in the *meta* position of the *p*-coumaric acid formed. Reed *et al.* (1973) using chick pea microsomes confirmed this observation and proposed a sequence of reactions which involved the isomerization of an arene oxide intermediate as shown in Fig. 16.4. A

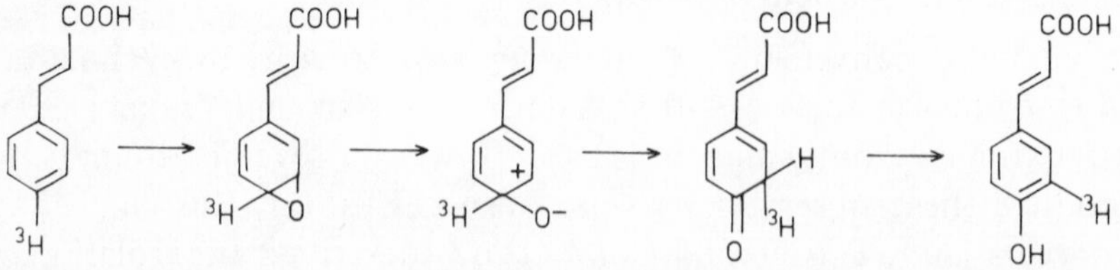

Figure 16.4 Hypothetic sequence of reactions involved in the conversion of (4-^{3}H)-cinnamic acid to (3-^{3}H)-*p*-coumaric acid catalyzed by a microsomal preparation from chick pea (*Cicer arietinum*) as proposed by Reed *et al.* (1973).

considerable reduction of the retention of tritium by the presence of adjacent deuteriums in the substrate was interpreted in favor of the proposed mechanism (Reed *et al.*, 1973).

16.2.2.4 *p*-Coumarate: CoA ligase

Early experiments on the formation of CoA thiol esters of cinnamic and *p*-coumaric acids using extracts from leaves of chick pea (*Cicer arietinum*) and parsley (*Petroselinum crispum*) were limited by the relatively low rates of activation, compared with a number of aliphatic acids, and by the lack of criteria indicating the presence of an enzyme (or enzymes) specific for the activation of cinnamic acids (Grisebach *et al.*, 1966). While the synthesis of CoA thiol esters of various cinnamic acids with a non-specific acid:CoA ligase (E.C. 6.2.1.2) from beef liver mitochondria activating a large number of aliphatic and aromatic acids (Mahler *et al.*, 1953; Schachter and Taggart, 1954) was described by Gross and Zenk (1966), Walton and Butt (1970, 1971) were the first to demonstrate the formation of cinnamoyl CoA in extracts from leaves of spinach beet (*Beta vulgaris*). Although this reaction was reported to be 'virtually specific' for cinnamic acid as substrate, a correlation between this enzymic activity and phenylpropanoid metabolism was not possible (Walton and Butt, 1971).

p-Coumarate:CoA ligase was first isolated from illuminated cell suspension cultures of *Petroselinum crispum* and shown to be specifically related to flavonoid biosynthesis (Hahlbrock and Grisebach, 1970). Subsequently, the enzyme was also demonstrated in cell suspension cultures of soya bean (*Glycine max*) during a short period of the growth cycle which was defined by concomitant large increases in the activities of several enzymes involved in phenylpropanoid metabolism (Hahlbrock *et al.*, 1971b; Hahlbrock and Kuhlen, 1972; Ebel and Grisebach, 1973).

The enzyme from soya bean cell cultures has been purified about twelvefold, and it has been used as a convenient source for the enzymic synthesis of *p*-coumaroyl CoA, since no loss of enzyme activity was observed either when the cells were kept frozen at $-20°C$ or when the partially purified *p*-coumarate: CoA ligase was stored at this temperature (Lindl *et al.*, 1973). A molecular weight of about 55 000 daltons was determined for *p*-coumarate: CoA ligase from soya bean. The enzyme requires CoASH, ATP and Mg^{2+} for

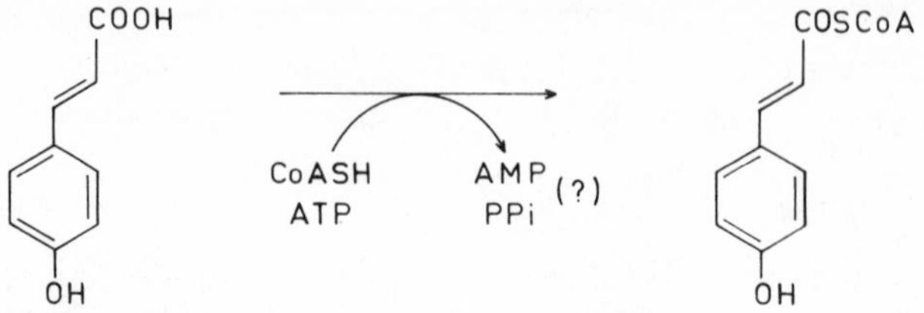

Figure 16.5 Formation of *p*-coumaroyl CoA from *p*-coumaric acid by *p*-coumarate : CoA ligase.

activity (see Fig. 16.5) and is stabilized in solution to some extent by the addition of thiol reagents such as mercaptoethanol or dithio-erythritol.

16.2.2.5 Phenolases

Vaughan and Butt (1969) described the isolation and purification of an enzyme from spinach beet (*Beta vulgaris*) which catalyses the conversion of *p*-coumaric acid to caffeic (3,4-dihydroxycinnamic) acid. While the enzyme exhibited only low substrate specificity, considerably high specificity for the position of hydroxylation was observed. Besides *p*-coumaric acid, the flavonoids naringenin, dihydrokaempferol and kaempferol were also hydroxylated in the 3'-position (Vaughan *et al.*, 1969). However, no conclusion as to whether cinnamic acids or flavonoids (or both) are natural substrates of this enzyme can be drawn from the available data on its substrate specificity (Roberts and Vaughan, 1971).

The enzyme from spinach beet showed both hydroxylase and catechol oxidase activities at a constant ratio throughout an extensive purification of about 1000-fold (Vaughan and Butt, 1969). Ascorbic acid was used as an effective reducing agent which caused reduction of *ortho*-quinones formed by the phenolase. The purified enzyme was shown to contain copper which, according to Vaughan and Butt (1970), might be converted by the further oxidation of the intermediate *ortho*-dihydric phenols into a form in turn catalytically active in the hydroxylation reaction. Sato (1969), on the other hand, ascribed to ascorbic acid, in the reaction catalysed by a phenolase from the mushroom *Agaricus campestris*, a dual function as substrate of the enzyme as well as an agent to prevent caffeic acid from being further oxidized.

A phenolase preparation catalyzing the hydroxylation of *p*-coumaric acid to caffeic acid with ascorbate as reducing agent was also obtained from cell suspension cultures of parsley (Schill and Grisebach, 1973). In contrast to the phenolase from *Beta vulgaris*, the enzyme preparation from parsley cells showed only weak hydroxylating activity with naringenin as substrate and no detectable reaction with dihydrokaempferol or apigenin. Whereas light increased the activity of all other enzymes involved in flavone glycoside biosynthesis in parsley (see Section 16.3.3), illumination of the cell cultures had no effect on the extractable enzyme activity for hydroxylation of *p*-coumaric acid or naringenin. It is therefore unlikely that the phenolase is specifically involved in flavone glycoside biosynthesis in parsley cell cultures.

16.2.2.6 Hydroxycinnamic acid 3-*O*-methyltransferases

Specific methylation in the 3-*O*-position of various hydroxycinnamic acids in cell-free extracts from plants has been reported by Finkle and Nelson (1963), Finkle and Masry (1964), Hess (1966), Finkle and Kelly (1970), Shimada *et al.* (1970), Mansell and Seder (1971), Ebel *et al.* (1972) and Shimada *et al.* (1972). The donor of the methyl groups was shown to be *S*-adenosylmethionine (Finkle and Masry, 1964).

While most of these reactions were not specific for cinnamic acids but also included, among other substrates, several flavonoids, Hess (1966) based his 'cinnamic acid starter hypothesis' (see Section 16.2.1) partially on the observation that cinnamic acids were more efficient substrates for 3-*O*-methylation in extracts from *Petunia hybrida* than were anthocyanins. By contrast, however, recent studies on the substrate specificity of an *O*-methyltransferase purified from cell suspension cultures of *Petroselinum crispum* (Ebel *et al.*, 1972) and on changes in the activity of this enzyme after treatment of the cells with light suggested that in this case methylation takes place preferentially at the flavonoid stage (see Section 16.3.3).

A final answer to the question, at which stages during flavonoid biosynthesis hydroxylation and methylation occurs will depend on the results of studies on the substrate specificities of all of the enzymes related to this pathway, especially of *p*-coumarate:CoA ligase and chalcone/flavanone synthetase, the two enzymes involved in reactions which interlock general phenylpropanoid metabolism with flavonoid biosynthesis.

16.3 Intermediates and end-products of flavonoid biosynthesis

16.3.1 The chalcone/flavanone intermediate

As an extension of Birch's hypothesis (Birch and Donovan, 1953) it was postulated (Grisebach, 1962) that the first specific reaction in flavonoid biosynthesis is the enzyme-mediated condensation of an activated cinnamic acid with three molecules of malonyl CoA to give a chalcone or flavanone (Fig. 16.6). This was further substantiated by feeding experiments with (3-^{14}C)-cinnamic acid using shoots of *Cicer arietinum* from which labelled 4,2′,4′-trihydroxychalcone could be isolated by dilution analysis (Grisebach and Brandner, 1962b). More recently, Endress (1972) isolated radioactive 3,4,2′,4′,6′-pentahydroxychalcone after administration of (1-^{14}C)-

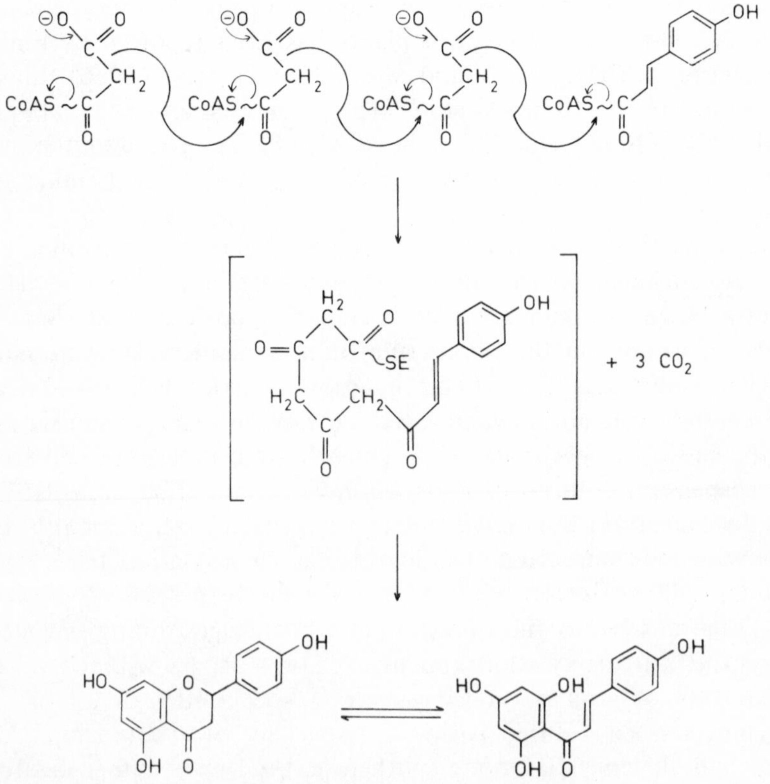

Figure 16.6 Proposed mechanism for the formation of chalcone/flavanone from CoA thiol esters of malonic and *p*-coumaric acids.

acetate to buds of *Petunia hybrida.* Direct evidence for the reaction mechanism formulated in Fig. 16.6 has been obtained from experiments with an enzyme preparation from cell suspension cultures of parsley which catalyses the formation of naringenin (5,7,4'-trihydroxyflavanone) from *p*-coumaroyl CoA and malonyl CoA (see Section 16.3.3.2).

Using specifically labelled chalcones and flavanones it has been repeatedly demonstrated that these compounds are central intermediates from which most, if not all, other flavonoids originate. Results of these experiments are summarized in Table 16.1. The following general conclusions should be noted.

(i) The incorporation of a chalcone/flavanone into other flavonoids seems to be specific for the hydroxylation pattern of ring A. Chalcones with a phloroglucinol-type ring A are exclusively incorporated into 5,7-dihydroxyflavonoids, while chalcones with a resorcinol-type ring A are selectively converted to 7-hydroxyflavonoids.

(ii) No such specificity was observed with regard to the substitution pattern of ring B. Thus, 4-hydroxychalcones (chalcone numbering) are not only incorporated into 4'-hydroxy- and 4'-methoxyflavonoids, but also into flavonoids further substituted in this ring. No significant differences in the incorporation rates and dilution values were found when the conversions of labelled 4,2',4',6'-tetrahydroxychalcone and 3,4,2',4',6'-pentahydroxychalcone 2'-*O*-glucosides to quercetin and cyanidin were compared (Table 16.1, No. 2). A particularly specific incorporation of (β-^{14}C)-2,2',4',6'-tetrahydroxychalcone into datiscetin was observed with *Datisca cannabina* (Table 16.1, No. 10); no label was incorporated into galangin (see Fig. 16.7). Cinnamic and *p*-coumaric acids were much less efficiently converted to datiscetin than the chalcone with the correct substitution pattern of ring B (Grisebach and Grambow, 1968).

(iii) It could be argued that labelled chalcone/flavanone intermediates administered to a plant are first degraded to cinnamic acids and subsequently enter the pool of the precursors of flavonoid biosynthesis. However, the intact incorporation of a multiply ^{14}C-labelled flavanone into cyanidin and biochanin A has been demonstrated by feeding experiments with seedlings of *Brassica oleracea* and *Cicer arietinum* (Table 16.1, No. 4). Conclusive evidence has not been obtained to answer the question whether

Table 16.1   Incorporation of the chalcone/flavanone precursor into various flavonoids

No.	Chalcone/Flavanone precursor	Product	Incorporation rate (%)	Dilution factor	Plant	Remarks	References
1	[β-¹⁴C]-4,2',4',6'-Tetrahydroxychalcone 2'-glucoside (I)	Cyanidin	n.d.		Brassica oleracea var. capitata seedlings	Radioactivity located in C-2 of the product	Grisebach and Patschke, 1961.
		Quercetin	0.6	480	Fagopyrum esculentum seedlings		
2	Chalcone I	Cyanidin	2.63	14	Fagopyrum esculentum seedlings	Parallel experiments; no significant difference in the incorporation of either chalcone	Patschke and Grisebach, 1965b.
		Quercetin	0.11	3040			
	[β-¹⁴C] 3,4,2',4',6'-Pentahydroxychalcone 2'-glucoside	Cyanidin	2.26	15			
		Quercetin	0.07	3930			
3	Chalcone I	Cyanidin	0.31	541	Haplopappus gracilis cell suspension cultures	For comparison with incorporation of dihydrokaempferol see text	Fritsch et al., 1971.
4	[2,6,8,10-¹⁴C₄]-5,7,4'-Trihydroxyflavanone	Cyanidin	n.d.	n.d.	Brassica oleracea seedlings	The radioactivity ratio ring A / C-2 remained unchanged	Patschke et al., 1964.
		Biochanin A	n.d.	n.d.	Cicer arietinum seedlings		
5	(2S)[2-¹⁴C]-5,7,4'-Trihydroxyflavanone 5-β-D-glucoside	Cyanidin	2.36	28	Fagopyrum esculentum seedlings		Patschke et al., 1966b.
		Quercetin	3.07	146			
		Biochanin A	0.21	163	Cicer arietinum seedlings		
		Formononetin	0.005	27 000			
	(2R)[2-¹⁴C]-5,7,4'-Trihydroxyflavanone 5-β-D-glucoside	Cyanidin	0.66	99	Fagopyrum esculentum seedlings	Comparison of incorporation of enantiomers	
		Quercetin	0.19	2514			
		Biochanin A	0.015	5440	Cicer arietinum Seedlings		
		Formononetin	0.0023	46 000			
6	[2-¹⁴C]-5,7,4'-Trihydroxyflavanone	Apigenin	1.98	1075	Petroselinum crispum	About 90% of the radioactivity were located in C-2 of apigenin; for comparison with incorporation of 3,5,7,4'-tetrahydroxy-flavanone see text.	Grisebach and Bilhuber, 1967.
7	Chalcone I	Phloridzin	n.d.	n.d.	Malus toringo leaves	Incorporation of I into phloridzin also with cell-free extract from freeze-dried leaves	Grisebach and Patschke, 1962.
8	[β-¹⁴C,β-³H]-4,2',4',6'-Tetrahydroxy chalcone 2'-glucoside	Taxifolin	n.d.	n.d.	Chamaecyparis obtusa leaves	¹⁴C/³H ratio was unchanged in taxifolin	Grisebach and Kellner, 1965.

No.	Chalcone/substrate	Product		Plant material	Remarks	Reference
	Trihydroxyflavanone 7-O-glucoside	7-O-glucoside (Cosmosine)		flowers		Chabannes et al., 1966.
10	[β-^{14}C]-2,2',4,6'-Tetrahydroxychalcone 4'-glucoside	Datiscetin 7-O-glucoside (Cosmosine) Galangin	0.71 no incorporation · 1728	Datisca cannabina plants	For comparison with other precursors see text	Grisebach and Grambow, 1968.
11	Chalcone I	Epicatechin	n.d. · n.d.	Thea sinensis leaves	All radioactivity was located in C-2 of epicatechin	Patschke and Grisebach, 1965a.
12	[β-^{14}C]-4,4',6'-Trihydroxychalcone 4'-glucoside (II)	Formononetin	n.d. · 417	Trifolium pratense and Cicer arietinum seedlings	All radioactivity was located in C-2 of formononetin; no incorporation into biochanin A	Grisebach and Patschke, 1960; Grisebach and Brandner, 1961.
13	Chalcone I	Biochanin A	n.d. · n.d.	Cicer arietinum seedlings	All radioactivity was located in C-2 of biochanin A; very low and non-specific incorporation into formononetin	Grisebach and Brandner, 1962a.
14	Chalcone II	Formononetin	n.d. · n.d.	Cell-free extract from Cicer arietinum seedlings		Grisebach and Brandner, 1962b.
15	[2,6,8,10-^{14}C]-5,7,4'-Trihydroxyflavanone	Biochanin A	0.05 · 46	Cicer arietinum seedlings	Radioactivity ratio ring A/C-2 remained unchanged	Patschke et al., 1964.
16	Chalcone II 4-[Me-^{3}H]-Methoxy-2',4'-dihydroxy-[β-^{14}C]-chalcone	Formononetin Formononetin	5.3 · 61 / 0.03 (^{14}C) · 13 360 / 0.02 (^{3}H) · 10 210	Cicer arietinum seedlings	Parallel experiments: no intact incorporation of 4-methoxychalcone	Barz and Grisebach, 1967.
17	Chalcone II	Coumestrol	0.05 · 182	Medicago sativa	All radioactivity was located in C-2 of coumestrol	Grisebach and Barz, 1964.
18	[β-^{14}C]-4,2',4'-Trihydroxychalcone	Coumestrol Daidzein	0.03 · 29 / 0.13 · 12	Aseptic Phaseolus aureus seedlings	For comparison with other precursors see text	Dewick et al., 1970.
19	[γ-^{14}C]-4,2',4'-Trihydroxychalcone	Formononetin 3,7,4'-Trihydroxyflavanone	>0.45 · 34 / >0.6 · n.d.	Cicer arietinum seedlings and cell free extracts		Wong, 1965.
20	Chalcone II	Amorphigenin	1.22 · 732	Amorpha fruticosa seeds	No incorporation of the corresponding 4-methoxy- and 2,4-dihydroxychalcones	Crombie et al., 1971b.
21	[γ-^{14}C]-4,2',4'-Trihydroxychalcone 4'-glucoside	6,4'-Dihydroxyaurone 6-glucoside	1 / 1.5	Glycine max cell free extracts		Wong, 1966a.
22	[γ-^{14}C]-4,2',4'-Trihydroxychalcone	7,4'-Dihydroxyflavone, daidzein and formononetin	1.9 / 7.1	Trifolium subterraneum seedlings		Wong, 1968.

n.d. = not determined.

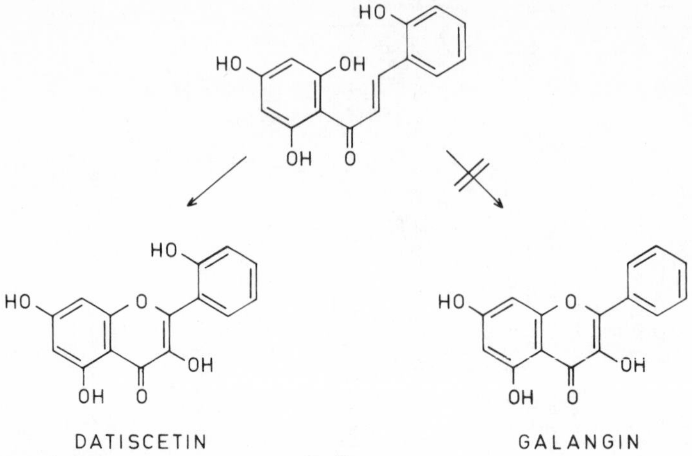

DATISCETIN   GALANGIN

Figure 16.7 Specific incorporation of 2,2',4',6'-tetrahydroxy-chalcone into datiscetin in *Datisca cannabina* (cf. Table 16.1, No. 10).

chalcones or flavanones are the more direct precursors of the various flavonoids. Since (2S)-5,7,4'-trihydroxyflavanone (the naturally occurring enantiomer) was shown to be a significantly more efficient precursor of cyanidin, quercetin, and biochanin A than the (2R) enantiomer (Table 16.1, No. 5), it was assumed that the flavanone is the immediate precursor of these flavonoids (Patschke *et al.*, 1966b). However, this is not an unequivocal interpretation, since it can also be assumed that only the (2S)-flavanone serves as substrate for the reverse reaction of the chalcone-flavanone isomerase (see Section 16.3.2) which then would supply a chalcone concentration sufficiently high for other enzymes to act on this substrate. Competitive feeding experiments with chalcones and flavanones (Wong, 1968) and kinetic studies on the conversion of the pair (4-^{14}C)-4,2',4'-trihydroxychalcone/(3',5'-^{3}H)-7,4'-dihydroxyflavanone to the corresponding dihydroflavonol, flavone, flavonol and 4'-methoxy-isoflavone (Wong and Grisebach, 1969) were interpreted as favouring a chalcone as the more immediate precursor. Certainly, a definite answer to this question will only be obtained from enzymic studies which include a complete separation of chalcone-flavanone isomerase activities from the other enzymes involved.

Other possible cyclization products of chalcones are dihydro-aurones, intermediates in the biosynthesis of aurones (see Chapter 9).

While no enzymic evidence for such a reaction sequence is available, Wong (1966a, b, 1967) reported the conversion of 4,2′,4′-trihydroxychalcone (1) to the corresponding 6,4′-dihydroxyaurone, hispidol (5), by cell-free extracts from soya bean seedlings, and the isolation of two diastereoisomers of 4′,6-dihydroxy-2-(α-hydroxybenzyl)coumaranone (4) which were proposed as intermediates in the biosynthesis of the aurone (Fig. 16.8).

Since the chalcone was also converted to both the corresponding dihydroflavonol (2) and flavonol (3) by cell-free extracts of chick pea and soya bean seedlings (Wong, 1965; Wong and Wilson, 1972) and to the flavonol (3) and aurone (5) either by extracts of *Phaseolus vulgaris* hypocotyls or by horseradish peroxidase (Rathmell and Bendall, 1972), it is possible that these reactions (Fig. 16.8) are

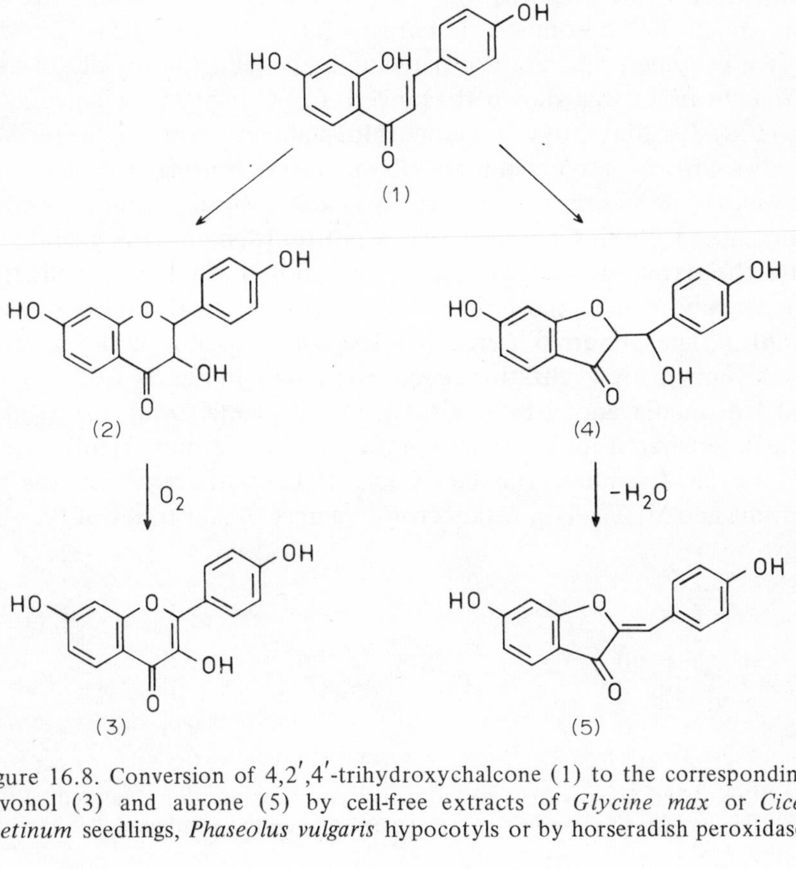

Figure 16.8. Conversion of 4,2′,4′-trihydroxychalcone (1) to the corresponding flavonol (3) and aurone (5) by cell-free extracts of *Glycine max* or *Cicer arietinum* seedlings, *Phaseolus vulgaris* hypocotyls or by horseradish peroxidase.

generally catalyzed by peroxidases via free radicals as intermediates. Similar conclusions have been drawn by Pelter *et al.* (1971) from studies on the oxidation of a chalcone by potassium ferricyanide to the corresponding aurone, flavone, dihydroflavonol, and isoflavone (see Section 16.3.7).

16.3.2 Chalcone-flavanone isomerase

Chalcone-flavanone isomerase was the first enzyme reported to catalyse a reaction specifically involved in the biosynthesis of flavonoids (Moustafa and Wong, 1967). The enzyme has no co-factor requirements and catalyses the formation of the 6-membered heterocyclic ring of flavanones from the corresponding chalcones. Although the chemical equilibrium of this reaction lies far on the side of the flavanone at physiological pH values, there is no unequivocal evidence as yet concerning the direction of the enzyme-mediated isomerization *in vivo* (cf. Section 16.3.1).

An enzyme preparation isolated and purified from soya bean seed (*Glycine max*) was shown to convert 4,2′,4′-trihydroxychalcone to (−)-(2S)-7,4′-dihydroxyflavanone (Moustafa and Wong, 1967). NMR studies on the stereochemistry of this reaction using two chalcone-flavanone isomerase isoenzymes from mung bean seedlings (*Phaseolus aureus* Roxb.) revealed that a proton (deuteron) is introduced into the flavanone molecule upon formation of the heterocyclic ring specifically in the axial position at C-3 (Fig. 16.9) (Hahlbrock *et al.*, 1970b). The preferred stereochemical conformation of the product of the enzymatic cyclization reaction is shown in Fig. 16.10.

Chalcone-flavanone isomerases from all plants so far investigated can be separated into varying numbers of isoenzymes (Hahlbrock *et al.*, 1970a; Grambow and Grisebach, 1971). Differences in the pH optima and Michaelis constants for a number of substrates of the two

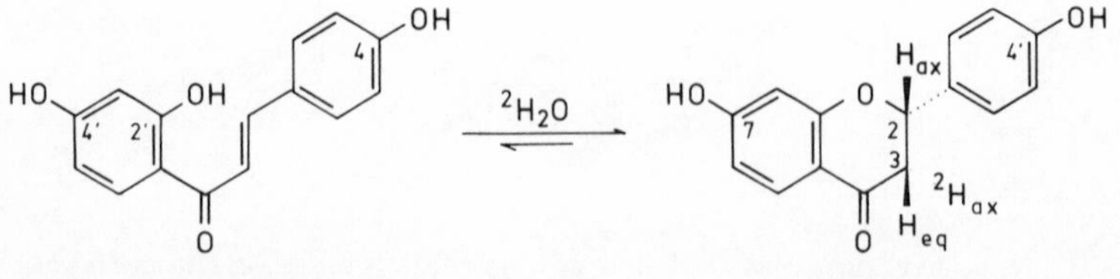

Figure 16.9 Stereospecificity of the isomerization reaction catalysed by chalcone-flavanone isomerase from *Phaseolus aureus* seedlings.

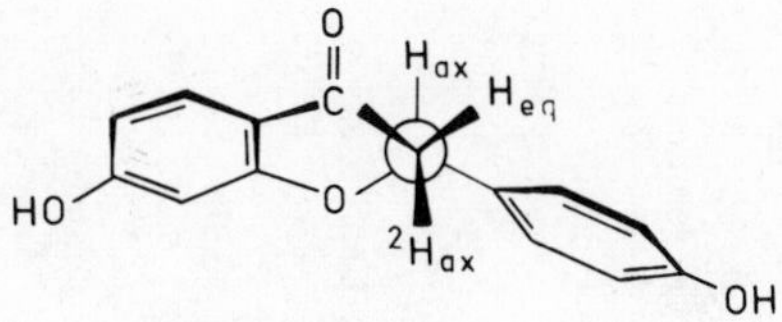

Figure 16.10 Preferred conformation of the (−)(2S)-7,4′-dihydroxyflavanone formed from the corresponding chalcone (see Fig. 16.9) by either one of the two chalcone-flavanone isomerase isoenzymes from *Phaseolus aureus* seedlings.

isomerases purified from mung bean seedlings suggest that these enzymes must be regarded as true isoenzymes and not as enzymes differing only in protein conformation. Furthermore, the K_m values measured for several chalcones suggest that substrate specificities of a chalcone-flavanone isomerase from a particular source correspond significantly to the substitution patterns of flavonoids occurring in that particular plant (Table 16.2 and Fig. 16.11). Thus, isomerases from parsley leaves only operate on 4,2′,4′,6′-tetrahydroxychalcone, the chalcone which corresponds in A-ring hydroxylation pattern with

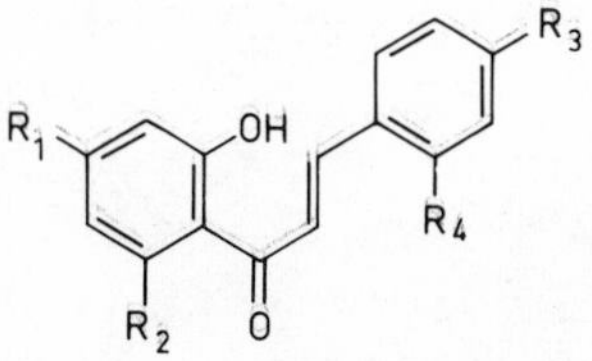

Figure 16.11 Chalcones tested as substrates of the chalcone-flavanone isomerases listed in Table 16.2.

(A) $R_1 = OH; R_2 = R_3 = R_4 = H$
(B) $R_1 = R_2 = R_4 = H; R_3 = OH$
(C) $R_1 = R_3 = OH; R_2 = R_4 = H$
(D) $R_1 = R_2 = OH; R_3 = R_4 = H$
(E) $R_1 = R_2 = R_3 = OH; R_4 = H$
(F) $R_1 = R_2 = R_4 = OH; R_3 = H$
(G) $R_1 = OGlc; R_3 = OH; R_2 = R_4 = H$
(H) $R_1 = OGlc; R_2 = OH; R_3 = R_4 = H$
(I) $R_1 = OGlc; R_2 = R_3 = OH; R_4 = H$
(K) $R_1 = OGlc; R_2 = R_4 = OH; R_3 = H$
(L) $R_2 = OGlc; R_1 = R_3 = OH; R_4 = H$

Table 16.2 *Comparison of some chalcone-flavanone isomerase isoenzymes and heteroenzymes*

Source of enzyme	Number of isoenzymes separated	tested	Chalcones tested (cf. Fig. 16.11)	Chalcones converted (K_m values, mole l^{-1})	Basis for distinction between isoenzymes	References
Phaseolus aureus (seedlings)		2	Isoenzyme I: A,B,C,E,G,I,L Isoenzyme II: A,B,C,E,G,I,L	A(2.6·10^{-4}),B(1.9·10^{-4}),C(5.7·10^{-5}),E(1.8·10^{-5}) A(4.1·10^{-5}),B(3.7·10^{-5}),C(1.4·10^{-5}),E(4.4·10^{-5})	Differences in electrophoretic mobility, K_m values, pH optimum	Hahlbrock *et al.*, 1970a.
Cicer arietinum (germs)	≥2	1	A,B,C,E,I,L	A(1.8·10^{-5}),B(7·10^{-6}),C(1.6·10^{-5}),E(4.4·10^{-5})	Differences in electrophoretic mobility	Hahlbrock *et al.*, 1970a.
Petroselinum crispum (leaves)	≥5	1	A,B,C,E,G,I,L	E(1.6·10^{-5})	Differences in electrophoretic mobility	Hahlbrock *et al.*, 1970a.
Datisca cannabina (leaves)	≥3	Mixture	C,D,E,F,I,K,L	D,E,F (K_m not det.)	Differences in electrophoretic mobility	Grambow and Grisebach, 1971.

the specific flavonoids found in parsley. Accordingly, the rather low specificities of isomerases from *Phaseolus aureus* and *Cicer arietinum* are in agreement with the occurrence of both phloroglucinol- and resorcinol-type substituted flavonoids in these plants. These implications are further supported by tracer studies which suggest that the removal of the 5-hydroxyl group (flavanone numbering) takes place at a stage prior to chalcone/flavanone formation (see Section 16.3.1).

Since none of the chalcone glucosides listed in Table 16.2 was isomerized by the enzymes tested, it must be assumed that chalcones, and not chalcone glucosides, are the natural intermediates. Investigations on the relationship between chalcone-flavanone isomerase activities and flavonoid biosynthesis in various plants and plant tissues will be discussed below (Sections 16.3.3 and 16.3.4).

16.3.3 Enzymology of flavonoid glycoside formation in *Petroselinum crispum*

16.3.3.1 Regulation of flavonoid glycoside formation in plants and cell suspension cultures

Parsley leaves contain flavone and flavonol glycosides with a phloroglucinol-type A-ring. Since relatively few enzymic steps are involved in the formation of such compounds, this plant is a good choice for enzymic studies of flavonoid biosynthesis. Recent progress in these investigations was largely stimulated by the observation that the rates of synthesis and accumulation of flavonoid glycosides in parsley varied greatly with plant growth and differentiation (Hahlbrock *et al.*, 1971c). Striking similarities among the curves obtained for changes in the activities of several enzymes of this pathway during the development of young parsley plants suggested that all of the enzymes involved in this biosynthetic sequence were regulated by the same mechanism (Fig. 16.12).

However, more detailed studies on this regulatory aspect and also the isolation of amounts of protein sufficient for the purification of enzymes were limited by the minute size of the young leaves when enzymic activities were at their maximum (Fig. 16.12). Cell suspension cultures originally derived from leaf petioles of this plant (Vasil and Hildebrandt, 1966) were therefore used for further experiments. Large increases in the activities of enzymes involved in the subsequent accumulation of flavone and flavonol glycosides were

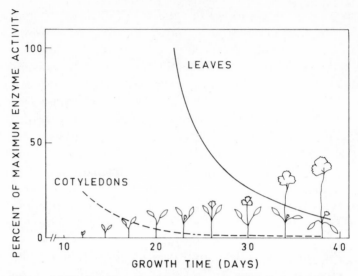

Figure 16.12 Relative changes in the activities of phenylalanine ammonia-lyase, chalcone-flavanone isomerase, UDP-glucose : flavonoid 7-O-glucosyltransferase and UDP-apiose : flavone glycoside apiosyltransferase in cotyledons (– – –) and leaves (———) of developing parsley plants (*Petroselinum crispum.*) For details see Hahlbrock *et al.*, 1971c.

shown to occur upon illumination of parsley cell cultures which had previously been grown in the dark for several days in liquid medium (Hahlbrock and Wellmann, 1970). The products formed were a number of malonylated and non-acylated flavone 7-O-glucosides and 7-O-apiosyl(1 → 2)-glucosides, and flavonol 7-O-glucosides and 3,7-O-diglucosides (Kreuzaler and Hahlbrock, 1973). Fig. 16.13 illustrates the close structural similarities between the flavone and flavonol aglycones.

Two groups of enzymes, which were differently regulated, could be distinguished on the basis of concomitant changes in their activities following the illumination of previously dark-grown cells (Fig. 16.14). The first group consists of the enzymes of general phenylpropanoid metabolism, while the second group comprises all the enzymes specifically related to flavonoid biosynthesis (Hahlbrock *et al.*, 1971a; Hahlbrock, 1972b). As a consequence of this observation, most of the previously unknown enzymes of the flavonoid glycoside pathway could be isolated from illuminated parsley cell suspension cultures. Apart from the substrate specificities

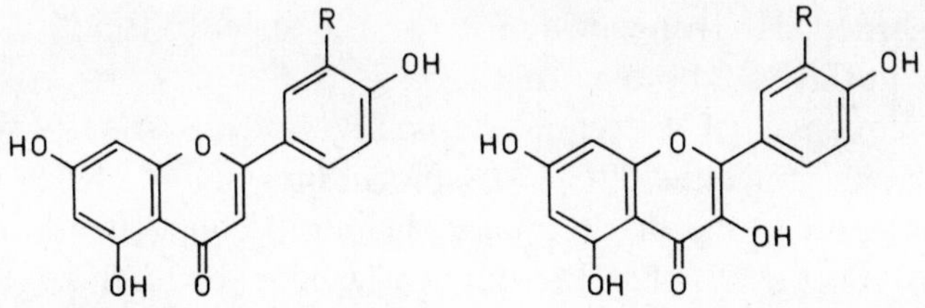

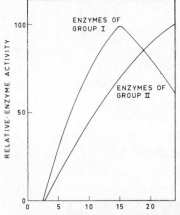

FLAVONES: R =   FLAVONOLS:

APIGENIN   H   KAEMPFEROL

LUTEOLIN   OH   QUERCETIN

CHRYSOERIOL   OCH$_3$ ISORHAMNETIN

Figure 16.13 Aglycones of flavone and flavonol glycosides formed upon illumination of cell suspension cultures of *Petroselinum crispum.*

Figure 16.14 Relative changes in the activities of the enzymes of general phenylpropanoid metabolism (group I) and of flavone glycoside biosynthesis (group II) following the illumination of cell suspension cultures of *Petroselinum crispum*. For details see Hahlbrock *et al.* (1971a), Hahlbrock (1972b) and Sections 16.2.2 and 16.3.3.

of these enzymes, the characteristic light-induced changes in activity were regarded as strong evidence for the specific function of these enzymes in flavonoid biosynthesis.

The sequence of enzymic steps involved in the formation of malonylated apigenin 7-O-[β-D-apiofuranosyl($1 \rightarrow 2$)]β-D-glucoside ('malonylapiin'), one of the major flavonoid constituents in parsley, is shown in Fig. 16.15. Present knowledge concerning the enzymology of this pathway as well as additional reactions specifically related to the formation of 3'-methoxyflavonoids and of flavonol diglucosides will be discussed below. All of the enzymes so far investigated are 'soluble' enzymes. The degrees of purification and some properties of these enzymes are summarized in Table 16.3.

16.3.3.2 Chalcone/flavanone synthetase

This enzyme catalyses the reaction depicted in Fig. 16.16. While no experimental evidence has been obtained so far regarding possible intermediates in the formation of the aromatic ring A from 'acetate units', chemical degradation of the overall reaction product proved that ^{14}C-labelled malonyl CoA was incorporated exclusively into this ring (Kreuzaler and Hahlbrock, 1972). Phenolic compounds which might possibly interfere with the synthetase activity were removed from buffered cell extracts with an anion exchanger. The enzyme has an optimum around pH 8 and does not require the addition of any co-factors for activity (Kreuzaler, unpublished).

No chalcone could be detected as a product of a partially purified synthetase which catalysed the conversion of about 10% of the added p-coumaroyl CoA to naringenin in 3 min (Kreuzaler, unpublished). Since this enzyme preparation was free of chalcone-flavanone isomerase activity, it seems very likely that the flavanone is in fact the immediate product of the synthetase reaction.

16.3.3.3 Chalcone-flavanone isomerase

Whereas a number of chalcone-flavanone isomerase isoenzymes from several plants including *Petroselinum crispum* have been partially purified and characterized (see Section 16.3.2), the properties of this enzyme from parsley cell suspension cultures have not been studied except for changes in activity upon illumination of the cells (Hahlbrock *et al.*, 1971a). The latter observation proves, however, that chalcone-flavanone isomerase is one of the enzymes specifically

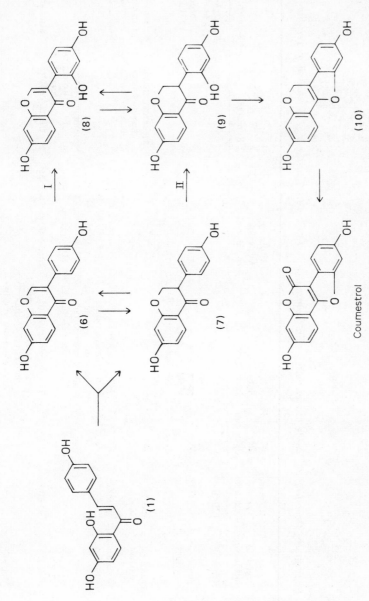

Figure 16.25 Proposed biosynthetic formation of coumestrol in *Phaseolus aureus* illustrating a 'metabolic grid' of isoflavanones and isoflavones, see text for details.

Table 16.3   *Enzymes of flavonoid glycoside biosynthesis from illuminated cell suspension cultures of Petroselinum crispum*

Enzyme	Degree of purification	Molecular weight (daltons)	pH optimum	Substrate specificity	Apparent K_m values of substrates	Remarks	References
Chalcone/flavanone synthetase	300-fold[a]	120 000[a]	8[a]				Kreuzaler and Hahlbrock, 1972. Hahlbrock et al., 1971a.
Chalcone-flavanone isomerase	crude extract	50 000[a]	7[a]	specific for phloroglucinol-type substitution pattern of ring A[a]			
Chalcone/flavanone oxidase	crude extract	n.d.[b]	7[a]			Demonstrated in leaves of young parsley seedlings	Sutter and Grisebach, unpublished.
UDP-Glucose: flavone/flavonol 7-O-glucosyltransferase	89-fold	50 000	7.5	7-O-glucosylation of various flavones, flavonols, flavanones; also conversion of quercetin 3-O-glucoside[a]	apigenin ($2.7 \cdot 10^{-6}$ M) luteolin ($1.5 \cdot 10^{-6}$ M) naringenin ($1 \cdot 10^{-5}$ M) UDP-glucose ($1.2 \cdot 10^{-4}$ M) TDP-glucose ($2.6 \cdot 10^{-4}$ M)	incomplete separation of the two glucosyltransferases on DEAE-cellulose	Sutter et al., 1972.
UDP-Glucose: flavonol 3-O-glucosyltransferase	40-fold[a]	50 000	8	3-O-glucosylation of various flavonols incl. quercetin 7-O-glucoside; no conversion of dihydroquercetin	quercetin ($<10^{-6}$ M)		Sutter and Grisebach, 1973.

Enzyme	Purification	Molecular weight	pI	Reaction	Substrate	Remarks	Reference
UDP-Apiose Synthase	1000-fold	101 000–115 000	8.2		UDP-D-Glucuronic acid ($2 \cdot 10^{-6}$ M)	exhibits also UDP-xylose synthase activity	Baron *et al.*, 1972, 1973.
UDP-Apiose: flavonoid 7-O-glucoside apiosyl-transferase	123-fold	55 000	7.0	Apiosylation of 7-O-glucosides of various flavones, flavanones, isoflavones; no conversion of flavonol 7-O-glucosides	apigenin 7-O-glucoside ($6.6 \cdot 10^{-5}$ M)		Ortmann *et al.*, 1972.
Malonyl CoA: flavonoid glycoside malonyl-transferase	crude extract	70 000[a]	n.d.	—	—		Hahlbrock, 1972b.
SAM: *ortho*-dihydric phenol *meta-O*-methyl-transferase	82-fold	48 000	9.7	Methylation in the *meta*-position of various flavonoids, caffeic acid, protocatechuic acid	luteolin ($4.6 \cdot 10^{-5}$ M), luteolin 7-O-glucoside ($3.1 \cdot 10^{-5}$ M), eriodictyol ($1.2 \cdot 10^{-3}$ M), caffeic acid ($1.6 \cdot 10^{-3}$ M)		Ebel *et al.*, 1972.

[a] Unpublished results.
[b] n.d. = not determined.

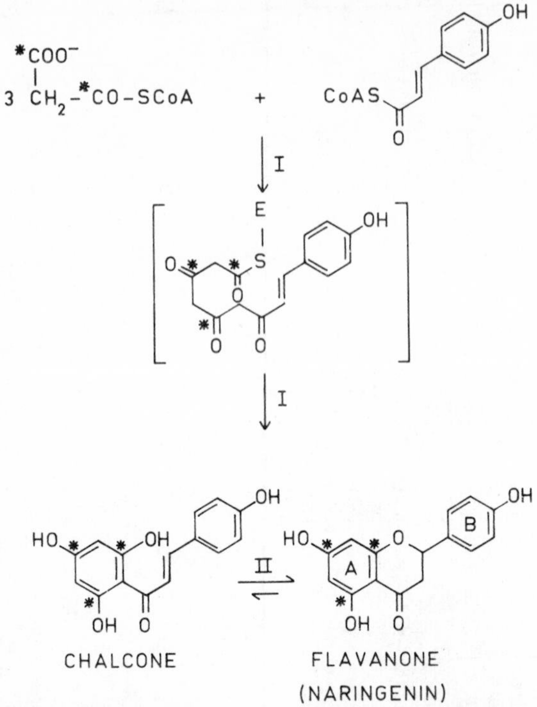

Figure 16.16 Hypothetical scheme for the enzymic formation of 4,2',4',6'-tetrahydroxychalcone and 5,7,4'-trihydroxyflavanone (naringenin) from malonyl CoA and p-coumaroyl CoA. I = Reactions catalysed by chalcone/flavanone synthetase, II = reaction catalysed by chalcone-flavanone isomerase; E = enzyme.

related to and probably essential for flavonoid biosynthesis, regardless of any discussion about its actual function (see Sections 16.3.1, 16.3.2 and 16.3.4).

16.3.3.4 Chalcone/flavanone oxidase

Evidence obtained for the highly co-ordinated expression of all other enzyme activities involved in the pathway illustrated in Fig. 16.15 strongly suggests that an enzyme converting the chalcone/flavanone intermediate to the corresponding flavone should also be active in illuminated parsley cell cultures. However, while attempts to isolate this hypothetical enzyme from the cell cultures have not led to reproducible results, its occurrence in young parsley leaves has recently been demonstrated (Sutter, 1972). The enzymatic reaction requires molecular oxygen, ferrous or ferric iron and a heat-stable co-factor(s) which can be removed by dialysis or gel filtration through Sephadex G-25. The nature of this co-factor(s) is still unknown.

Since chalcone-flavanone isomerase was also active in the crude extracts used for the demonstration of chalcone/flavanone oxidase activity, it has not been possible to decide whether the chalcone or the flavanone is the actual substrate of this enzyme. Further information will be required for its precise identification.

16.3.3.5 Glucosyltransferases

Flavone 7-*O*-glycosides as well as flavonol 7-*O*-glycosides and 3,7-di-*O*-glycosides were shown to be formed by illuminated parsley cell suspension cultures (see Section 16.3.3.1). The corresponding enzymes glucosylating flavonoids specifically in the 7-*O* and in the 3-*O*-position, respectively, were isolated from these cells. The reactions catalysed by the two glucosyltransferases are shown in Fig. 16.17. Both enzymes have a similar molecular weight of about 55 000 daltons as calculated from the elution volumes after gel chromatography (Sutter *et al.*, 1972; Sutter and Grisebach, 1973). They can be separated according to their different charge by DEAE-cellulose column chromatography or by disc gel electrophoresis. Neither requires co-factors, thiol reagents or divalent cations for activity.

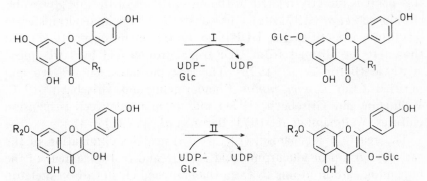

Figure 16.17 Reactions catalysed by two glucosyltransferases, respectively, which were isolated from cell suspension cultures of *Petroselinum crispum*. I = UDP-glucose : flavone/flavonol 7-*O*-glucosyltransferase (R_1 = H or OH), II = UDP-glucose : flavonol 3-*O*-glucosyltransferase (R_2 = H or glucose).

The first enzyme, UDP-glucose:flavone/flavonol 7-O-glucosyltrans-ferase, was purified about 90-fold and is specific for the 7-O-position of various flavones, flavonols (but not flavonol 3-O-glycosides), and flavanones as acceptors and for UDP-glucose or TDP-glucose as glucosyl donors (Sutter *et al.*, 1972). The substrate specificities (see Table 16.3) are consistent with the occurrence of both flavone and flavonol 7-O-glycosides in the cells used as the source of this enzyme.

More recently, the UDP-glucose:flavonol 3-O-glucosyltransferase has also been isolated from cell suspension cultures of parsley (Sutter and Grisebach, 1973). This enzyme was completely separated from the 7-O-glucosyltransferase. While the 3-O-glucosyltransferase is specific for the position of glucosylation, several flavonols, including quercetin 7-O-glucoside, can serve as substrates. However, no glucosylation of dihydroquercetin was observed, supporting the conclusion drawn from tracer experiments that glycosylation is the last step in flavonoid glycoside formation (Grisebach, 1965). It seems likely that this enzyme is involved in the formation of both 3-O-glucosides and 3,7-di-O-glucosides of flavonols in parsley, since no diglucosides are formed by the 7-O-glucosyltransferase.

16.3.3.6 UDP-Apiose synthase

In parsley, the branched-chain sugar apiose occurs only in flavone 7-O-[β-D-apiofuranosyl(1 → 2)]β-D-glucosides (Sandermann, 1969; Kreuzaler and Hahlbrock, 1973). The regulation of its enzymic formation in young leaves and illuminated cell suspension cultures of this plant is directly related to the biosynthesis of flavone glycosides (Hahlbrock *et al.*, 1971a, c). The substrate for the apiosyltransferase reaction was identified as UDP-D-apiose (Ortmann *et al.*, 1972) and shown to be formed from UDP-D-glucuronic acid by UDP-apiose synthase (Baron *et al.*, 1973). This enzyme has been isolated and purified from *Lemna minor* (Sandermann and Grisebach, 1970; Wellmann and Grisebach, 1971) and from parsley cell suspension cultures (Wellmann *et al.*, 1971; Baron *et al.*, 1972, 1973).

The reaction mechanism (Fig. 16.18) involves elimination of the carboxyl group of glucuronic acid as CO_2 and rearrangement of the remaining carbon atoms to form the branched-chain carbon skeleton of apiose (Grisebach and Schmid, 1972). The formation of a second product, UDP-D-xylose, is also catalysed by the same enzyme. The relative rates of formation of the two sugar nucleotides, UDP-D-

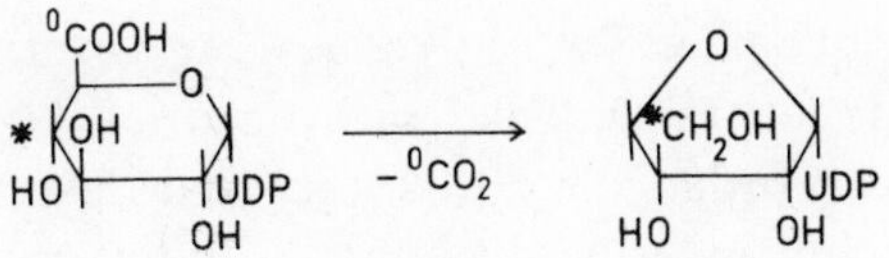

Figure 16.18 Formation of UDP-apiose from UDP-glucuronic acid.

apiose and UDP-D-xylose, remained constant throughout a 1000-fold purification of the enzyme. This suggests that the synthase is either an enzyme complex composed of at least two subunits with different catalytic activities or a multifunctional protein with two catalytic sites.

The molecular weight of the UDP-apiose/UDP-xylose synthase is 100 000–115 000 daltons. The enzyme requires NAD and thiol reagents for activity and is stimulated by NH_4^+. In these properties it differs markedly from a UDP-xylose synthase also present in parsley cell suspension cultures (Wellmann et al., 1971). While the apiose moiety of UDP-apiose is efficiently transferred to flavone 7-O-glucosides, UDP-xylose is not utilized as substrate for the formation of flavonoid glycosides in parsley.

16.3.3.7 Apiosyltransferase

UDP-D-apiose:flavone 7-O-[β-D]-glucoside apiosyltransferase cataly-ses the formation of flavone 7-O-[β-D-]-apiofuranosyl(1 → 2)[β-D]-glucosides as shown in Fig. 16.19 (Ortmann et al., 1970, 1972). The enzyme is highly specific for UDP-D-apiose as glycosyl donor. By contrast, 7-O-glucosides of a large variety of flavones, flavanones and isoflavones, apigenin 7-O-glucuronide and glucosides of p-substituted

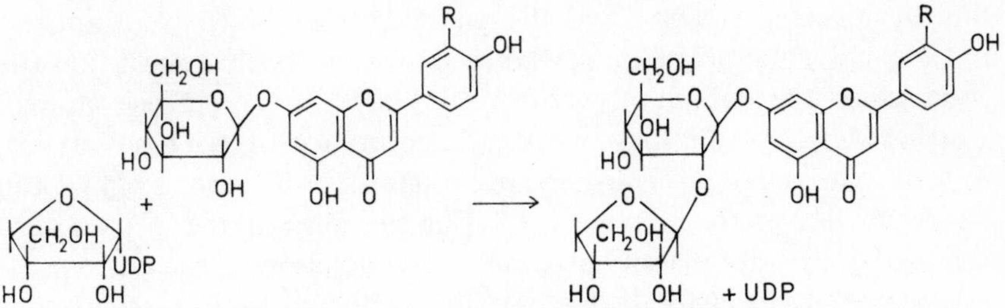

Figure 16.19 Formation of apiin (apigenin 7-O-[β-D]-apiofuranosyl91→2)[β-D]-glucoside) from apigenin 7-O-[β-D]-glucoside and UDP-D-apiose.

phenols (but not phenolic aglycones) can serve as acceptors. However, no reaction takes place with flavonol 7-*O*-glucosides. This observation is in accordance with the fact that flavonols do not occur in parsley as apiosylglucosides but only as mono- or diglucosides (see Section 16.3.3.1).

No co-factors are required for apiosyltransferase activity. The enzyme is rather unstable under various conditions unless partly purified and stored in the presence of thiol reagents such as dithioerythritol. Partial purification of the apiosyltransferase resulted in its complete separation from the 7-*O*-glucosyltransferase (see Section 16.3.3.5). The molecular weight of 50 000 daltons is about the same as that determined for the two glucosyltransferases.

All the data thus far available are consistent with the assumption that flavone apiosylglucosides are formed by a stepwise transfer of glucose and apiose from the corresponding UDP-sugars to the aglycones as shown in Fig. 16.15

16.3.3.8 Malonyltransferase

Malonylation presumably represents the last step of flavonoid glycoside biosynthesis in parsley. The enzyme catalysing this reaction, malonyl CoA:flavonoid glycoside malonyltransferase (Fig. 16.15, No. 7), was demonstrated in crude extracts of parsley cell suspension cultures (Hahlbrock, 1972b). Malonyl CoA functions as donor of the acyl residue. Of the number of possible acceptors (cf. Kreuzaler and Hahlbrock, 1973) only apigenin 7-*O*-apiosylglucoside (apiin) and 7-*O*-glucoside (cosmosiin) and chrysoeriol 7-*O*-apiosylglucoside (gravebioside B) have been tested and shown to be efficiently converted to the corresponding monomalonyl glycosides. Although the exact position of malonylation has not been determined, spectral evidence and the fact that apiin and cosmosiin, but not the aglycone apigenin, served as substrates, both suggest that the product of the enzymic reaction is acylated in the glucose moiety. Furthermore, by anology with the occurrence of 6'-*O*-malonyliso-flavone glucosides in *Trifolium* species (Beck and Knox, 1971) it can be assumed that the enzyme catalysing the formation of flavone and flavonol glycosides malonylates in the 6-*O*-position of the glucose.

The malonyltransferase has not been purified to any extent, and detailed studies on its substrate specificity and other properties have not yet been carried out. However, the enzyme was shown to be specifically related to flavonoid biosynthesis by the characteristic changes in its activity after illumination of the cell cultures (Fig. 16.14).

16.3.3.9 Methyltransferase

As already mentioned (see Section 16.2.2), it is not yet certain whether *O*-methyl groups are introduced into flavonoids at the phenylpropanoid stage or at a later stage during flavonoid biosynthesis, or both. Several lines of evidence suggest that a methyltransferase which has been isolated from illuminated cell suspension cultures of parsley is specifically involved in the methylation of flavonoids (Ebel *et al.*, 1972). Thus, the enzyme catalyses the transfer of the methyl group of *S*-adenosyl-L-methionine (SAM) to 3',4'-dihydroxyflavonoids exclusively in the *meta* position (Fig. 16.20). This is in agreement with the occurrence of a number of glycosides of chrysoeriol and isorhamnetin in these cells (Fig. 16.13). Furthermore, as concluded from the changes in the activity of the SAM: *ortho*-dihydric phenol *meta-O*-methyltransferase after illumination of the cell cultures, this methyltransferase is an enzyme of group II, as shown in Fig. 16.14, like all the other enzymes described in this section. Although ferulic acid is also readily formed from caffeic acid and SAM in the presence of the methyltransferase, the affinity of the enzyme for this acid ($K_m = 1.6 \times 10^{-3}$) is considerably lower than that for luteolin ($K_m = 4.6 \times 10^{-5}$) and luteolin 7-*O*-glucoside ($K_m = 3.1 \times 10^{-5}$).

The partially purified methyltransferase requires Mg^{2+} for optimal activity. In contrast to mammalian catechol *O*-methyltransferases (Molinoff and Axelrod, 1971), the enzyme from parsley cell cultures is not affected by the addition of inhibitors of thiol groups such as

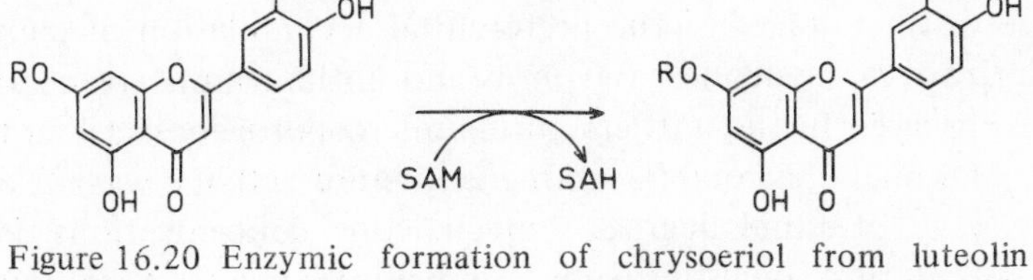

Figure 16.20 Enzymic formation of chrysoeriol from luteolin (R = H) and chrysoeriol 7-*O*-glucoside from luteolin 7-*O*-glucoside (R = glucose). SAM = *S*-adenosylmethionine, SAH = *S*-adenosylhomocysteine.

p-chloromercuribenzoate or iodoacetamide. The molecular weight of the methyltransferase was estimated from the elution volume on a Sephadex G-100 column to be about 48 000 daltons.

16.3.4 Enzymology and regulation of flavonoid biosynthesis in other plants

Certain aspects of this topic will be discussed later in Chapter 18. Investigations of the relationship between flavonoid production and the activities of enzymes related to this pathway have so far largely been confined to changes in PAL activity and the accumulation of anthocyanins, flavonols and phytoalexins in developing plants (Maier and Hasegawa, 1970; Amrhein and Zenk, 1970b; Weissenböck, 1971; Wiermann, 1972a) or in response to wounding or treatment of plants or plant tissues with light, hormones, microorganisms or various chemicals (see Chapter 18).

Amrhein and Zenk (1970a) reported a close relationship between concomitant changes in PAL and cinnamic acid 4-hydroxylase activity and the accumulation of cyanidin in hypocotyls of *Fagopyrum esculentum*. Also the production of apigenin in illuminated cell suspension cultures of *Glycine max* (Hahlbrock, 1972a) seems to be closely related to simultaneous increases in the activities of PAL and *p*-coumarate:CoA ligase (Hahlbrock *et al.*, 1971b).

Of the large number of enzymes more directly involved in flavonoid biosynthesis, only chalcone-flavanone isomerase has been studied in relation to the formation of flavonoids in plants other than parsley. Wiermann (1970, 1972a, b, 1973) investigated the variations in the activity of this enzyme in relation to the accumulation and further metabolism of chalcones and flavonols in developing anthers of *Narcissus pseudonarcissus, Lilium candidum* and *Tulipa* cv. Apeldoorn. The latter plant is an especially interesting one for studying the biosynthesis and metabolism of flavonoids, because distinct stages of the preferential accumulation of cinnamic acid derivatives, chalcones, flavonols and anthocyanins, respectively, can be observed in the anthers during microsporogenesis (Wiermann, 1970). Maximal chalcone-flavanone isomerase activity was measured at the stage of rapid decreases in chalcone concentrations and of concomitant and quantitatively comparable increases in flavonol concentrations (Wierman, 1972b).

Despite some ambiguity regarding the actual role of chalcone-flavanone isomerase in flavonoid biosynthesis, this observation adds further support to the assumption that this enzyme is, indeed, directly involved in this pathway (see Sections 16.3.2 and 16.3.3.3).

16.3.5 Glycosyltransferases from other sources

The formation of flavonol glycosides from the corresponding aglycones and sugar nucleoside diphosphates, besides that described for the parsley system, has been reported in cell-free extracts from *Phaseolus vulgaris* (Jacobelli *et al.*, 1958; Marsh, 1960), *Phaseolus aureus* (Barber and Neufeld, 1961; Barber, 1962), *Leucaena glauca* (Barber and Chang, 1968), *Impatiens balsamina* (Miles and Hagen, 1968) and *Zea mays* (Larson, 1971). Table 16.4 includes a list of the compounds tested as possible substrates and of the products reported to be formed by the action of glycosyltransferases. Of various possible activated sugars only TDP and UDP derivatives serve as glycosyl donors; flavonols were the only flavonoid substrates tested. Since none of the enzymes was purified to any considerable extent, it must be considered possible that the cell-free extracts contained mixtures of more than one glycosyltransferase of different substrate specificities. A comparison with the substrate specificities of the two glucosyltransferases and the apiosyltransferase from parsley (see Sections 16.3.3.5 and 16.3.3.7) is therefore not possible.

Table 16.4 *Glycosylation of flavonols by cell-free extracts from various plants (For enzymes from parsley see Section 16.3.3).*

Source of cell-free extract	Flavonoid compounds tested as possible substrates	Glycosyl donor	Products isolated	References
Phaseolus vulgaris	Quercetin 3-*O*-glucoside	TDP-Rhamnose	Rutin	Jacobelli *et al.*, 1958.
Phaseolus vulgaris	Quercetin	UDP-Glucuronic acid	Quercetin glucuronide	Marsh, 1960.
Phaseolus aureus	Quercetin	UDP-Glucose, TDP-Glucose	Quercetin 3-*O*-glucoside	Barber and Neufeld, 1961;
	Quercetin 3-*O*-glucoside	TDP-Rhamnose	Rutin	Barber, 1962.
Leucaena glauca	Quercetin	UDP-Rhamnose TDP-Rhamnose	Quercetin 3-*O*-glucoside	Barber and Chang, 1968.
Impatiens balsamina	Kaempferol, Quercetin, Myricetin, Pelargonidin	UDP-Glucose TDP-Glucose ADP-Glucose	3-*O*-glucosides of Kaempferol, Quercetin, Myricetin	Miles and Hagen, 1968.
Zea mays	Quercetin	UDP-Glucose	Quercetin 3-*O*-glucoside	Larson, 1971.

Like apiin and graveobioside B in parsley, however, the diglycoside rutin (quercetin 3-O-rhamnoglucoside) is formed by the stepwise addition of the sugar moieties to the aglycone by two different enzymes in extracts from *Phaseolus aureus* (Barber, 1962).

16.3.6 Flavone C-glycosides

A number of flavone O- and C-glycosides with hydroxylation patterns of ring B corresponding to those of apigenin and luteolin occur in *Spirodela polyrhiza, S. oligorhiza*, and in *Lemna minor* (for details see Chapter 12). When ^{14}C-labelled apigenin or luteolin were administered to these plants, radioactivity in the isolated flavonoid products was detected only in O-glycosylated and O-methylated, but not in C-glycosylated flavones (Wallace *et al.*, 1969). In contrast, $(2\text{-}^{14}\text{C})$-5,7,4'-trihydroxyflavanone (naringenin) was incorporated by *S. polyrhiza* in a parallel manner into the 7-O-glucosides of apigenin and luteolin and into apigenin 8-C-glucoside (vitexin) and luteolin 8-C-glucoside (orientin) (Wallace and Grisebach, 1973). Based on these results and on enzymic studies of flavonoid O-glucosylation (see Section 16.3.3 and 16.3.4, the pathway shown in Fig. 16.21 is proposed for the biosynthesis of flavone O- and C-glycosides in *Spirodela polyrhiza.*

16.3.7 Dihydroflavonols, flavonols, anthocyanins, catechins

Competitive feeding experiments, as described in Section 16.3.1 (Wong, 1968; Wong and Grisebach, 1969), can be cited in support of the mechanism proposed by Pelter *et al.* (1971) for the formation of a dihydroflavonol by oxidation of the corresponding chalcone (Fig. 16.22). Incorporation of $[\beta\text{-}^{14}\text{C},\beta\text{-}^{3}\text{H}]$-4,2',4',6'-tetrahydroxychalcone into taxifolin (3,5,7,3',4'-pentahydroxyflavanone) without significant change in the $^{14}\text{C}/^{3}\text{H}$ ratio (Table 16.1, No. 8) suggests that an intermediate with a central heterocyclic ring and a 2,3-double bond (flavone) or a paraquinoid structure of ring B (cf. Pelter *et al.*, 1971) can be excluded. So far, conclusive evidence for the origin of the oxygen atom in the 3-position of dihydroflavonols has not been obtained. Further experimental work will have to show whether it originates from water or from molecular oxygen.

The conversion of dihydroflavonols into flavonols, anthocyanins

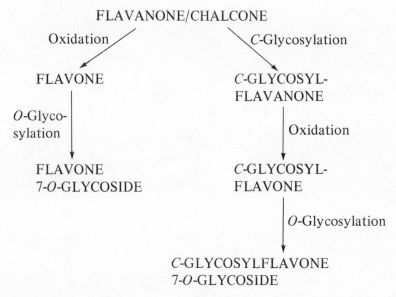

Figure 16.21 Proposed pathway for the biosynthesis of flavone O-glycosides and C-glycosides in *Spirodela polyrhiza*.

and catechins has been demonstrated by tracer studies. Whereas labelled dihydrokaempferol is incorporated into both kaempferol and quercetin in buckwheat seedlings (Patschke *et al.*, 1966a) and in pea plants (Patschke and Grisebach, 1968), dihydroquercetin is incorporated only into quercetin when fed to pea plants.

Earlier reports on the incorporation of [G-³H]-dihydrokaempferol into cyanidin in buckwheat seedlings (Patschke *et al.*, 1966a) were more recently confirmed by feeding experiments with cell suspension cultures of *Haplopappus gracilis* (Fritsch *et al.*, 1971). Dihydrokaempferol was converted to cyanidin at a much higher incorporation rate (up to 12%) and with a much lower dilution value, 20, than the corresponding chalcone or L-phenylalanine. Attempts to demonstrate the formation of cyanidin from dihydrokaempferol in cell-free extracts from *H. gracilis* have so far been unsuccessful.

[G-³H]-Dihydrokaempferol was also incorporated into catechins in young tea shoots with about the same efficiency as [¹⁴C]-shikimic acid or L-[¹⁴C]-phenylalanine (Zaprometov and Grisebach, 1973).

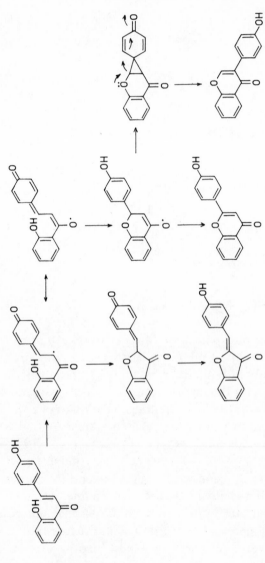

Figure 16.22 Proposed mechanisms for the conversion of chalcones to dihydroflavonols, aurones, flavones and isoflavones (after Pelter *et al.*, 1971).

16.3.8 Isoflavonoids

Extensive tracer studies with phenylpropanoid and chalcone precursors have provided evidence for the formation of the isoflavone skeleton by an aryl migration (Grisebach and Doerr, 1960) occurring at or after the chalcone/flavanone stage (Table 16.1, No. 5, 12–15; Fig. 16.23). It is uncertain, however, whether isoflavanones are intermediates in the biosynthesis of isoflavones. Although dehydrogenation of an isoflavanone to the corresponding isoflavone was demonstrated in chick pea seedlings (*Cicer arietinum*), it seems more likely that, conversely, isoflavanones are formed from isoflavones (Grisebach and Zilg, 1968). Dihydroflavonols do not function as precursors of isoflavones (Wong, 1965; Barz and Grisebach, 1966). Oxidation of a chalcone as shown in Fig. 16.22 has been proposed as one possible mechanism for aryl migration (Pelter *et al.*, 1971).

Feeding experiments with p-[^{14}C,^{3}H]-methoxy-[β-^{14}C]-cinnamic acid (Barz and Grisebach, 1967), p-[Me-^{2}H$_3$]-methoxycinnamic acid (Ebel *et al.*, 1970a) and 4-[Me-^{3}H]-methoxy-2',4'-dihydroxy-[β-^{14}C]-chalcone (Table 16.1, No. 15) were carried out with chick pea seedlings in order to determine the stage during the biosynthesis of 4'-methoxyisoflavones at which methylation of the 4'-hydroxy group occurs. However, because of rapid demethylation reactions, the results did not permit a conclusive interpretation. Further investigations were therefore concentrated on the introduction of the 4'-O-methyl group at an enzymic level. According to preliminary results obtained with a crude enzyme preparation from cell suspension cultures of *Cicer arietinum* (Wengenmayer, 1972), an enzyme specific for the methylation of isoflavones seems to be involved in this reaction. Daidzein and genistein were efficiently converted to formononetin and biochanin A, respectively, by these cell-free extracts in the presence of *S*-adenosylmethionine (Fig. 16.24). No reaction was detectable with various concentrations of *p*-coumaric or

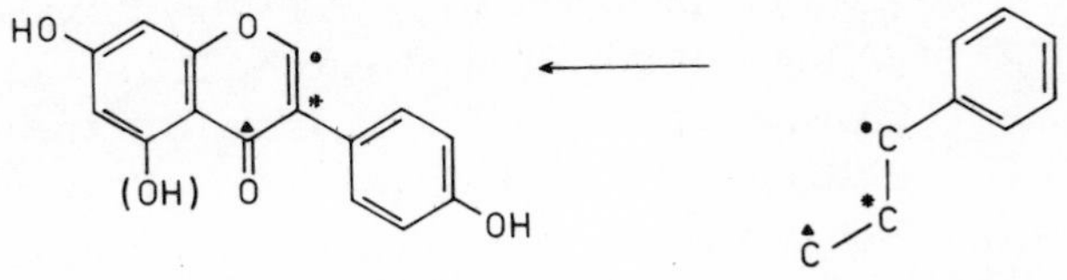

Figure 16.23 Incorporation of phenylpropanoid precursors into isoflavones.

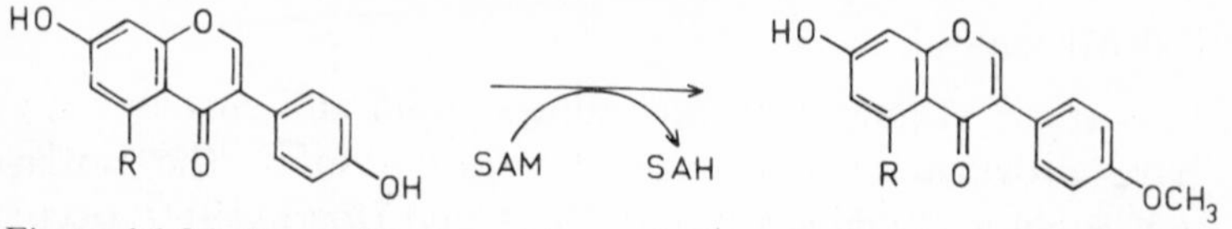

Figure 16.24 Enzymic formation of formononetin from daidzein (R = H) and of biochanin A from genistein (R = OH) in cell-free extracts of cell suspension cultures from *Cicer arietinum*. SAM = S-adenosylmethionine, SAH = S-adenosylhomocysteine.

caffeic acid, with the flavones apigenin and luteolin, or with 4,2',4'-trihydroxychalcone as substrates. Thus methylation apparently takes place only at the isoflavone stage, catalysed by a rather specific S-adenosylmethionine:isoflavone 4'-O-methyltransferase.

Coumestones, e.g. coumestrol, are biosynthetically related to the flavones (Fig. 16.25) and are correspondingly labelled when cinnamic acids (Grisebach and Barz, 1963) and chalcones (Table 16.1, No. 16, 17) are applied as precursors. Experiments designed to obtain information on later steps in coumestone biosynthesis show that daidzein and dihydrodaidzein are very efficient precursors of coumestrol in sterile-grown mung bean seedlings (Table 16.1, No. 17) and in suspension cultures of mung bean root cells (Berlin *et al.*, 1972). On the other hand, 3(4-hydroxyphenyl)-4,7-dihydroxy-coumarin can be excluded as an intermediate between daidzein and coumestrol (Dewick *et al.*, 1970). A pathway involving 2'-hydroxylation of daidzein (Fig. 16.25, route I) seems unlikely because daidzein is a more efficient precursor than 7,2',4'-trihydroxyisoflavone (8). An alternative route (Fig. 16.25, route II) based upon chemical analogy (Dewick *et al.*, 1969) would be dehydration/cyclization of 7,2',4'-trihydroxyisoflavanone (9) to 3,9-dihydropterocarp-6a-en (10) followed by allylic oxidation to coumestrol. Pterocarp-6a-ens are extremely easily oxidized to coumestones (Ferreira *et al.*, 1971).

Since conversions of daidzein to 7,2',4'-trihydroxyisoflavone (Dewick *et al.*, 1969) and of dihydrodaidzein and 7,2',4'-trihydroxy-isoflavone to coumestrol do occur, it appears that the situation existing in *Phaseolus aureus* is best considered to be a 'metabolic grid' (Bu'Lock, 1965). Another possibility is the formation of the pterocarpene from the isoflavanone by direct oxidative cyclization (Wong, 1970). [G-³H]-Daidzein was also incorporated into phase-

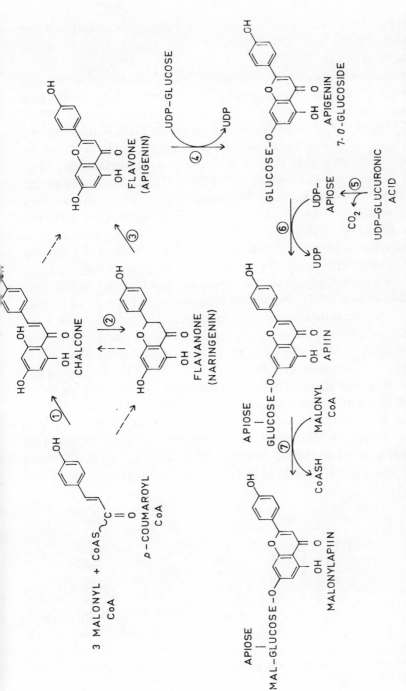

Figure 16.15 Proposed scheme for the enzymatic formation of apigenin glycosides in illuminated cell suspension cultures of *Petroselinum crispum*. The enzymes involved in this sequence of reactions are ① chalcone/flavanone synthetase, ② chalcone-flavanone isomerase, ③ chalcone/flavanone oxidase, ④ UDP-glucose : flavone/flavonol 7-*O*-glucosyltransferase, ⑤ UDP-apiose synthase, ⑥ UDP-apiose : flavone 7-*O*-glucoside apiosyltransferase, ⑦ malonyl CoA : flavonoid glycoside malonyltransferase. Possible alternative routes for reactions 1 to 3 are indicated by dotted and solid arrows, respectively.

ollin (7-hydroxy-3',4'-dimethylchromenylchromanocoumarane) in *Phaseolus vulgaris* with about the same efficiency as L-phenylalanine (Hess *et al.*, 1971).

Investigations on the formation of rotenoids by Crombie and collaborators led to the discovery of a direct biosynthetic link between these compounds and 2'-methoxyisoflavones. When [1-^{14}C]- and [2-^{14}C]-phenylalanine were fed to *Derris elliptica* plants, or [2-^{14}C]-phenylalanine to germinating seeds of *Amorpha fruticosa*, labelling patterns of rotenone and amorphigenin (Fig. 16.26) were consistent with a 1,2-shift of the aromatic ring to the 12a-position (Crombie and Thomas, 1967). The incorporation pattern of [3-^{14}C]-phenylalanine (Fig. 16.26) confirmed this conclusion (Crombie *et al.*, 1968). When [Me-^{14}C]-methionine was fed to these plants the isolated rotenoid was almost exclusively labelled at C-6 and in the methoxy groups (Fig. 16.27) proving that the 'extra' methylene group at C-6 could be provided by methionine (Crombie *et al.*, 1968). 9-Demethylmunduserone (12), an intermediate in the biosynthesis of amorphigenin (17), is formed from 7-hydroxy-2',4',5'-trimethoxyisoflavone (11) with the 'extra' methylene group at C-6 originating from the 2'-methoxy group as shown in Fig. 16.28 (Crombie *et al.*, 1970, 1971b). The corresponding isoflavanone was a much poorer precursor of (17). Formononetin, but not daidzein, was incorporated into (17) (Crombie *et al.*, 1971b). Since an isoflavone bearing the E-ring of amorphigenin or rotenone is not an acceptable precursor, the conclusion can be drawn that prenylation is a late step in rotenoid biosynthesis. This is confirmed by the observation that (±)-[6-^{3}H]-9-demethylmunduserone (12) is

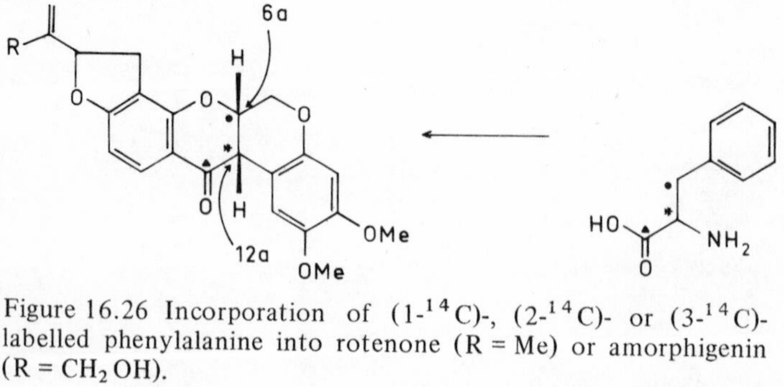

Figure 16.26 Incorporation of (1-^{14}C)-, (2-^{14}C)- or (3-^{14}C)-labelled phenylalanine into rotenone (R = Me) or amorphigenin (R = CH$_2$OH).

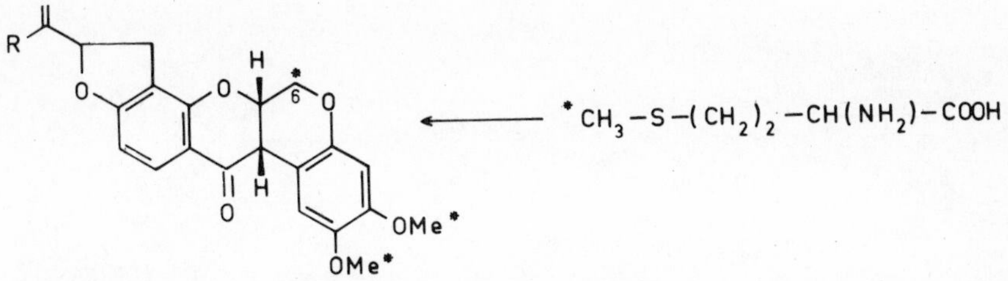

Figure 16.27 Incorporation of radioactive label from (Me-^{14}C)-methionine into rotenone (R = Me) or amorphigenin (R = CH$_2$OH).

an efficient precursor of (17) (Crombie *et al.*, 1971a). Clues to the formation of ring E were obtained by the efficient incorporation of rotenonic acid (13) and rotenone (16) into (19). These observations suggest that the post-isoflavonoid stages in amorphigenin biosynthesis are likely to be (a) formation of 9-demethylmunduserone (12), (b) 8-dimethylallylation (13), (c) epoxidation (14), and (d) cyclization either via the epoxide or via the corresponding diol to give dalpanol (15). Dehydration (e) then leads to rotenone (16) and 8'-hydroxylation (f) to amorphigenin (17). Accordingly, the biosynthesis of rotenoids can be formulated as shown in Fig. 16.28. Details concerning the pathway from the chalcone (1) to the isoflavone (11) are still lacking. From the fact that formononetin but not daidzein is a precursor of (17), Crombie *et al.* (1971b) concluded that the spirodienone (cf. Pelter *et al.*, 1971) could be decomposed by S-adenosylmethionine followed by loss of a proton. However, in the light of the recent discovery of a methyltransferase specific for the 4'-O-position of isoflavones (see above) this explanation seems unlikely. More work with cell-free systems will be necessary to clarify the mechanism.

16.3.9 Neoflavonoids

Radioactive calophyllolide was obtained when DL-[3-^{14}C]-phenylalanine had been fed to young shoots of *Calophyllum inophyllum* (Kunesch and Polonsky, 1967). By degradation and analysis of the labelled product these investigators demonstrated that about 80% of the radioactivity was located in the C-4 position (Fig. 16.29). This result disproves the suggestion (Grisebach and Ollis, 1961) that the biosynthesis of the neoflavonoid skeleton necessarily involves two successive 1,2-aryl shifts. Two possible mechanisms which also were

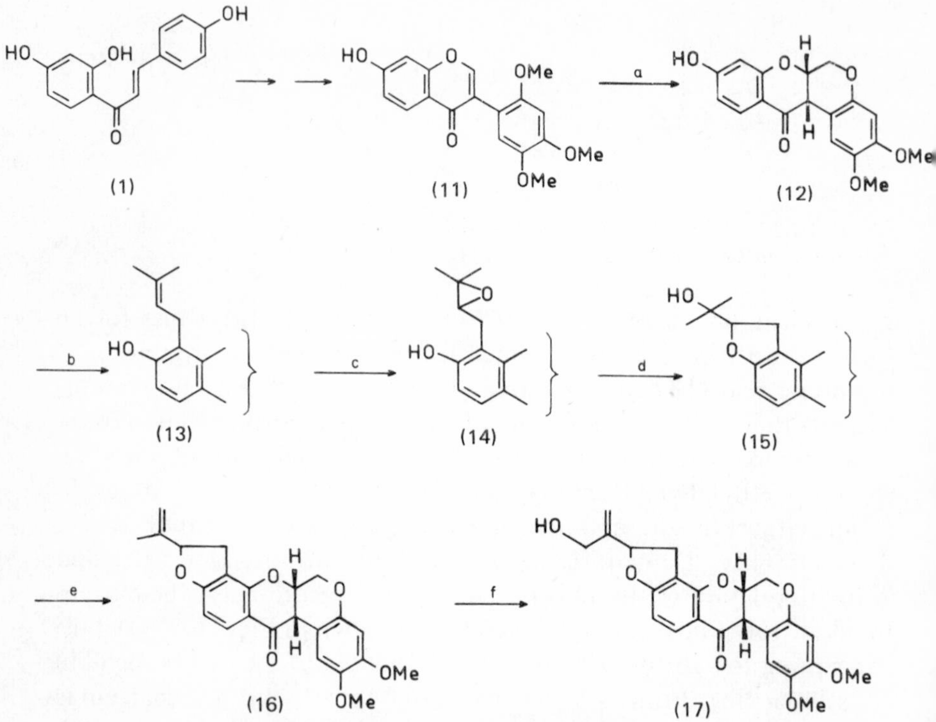

Figure 16.28 Proposed biosynthetic formation of rotenone (16) and amorphigenin (17).

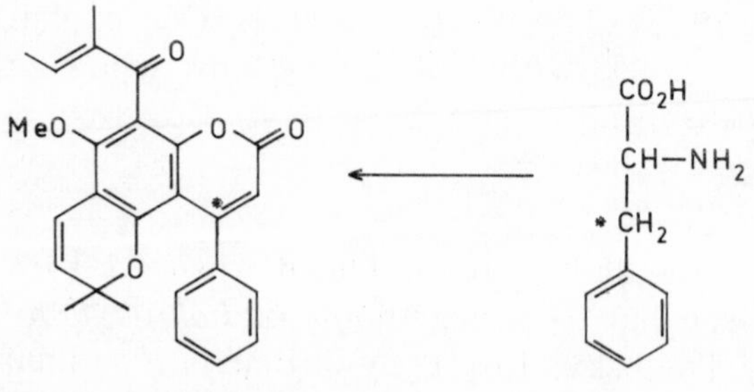

Figure 16.29 Incorporation of (3-^{14}C)-phenylalanine into calophyllolide.

proposed for the biosynthesis of neoflavonoids (Ollis and Gottlieb, 1968) are compatible with the result of the tracer experiment: (a) β-addition of a phenolic unit to a cinnamic acid leading to 4-aryl-coumarins as intermediates, or (b) alkylation of a phenolic unit by, for example, cinnamoyl pyrophosphate or an equivalent activated compound.

The biosynthesis of neoflavonoids is further discussed in Chapter 15 (p. 856) but as yet no enzymic studies have been carried out with these compounds.

References

Ahmed, S. I. and Swain, T. (1970), *Phytochemistry* **9**, 2287.

Alibert, G., Ranjeva, R. and Boudet, A. (1972), *Biochim. Biophys. Acta* **279**, 282.

Amrhein, N. and Zenk, M. H. (1970a), *Naturwissenschaften* **57**, 312.

Amrhein, N. and Zenk, M. H. (1970b), *Ztsch, Pflanzenphysiol.* **63**, 384.

Amrhein, N. and Zenk, M. H. (1971), *Ztsch. Pflanzenphysiol.* **64**, 145.

Attridge, T. H., Stewart, G. R. and Smith, H. (1971), *FEBS Letters* **17**, 84.

Barber, G. A. (1962), *Biochemistry* **1**, 463.

Barber, G. A. and Chang, M. T. Y. (1968), *Phytochemistry* **7**, 35.

Barber, G. A. and Neufeld, E. F. (1961), *Biochem. biophys. Res. Comm.* **6**, 44.

Baron, D., Streitberger, U. and Grisebach, H. (1973), *Biochim. Biophys. Acta* **293**, 526.

Baron, D., Wellman, E. and Grisebach, H. (1972), *Biochim. Biophys. Acta* **258**, 310.

Barz, W. and Grisebach, H. (1966), *Ztsch. Naturf.* **21b**, 47.

Barz, W. and Grisebach, H. (1967), *Ztsch. Naturf.* **22b**, 627.

Beck, A. B. and Knox, J. R. (1971), *Aust. J. Chem.* **24**, 1509.

Berlin, J., Dewick, P. M., Barz, W. and Grisebach, H. (1972), *Phytochemistry* **11**, 1689.

Birch, A. J. and Donovan, F. W. (1953), *Aust. J. Chem.* **6**, 360.

Boudet, A., Ranjeva, R. and Gadal, P. (1971), *Phytochemistry* **10**, 997.

Bu'Lock, J. D. (1965), *The Biosynthesis of Natural Products*, p. 82. McGraw-Hill, London.

Camm, E. L. and Towers, G. H. N. (1973), *Phytochemistry* **12**, 961.

Chabannes, B., Ville, A. and Pachéco, H. (1966), Pli cacheté déposé à la Soc. chim. de France No. 01628.

Crombie, L., Dewick, P. M. and Whiting, D. A. (1970), *Chem. Commun.* 1469.

Crombie, L., Dewick, P. M. and Whiting, D. A. (1971a), *Chem. Commun.* 1182.

Crombie, L., Dewick, P. M. and Whiting, D. A. (1971b), *Chem. Commun.* 1183.

Crombie, L., Green, C. L. and Whiting, D. A. (1968), *J. chem. Soc.* (C) 3029.

Crombie, L. and Thomas, M. B. (1967), *J. chem. Soc.* (C) 1796.

Dewick, P. M., Barz, W. and Grisebach, H. (1969), *Chem. Commun.* 466.

Dewick, P. M., Barz, W. and Grisebach, H. (1970), *Phytochemistry* **9**, 775.

Ebel, J., Achenbach, H., Barz, W. and Grisebach, H. (1970a), *Biochim. Biophys. Acta* **215**, 203.

Ebel, J., Barz, W. and Grisebach, H. (1970b), *Phytochemistry* **9**, 1529.

Ebel, J. and Grisebach, H. (1973), *FEBS Letters* **30**, 141.

Ebel, J., Hahlbrock, K. and Grisebach, H. (1972), *Biochim. Biophys. Acta* **269**, 313.

Endress, R. (1972), *Ztsch. Pflanzenphysiol.* **67**, 188.

Ferreira, D., Brink, C. v. D. M. and Roux, D. G. (1971), *Phytochemistry* **10**, 1141.

Finkle, B. J. and Kelly, S. H. (1970), *Pl. Physiol.* **46**, (Suppl.) 232.

Finkle, B. J. and Masry, M. S. (1964), *Biochim. Biophys. Acta* **85**, 167.

Finkle, B. J. and Nelson, R. F. (1963), *Biochim. Biophys. Acta* **78**, 747.

Fritsch, H. J., Hahlbrock, K. and Grisebach, H. (1971). *Ztsch. Naturf.* **26b**, 581.

Grambow, H. J. and Grisebach, H. (1971), *Phytochemistry* **10**, 789.

Grisebach, H. (1962), *Planta med.* **10**, 385.

Grisebach, H. (1965). In *Chemistry and Biochemistry of Plant Pigments* (ed. T. W. Goodwin), p. 279. Academic Press, London and New York.

Grisebach, H. (1967), *Biosynthetic Patterns in Microorganisms and Higher Plants*, pp. 1–31. J. Wiley, New York.

Grisebach, H. (1968). In *Recent Advances in Phytochemistry* (eds. T. J. Mabry, R. E. Alston and V. C. Runeckles), Vol. 1, pp. 379–406. Appleton-Century-Crofts, New York.

Grisebach, H. and Barz, W. (1963), *Ztsch. Naturf.* **18b**, 466.

Grisebach, H. and Barz, W. (1964), *Ztsch. Naturf.* **19b**, 569.

Grisebach, H. and Barz, W. (1969), *Naturwissenschaften* **56**, 538.

Grisebach, H., Barz, W., Hahlbrock, K., Kellner, S. and Patschke, L. (1966), *Proc. 2nd Meet. Fed. Eur. Biochem. Soc., Vienna* **3**, 25.

Grisebach, H. and Bilhuber, W. (1967), *Ztsch. Naturf.* **22b**, 746.

Grisebach, H. and Brandner, G. (1961), *Ztsch. Naturf.* **16b**, 2.

Grisebach, H. and Brandner, G. (1962a), *Experientia* **18**, 400.

Grisebach, H. and Brandner, G. (1962b), *Biochim. Biophys. Acta* **60**, 51.

Grisebach, H. and Doerr, N. (1960), *Ztsch. Naturf.* **15b**, 284.

Grisebach, H. and Grambow, H. J. (1968), *Phytochemistry* **7**, 51.

Grisebach, H. and Kellner, S. (1965), *Ztsch. Naturf.* **20b**, 446.

Grisebach, H. and Ollis, W. D. (1961), *Experientia* **17**, 4.

Grisebach, H. and Patschke, L. (1960), *Chem. Ber.* **93**, 2326.

Grisebach, H. and Patschke, L. (1961), *Ztsch. Naturf.* **16b**, 645.

Grisebach, H. and Patschke, L. (1962), *Ztsch. Naturf.* **17b**, 857.

Grisebach, H. and Schmid, R. (1972), *Angew. Chem.* **84**, 192; *Angew. Chem. (internatl. Ed.)* **11**, 159.

Grisebach, H. and Zilg, H. (1968), *Ztsch. Naturf.* **23b**, 499.

Gross, G. G. and Zenk, M. H. (1966), *Ztsch. Naturf.* **21b**, 683.

Hahlbrock, K. (1972a), *Phytochemistry* **11**, 165.

Hahlbrock, K. (1972b), *FEBS Letters* **28**, 65.

Hahlbrock, K., Ebel, J., Ortmann, R., Sutter, A., Wellmann, E. and Grisebach, H. (1971a), *Biochim. Biophys. Acta* **244**, 7.

Hahlbrock, K. and Grisebach, H. (1970), *FEBS Letters* **11**, 62.

Hahlbrock, K. and Kuhlen, E. (1972), *Planta* **108**, 271.

Hahlbrock, K., Kuhlen, E. and Lindl, T. (1971b), *Planta* **99**, 311.

Hahlbrock, K., Sutter, A., Wellmann, E., Ortmann, R. and Grisebach, H. (1971c), *Phytochemistry* **10**, 109.

Hahlbrock, K. and Wellmann, E. (1970), *Planta* **94**, 236.

Hahlbrock, K. and Wellmann, E. (1973), *Biochim. Biophys. Acta* **304**, 702.

Hahlbrock, K., Wong, E., Schill, L. and Grisebach, H. (1970a), *Phytochemistry* **9**, 949.

Hahlbrock, K., Zilg, H. and Grisebach, H. (1970b), *Eur. J. Biochem.* **15**, 13.

Hanson, K. R. and Havir, E. A. (1970). *Arch. Biochem. Biophys.* **141**, 1.

Hanson, K. R. and Havir, E. A. (1972a). In *The Enzymes*, 3rd Ed. (ed. P. D. Boyer), Vol. 7, pp. 75–166. Academic Press, New York.

Hanson, K. R. and Havir, E. A. (1972b). In *Recent Advances in Phytochemistry* (eds. V. C. Runeckles and J. E. Watkin), Vol. 4, pp. 45–85. Appleton-Century-Crofts, New York.

Harborne, J. B. (1967), *Comparative Biochemistry of the Flavonoids*, pp. 250–266. Academic Press, London.

Havir, E. A. and Hanson, K. R. (1968), *Biochemistry* **7**, 1896.

Havir, E. A. and Hanson, K. R. (1972), *Pl. Physiol.* **49** (Suppl.), 25.

Havir, E. A., Reid, P. D. and Marsh, H. V. (1971), *Pl. Physiol.* **48**, 130.

Hess, D. (1964), *Planta* **60**, 568.

Hess, D. (1966), *Ztsch. Pflanzenphysiol.* **55**, 374.

Hess, D. (1968), *Biochemische Genetik*, pp. 89–106. Springer-Verlag, Berlin.

Hess, S. L., Hadwiger, L. A. and Schwochau, M. E. (1971), *Phytopathology* **61**, 79.

Jacobelli, G., Tabone, M. J. and Tabone, D. (1958), *Bull. Soc. Chim. biol.* **40**, 955.

Koukol, J. and Conn, E. (1961), *J. biol. Chem.* **236**, 2692.

Kreuzaler, F. and Hahlbrock, K. (1972), *FEBS Letters* **28**, 69.

Kreuzaler, F. and Hahlbrock, K. (1973), *Phytochemistry* **12**, 1149.

Kunesch, G. and Polonsky, J. (1967), *Chem. Commun.* 317.

Larson, R. L. (1971), *Phytochemistry* **10**, 3073.

Lindl, T., Kreuzaler, F. and Hahlbrock, K. (1973), *Biochim. Biophys. Acta* **302**, 457.

Macleod, N. J. and Pridham, J. B. (1963), *Biochem. J.* **88**, 45 P.

Mahler, H. R., Wakil, S. J. and Bock, R. M. (1953), *J. biol. Chem.* **204**, 453.

Maier, V. P. and Hasegawa, S. (1970), *Phytochemistry* **9**, 139.

Mansell, R. L. and Seder, J. A. (1971), *Phytochemistry* **10**, 2043.

Marsh, C. A. (1960), *Biochim. Biophys. Acta* **44**, 359.

Meier, H. and Zenk, M. H. (1965), *Ztsch. Pflanzenphysiol.* **53**, 415.

Miles, C. D. and Hagen, C. W. (1968), *Pl. Physiol.* **43**, 1347.

Molinoff, P. B. and Axelrod, J. (1971), *Annual Rev. Biochem.* **40**, 465.

Moustafa, E. and Wong, E. (1967), *Phytochemistry* 6, 625.
Nair, P. M. and Vining, L. C. (1965), *Phytochemistry* 4, 161.
Neish, A. C. (1961), *Phytochemistry* 1, 1.
Neish, A. C. (1964). In *Biochemistry of Phenolic Compounds* (ed. J. B. Harborne), pp. 295–360. Academic Press, London and New York.
Ollis, W. D. and Gottlieb, O. R. (1968), *Chem. Commun.* 1396.
Ortmann, R., Sandermann, H. and Grisebach, H. (1970), *FEBS Letters* 7, 164.
Ortmann, R., Sutter, A. and Grisebach, H. (1972), *Biochim. Biophys. Acta* 289, 293.
Pachéco, H. (1969), *Bull. Soc. franc. Physiol. veg.* 15, 3.
Patschke, L., Barz, W. and Grisebach, H. (1964), *Ztsch. Naturf.* 19b, 1110.
Patschke, L., Barz, W. and Grisebach, H. (1966a), *Ztsch. Naturf.* 21b, 45.
Patschke, L., Barz, W. and Grisebach, H. (1966b), *Ztsch. Naturf.* 21b, 201.
Patschke, L. and Grisebach, H. (1965a), *Ztsch. Naturf.* 20b, 399.
Patschke, L. and Grisebach, H. (1965b), *Ztsch. Naturf.* 20b, 1039.
Patschke, L. and Grisebach, H. (1968), *Phytochemistry* 7, 235.
Pelter, A., Bradshaw, J. and Warren, R. F. (1971), *Phytochemistry* 10, 835.
Pla, J., Ville, A. and Pachéco, H. (1967), *Bull. Soc. Chim. biol.* 49, 395.
Potts, J. R. M., Weklych, R. and Conn, E. E. (1974), *J. biol. Chem.* 249, 5019.
Rathmeli, W. G. and Bendall, D. S. (1972), *Biochem. J.* 127, 125.
Reed, D. J., Vimmerstedt, J., Jerina, D. M. and Daly, J. W. (1973), *Arch. Biochem. Biophys.* 154, 642.
Roberts, R. J. and Vaughan, P. F. T. (1971), *Phytochemistry* 10, 2649.
Robinson, R. (1955), *The Structural Relations of Natural Products*, p. 41. Clarendon Press, Oxford.
Russell, D. W. (1971), *J. biol. Chem.* 246, 3870.
Russell, D. W. and Conn, E. E. (1967), *Arch. Biochem. Biophys.* 122, 256.
Russell, D. W., Conn, E. E., Sutter, A. and Grisebach, H. (1968), *Biochim. Biophys. Acta* 170, 210.
Sandermann, J., Jr. (1969), *Phytochemistry* 8, 1571.
Sandermann, H., Jr. and Grisebach, H. (1970), *Biochim. Biophys. Acta* 208, 173.
Sato, M. (1969), *Phytochemistry* 8, 353.
Schachter, D. and Taggart, J. V. (1954), *J. biol. Chem.* 208, 263.
Schill, L., and Grisebach, H. (1973), *Ztsch. physiol. Chem.* 354, 1555.
Schopfer, P. (1971), *Planta* 99, 339.
Shimada, M., Fushiki, H. and Higushi, T. (1972), *Phytochemistry* 11, 2657.
Shimada, M., Ohashi, H. and Higushi, T. (1970), *Phytochemistry* 9, 2463.
Steiner, A. M. (1970), *Ztsch. Pflanzenphysiol.* 63, 370.
Sutter, A. (1972), Ph.D. thesis, University of Freiburg/Br., Germany.
Sutter, A. and Grisebach, H. (1973), *Biochem. Biophys. Acta* 309, 289.
Sutter, A., Ortmann, R. and Grisebach, H. (1972), *Biochim. Biophys. Acta* 258, 71.
Vasil, I. K. and Hildebrandt, A. D. (1966), *Planta* 68, 69.
Vaughan, P. F. T. and Butt, V. S. (1969), *Biochem. J.* 113, 109.
Vaughan, P. F. T. and Butt, V. S. (1970), *Biochem. J.* 119, 89.

Vaughan, P. F. T., Butt, V. S., Grisebach, H. and Schill, L. (1969), *Phytochemistry* **8**, 1373.

Wallace, J. W. and Grisebach, H. (1973), *Biochim. Biophys. Acta* **304**, 837.

Wallace, J. W., Mabry, T. J. and Alston, R. E. (1969), *Phytochemistry* **8**, 93.

Walton, E. and Butt, V. S. (1970), *J. exp. Bot.* **21**, 887.

Walton, E. and Butt, V. S. (1971), *Phytochemistry* **10**, 295.

Weissenböck, G. (1971), *Ztsch. Pflanzenphysiol.* **66**, 73.

Wellmann, E., Baron, D. and Grisebach, H. (1971), *Biochim. Biophys. Acta* **244**, 1.

Wellmann, E. and Grisebach, H. (1971), *Biochim. Biophys. Acta* **235**, 389.

Wengenmayer, H. (1972), Diplomarbeit, Freiburg/Br., Germany.

Wiermann, R. (1970), *Planta* **95**, 133.

Wiermann, R. (1972a), *Ztsch. Pflanzenphysiol.* **66**, 215.

Wiermann, R. (1972b), *Planta* **102**, 55.

Wiermann, R. (1973), *Planta* **110**, 353.

Wong, E. (1964), *Chem. Ind.* 1985.

Wong, E. (1965), *Biochim. Biophys. Acta* **111**, 358.

Wong, E. (1966a), *Phytochemistry* **5**, 463.

Wong, E. (1966b), *Chem. Ind.* 598.

Wong, E. (1967), *Phytochemistry* **6**, 1227.

Wong, E. (1968), *Phytochemistry* **7**, 1751.

Wong, E. (1970). In *Progress in the Chemistry of Organic Natural Products* (eds. W. Herz, H. Grisebach and A. I. Scott) Vol. 28, p. 57. Springer-Verlag, Wien.

Wong, E. and Grisebach, H. (1969), *Phytochemistry* **8**, 1419.

Wong, E. and Wilson, J. M. (1972), *Phytochemistry* **11**, 875.

Zaprometov, M. N. and Bukhlaeva, V. Ya. (1968), *Biokhimiya* **33**, 383.

Zaprometov, M. N. and Bukhlaeva, V. Ya. (1971), *Biokhimiya* **36**, 270.

Zaprometov, M. N. and Grisebach, H. (1973), *Ztsch. Naturf.* **28c**, 113.

Chapter 17

Metabolism of Flavonoids

WOLFGANG BARZ and WOLFGANG HÖSEL

17.5 Flavonoid metabolism in microorganisms

17.5.1 Flavanones, chalcones and dihydrochalcones
17.5.2 Flavonols, dihydroflavonols, flavones and catechins
17.5.3 Anthocyanins and proanthocyanidins
17.5.4 Isoflavonoids
17.5.5 Aromatic acids and phenols

17.6 Conclusions

17.1 Introduction

This chapter will deal with various aspects of flavonoid turnover and degradation in higher plants and microorganisms. The microbial dissimilation of flavonoids has been a well understood area of biochemical research for many decades. It has been part of the more general field of microbial transformations of natural aromatic compounds, where it has been shown that all organic molecules are degraded by some form of life to maintain the carbon cycle of nature. Soil and sewage microflora are specifically adapted to perform this task. The results of microbial metabolism will be discussed here to outline the chemical principles governing flavonoid dissimilation and to relate them to the fate of flavonoids in higher plants.

Turnover and degradation of flavonoids in higher plants represents a new aspect of plant biochemistry. For a long time, flavonoids as with other 'secondary' plant products have been regarded as metabolically inactive end products stored as waste material in various plant tissues (Schwarze, 1958; Reznik, 1960; Mothes, 1969). This concept, based on the assumption that 'these plant constituents represent no biochemically usable energy potential' (Reznik, 1960), completely neglects the dynamic aspect and was largely based on insufficient experimental data. Now, we know that not only in ageing and senescent tissues but also in all actively growing cells the continual synthesis and turnover of polyphenols occur. We are therefore dealing with steady-state concentrations of flavonoids which are subject to varying rates of influx and efflux.

Since the term *metabolism* may be defined as 'the totality of the chemical processes that a cell is capable of performing' (Mahler and Cordes, 1966), the major subject of this chapter will be 'catabolism of flavonoids'. We thus refer to 'essentially degradative processes in which large organic molecules are broken down to simpler cellular constituents' (Mahler and Cordes, 1966). We will use the term 'catabolism' regardless of whether or not chemical free energy is being released which can be utilized by the plant cell.

Flavonoid interconversions (chalcone — flavanone, flavanone — flavone, dihydroflavonol — anthocyanin) representing anabolic routes are beyond the scope of this chapter (see Chapter 16). We will, however, briefly mention those oxidative synthetic processes which may lead from flavonoid monomers to various kinds of polymeric substances, because in some cases such oxidative reactions can explain an observable turnover of flavonoids.

17.2 General aspects of flavonoid metabolism

17.2.1 In plants

That the phenomena of flavonoid turnover and catabolism in higher plants have until very recently been neglected or underestimated seems to be largely a result of the highly differentiated structure of plant tissues. All observations point to the fact that, in general, synthesis, accumulation and dissimilation of flavonoids occur in one tissue or even in closely related compartments at the same time (cf. Chapter 18). Therefore, determination of both simultaneous synthesis and turnover require experimental techniques (tracer studies) which allow the characterization of metabolic rates. This has in many cases not yet been done. Furthermore, experiments to be described in this chapter have revealed that the stationary concentrations of flavonoid catabolites turn out to be very low. This indicates that, of a large number of intermediates associated with any one particular pathway, only a few are likely to accumulate to any measurable extent as glycosides, esters, vacuole-stored compounds or otherwise 'protected' substrates. These accumulated compounds are the ones isolated by the plant chemist and have incorrectly been termed 'secondary end products'.

The low concentrations of many flavonoid catabolites encountered so far seem to be the result of a complete lack of any significant excretory system in higher plants. Once an endogenous compound has been transferred from the metabolically inactive pool (see Section 17.6) to an active form or to suitable reaction centres (metabolically active pool), inactivation (polymerization or catabolism) must occur. Since excretion or accumulation of intermediates is excluded, the degradative pathways must go to completion, yielding compounds of primary metabolism or CO_2. Apart from the well known seasonal variations of flavonoid composition of plants (cf. Tissut, 1968; Tissut and Egger, 1972; Staude and Reznik, 1973) and the phenomenon of 'juvenile pigmentation' (see Section 17.3.7), both indicating flavonoid turnover, various other experimental observations indicate that turnover or degradative reactions must take place in plants.

Fig. 17.1 presents some accumulation curves of chlorogenic acid, rutin and isoflavones, randomly chosen from the literature. It can be seen that during ontogenesis of various plants the amounts of flavonoids per plant organ or per seedlings never rise above a certain level, but rather remain constant over a longer period of growth. Since flavonoid synthesis in such cases undoubtedly proceeds unaltered, the constant amounts can only be interpreted by assuming turnover or catabolism. Such typical sigmoidal accumulation curves with a pronounced plateau therefore describe systems of comparable intensities of synthesis and turnover. Whenever pulse labelling experiments have been carried out, turnover of flavonoids has been indicated (see Sections 17.3.6, 17.3.7, 17.3.9 and 17.4.2). Furthermore, in the case of flavonol turnover (Hösel *et al.*, 1972a), the enzyme(s) for turnover are known to be formed after the onset of flavonoid biosynthesis and shortly before the accumulation curves reach a plateau. Also, many feeding experiments with [14]C- or ^{3}H-labelled compounds directed towards elucidation of biosynthetic pathways have resulted in substantial randomization of label. These

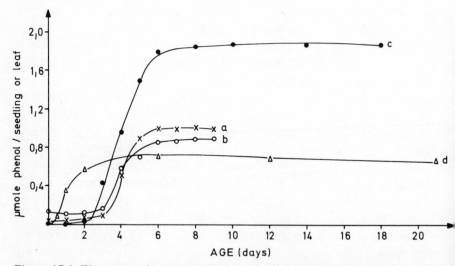

Figure 17.1 The curves show accumulation of (a) chlorogenic acid and (b) rutin in buckwheat seedlings (Amrhein and Zenk, 1970), (c) isoflavone formononetin in *Cicer arietinum* seedlings (Barz and Adamek, 1970) and (d) isoflavone genistein in subterranean clover leaves (Rossiter and Beck, 1967a,b).

'unspecific side-reactions' are presumably the result of the ability of higher plants to degrade a wide variety of endogenous and exogenous compounds (for suitable examples, see Barz and Grisebach, 1967; Berlin, 1972).

Numerous physiological studies on the effect of the environment on the level of flavonoids in plant tissues have been conducted (see Chapter 18); any alterations in the composition and in the level of flavonoids have been interpreted as solely being the result of changes in the biosynthetic machinery. However, the rates of turnover or degradation could equally well have been effected; possibly both synthesis and turnover are regulated independently of each other. Taking any given flavonoid pool, we may be dealing with systems of varying intensities of influx *and* efflux. One example of flavonol turnover has clearly shown that though synthesis remains unaltered an increase in turnover leads to a general decrease in flavonol concentration (Barz *et al.*, 1971).

Though there can be no doubt any longer about the dynamic role of flavonoids in higher plants, it should always be remembered that an observed turnover of a particular compound may be the result of *either* anabolism *or* catabolism. Turnover of a given flavonoid is usually determined by conducting pulse labelling experiments with suitable precursors (acetate, phenylalanine or cinnamic acid). The velocity of transfer of label into and eventually through the pool of the flavonoid under investigation will clearly reveal the rate of turnover under the actual experimental conditions. To determine whether anabolic or catabolic routes are responsible, intensive studies with the aid of labelled flavonoids are usually required.

Anabolic routes transfer the flavonoids under investigation into other monomers with a higher degree of substitution or (by means of oxidation reactions) to polymers. Such routes are normally easily detected by radiochemical analyses. These polymers such as lignin and condensed tannins are unlikely to undergo catabolism within the plant (cf. Neish, 1965) and their production probably represents an efficient detoxification mechanism.

Turnover of flavonoids in plants may in many cases be the result of both anabolism and catabolism (see Berlin, 1972). Determination of catabolic pathways therefore requires special experiments using labelled substrates. However, feeding of compounds to plants is often severely hampered by several factors such as (a) permeability barriers, (b) lack of transport, (c) unphysiological polymerization reactions, (d) metabolism by other than the normal pathways, (e) compartmentalization of metabolic centres and (f) microbial contamination. Harborne (1967) has given a detailed description of the possible pitfalls; possible experimental systems which overcome some of these difficulties are evaluated in Chapter 18, Section 18.2.6.

A combination of experimental approaches, based on enzymic studies and use of plant cell suspension cultures, seems to offer a solution. The advantages of plant cell suspension cultures and the avoidance of microbial contamination have been discussed by Street (1973). A suitable example in connection with successful determination of flavonol catabolism in plants has been published by Hösel *et al.* (1972b).

Despite the various difficulties discussed, catabolism of flavonoids has been observed when it has been deliberately looked for (see Sections 17.3.1, 17.3.2, 17.3.3, 17.3.6 and 17.3.9). Though the pathways involved are largely still unknown, these degradative reactions form a link between secondary and primary plant metabolism. Some of the degradation products of flavonoids are undoubtedly removed by respiration, the chemical energy laid down in plant polyphenols thus being released. It is, however, doubtful whether this significantly contributes to the plant's energy balance.

17.2.2 In microorganisms

When compared with flavonoid catabolism in higher plants, the equivalent studies with microorganisms are much less complicated. Since microorganisms do not produce and normally contain no flavonoids, dissimilation of these plant pigments by microbes shows all the characteristics of normal microbial degradation of exogenous compounds. Rarely if ever are problems due to cellular or organ differentiation, permeability barriers or polymerization reactions encountered. These facts, combined with the observation that microbial catabolites are often excreted into the culture medium, greatly facilitate the elucidation of pathways.

Flavonoids are degraded by bacteria or fungi as sources for energy and carbon. Therefore, much greater quantities of flavonoid are consumed by microbes as compared with higher plants. To obtain carbon and energy for growth and differentiation, the flavonoid substrates must be transformed in such a way that aliphatic ring fission products are obtained. These aliphatic intermediates (acetate, pyruvate, oxaloacetate, succinate), are eventually funnelled into the high energy release reactions of primary metabolism. The degradation of aromatic substrates to aliphatic intermediates is carried out with great economy using surprisingly few reaction steps. While the enzymes for energy release and synthesis of cell material are constitutive, the bulk of the enzymic machinery for flavonoid or aromatic dissimilation is inducible. As a result, these systems have been used to study the phenomena of substrate induction of enzymes, repression of enzyme synthesis, catabolite repression and regulation of branched and unbranched catabolic pathways (see Gibson, 1968; Evans, 1969; Dagley, 1971; Ornston, 1971). A further proof of the economy and efficiency of microbial catabolic pathways is the evolution of branched, interrelated catabolic pathways. Thus, different classes of substrates can eventually be dissimilated along a single pathway with minimum enzymic machinery. This has also been shown to occur in the case of flavonoid degradation (see Section 17.5.2). It is thus possible to classify flavonoid degradative pathways of microorganisms, at least in part, into reactions types according to the pathway chosen.

17.3 Flavonoid metabolism in plants

17.3.1 Chalcones and flavanones

Since chalcones and flavanones are readily inter-convertible in higher plants by action of various chalcone-flavanone isomerases (see Chapter 16), catabolic reactions which affect either class affect both. Catabolic reactions involving flavanones/chalcones may be of interest from a regulatory point of view; any efflux from the C_{15}-pool representing the starting point for flavonoid biosynthesis (Chapter 16) could result in a constant drain of essential intermediates. Because of their role as intermediates in flavonoid biosynthesis, variations in the concentration of flavanones/chalcones have always been correlated with regulatory properties of biosynthetic pathways only; for work with *Citrus* and *Prunus* species involving naringenin glycosides (Fig. 17.2 (2)) see Erez and Lavee (1969) and Maier and Hasegawa (1970). Patschke *et al.* (1964a) first reported a degradation of chalcones, when they observed that both 4,2',4',6'-tetrahydroxychalcone (1) and 4,2',4'-trihydroxychalcone (3), when fed to red cabbage seedlings or *Petunia hybrida* stems, yielded p-coumaric acid which in turn was assumed to be converted to ferulic and sinapic acid (see Fig. 17.2).

Later studies in the author's laboratory (Janistyn *et al.*, 1971; Kiss, 1972; Berlin *et al.*, 1974) using plant suspension cultures of *Pisum sativum* and *Glycine max* provided further information on this catabolic sequence; (1), (2) and (3) are first split to yield p-coumaric acid, a good portion of which is then catabolized via p-hydroxy-benzoic acid (see Fig. 17.2). The exact nature of the fragments of ring A is still obscure (phloroglucinol and resorcinol have not yet been isolated) though experiments with (2) and (3), specifically ^{14}C-labelled in ring A, showed an unusually high rate of respiratory dissimilation of the A-ring carbon atoms. Chalcones with a phloro-glucinol-type ring A were more rapidly degraded than chalcones with a resorcinol-type substitution pattern. The data on chalcone ring A degradation support earlier findings with intact *Cicer arietinum*

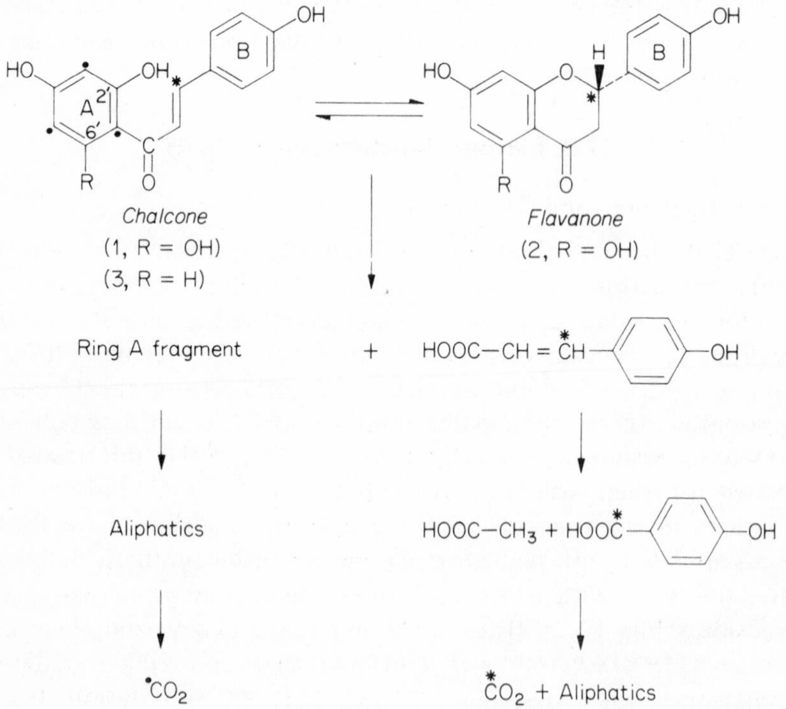

Figure 17.2 Chalcone/flavanone degradation in plants.

plants (Patschke *et al.*, 1964b). Fission of (1), (2) and (3) to yield *p*-coumaric acid is highly reminiscent of the plant and microbial *C*-acylhydrolases shown to attack dihydrochalcones (see Sections 17.3.2 and 17.5.1). Though chalcones are not substrates for these *C*-acylhydrolases, higher plants may well possess such enzymes. However, enzymic studies presently under investigation have revealed that the naringenin degrading enzyme from *Glycine max* cell suspension cultures is a peroxidase type of enzyme, also involved in aurone (see Section 17.3.3) and flavonol (see Section 17.3.6) degradation. Ring A of naringenin is split to unsaturated aliphatic acids of yet unknown structure which are then further metabolized (Barz and Patzlaff, unpublished).

17.3.2 Dihydrochalcones

Of the few known dihydrochalcones, only phloretin (Fig. 17.3 (4)) seems to have been investigated catabolically. This phenol, widely distributed in *Malus* as the 2'-β-D-glucoside phloridzin (Fig. 17.3 (5)), is known to have various physiological effects in animals (Ramwell and Sherratt, 1964) and plant tissues (Neumann and Avron, 1967; Podstolski and Lewak, 1970). In apple leaves, transformation products of phloridzin are involved in the defence mechanism against various fungi (Noveroske *et al.*, 1964a, 1964b). Recent investigations (Neumann and Avron, 1967; Raa and Overeen, 1968; Sarapuu, 1971) provide a chemical explanation (Fig. 17.3) for the nature of these transformation products. With the aid of apple phenolase, first 3-hydroxyphloretin (6) is formed ,which via the *ortho*-quinone (7) is eventually polymerized. Both the *ortho*-quinone and the polymers show fungistatic properties and furthermore are supposed to block the infection centre from the uninfected tissue. Three phloridzin-specific glucosidases, long known to occur in apple tissue (Lewak and Podstolski, 1966), have been purified and shown to be involved in phloridzin metabolism (Podstolski and Lewak, 1970).

Based on earlier reports of seasonal fluctuations in phloridzin content in apple leaves (Sarapuu, 1965), more recent experiments (Grochowska, 1967) demonstrated that phloridzin is also subject to catabolism. Crude protein extracts from the sap of apple spurs hydrolyzed phloridzin and phloretin to phloroglucinol and *p*-hydroxydihydrocinnamic acid. This is reminiscent of comparable reactions found with fungal enzymes (see Section 17.5.1). The

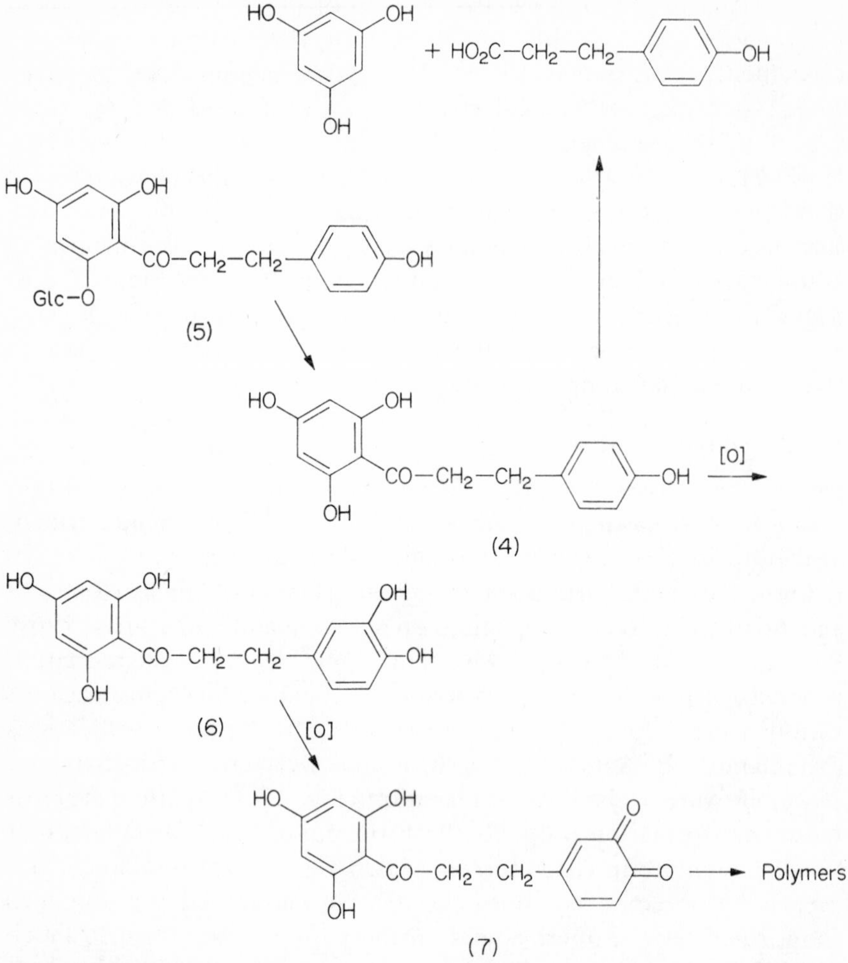

Figure 17.3 Metabolic pathways of the dihydrochalcone phloretin in apple tissue.

alternative reactions of phloridzin, catabolism versus polymerization, shown to occur in apple tissue (Fig. 17.3) are typical of the natural fate of most plant polyphenols.

17.3.3 Aurones

In view of their restricted occurrence as yellow flower pigments, naturally occurring aurones are presumably end-products derived from the corresponding chalcones, at least under normal conditions of growth and flowering (Wong, 1966, 1967; Rathmell and Bendall, 1972). Such an assumption is, for example, indicated by an observation (Puri and Seshadri, 1955) that young buds of *Butea frondosa* contain the chalcone isobutrin while the flowers contain predominantly the analogous aurone, palasitrin.

However, recent investigations in the authors' laboratory indicate that aurones can be degraded (Barz *et al.*, 1974). 6,4'-Hydroxy-aurone 6-β-D-glucoside-(α-^{14}C) (hispidol 6-glucoside) (Fig. 17.4 (8)) when fed to cell suspension cultures of *Glycine max*, *Phaseolus aureus*, *Cicer arietinum* and *Petroselinum crispum* was readily catabolized in each culture as measured by $^{14}CO_2$ production (ranging from 1.7% to 0.5% within 72 h) and randomization of label (standard fractionation according to Barz *et al.*, 1970). *p*-Hydroxy-benzoic acid–(carboxyl-^{14}C) was isolated as the main catabolite from all cultures investigated (Fig. 17.4).

While hispidol 6-glucoside has only been shown to be a natural constituent of *Glycine max* (Wong, 1966), the cell cultures of the other three plants mentioned were included in the experiments, in order to obtain an indication as to variation in aurone metabolism. Though the soyabean cultures when compared with the other three showed by far the highest degree of aurone degradation, the question of the specificity of the degradative reactions remains to be solved. Nevertheless, these studies show that aurones are degradable compounds under *in vivo* conditions, with the B-ring liberated as the equivalent benzoic acid.

17.3.4 Flavones

Little is known of the catabolism of flavones. The label of (2-^{14}C)-7,4'-dihydroxyflavone, when fed to *Cicer arietinum* seedlings, underwent substantial randomization in both the soluble and insoluble fractions (Brandner, 1962). While incorporation into

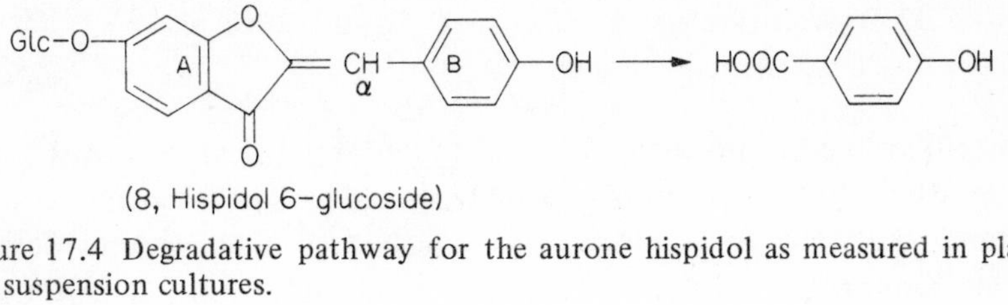

(8, Hispidol 6–glucoside)

Figure 17.4 Degradative pathway for the aurone hispidol as measured in plant cell suspension cultures.

isoflavones could not be determined, the results point to some sort of degradation. 5,7,4'-Trihydroxyflavone (apigenin) is efficiently dissimilated by cell suspension cultures of soyabean and mungbean, though p-hydroxybenzoic acid could not be detected as a catabolite (Hösel and Barz, unpublished). McClure and Wilson (1970), in their studies on photo-control of C-glycosylflavones in barley seedlings, have described rapid losses of saponarin from plumules upon red-light treatment. Though the authors discuss translocation or the formation of insoluble complexes as possible explanations for this loss, catabolic reactions could well have been involved.

17.3.5 Dihydroflavonols

The majority of dihydroflavonols (flavanonols) are known as wood constituents and are found in the free state (see Chapter 11); no relevant report has appeared so far describing their metabolic fate. However, dihydroflavonols, such as dihydrokaempferol (9) and dihydroquercetin (10) have recently been shown to be intermediates in the biosynthesis of flavonols, anthocyanins and catechins (see Chapter 16, Section 16.3.4). Apart from this function, dihydro-flavonols must also be subject to catabolic reactions. In the course of their studies on isoflavone biosynthesis, Barz and Grisebach (1966a) conclusively demonstrated the high degree of dihydrokaempferol degradation in *Cicer arietinum*. The catabolites were not isolated, but the radioactivity was randomized over all fractions (Barz, 1964).

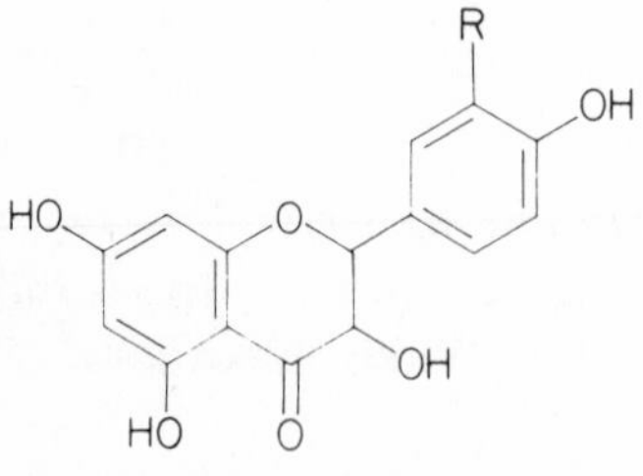

(9, Dihydrokaempferol R = H)   (11, Garbanzol)
(10, Dihydroquercetin R = OH)

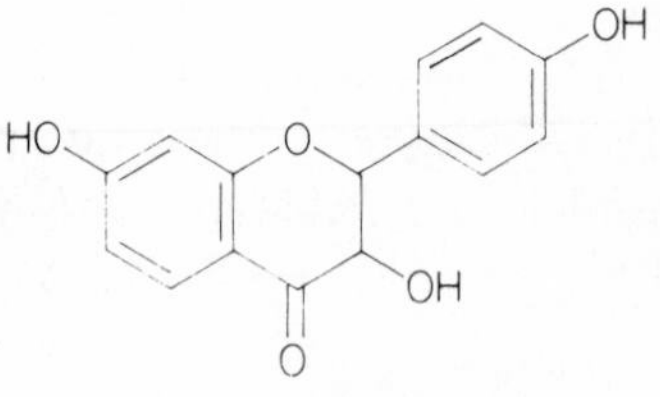

Comparable results have been obtained by Imaseki *et al.* (1965) using the dihydroflavonol garbanzol (11). These authors have claimed specific incorporation of garbanzol into the isoflavone formononetin (21). However, in view of all other data on isoflavone biosynthesis (see Chapter 16, Section 16.3.8) these results can only be interpreted by assuming degradation of (11) (cf. Barz and Grisebach, 1966a), with the catabolites finally being funnelled into anabolic routes.

17.3.6 Flavonols

Much information is available showing that flavonols such as kaempferol (12), quercetin (13) or isorhamnetin (14) are subject to turnover and catabolism in higher plants. There are, for example, many reports describing diurnal and seasonal variations in flavonol content in various plant species. Nöll (1955) measured diurnal fluctuations of rutin concentration in buckwheat and Ahlgrim (1956) found a substantial decrease in the amount of rutin in buckwheat plants after blooming, in ageing plants or in plants growing in the shade. Comparable seasonal variations have been found in leaves of various trees (Tissut, 1968; Tissut and Egger, 1972), in the cell sap of water plants (Burian and Aichmair, 1968), in *Prunus domestica* leaves (Hillis and Swain, 1959) as well as in ferns (Voirin and Lebreton, 1972). In general, flavonol metabolism is most pronounced at times of intensive plant growth or differentiation (Stafford, 1969; Weissenböck and Reznik, 1970). For example, the main flavonol of *Corylus avellana* winter buds, kaempferol 3-(p-coumarylglucoside), gradually disappears upon opening of the buds and is totally absent from mature leaves (Staude and Reznik, 1973).

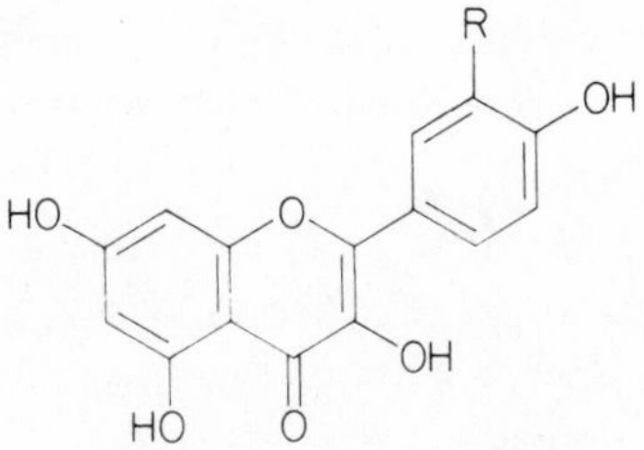

(12, Kaempferol R = H)
(13, Quercetin R = OH)
(14, Isorhamnetin R = OMe)

More precise information on the dynamics of flavonol turnover has been obtained with pulse labelling experiments. Using $^{14}CO_2$ (Dittrich and Kandler, 1971; Zaprometov *et al.*, 1971) or

[14] C-phenylalanine (Grisebach and Bopp, 1959), turnover of flavonols and their glycosides has been determined in *Picea abies* L., apple leaves and buckwheat seedlings, respectively.

The experiments with *Picea abies* L. (Dittrich and Kandler, 1971) further showed that kaempferol and quercetin 3-glucosides are much more rapidly metabolized than the equivalent 7-glucosides. This is in agreement with other data (Galston, 1969) which demonstrate varying rates of metabolism for different glycosides of the same aglycone.

Kaempferol, quercetin and isorhamnetin, present as various glycosides in stems and leaves of *Cicer arietinum*, have been shown by pulse labelling experiments to be metabolized with biological half-lives in the order of 7 to 12 days (Barz and Hösel, 1971). The different rates of turnover measured for these flavonols under various light conditions and in darkness (Barz *et al.*, 1971) again show that stationary concentrations of plant constituents are the result of both synthesis and degradation (cf. Zucker, 1972). The differences between light and darkness with regard to metabolism also possibly explain the diurnal variations found in *Impatiens balsamina* (Weissenböck, 1971) or buckwheat (Nöll, 1955).

In contrast to the rather slow turnover of flavonols discussed above, more rapid changes in flavonol concentration have been reported in pea tendrils upon coiling or mechanical stimulation (Jaffe and Galston, 1967). In pea tendrils quecetin 3-(*p*-coumaryltriglucoside) concentration was apparently lowered to some 30% of the original value within 30 to 60 min. Attempts to repeat these experiments have, however, not been successful (see p. 1029, Chapter 18) so there is some doubt as to whether such rapid changes do occur.

In 1969, Noguchi and Mori first reported a plant enzyme which degraded flavonols. Crude protein preparations from buckwheat were supposed to break down rutin to phloroglucinol and protocatechuic acid. However, this reaction, reminiscent of the fungal enzymes degrading flavonols (see Section 17.5.2) has not so far been detected again, in spite of several attempts in the authors' laboratory. A more reliable flavonol degrading enzyme system has been isolated from *Cicer arietinum* (Hösel and Barz, 1972) and this converts flavonol aglycones to the corresponding 2,3-dihydroxyflavanones ((15) in Fig. 17.5). This reaction requires molecular oxygen probably in stoichio-

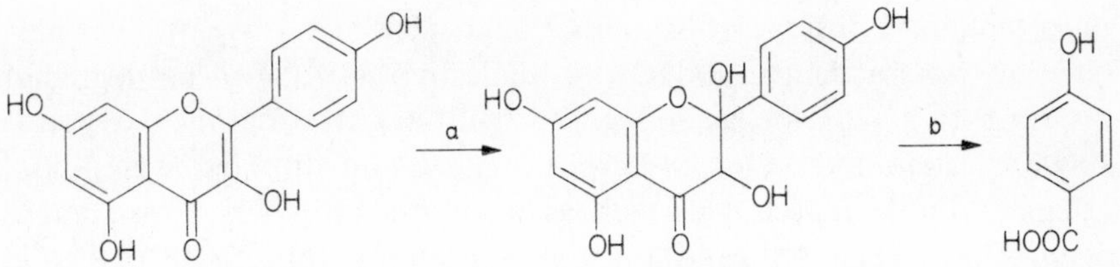

(15)

Figure 17.5 Degradative pathway of flavonols, in plants and plant cell suspension cultures (Hösel and Barz, 1972; Hösel *et al.*, 1972b).

metric amounts. In a second enzyme reaction the 2,3-dihydroxy-flavanones are further degraded to substituted benzoic acids derived from ring B. Phloroglucinol, a probable catabolite of ring A, could not be isolated from the reaction mixture; other degradation products still have to be elucidated (Hösel *et al.*, 1972b).

Later experiments (Hösel and Barz, unpublished) have revealed that the flavonol degrading enzyme (both for reaction (a) and (b) in Fig. 17.5) from *Cicer arietinum* is probably a peroxidase. All peroxidase isoenzymes from *Cicer arietinum* and *Sinapis alba* separated by isoelectric focusing techniques were shown to degrade flavonols with comparable rates according to the pathway shown in Fig. 17.5 (Frey, 1973). The identity of peroxidase and flavonol-degrading enzyme also explains why the same flavonol degradation pathway (Fig. 17.5) has been found in all plants tested so far (Hösel *et al.*, 1972a). In addition, it explains the observed inhibitory action of quercetin on IAA oxidase (cf. Chapter 18, Section 18.4.3), though quercetin should not be regarded as a noncompetitive inhibitor but rather as a substrate for IAA oxidase.

In recent investigations with *Tulipa* anthers, Wiermann (1973) has established an interesting correlation between flavonol accumulation and the level of enzyme activity causing flavonol degradation. In this tissue enzyme activity for flavonol degradation is high as long as flavonols are absent, but upon formation and accumulation of flavonols a marked decrease of enzyme activity was observed.

Apart from these enzyme studies, various *in vivo* investigations in several laboratories have proved that flavonols undergo catabolism. Dietermann *et al.* (1969) applied (U-^{14}C)-quercetin to tobacco leaves and measured rapid incorporation of radioactivity into polymeric material ('lignin'). Ethanol extracts of the leaves contained several radioactive catabolites, two of which were tentatively identified as phloroglucinol and protocatechuic acid. (U-^{14}C)-quercetin has also been shown by Durmishidze and Shalashvili (1968) to be degraded by root tissue of maize and grape seedlings; the radioactivity was finally distributed between respiratory carbon dioxide, water- and ethanol-soluble material as well as insoluble polymers. Comparable results have been reported for the degradation of (U-^{14}C) labelled kaempferol (12), quercetin (13) and isorhamnetin (14) applied to various plant cell suspension cultures (Hösel *et al.*, 1972b). The cell suspension cultures were very suitable experimental systems because the large percentages of $^{14}CO_2$ obtained (up to 43% within 72 h) and the high degree of randomization of radioactivity demonstrate high efficiency for catabolic reactions in these cells.

17.3.7 Anthocyanins

Anthocyanins (for their description, see Chapter 5) are mostly located in vacuoles of plant cells and have therefore been regarded as inactive end-products of biosynthesis. Evidence for turnover of anthocyanins has only fairly recently been reported. Leaves of tropical trees, such as *Mangifera indica*, contain petunidin glucoside as 'juvenile pigmentation'; this reaches its highest concentration at the early stages of leaf growth and afterwards completely disappears (Paris and Jacquemin, 1970). The same phenomenon has been found in *Catalpa ovata* (Zenk, 1967), or in violet-coloured chrysanthemum flowers (Stickland, 1972). Drastic changes in the spectrum of anthocyanin pigments (aglycones and glycosides) during growth and development of leaves or petals have repeatedly been reported (Proctor and Creasy, 1969; Ishikura, 1972). These observations are best interpreted by assuming catabolism.

Pulse labelling experiments with ^{14}C-acetate led Steiner (1971) to the assumption that the anthocyanin content of *Petunia hybrida* petals is the result of both concomitant synthesis and turnover. The biological half-life of the 3-monoglucosides of delphinidin, petunidin and malvidin, respectively, were thus determined to be 25 to 31 hours. Malvidin 3-glucoside with the highest degree of methylation in ring B showed the slowest rate of turnover. Turnover of the petunia anthocyanin pigments was measurable both under white light and in darkness (Steiner, 1971). Further investigations showed that light of various wavelengths (far red, red, blue) influenced biosynthesis or turnover of pigments to varying degrees (Steiner, 1972). Recent investigations on anthocyanin pigments in plant cell suspension cultures (Fritsch *et al.*, 1971; Constabel *et al.*, 1971; Lackman, 1971; Stickland and Sunderland, 1972a, 1972b) have also demonstrated that such compounds can be degraded by the same cells which produce them. This is very evident from the anthocyanin accumulation curves obtained with *Haplopappus gracilis* (Fritsch *et al.*, 1971; Stickland and Sunderland, 1972a, 1972b) because cell number as well as fresh and dry weight of cells remained constant while the anthocyanin content rapidly fell. The decrease of anthocyanin pigments in such *Haplopappus gracilis* cell cultures was markedly inhibited under red or green light when compared with white or blue illumination.

Anthocyanin degradation in plant cells should be accompanied by removal of the glycosidically bound sugar moieties to give the free aglycones. Such glycosidases from various plant sources have repeatedly been reported (Suzuki, 1962; Harborne, 1965; Mansell and Hagen, 1966; Boylen *et al.*, 1969) and in at least one case, *Theobroma cacao*, the enzyme is known to be specific for the particular anthocyanidin glycosides produced by the plant.

Higher plants just as fungi (Huang, 1955, 1956a, 1956b; Peng and Markarkis, 1963) have occasionally been reported to contain anthocyanin decolourizing enzyme systems. The discolouration of anthocyanins can, however, be attributed to purely chemical factors (Boylen *et al.*, 1969), to oxidative reactions by phenolases (van Buren *et al.*, 1960; Schumacker and Bastin, 1965; Sakamura *et al.*, 1965; Proctor and Creasey, 1969), or to enzymic destruction by the action of peroxidase (Jürgensmeyer and Bopp, 1961).

17.3.8 Catechins and proanthocyanidins

Catechins, 3-hydroxyflavans (16) and flavan-3,4-diols (17) are fundamental units for the 'condensed tannins' found in many plants. While evidence for turnover and catabolism of catechin and flavan-3,4-diols is available, much less is known of the metabolic fate of oligomeric or polymeric material.

In 1959, Zaprometov in connection with studies on catechins in the young tea plant, was one of the first to report substantial evidence for turnover and degradation of secondary plant constituents. Photosynthetically prepared ^{14}C-labelled catechins when applied to cut tea shoots led to a dramatic formation of respiratory

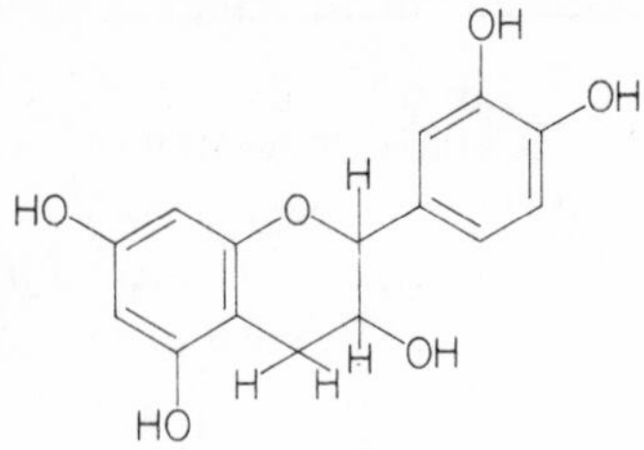

(16, Catechin) 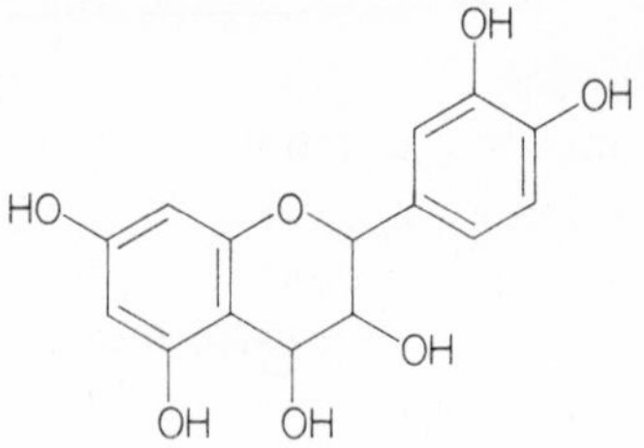 (17, 5,7,3',4'-tetrahydroxyflavan-3,4-diol)

$^{14}CO_2$ both in light and in darkness. After 70 h, some 80% of the applied radioactivity appeared in CO_2. In later studies (Zaprometov and Bukhlaeva, 1967) using uniformly labelled (−)-epicatechin, (−)-epigallocatechin, (−)-epicatechin gallate and (−)-epigallocatechin gallate the distribution of radioactivity in the tea shoots was measured. After 55 h the following values were obtained: carbon dioxide (15%), organic acids (20%, predominantly shikimic and quinic acid), carbohydrates (5%), protein (2%), chlorophyll and terpenes (3%), phenols (10%) and polymeric material (40%, lignin etc.). Though the possibility of microbial contamination cannot be ruled out, these data provide evidence for a complete breakdown of the applied catechins and are in total agreement with aseptically conducted investigations on flavonol degradation in plant tissue cultures (Hösel et al., 1972b, see Section 17.3.6). During pulse-labelling experiments, Zaprometov and Bukhlaeva (1968) further-more determined the biological half-lives of (−)-epicatechin and (−)-epigallocatechin to be in the order of 50 h and those of (−)-epicatechin gallate and (−)-epigallocatechin gallate to be approx-imately 70 h. Such experiments indicate that catechins in the tea plant are subject to various metabolic reactions and that, depending on the physiological conditions of the cell, such polyphenols are either catabolized or converted into polymers.

With regard to the proanthocyanidins little is known from a catabolic point of view. Combier and Lebreton (1968) reported on diurnal proanthocyanidin fluctuations in Sedum album with maxi-mum concentrations at midnight and minimum values at noon. The reasons for these fluctuations (catabolism, polymerization, transport) have, however, not been reported. Further information on proantho-cyanidin and catechin metabolism with regard to polymer formation can be found in Chapter 10.

17.3.9 Isoflavones and isoflavanones

Interest in isoflavone metabolism is largely due to their oestrogenic properties and to their connection with sheep infertility (Bradbury and White, 1951; Curnow, 1954); numerous physiological and ecological studies on isoflavones have been carried out with various clover species. The results show that environmental changes (i.e. temperature, light, metal ions and phosphate supply) greatly influence the isoflavone content of clover (Beck, 1964; Francis and Millington, 1965a, b; Rossiter and Beck, 1966a, b, 1967a, b; Rossiter, 1967, 1970; Schultz, 1965, 1967; Dedio and Clark, 1968). Furthermore, the data conclusively show that the isoflavones biochanin A (19), genistein (18), daidzein (20) and formononetin (21) in clover species are subject to active metabolism. This is especially corroborated by the observations (Dedio and Clark, 1968) that isoflavone content rapidly falls off after flowering and that genistein rapidly disappears from leaf homogenates upon standing (Francis and Millington, 1965b).

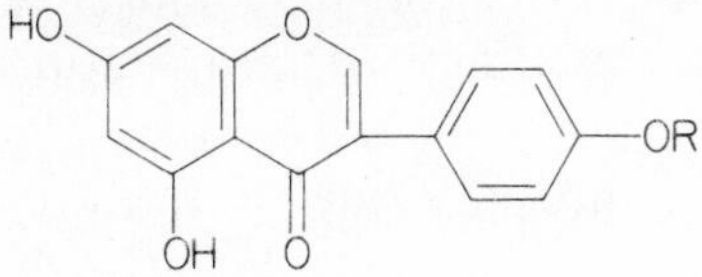

(18, Genistein R = H)
(19, Biochanin A R = Me)

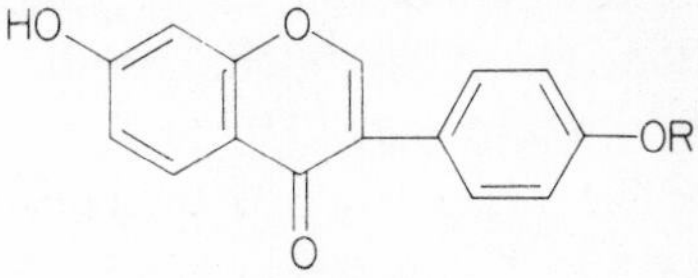

(20, Daidzein R = H)
(21, Formononetin R = Me)

Later experiments (Barz, 1969; Barz and Roth-Lauterbach, 1969; Barz et al., 1971; Barz and Hösel, 1971) have revealed that biochanin A and formononetin in *Cicer arietinum* as well as daidzein in *Phaseolus aureus* are simultaneously synthesized and turned over with biological half-lives in the order of 50–70 hours. Some basic ideas of the relationship and consequences of concomitant synthesis and turnover elucidated with the *Cicer* isoflavone system have already been discussed in Section 17.2.1 (Fig. 17.1) (see also Grisebach and Barz, 1969; Barz and Adamek, 1970). Feeding experiments with various ^{14}C-labelled samples of formononetin, daidzein and biochanin A in aseptically grown *Cicer arietinum* and *Phaseolus aureus* seedlings have demonstrated (formation of [14]CO_2) that the observed turnover can at least in part be explained by assuming catabolic reactions to be involved (Barz et al., 1970; Berlin, 1972; Berlin et al., 1974).

However, elucidation of the degradative pathways have severely been hampered by the very low concentrations of all catabolites. To improve the experimental system, studies on isoflavone degradation have lately been carried out with cell suspension cultures of mungbean, soyabean and *Cicer arietinum* (Berlin and Barz, 1971; Berlin, 1972). It appears that essentially all carbon atoms of biochanin A, formononetin and daidzein can be funnelled into catabolic reactions and finally yield CO_2 (up to 15%); cleavage of the aromatic rings has also been detected.

In elucidating flavonoid catabolic pathways, one is confronted with the difficulty that one particular compound is likely to be directed into different metabolic routes. Daidzein and mungbean cell suspension cultures have been used as an experimental system to clarify some of the aspects involved. Fig. 17.6 shows what happens in a mungbean cell to which daidzein is exogenously fed. Four reactions are possible: (a) glucosylation to daidzein 7-O-glucoside (Berlin, 1972); (b) phenolase catalyzed polymerization via 7,3'4'-trihydroxy-isoflavone (Berlin *et al.*, 1974); (c) transformation to coumestrol (Barz and Grisebach, 1966b, see Chapter 16); and (d) degradation (Berlin and Barz, 1971).

The quantities of daidzein introduced into any one of the four

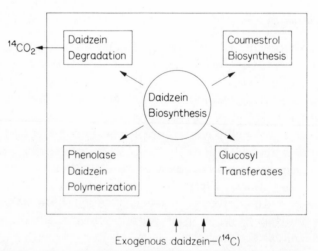

Figure 17.6 Possible metabolic routes for the isoflavone daidzein exogeneously applied to mungbean cell suspension cultures (Berlin, 1972).

pathways will depend on a variety of parameters. If, for instance, mungbean cell suspension cultures are kept under anaerobic conditions, the three oxygen dependent pathways are blocked so that the bulk of exogenous daidzein is transferred to daidzein 7-O-glucoside (Berlin, 1972). Degradative processes are, therefore, a question of quantity depending on the rates of other possible alternatives.

Biochanin A and formononetin in *Cicer arietinum* are mainly found as glycosides (Barz and Grisebach, 1966a) and are located in root tissue in the vacuoles of large cortex cells (Gierse and Barz, unpublished). For degradation reactions to occur, glycosidases are necessary. Two β-glycosidases from root and leaf tissue have been purified and are highly specific for isoflavone 7-O-glucosides (Hösel and Barz, unpublished).

In connection with studies on rotenoid biosynthesis, Crombie *et al.* (1970, 1971a, 1971b; see Section 17.3.11) presented evidence that of many isoflavones tested two, namely formononetin (21) and 7-hydroxy-2',4',5'-trimethoxyisoflavone (22), can act as intermediates in rotenoid formation. Apart from the demonstration of an anabolic function of formononetin, the data given are of further interest. One could argue that the incorporation values obtained with some of the other isoflavones and isoflavanones tested are misleading, due to the effects of catabolic reactions with consequent randomization of label. This is further indicated by our observation that isoflavanones such as 7,4'-dihydroxyisoflavanone (23) are more rapidly degraded in plants than the equivalent isoflavones (Berlin, 1972; Berlin *et al.*, 1972).

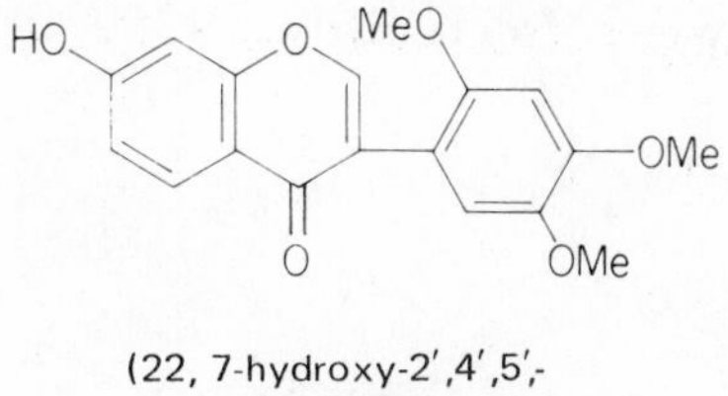

(22, 7-hydroxy-2',4',5',-
trimethoxyisoflavone)

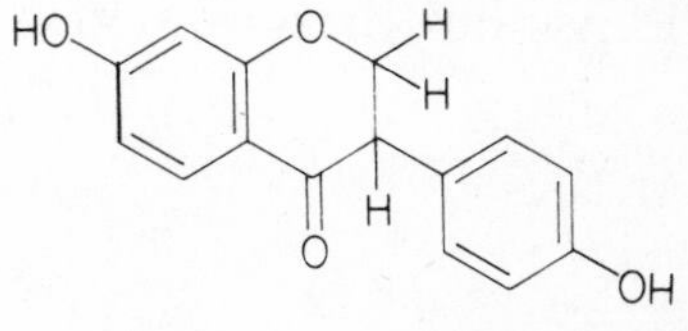

(23, 7,4'-dihydroxyisoflavanone)

17.3.10 Coumestans and pterocarpans

Of the few known naturally occurring benzofurocoumarins (coumestans) only coumestrol (24) has been intensively studied for metabolic reactions; it occurs in alfalfa and other legumes and it has been shown to accumulate to significant levels in alfalfa infected with foliar diseases and that its concentration is directly correlated with the degree of infection (for reviews see Hanson *et al.,* 1965; Bickoff, 1967). Though no direct evidence for coumestrol turnover has been obtained in relation to pathogen infection of alfalfa, the various seasonal fluctuations encountered (Bickoff, 1967) support this assumption. Pulse labelling experiments conducted with *Phaseolus aureus* seedlings have, however, clearly shown that coumestrol is rapidly metabolized in uninfected tissue (Barz, 1969) and that it can not thus be considered an end-product. Studies on coumestrol biosynthesis (Berlin *et al.,* 1972) have also provided evidence for a transitory role for this plant phenol.

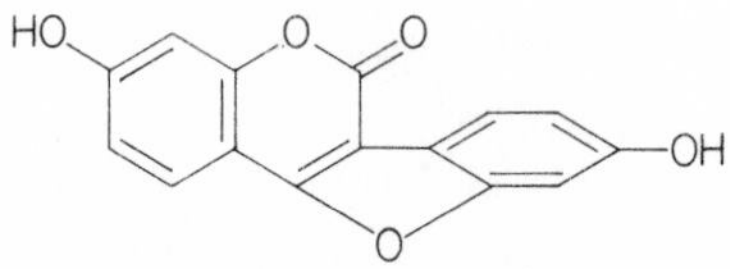

(24, Coumestrol)

Among the naturally occurring pterocarpans, the two phytoalexins pisatin (25) and phaseollin (26) have attracted considerable interest, especially by phytopathologists (see Deverall, 1972). Pisatin is formed upon fungal infection in pea (*Pisum sativum* L.) pods and phaseollin, comparably, in bean (*Phaseolus vulgaris* L.). tissue. While uninfected tissues contain no or very little phytoalexin (Hadwiger, 1967; Hadwiger *et al.,* 1970; Hess *et al.,* 1971) dramatic accumulation is observed either upon infection or upon challenging the tissue with various chemicals (Hess and Hadwiger, 1971; Hadwiger, 1972).

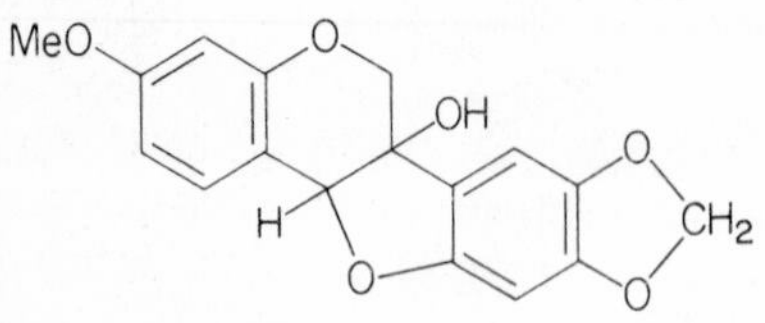

(25, Pisatin) 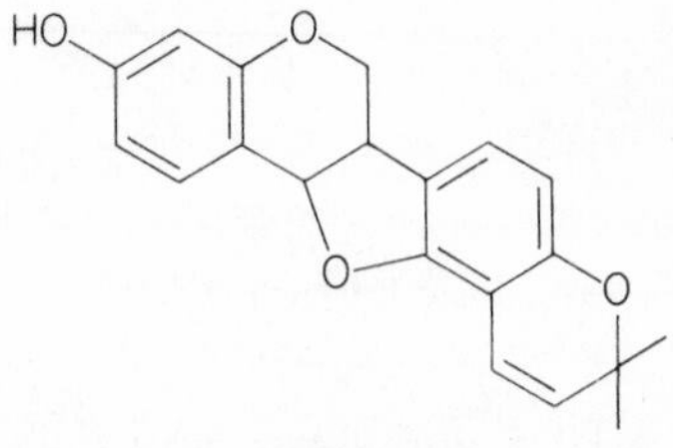   (26, Phaseollin)

One of the basic questions in phytoalexin accumulation has been whether the low levels of anti-fungal material are due to such rapid turnover in the untreated tissue that no significant accumulation occurs. However, using various radioactive precursors Hadwiger *et al.* (1970) elegantly demonstrated the absence of phytoalexin production in uninfected tissue. Furthermore, ^{14}C-pisatin when fed to uninfected pea tissue for a period of up to 52 hours was quantitatively recovered from both pea seedlings and immature pod tissue. The authors thus concluded that 'the turnover of pisatin appears to be minimal or nil'. The same seems to hold true for phaseollin in bean tissue. Substantial degradation of the related isoflavone daidzein (20) has however been observed in parallel experiments using both infected and uninfected pea tissue (Schwochau, 1968). Within 24—36 hours, some 10% of the radioactivity of daidzein were found in the ethanol insoluble protein-lignin fraction, 15—20% appeared as water soluble catabolites with only approximately 10% of unchanged daidzein being recovered. The fate of daidzein thus contrasts with that of pisatin and indicates that flavonoid catabolic reactions can be very specific.

17.3.11 Rotenoids

The biosyntheses of rotenone (27) in *Derris elliptica* plants and amorphigenin (28) in germinating seeds of *Amorpha fruticosa* follow the characteristics of isoflavonoid formation involving an aryl migration (Crombie and Thomas, 1967; Crombie *et al.*, 1968). Rotenone and amorphigenin are both metabolically active products. Thus, Crombie and co-workers (1971a) demonstrated the conversion of (27) into (28) by 8'-hydroxylation in *Amorpha fruticosa* seed, proving the transitory role of rotenone. Considering the present data on amorphigenin biosynthesis (a sequence involving as key intermediates 2',4',4-trihydroxychalcone (3), 7-hydroxy-4'-methoxy-isoflavone (21), 7-hydroxy-2',4',5'-trimethoxyisoflavone (22), 9-demethylmunduserone (30), rotenonic acid (29) and rotenone (27)) (Crombie *et al.*, 1970; 1971a, b), (28) can be regarded as the most fully substituted rotenoid known so far. Yet, amorphigenin is metabolically active, since recent pulse-labelling experiments with (1-^{14}C)-phenylalanine in *A. fruticosa* seedlings (Crombie *et al.*, 1973) show a definite turnover of the rotenoid. The data further show that the observed turnover of (28) cannot be explained by assuming conversion of amorphigenin into the known vicianoside amorphin in this plant system. One must assume catabolic reactions to be involved because older *A. fruticosa* plants are completely devoid of amorphigenin or amorphin.

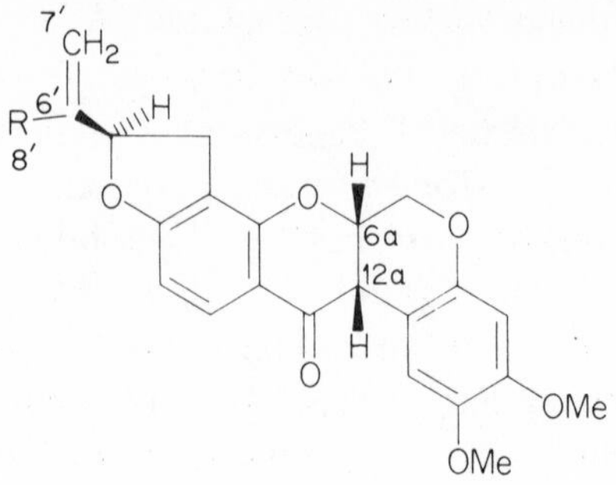

(27, R = Me Rotenone)
(28, R = CH₂OH Amorphigenin)

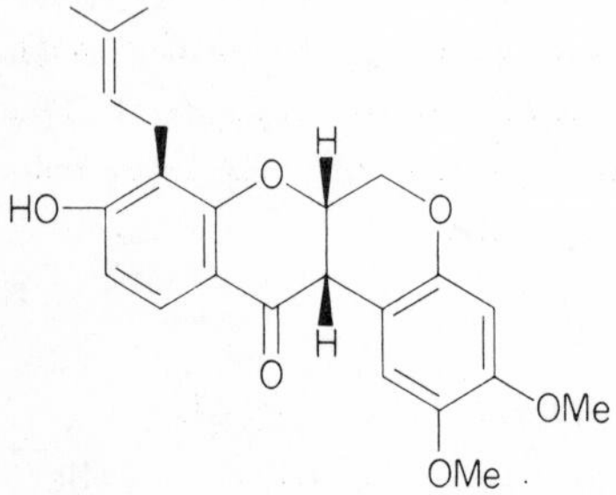

(29, Rotenonic acid)

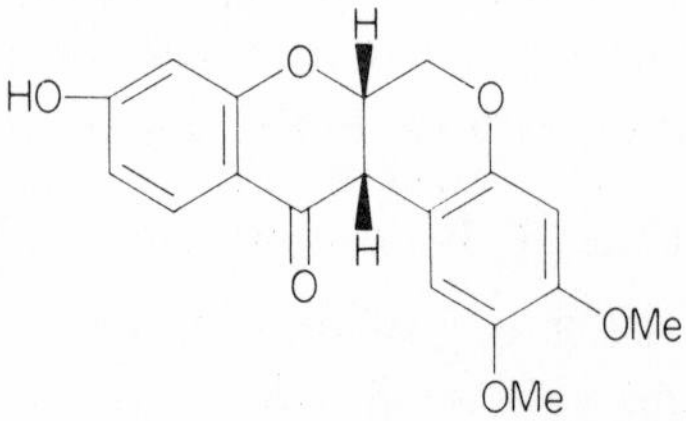

(30, 9–demethylmunduserone)

Though the fate of amorphigenin in plants is as yet undetermined, hints as to possible reactions stem from work on oxidative rotenoid metabolism in the microsome mixed function oxidase systems of mammalian liver, fish liver and insect tissues (Fukami *et al.*, 1967, 1969; Yamamoto, 1969). The reactions observed so far with rotenone (27) as substrate are hydroxylation of the 12a-position yielding rotenolones, oxidation of the isopropenyl side chain, yielding (28) and 6',7'-dihydro-6',7'-dihydroxyrotenone and formation of several unknown water-soluble catabolites.

17.4 Metabolism of aromatic acids and phenols in plants

Ring cleavage reactions form a critical point in the catabolism of aromatic compounds and flavonoids in higher plants. Since the catabolism of simple aromatic acids has well been substantiated for microorganisms (see Section 17.5.5) and, furthermore, since flavonoid structures in plants are degraded to cinnamic and benzoic acids (see Sections 17.3.1–17.3.4 and 17.3.6) a summary of our present knowledge of ring fission reactions of aromatic acids in plant systems deserves presentation. A review of ring cleavage reactions in connection with biosynthetic pathways has been given by Thomas (1965).

17.4.1 Amino acids and phenylacetic acids

Structural modifications of the side chain of tryptophan in plants have repeatedly been observed, especially in connection with auxin and indolalkaloid biosynthesis. The formation of tryptamine, indole-3-acetaldehyde, indole-3-ethanol and indole-3-methanol being suitable examples for such modifications (Gross, 1969; Libbert *et al.,* 1970; Magnus *et al.,* 1971, 1973). By contrast, evidence for ring cleavage reactions of tryptophan is meagre. In root tip tissue of pea seedlings, the conversion of indolacetic acid to *o*-formamidoacetophenone has been observed (Collet, 1968), a reaction obviously comparable to the well known tryptophan-pyrrolase fission of the indole nucleus (Ribbons, 1965). The only unequivocal demonstration of the ability of plant tissue to degrade the benzene ring of tryptophan stems from work by Ellis and Towers (1970) with tissue cultures of *Ruta graveolens* and *Melilotus alba.* Over a 7-day period, some 0.26% of the radioactivity of (benzene-ring-^{14}C)-tryptophan appeared in CO_2. The levels of ^{14}C in respiratory carbon dioxide observed in these and other studies, though comparatively small, are probably indicative of considerably more radioactivity circulating in the general metabolic pathways.

A much larger body of information is available concerning the degradation and ring fission of phenylalanine and tyrosine. The relevant data, together with those for phenylacetic and benzoic acids, are summarized in Fig. 17.7. Various tracer studies using intact plants, leaf discs and plant tissue cultures have shown that phenylalanine and tyrosine are completely degraded to various primary metabolites and carbon dioxide (Ibrahim *et al.,* 1961; Rosa, 1966; Ellis and Towers, 1970; Berlin and Barz, 1971; Zaprometov and Bukhlaeva, 1971; Zenk, 1972; Ellis, 1973; Durand and Zenk, 1974).

Three major pathways, depending on the plant species investigated, seem to exist for phenylalanine: (i) phenylamine → phenylpyruvic acid → 2-hydroxyphenylacetic acid → 2,3-dihydroxyphenylacetic acid (Kindl, 1969a); (ii) phenylalanine → *trans*-cinnamic acid → 2-hydroxycinnamic acid → salicylic acid → catechol (Grise-

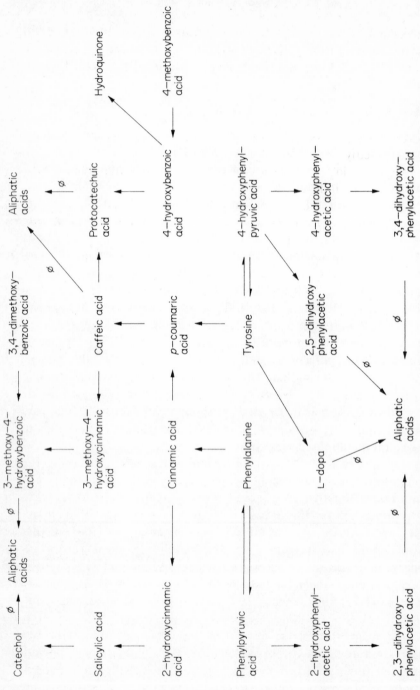

Figure 17.7 Metabolic grid depicting the degradative pathways for phenylalanine, tyrosine, cinnamic acids, phenylacetic acids and benzoic acids. (The symbol Ø indicates ring fission of aromatic compounds).

bach and Vollmer, 1963; Zenk, 1966; Ellis and Towers, 1969; Kindl, 1969b); and (iii) phenylalanine → *trans*-cinnamic acid → *p*-coumaric acid → 4-hydroxybenzoic acid (Zenk and Müller, 1964; Vollmer *et al.*, 1965; Zenk, 1966; Kindl, 1969a).

Except for phenylalanine ammonia lyase, the *trans*-aminase and *trans*-cinnamic acid 4-hydroxylase (see Chapter 16), the reactions involved in the modification and shortening of the side chain of phenylalanine have hardly been studied.

Comparable studies with tyrosine in various different plant species also reveal the existence of two possible routes, namely: (i) tyrosine → *p*-coumaric acid → *p*-hydroxybenzoic acid (Vollmer *et al.*, 1965; Zenk and Müller, 1964; Zenk, 1966); and (ii) tyrosine → 4-hydroxyphenylpyruvic acid, which in turn can be converted to either 3,4-dihydroxyphenylacetic or 2,5-dihydroxyphenylacetic acid (homogentisic acid) (Kindl and Hoffmann-Ostenhof, 1966; Whistance and Threlfall, 1968a, b; Ellis and Towers, 1970; Zenk, 1972; Ellis, 1973; Durand and Zenk, 1974).

The three substituted phenylacetic aids shown in Fig. 17.7 are suitable substrates for ring cleavage reactions. Thus, when ^{14}C-labelled samples of these phenylacetic acids were fed to *Sinapis alba* and *Astilbe chinensis* plants, considerable radioactivity was measured in CO_2 and D-glucose (Kindl, 1969a). Earlier assumptions as to homogenistic acid participation in tyrosine degradation (Ibrahim *et al.*, 1961; Ellis and Towers, 1970) have recently been proved unequivocally by Zenk and co-workers using plant tissue cultures (Zenk, 1972; Ellis, 1973). All evidence supports the concept that the homogentisic acid pathway, well known in microorganisms and mammals (Meister, 1965), is also operating in a large number of plant systems (Durand and Zenk, 1974).

Biosynthetic studies on the betalain pigments have recently provided conclusive evidence for an extradiol cleavage of the aromatic ring of 3,4-dihydroxyphenylalanine (DOPA) (Fischer and Dreiding, 1972). While the tracer studies on betalains provide precise evidence as to the position of ring fission, the extensive catabolism of (ring-^{14}C)-DOPA observed in plant tissue cultures by measuring $^{14}CO_2$-production further strengthen other observations (see Sections 17.4.2 and 17.4.3) that *ortho*-dihydroxy aromatics are suitable substrates for ring cleavage in plants, too (Ellis, 1973).

17.4.2 Cinnamic acids

Shortening of the side chain of cinnamic acids by β-oxidation is the main biosynthetic pathway leading to benzoic acids in plants (Zenk, 1966; Towers and Subba Rao, 1972) and the characteristic substitution pattern of benzoic acids is largely determined at the cinnamic acid stage. Considering the available evidence for benzoic acid catabolism (see Section 17.4.3), the conversion of cinnamic to benzoic acids is one catabolic step in the degradative pathway of phenylalanine and tyrosine (see Fig. 17.7). While ring cleavage reactions with cinnamic, p-coumaric or o-coumaric acid as substrates are not likely to occur, fission of the aromatic ring of caffeic acid has recently been indicated in studies with plant tissue cultures (Berlin *et al.*, 1971). The evidence for a direct fission of the o-dihydroxy ring structure of caffeate relies largely on the observation that the degree of degradation of caffeic acid was well correlated with those of other o-dihydroxy compounds (catechol, protocatechuate) in the cultures tested; thus the enzyme(s) involved are not very specific (compare also Ellis, 1973).

One of the earliest pulse labelling studies proving polyphenol turnover in plants was carried out by Taylor and Zucker (1966) on chlorogenic acid (3-*O*-caffeylquinic acid) in *Xanthium pennsylvanicum* leaves and potato tubers. The authors determined the biological half-life of chlorogenic acid in *Xanthium* leaves to be in the order of 14 hours and further showed that both chlorogenic acid and 3,5-dicaffeylquinic acid, apart from being interconvertible, are subject to other metabolic conversions as well. While a considerable portion of these conversions can be accounted for as incorporation of the caffeyl moieties into insoluble polymeric material ('lignin'), catabolism of the caffeic acid portion of the conjugates must also be taken into consideration.

17.4.3 Benzoic acids

Various quantitative determinations of benzoic acids have revealed significant fluctuations and variations in the concentrations of these compounds in plants and thus demonstrated their dynamic status (e.g. Kuc and Nelson, 1964; Glass and Bohm, 1969). Recent feeding experiments with various labelled benzoic acids in normal and aseptically grown plants as well as in plant suspension cultures demonstrated the following catabolic reactions to be operating (Bolkart and Zenk, 1968; Kindl, 1969b; Harms et al., 1969a, b, 1971; Ellis and Towers, 1970; Berlin et al., 1971; Ellis, 1971; Harms et al., 1972; Harms and Prieß, 1973): decarboxylation, O-demethylation and ring fission.

Decarboxylation. In an hydroxylation-induced decarboxylation, benzoic acids are converted to simple phenols. The formation of hydroquinone, methoxyquinone and catechol from p-hydroxybenzoic acid, vanillic acid and salicylic acid, respectively, are good examples. The rate of decarboxylation is especially pronounced in the case of p-hydroxy-substituted acids with m-substituents exercising a strong stimulatory effect. The formation of $^{14}CO_2$ from carboxyl-labelled benzoic acids by cell suspension cultures of soyabean increased in the following sequence: salicylic acid 1.4%, anisic acid 4%, p-hydroxybenzoic acid 21%, protocatechuic acid 66%, vanillic acid 83% and syringic acid 88%.

O-demethylation. The pronounced ability of higher plants to split alkylaryl-ether bonds earlier observed with cinnamic acids (El-Basyouni et al., 1964; Barz and Grisebach, 1967; Ebel et al., 1970; Berlin et al., 1971) and isoflavones (Barz et al., 1970) also occurs with benzoic acids. O-demethylation with the plant systems studied so far (mungbean, soyabean, garbanzo bean, wheat) seems to be very specific for the p-position because m-methoxy groups are oxidized by almost two orders of magnitude less efficiently. Anisic acid, veratric acid and 3,4,5-trimethoxybenzoic acid yield p-hydroxybenzoic acid, vanillic acid and syringic acid, respectively (see Fig. 17.7). By contrast, isovanillic acid and 5-hydroxyveratric acid were not formed in tissue cultures inoculated with veratric or 3,4,5-trimethoxybenzoic acid. Results obtained with aseptically grown plants and with plant suspension cultures were essentially identical.

Ring fission. Unequivocal evidence for ring cleavage reactions by measuring $^{14}CO_2$-production from specifically ring labelled benzoic acids has been obtained with at least six different acids and a wide range of plant systems. The extent of degradation decreased in the order: protocatechuic acid $\gg$ p-hydroxybenzoic acid > vanillic acid > anisic acid > salicylic acid > benzoic acid. The superiority of o-dihydroxy compounds as substrates for ring fission (protocatechuate) was again demonstrated and this parallels the results obtained with microorganisms. These findings do, however, require that monohydroxybenzoic acids such as p-hydroxybenzoate are further hydroxylated to form suitable substrates for ring cleavage reactions to occur. Recent experiments with *Phaseolus aureus* suspension cultures (Barz and Mohr, unpublished) show that p-hydroxybenzoic acid catabolism proceeds via protocatechuic acid.

All the above data on decarboxylation, O-demethylation and ring fission of benzoic acids have been obtained exclusively during tracer studies under *in vivo* conditions and do not allow any conclusions to be drawn as to the enzymes involved or to the ring cleavage mechanisms operating. Two recent cell-free studies in connection with ring fission reactions are, therefore, of interest. Using cell free preparations of hypocotyls of mungbean seedlings, Tateoka (1970) observed an oxygen-dependent reaction with protocatechuic acid and claimed the identification of ring fission products such as β-carboxy-*cis,cis*-muconic acid, β-carboxy-muconolactone and β-ketoadipic acid. Such a pathway would require an intradiol cleavage of protocatechuic acid and establish a degradation scheme similar to that of *Neurospora crassa* (Gross *et al.,* 1956). The experimental evidence given by Tateoka (1970) does, however, cast some doubt as to the validity of the results presented and furthermore, independent attempts to repeat the experiments have so far totally failed (Shaw and Barz, unpublished). Though the mere fact of protocatechuate dissimilation by mungbean tissue seems to be beyond any doubt (Berlin *et al.,* 1971), elucidation of the degradative pathway involved warrants further studies.

The only other report of a cell-free study showing ring cleavage of a benzoic acid has quite recently been published by Sharma *et al.* (1972). An enzyme system obtained from tender leaves of *Tecoma stans* converted 2,3-dihydroxybenzoic acid into a γ-dilactone carboxylic acid which suggests an initial intradiol cleavage of the substrate with α-carboxy-*cis,cis*-muconic acid as an intermediate. Whether the γ-dilactone carboxylic acid is a true intermediate in the catabolic pathway or an artifact due to the isolation procedure requires further investigations.

17.4.4 Simple phenols

Simple phenols such as hydroquinone, methoxyhydroquinone or catechol arise in higher plants by hydroxylation induced decarboxylation of the equivalent benzoic acids. Catechol is also an important ring-cleavage substrate in known pathways of microbiol metabolism. It is, therefore, not surprising that all of a large number of plant species from some eleven families tested showed a considerable ability to convert catechol to aliphatic metabolites and carbon dioxide (Ellis, 1971; Berlin et al., 1971). This is in accordance with other studies on similar 1,2-dihydroxy aromatic compounds such as protocatechuic acid, caffeic acid or L-DOPA (see Fig. 17.7). Hydroquinone, as the β-glucoside arbutin, and various galloyl-arbutin derivatives (Britton and Haslam, 1965) are obviously subject to considerable turnover in plants, for example as has been measured in *Bergenia crassifolia* leaves (Danner, 1940; Friedrich, 1954; Moritz and Morf, 1956; Wehnert, 1970). Turnover of the hydroquinone moiety can be explained by phenoloxidase catalysed polymerization (Frohne, 1964) without any ring-fission. This is corroborated by the observation that feeding of ^{14}C-hydroquinone to *Avena* coleoptiles resulted in extensive incorporation into cell wall material without any $^{14}CO_2$ formation (Reuter, 1970).

17.5 Flavonoid metabolism in microorganisms

The basic biological aspects of microbial metabolism of flavonoids have been dealt with in Section 17.2.2. Intensive investigations in the last decade have provided evidence that only a few degradative pathways have been evolved in microorganisms and that this limited number of alternative pathways is sufficient for the catabolism of a large variety of different structures. Only monomeric flavonoids will be considered here, since the study of condensed tannins and other polymers has been neglected (see Evans, 1969).

17.5.1 Flavanones, chalcones and dihydrochalcones

An important degradative pathway of flavanones carried out by the intestinal microflora of mammals will be discussed in the next section. Here, we concentrate on various transformation reactions of the (±)-flavanones (31) and (32) (Fig. 17.8) that are carried out by the fungus *Gibberella fujikuroi* (Udupa et al., 1968, 1969). The following products have been identified: (−)-flavan-4α-ol (33), 2′-hydroxychalcone (34), 4,2′-dihydroxydihydrochalcone (35), 4,2′-dihydroxychalcone (36), (±)-4′-hydroxyflavanone (32) and (−)-4′-

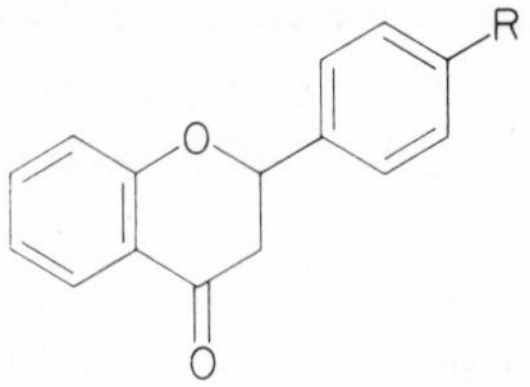

(31, R = H; 32, R = OH) 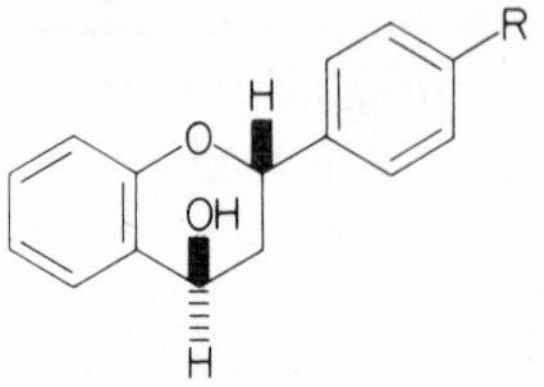   (33, R = H; 37, R = OH)

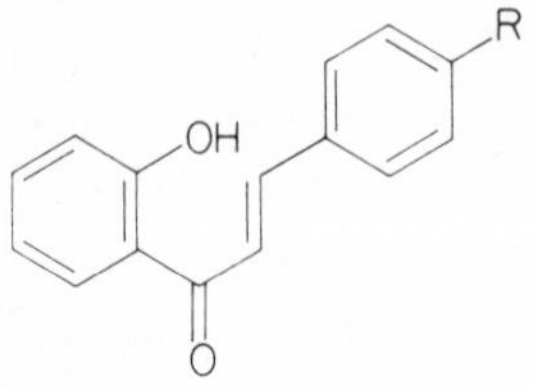

(34, R = H; 36, R = OH) 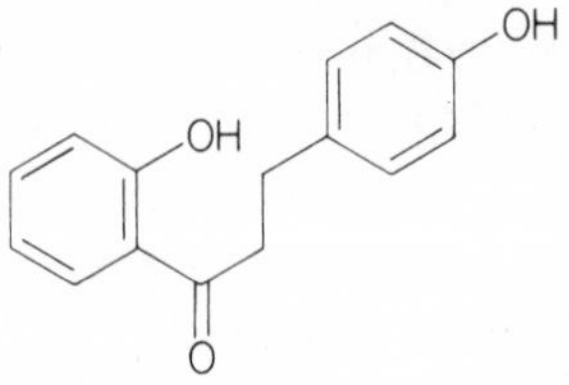   (35)

Figure 17.8 Transformation products of (±)-flavanones by the fungus *Gibberella fujikuroi.*

hydroxyflavan-4α-ol (37). Reduction of the carbonyl groups of the flavanones (31) and (32) occurs stereospecifically.

Ciegler *et al.* (1971) investigated various fungi for naringin (naringenin 7-*O*-rhamnosylglucoside) degradation and observed as main reactions the stepwise removal of first rhamnose, yielding prunin (naringenin 7-*O*-glucoside), then glucose. The two glycosidases necessary are sequentially induced by their substrates. Identical results have been obtained with a strain of *Fusarium oxysporum* which eventually degrades flavanones to benzoic acids, derived from the B-ring (Barz, 1971). Though chalcones have been shown to be degraded by species of *Fusarium* (Barz, 1971; Barz and Schlepphorst, unpublished), details of the pathways involved remain to be elucidated. However, chalcone degradation is presumably carried out along pathways different from those used for dihydrochalcone dissimilation. The *C*-acylphenol acylhydrolases involved in dihydrochalcone degradation do not act upon chalcones or flavanones (Chatterjee and Gibbins, 1969; Minamikawa *et al.,* 1970) (Fig. 17.9).

With regard to dihydrochalcone degradation, interest has focused on phloridzin dissimilation (Fig. 17.9). Various fungi such as *Venturia inaequalis* (Holowczak *et al.,* 1960) *Penicillium* species (Towers, 1964) and *Aspergillus* species (Jayasankar *et al.,* 1969)

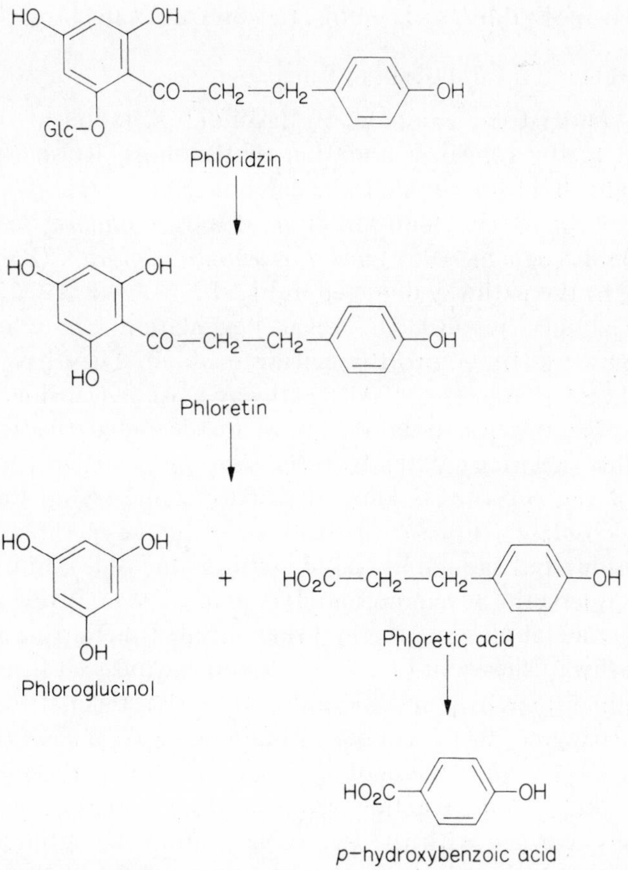

Figure 17.9 Degradative pathway of phloridzin occurring in fungi and *Erwinia herbicolor*.

hydrolyse phloridzin to phloretin which in turn is degraded to phloroglucinol, phloretic acid and *p*-hydroxybenzoic acid, respectively (Fig. 17.9). Identical results have been obtained with numerous species of *Fusarium* (Barz, 1971 and unpublished). While the bacterium *Erwinia herbicolor* (Chatterjee and Gibbins, 1969) also degrades phloridzin according to Fig. 17.9, phloretic acid is not further dissimilated to *p*-hydroxybenzoic acid by this organism. The inducible bacterial (Chatterjee and Gibbins, 1969) and fungal (Minamikawa *et al.*, 1970) enzymes cleaving phloretin turned out to be *C*-acylphenol acylhydrolases strictly specific for dihydrochalcones and phloracetophenones.

17.5.2 Flavonols, dihydroflavonols, flavones and catechins

17.5.2.1 Flavonol catabolism in fungi

The best understood example of flavonoid metabolism in micro-organisms is the fungal degradation of flavonols. Rutin (quercetin 3-rutinoside) has been shown to be catabolized by *Aspergillus* species (Westlake *et al.*, 1959; Simpson *et al.*, 1960), *Pullularia fermentans* (Hattori and Noguchi, 1959) and *Fusarium oxysporum* (Barz, 1971) according to the pathway depicted in Fig. 17.10.

The enzymes involved in rutin degradation are exoenzymes excreted by the fungi into the culture medium. They have all been separated and characterized. The pathway as shown in Fig. 17.10 is very specific, because only flavonols are degraded (Padron *et al.*, 1960). This specificity extends to the flavonol as well as to the sugar moiety of the substrates. Thus, the first enzyme (A) in Fig. 17.10, the β-glycosidase rutinase hydrolyses rutin, hyperin (quercetin 3-galactoside) and sakuranin (sakuranetin 5-glucoside), but does not act upon quercetin 3-rhamnoside (Hay *et al.*, 1961). A few strains of *Aspergillus* metabolize quercetin 3-rhamnoside (quercitrin) according to the pathway shown in Fig. 17.10 due to the induced formation of an additional quercitrin specific glycosidase (Westlake, 1963). Quercetinase (enzyme B), a copper containing dioxygenase (Oka and Simpson, 1971), splits flavonol aglycones yielding carbon monoxide and a depside. This reaction requires stoichiometric amounts of atmospheric oxygen with the two oxygen atoms being incorporated into the carbonyl and the carboxyl group of the depside (Krishnamurty and Simpson, 1970). Quercetinase shows absolute specificity for the flavonol nucleus requiring a carbon double bond between C-2 and C-3 and a hydroxyl substituent at C-3. Hydroxyl groups at the aromatic rings A and B both influence maximum velocity and the Michaelis constants of the enzyme reaction. The glycoprotein quercetinase contains two atoms of divalent copper per enzyme molecule, and for binding of substrate the hydroxyl group at C-3 and the carbonyl group at C-4 are essential. A detailed mechanism for this enzyme reaction has been proposed (Oka *et al.*, 1972).

The esterase (enzyme C in Fig. 17.10) is specific for flavonol derived depsides and digalloyl depside, while ester bonds of a large variety of aromatic substrates cannot be hydrolysed (Child *et al.*, 1963). The enzymes A – D are concomitantly induced in *Aspergillus*

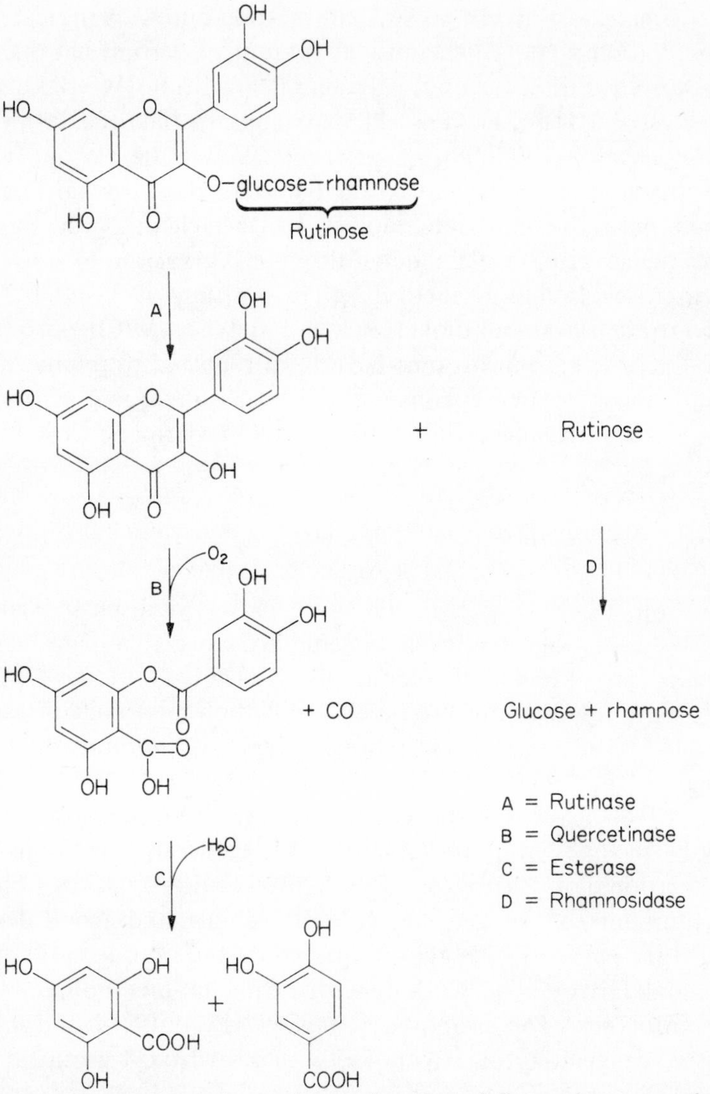

Figure 17.10 Degradative pathway of rutin and flavonols as occurring in fungi.

by rutin or quercetin. However, independent genetic control of at least enzyme C must be assumed, because tannic acid solely induced the formation of the esterase but not ot rutinase and quercetinase (Child *et al.*, 1963).

In addition to the important flavonol degradative pathway (Fig.

17.10), fungi most likely possess alternative routes, as indicated by the observation that the theoretical amount of carbon monoxide is rarely evolved during flavonol aglycone degradation (Westlake *et al.*, 1961; Westlake, 1963). Thus, Pickard and Westlake (1969, 1970) found a laccase in *Polyporus versicolor* converting flavonols into brown pigments. Furthermore, flavonol 3-glycosides can be dissimilated with the glycoside moiety still attached to the flavonol nucleus. Quercetin, when metabolized by *Aspergillus niger* van Tieghem (Fig. 17.11) is methylated in positions C-7 and C-3 and reduced to the flavan-3,4-diol (Haluk and Metche, 1970).

Wood-destroying mushrooms have been reported to remove rhamnose and glucose stepwise from rutin with the subsequent addition of ribosyl groups to quercetin at C-7 and C-4 (Armand-Fraysse and Lebreton, 1969).

17.5.2.2 Flavonol, flavone, dihydroflavonol and catechin catabolism in microorganisms

In addition to the specific flavonol pathway discussed above, alternative metabolic routes applicable to more than one class of flavonoids are known. Bacteria, such as *Pseudomonas* species, degrade flavonols without production of carbon monoxide (Westlake *et al.*, 1961). Recent investigations using a *Pseudomonas* species (Jeffrey *et al.*, 1972a, 1972b; Schultz and Wood, 1972; Schultz *et al.*, 1974) have led to the discovery that the first step in flavonol, flavone, dihydroflavonol and catechin metabolism is hydroxylation at C-8 (ring A). This hydroxylation requires stoichiometric amounts of oxygen and NADH. Schultz and Wood demonstrated this pathway using protein extracts of a quercetin-grown species and Jeffrey *et al.* working with a catechin-metabolizing strain of *Pseudomonas.* The proposed pathways (Figs. 17.12 and 17.13) seem to be specific for flavonoids with a hydroxyl group at C-7.

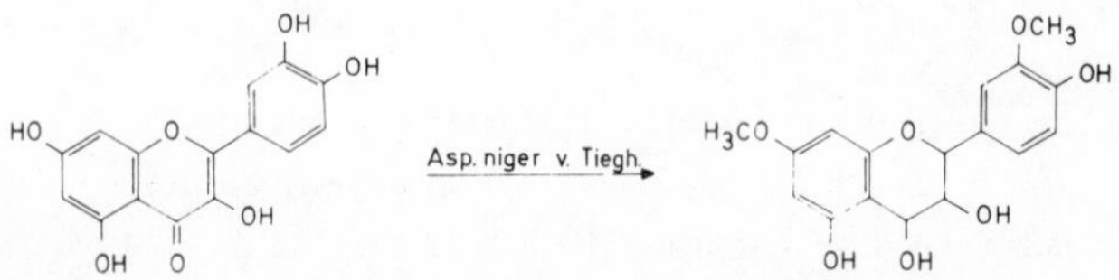

Figure 17.11 Transformation of quercetin by *Aspergillus niger* van Tieghem (Haluk and Metche, 1970).

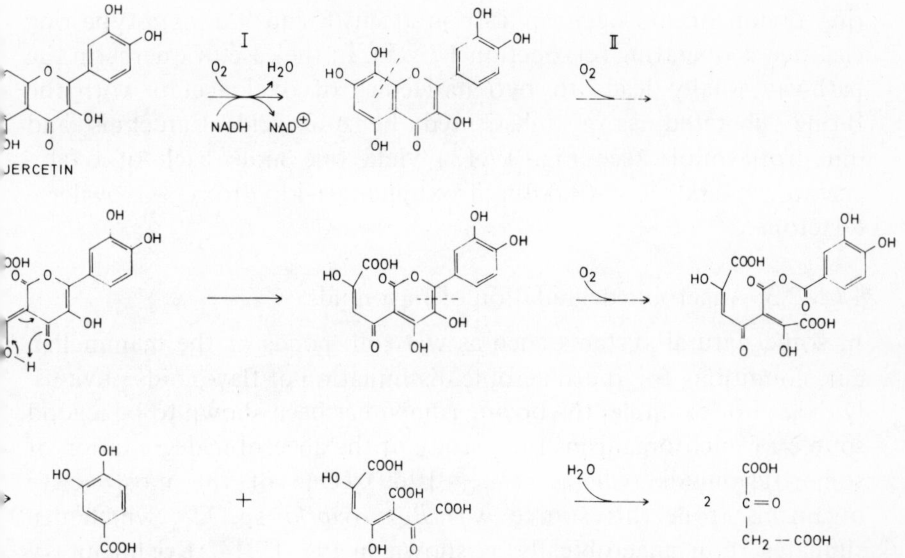

Figure 17.12 Degradation of flavonols by *Pseudomonas* species according to Schultz and Wood (1972).

Methyl or methoxyl substituents at this position prevent the hydroxylation at C-8. The monooxygenases responsible for hydroxylation at position 8 of flavonoids have both been shown to be flavoproteins.

The intermediate 7,8-dihydroxyflavonoids are further metabolized by the action of oxygen-requiring dioxygenases. Since in all cases,

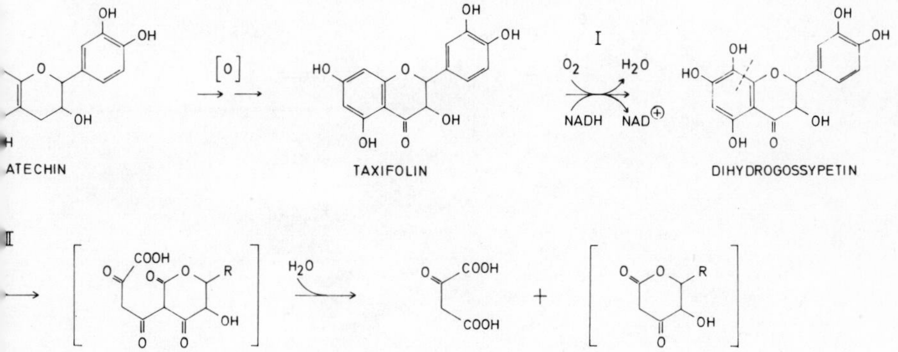

Figure 17.13 Degradation of catechin/taxifolin by *Pseudomonas* species according to Jeffrey *et al.* (1972a,b).

ring fission occurs between carbon atoms 8 and 9 a *meta*-type ring cleavage is operating (cf. Section 17.5.5). In the case of quercetin the pathway finally leads to two molecules of oxaloacetate with the B-ring liberated as a substituted benzoic acid. Catechins and dihydroflavonols (see Fig. 17.13) yield one mole each of oxalo-acetate and 5-(3,4-dihydroxyphenyl)-4-hydroxy-3-oxovalero-δ-lactone.

17.5.2.3 Anaerobic degradation of flavonoids

In some natural systems such as wet soil, ponds or the mammalian gut, conditions for the anaerobic dissimilation of flavonoids obvious-ly exist. For example, the bovine rumen has been shown to be a good source of microorganisms for a study of the anaerobic degradation of some flavonoids (Cheng *et al.*, 1969). One of the most active organisms from this source was *Butyrivibrio* sp. C_3, which dis-similates rutin anaerobically as shown in Fig. 17.14 (Krishnamurty *et al.*, 1970). Since only the glycosides rutin and quercitrin but not the aglycone quercetin are dissimilated, the glycosides were assumed to be the essential transport form for uptake of substrate by *Butyrivibrio* sp. C_3. This is in agreement with the observation that exoenzymes of this organism could not be detected.

Fig. 17.15 summarized the results of various experiments with

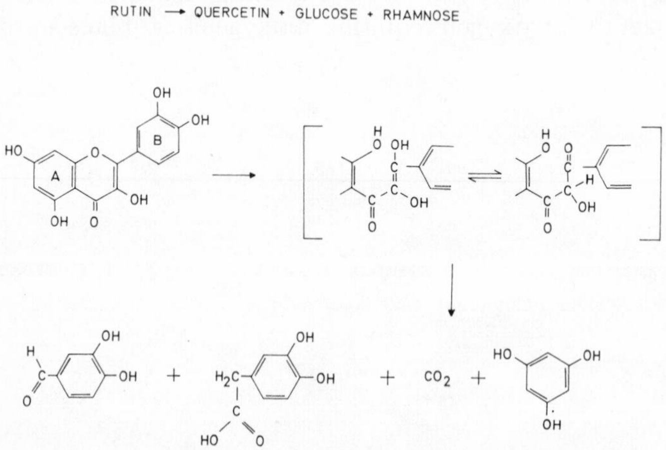

Figure 17.14 Anaerobic degradation of rutin by *Butyrivibrio* sp. C_3 (Krishnamurty *et al.*, 1970).

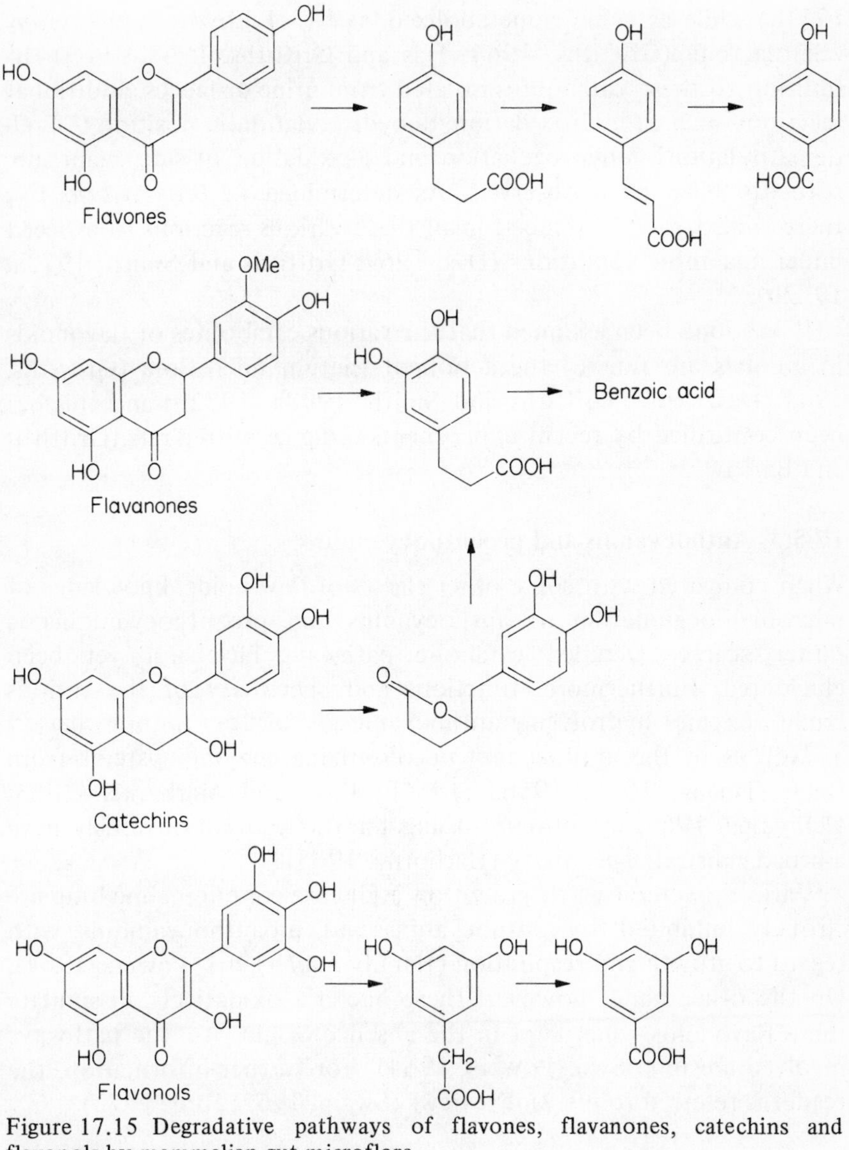

Figure 17.15 Degradative pathways of flavones, flavanones, catechins and flavonols by mammalian gut microflora.

flavonoids orally applied to mammals (rat, guinea pig, rabbit). Flavonols are degraded to phenylacetic acids (Booth *et al.*, 1956; Griffiths and Smith, 1972a), flavones and flavanones give rise to phenylpropionic acids (Booth *et al.*, 1958; Griffiths and Smith,

1972b) while catechin is metabolized via δ-(3,4-dihydroxyphenyl)-γ-valerolactone (Griffiths, 1964; Das and Griffiths, 1968, 1969). In addition to these catabolites isolated from urine or faeces, additional reactions such as hydroxylation, dehydroxylation in position C-4′,O-demethylation, dehydrogenation and β-oxidation of side chain substituents have been observed. As determined in *Butyrivibrio* C_3, there is also good evidence for all these various reactions to proceed under anaerobic conditions (Das, 1969; Griffiths and Smith, 1972a, 1972b).

It has long been assumed that the various catabolites of flavonoids in animals are due to the action of the gut microflora (Griffiths, 1964; Das, 1969; Griffiths and Smith, 1972a, 1972b) and this has been confirmed by recent experiments using germ-free rats (Griffiths and Barrow, 1972).

17.5.3 Anthocyanins and proanthocyanidins

When compared with some other classes of flavonoids, knowledge of microbial degradation of anthocyanins and proanthocyanidins is rather scarce. Detailed catabolic pathways have not yet been elucidated. Furthermore, functions and specificity of the various crude enzymes hydrolysing anthocyanin glycosides ('anthocyanase') as well as of the anthocyanin decolourizing enzyme systems from fungi (Huang, 1955, 1956a, 1956b; Peng and Markarkis, 1963; Uchiyama, 1969) are obscure. Fungal anthocyanases obviously have a broad substrate specificity (Harborne, 1965).

Various bacteria when grown on a glucose-containing medium are strongly inhibited by anthocyanins and proanthocyanidins with regard to growth and respiration (Hamdy *et al.*, 1961; Powers, 1964). On the other hand, however, these bacteria oxidatively dissimilate these flavonoids when kept in the absence of glucose, the pathways involved are unknown (Powers, 1964). For further information, the reader is referred to relevant reviews (Swain, 1962; Jurd, 1972).

17.5.4 Isoflavonoids

Isoflavonoids have less frequently been investigated for microbial dissimilation. Most of the work has been done in connection with metabolism of forage estrogens (isoflavones and coumestrol) in an attempt to clarify their role in the infertility syndrome of mammals (Bickoff, 1968). As pointed out in Section 17.5.2, most of the results obtained with animals are essentially due to the action of the intestinal microflora.

The isoflavones daidzein, formononetin, genistein and biochanin A, all constituents of legume forage plants, are efficiently catabolized by the intestinal microflora of sheep (Batterham *et al.*, 1965, 1971; Braden *et al.*, 1967; Shutt and Braden, 1968), the domestic fowl (Cayen *et al.*, 1965), the rat (Griffiths and Smith, 1972a) and the guinea pig (Shutt and Braden, 1968). The main catabolite (Fig. 17.16) is the isoflavan equol, long known to occur in the urine of mares and goats (Marrian and Haslewood, 1932; Klyne and Wright, 1957). The remarkable dehydroxylation of genistein in position 5 and the reduction of the heterocyclic ring point to the anaerobic conditions. Among other catabolites, *p*-ethylphenol, an α-methyl-substituted deoxybenzoic and two substituted phenylbenzyl ketones have been recovered in some of these experiments (Tang and Common, 1968).

An investigation of urinary conversion products of the coumestan coumestrol (24) in the fowl showed no conversion of this compound to equol (Cayen and Common, 1965).

While the flavonol degrading species of *Aspergillus niger* (see Section 17.5.2) cannot degrade isoflavones (Padron *et al.*, 1960), dissimilation of the four isoflavones of Fig. 17.16 seems to be a widespread property in the genus *Fusarium* (Barz, 1971; Barz and Schlepphorst, unpublished). With the aid of inducible enzymes, the four isoflavones are completely degraded and they serve as sole source of carbon and energy. Intermediates of the pathway are presently under investigation. A remarkable capacity of *Fusarium* species for dissimilation of aromatic compounds is corroborated by the observation that even the fungistatic phytoalexin pisatin (25) is degraded by *Fusarium oxysporum* Schlecht (Nonaka, 1967; Dewit-Elshove, 1968).

A *Nocardia* bacterium isolated from the rhizosphere of isoflavone bearing *Cicer arietinum* has been shown to completely degrade flavones, flavonols and isoflavones (Barz, 1970). The basic skeletons of flavonoids degraded by this bacterium are shown in Fig. 17.17; oxygen or carbon substituents in the (−)-labelled position inhibit degradation while substituents in the (+)-labelled positions do not. Isoflavone catabolism is started by oxidative attack of ring A along a pathway most likely comparable to the one shown in Fig. 17.12 (Gierse, 1971; Barz and Müller-Enoch, unpublished).

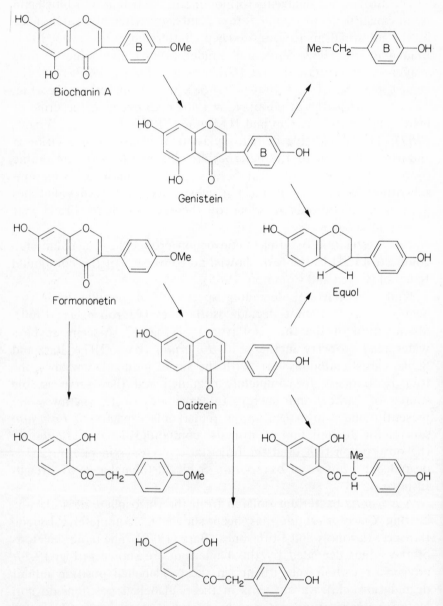

Figure 17.16 Transformation of isoflavones to equol and various catabolites by the intestinal microflora of animals.

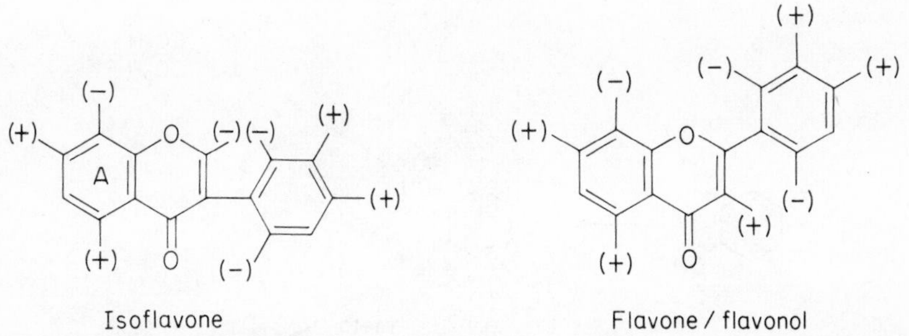

Isoflavone   Flavone / flavonol

Figure 17.17 Basic skeletons of flavonoids degraded by a *Nocardia* species ((+): substituent does not prevent degradation; (−): substituent completely inhibits degradation) (Barz, 1970).

17.5.5 Aromatic acids and phenols

The astonishing ability of microorganisms to catabolize a wide variety of aromatic acids and phenols is one of the best understood research areas of biochemistry. An impressive body of information has been accumulated and a detailed presentation is beyond the scope of this chapter. The basic principles are illustrated in Fig. 17.18. All aromatic structures must be capable of being modified into either *o*- or *p*-dihydroxyphenols before cleavage can occur. *O*-Dihydroxyphenols are oxidatively dissimilated by two basically different pathways: (a) fission of the bond between the carbon atoms bearing the hydroxyl groups to produce a *cis-cis*-muconic acid (*ortho*-or intradiol cleavage; 'a' in Fig. 17.18; and (b) fission of the bond between the carbon atom carrying a hydroxyl and its neighbour that carries a hydrogen atom or a side chain substituent to produce a muconic acid semialdehyde or a keto acid (*meta*- or extradiol cleavage; 'b' in Fig. 17.18).

In the case of *p*-dihydroxyphenols, rupture of the bond between the carbon atom bearing a hydroxyl and an adjacent carbon atom carrying a hydrogen, carboxyl or a side chain occurs ('c' in Fig. 17.18), yielding a substituted *cis-cis*-muconic acid derivative. Depending on the species studied as well as on the pathway and cleavage enzyme involved, the various possible substituents of aromatic compounds are either shortened or oxidatively removed prior to ring cleavage or they remain on the ring and are eliminated after ring fission at the stage of aliphatics. For more information the reader is referred to useful recent reviews (Evans, 1963; Ribbons,

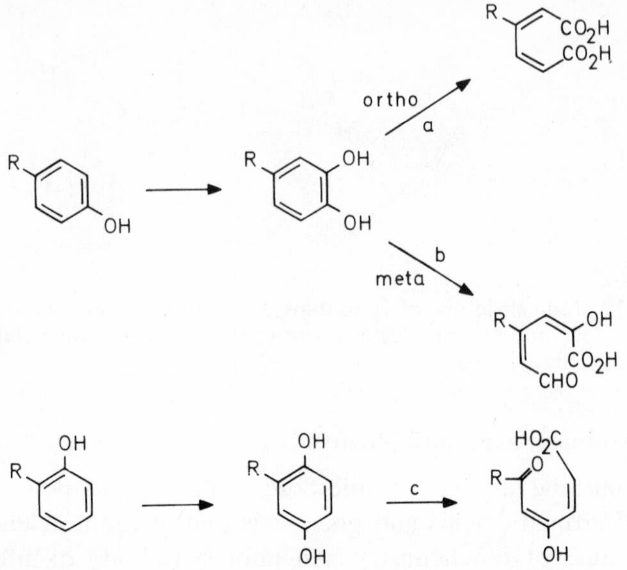

Figure 17.18 Fundamental mechanisms of ring cleavage reactions in microorganisms.

1965; Canovas *et al.*, 1967; Gibson, 1968; Evans, 1969; Dagley, 1971; Towers and Subba Rao, 1972).

17.6 Conclusions

Recent investigations have provided ample evidence for flavonoid turnover and degradation in higher plants. The observation of simultaneous synthesis, accumulation and turnover of flavonoids, especially in rapidly growing and differentiating tissues, indicates that the systems involved are of much greater complexity than so far expected. On the other hand, though turnover can be expected to occur, it does not always happen (Section 17.3.10). Turnover of flavonoids comprises reactions which transfer the plant products partly into polymers and partly into catabolic pathways. The polymerization reactions are mainly catalyzed by peroxidase and phenolase. It will depend on the physiological situation of the cells involved and on the availability of enzymes how much flavonoid is metabolized along the two pathways.

Catabolism versus polymerization is, therefore, a quantitative matter and depends on the relevant enzymes, the localization of both substrates and enzymes, direction of transport and the pool sizes of the substrates under investigation. The situation is further complicated by the fact that under certain physiological conditions, flavonoids can for various intervals of time either partly or totally, be withdrawn from further metabolism (see Sections 17.3.9 and 17.3.10); they are then in a metabolically inactive pool. Various endogenous or exogenous regulatory factors which lead for example to *de novo* enzyme synthesis or transport of substrates will convey these 'protected' compounds to a metabolically active pool. From this metabolically active pool, turnover and catabolism will proceed. This scheme is summarised in Fig. 17.19.

Whenever flavonoid catabolism in higher plants has successfully been demonstrated, benzoic acids are among the main degradation products. There is thus a great similarity with the situation in microorganisms. A more detailed analysis of benzoic acid metabolism

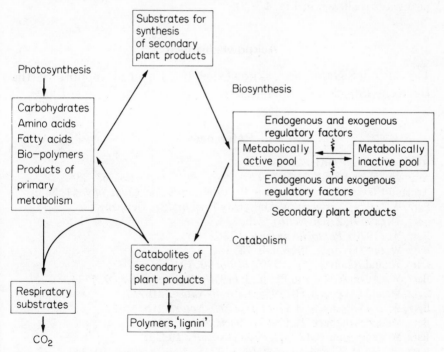

Figure 17.19 Relationships between primary and secondary metabolism and description of two, metabolically different pools for secondary plant products.

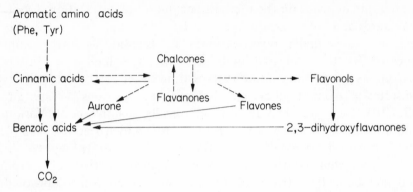

Figure 17.20 Metabolic grid depicting anabolic (– – – –➤) and catabolic (————➤) pathways of some flavonoids.

will therefore contribute to our understanding of flavonoid degradation. As a summary of our present knowledge of flavonoid degradation, a metabolic grid depicting anabolic and catabolic pathways is shown in Fig. 17.20.

Acknowledgement

The authors gratefully acknowledge the financial support of the Deutsche Forschungsgemeinschaft.

References

Ahlgrim, E.-D. (1956), *Planta* **47**, 268.
Amrhein, N. and Zenk, M. H. (1970), *Ztsch. Pflanzenphysiol.* **63**, 384.
Armand-Fraysse, D. and Lebreton, P. (1969), *Bull. Soc. Chim. biol.* **51**, 563.
Barz, W. (1964), Dissertation, University Freiburg/Br., Germany.
Barz, W. (1969), *Ztsch. Naturf.* **24b**, 234.
Barz, W. (1970), *Phytochemistry* **9**, 1745.
Barz, W. (1971), *Arch. Mikrobiol.* **78**, 341.
Barz, W. and Adamek, Ch. (1970), *Planta* **90**, 191.
Barz, W., Adamek, Ch. and Berlin, J. (1970), *Phytochemistry* **9**, 1735.
Barz, W. and Grisebach, H. (1966a), *Ztsch. Naturf.* **21b**, 47.
Barz, W. and Grisebach, H. (1966b), *Ztsch. Naturf.* **21b**, 1113.
Barz, W. and Grisebach, H. (1967), *Ztsch. Naturf.* **22b**, 627.
Barz, W. and Hösel, W. (1971), *Phytochemistry* **10**, 335.
Barz, W., Hösel, W. and Adamek, Ch. (1971), *Phytochemistry* **10**, 343.
Barz, W., Mohr, I. and Teufel, E. (1974), *Phytochemistry* **13**, 1785.
Barz, W. and Roth-Lauterbach, B. (1969), *Ztsch. Naturf.* **24b**, 638.

Batterham, T. J., Hart, N. K., Lamberton, J. A. and Braden, A. W. H. (1965), *Nature* **206**, 509.

Batterham, T. J., Shutt, D. A., Hart, N. K., Braden, A. W. H. and Tweeddale, H. J. (1971), *Aust. J. agric. Res.* **22**, 131.

Beck, A. B. (1964), *Aust. J. agric. Res.* **15**, 223.

Berlin, J. (1972), Dissertation, University Freiburg/Br., Germany.

Berlin, J. and Barz, W. (1971), *Planta* **98**, 300.

Berlin, J., Barz, W., Harms, H. and Haider, K. (1971), *FEBS Letters* **16**, 141.

Berlin, J., Dewick, P. M., Barz, W., Grisebach, H. (1972), *Phytochemistry* **11**, 1689.

Berlin, J., Kiss, P., Müller-Enoch, D., Gierse, D., Barz, W. and Jauistyu, B. (1974), *Ztsch. Naturf.* **29c**, 374.

Bickoff, E. M. (1967), Studies on chemical and biological properties of coumestrol and related compounds. United States Department of Agriculture. Tech. Bull. No. 1408, p. 95.

Bickoff, E. M. (1968), Oestrogenic constituents of forage plants. Review Series No. 1/1968. Commonwealth Bur. of Pastures and Field Crops. Hurley, Berkshire, England.

Bolkart, K. H. and Zenk, M. H. (1968), *Ztsch. Pflanzenphysiol.* **59**, 439.

Booth, A. N., Jones, F. T. and DeEds, F. (1958), *J. biol. Chem.* **230**, 661.

Booth, A. N., Murray, L. W., Jones, F. T. and DeEds, F. (1956), *J. biol. Chem.* **223**, 251.

Boylen, C. W., Hagen, C. W., Jr., Mansell, R. C. (1969), *Phytochemistry* **8**, 2311.

Bradbury, R. B. and White, D. E. (1951), *J. chem. Soc.* 3447.

Braden, A. W. H., Hart, N. K. and Lamberton, J. A. (1967), *Aust. J. agric. Res.* **18**, 335.

Brandner, G. (1962), Dissertation, University Freiburg/Br., Germany.

Britton, G. and Haslam, E. (1965), *J. chem. Soc.* 7312.

Burian, K. and Aichmair, M. I. (1968), *Protoplasma* **66**, 451.

Canovas, J. L., Ornston, L. N. and Stanier, R. Y. (1967), *Science* **156**, 1695.

Cayen, M. N. and Common, R. H. (1965), *Biochim. Biophys. Acta* **100**, 567.

Cayen, M. N., Tang, G. and Common, R. H. (1965), *Biochim. Biophys. Acta* **111**, 349.

Chatterjee, A. K. and Gibbins, L. N. (1969), *J. Bacteriol.* **100**, 594.

Cheng, K. J., Jones, G. A., Simpson, F. J. and Bryant, M. P. (1969), *Can. J. Microbiol.* **15**, 1356.

Child, J. J., Simpson, F. J. and Westlake, D. W. S. (1963), *Can. J. Microbiol.* **9**, 653.

Ciegler, A., Lindenfelser, L. A. and Nelson, G. E. N. (1971), *Appl. Microbiol.* **22**, 974.

Collet, C. F. (1968), *Can. J. Bot.* **46**, 969.

Combier, H. and Lebreton, P. (1968), *C. r. Acad. Sci.* **267**, 421.

Constabel, F., Shyluk, J. P. and Gamborg, O. L. (1971), *Planta* **96**, 306.

Crombie, L., Dewick, P. M. and Whiting, D. A. (1970), *Chem. Commun.* 1469.

Crombie, L., Dewick, P. M. and Whiting, D. A. (1971a), *Chem. Commun.* 1182.

Crombie, L., Dewick, P. M. and Whiting, D. A. (1971b), *Chem. Commun.* 1183.

Crombie, L., Dewick, P. M. and Whiting, D. A. (1973), *J. chem. Soc. Perkin I* 1285.

Crombie, L., Green, C. L. and Whiting, D. A. (1968), *J. chem. Soc.* (C) 3029.

Crombie, L. and Thomas, M. B. (1967), *J. chem. Soc.* (C) 1796.

Curnow, D. H. (1954), *Biochem. J.* **58**, 283.

Dagley, S. (1971). In *Advances in Microbial Physiology* (A. H. Rose and J. F. Wilkinson, eds.), Vol. 6, pp. 1—46. Academic Press, London, New York.

Danner, H. (1940), *Bot. Arch.* **41**, 168.

Das, N. P. (1969), *Biochim. Biophys. Acta* **177**, 668.

Das, N. P. and Griffiths, L. A. (1968), *Biochem. J.* **110**, 449.

Das, N. P. and Griffiths, L. A. (1969), *Biochem. J.* **115**, 831.

Dedio, W. and Clark, K. W. (1968), *Can. J. Pl. Sci.* **48**, 175.

Deverall, B. J. (1972). In *Phytochemical Ecology* (J. B. Harborne, ed.), pp. 217—233. Academic Press, London and New York.

Dewit-Elshove, A. (1968), *Neth. J. Pl. Pathol.* **74**, 44.

Dietermann, L. J., Wender, S. H., Chorney, W. and Skok, K. (1969), *Phytochemistry* **8**, 2321.

Dittrich, P. and Kandler, O. (1971), *Ber. dt. bot. Ges.* **84**, 465.

Dürand, R. and Zenk, M. H. (1974), *FEBS Letters* **39**, 218.

Durmishidze, V., Shalashili, A. G. (1968), *Dokl. Akad. Nauk USSR* **181**, 1489.

Ebel, J., Achenbach, H., Barz, W. and Grisebach, H. (1970), *Biochim. Biophys. Acta* **215**, 203.

El Basyouni, S. Z., Chen, D., Ibrahim, R. K., Neish, A. C. and Towers, G. H. N. (1964), *Phytochemistry* **3**, 485.

Ellis, B. E. (1971), *FEBS Letters* **18**, 228.

Ellis, B. E. (1973), *Planta* **111**, 113.

Ellis, B. E. and Towers, G. H. N. (1969), *Phytochemistry* **8**, 1415.

Ellis, B. E. and Towers, G. H. N. (1970), *Phytochemistry* **9**, 1457.

Erez, A. and Lavee, S. (1969), *Pl. Physiol.* **44**, 342.

Evans, W. C. (1963), *J. gen. Microbiol.* **32**, 177.

Evans, W. C. (1969). In *Fermantation Advances* (D. Perlman, ed.), pp. 649—687. Academic Press, London, New York.

Fischer, N. and Dreiding, A. S. (1972), *Helv. chim. Acta* **55**, 649.

Francis, C. M. and Millington, A. J. (1965a), *Aust. J. agric. Res.* **16**, 23.

Francis, C. M. and Millington, A. J. (1965b), *Aust. J. agric. Res.* **16**, 557.

Frey, G. (1973), Diplomarbeit, University Freiburg/Br., Germany.

Friedrich, H. (1954), *Pharmazie* **9**, 138.

Fritsch, H. J., Hahlbrock, K. and Grisebach, H. (1971), *Ztsch. Naturf.* **26b**, 581.

Frohne, D. (1964), *Planta med.* **12**, 140.

Fukami, J., Shishido, T., Fukunaga, K. and Casida, J. E. (1969), *Agric. Fd. Chem.* **17**, 1217.

Fukami, J., Yamamoto, J. and Casida, J. E. (1967), *Science* **155**, 713.

Galston, A. W. (1969). In *Perspectives in Phytochemistry* (J. B. Harborne and T. Swain, eds.), pp. 193—204. Academic Press, London, New York.

Gibson, D. T. (1968), *Science* **161**, 1093.

Gierse, H. D. (1971), Diplomarbeit, University Freiburg/Br., Germany.

Glass, A. D. M. and Bohm, B. A. (1969), *Phytochemistry* **8**, 371.

Griffiths, L. A. (1964), *Biochem. J.* **92**, 173.

Griffiths, L. A. and Barrow, A. (1972), *Biochem. J.* **130**, 1161.

Griffiths, L. A. and Smith, G. E. (1972a), *Biochem. J.* **130**, 141.

Griffiths, L. A. and Smith, G. E. (1972b), *Biochem. J.* **128**, 901.

Grisebach, H. and Barz, W. (1969), *Naturwissenschaften* **56**, 538.

Grisebach, H. and Bopp, M. (1959), *Ztsch. Naturf.* **14b**, 485.

Grisebach, H. and Vollmer, K. O. (1963), *Ztsch. Naturf.* **18b**, 753.

Grochowska, M. J. (1967), *Bull. Acad. Polon. Sci.* **15**, 455.

Gross, D. (1969). In *Biosynthese der Alkaloide* (eds. K. Mothes and H. R. Schütte), pp. 439–458. VEB Deutscher Verlag der Wissenschaften, Berlin.

Gross, S. R., Gafford, R. D. and Tatum, E. L. (1956), *J. biol. Chem.* **219**, 781.

Hadwiger, L. A. (1967), *Phytopathology* **57**, 1258.

Hadwiger, L. A. (1972), *Biochem. biophys. Res. Comm.* **46**, 71.

Hadwiger, L. A., Hess, S. L. and von Broembsen, S. (1970), *Phytopathology* **60**, 332.

Haluk, J. P. and Metche, M. (1970), *Bull. Soc. Chim. biol.* **52**, 667.

Hamdy, M. K., Pratt, D. E., Powers, J. J. and Somaatmadjy, D. (1961), *J. Fd. Soc.* **26**, 457.

Hanson, C. H. (1965), Variation in coumestrol content of alfalfa. United States Department of Agriculture. Tech. Bull. No. 1333.

Harborne, J. B. (1965), *Phytochemistry* **4**, 107.

Harborne, J. B. (1967), *Comparative Biochemistry of the Flavonoids*, p. 267. Academic Press, London and New York.

Harms, H., Haider, K., Berlin, J., Kiss, P. and Barz, W. (1972), *Planta* **105**, 342.

Harms, H. and Prieß, J. (1973), *Planta* **109**, 307.

Harms, H., Söchtig, H. and Haider, K. (1969a), *Pl. Soil* **31**, 129.

Harms, H., Söchtig, and Haider, K. (1969b), *Pl. Soil* **31**, 257.

Harms, H., Söchtig, H. and Haider, K. (1971), *Ztsch. Pflanzenphysiol.* **64**, 437.

Hattori, S. and Noguchi, I. (1959), *Nature* **184**, 1145.

Hay, G. W., Westlake, D. W. S. and Simpson, F. J. (1961), *Can. J. Microbiol.* **7**, 921.

Hess, S. L. and Hadwiger, L. A. (1971), *Pl. Physiol.* **48**, 197.

Hess, S. L., Hadwiger, L. A. and Schwochau, M. E. (1971), *Phytopathology* **61**, 79.

Hillis, W. E. and Swain, T. (1959), *J. Sci. Fd. Agric.* **10**, 135.

Holowczak, J., Kuc, J. and Williams, E. G. (1960), *Phytopathology* **50**, 640.

Hösel, W. and Barz, W. (1972), *Biochim. Biophys. Acta* **261**, 294.

Hösel, W., Frey, G., Teufel, E. and Barz, W. (1972a), *Planta* **103**, 74.

Hösel, W., Shaw, P. D. and Barz, W. (1972b), *Ztsch. Naturf.* **27b**, 946.

Huang, H. T. (1955), *J. agric. Fd. Chem.* **3**, 141.

Huang, H. T. (1956a), *J. Am. chem. Soc.* **78**, 2390.

Huang, H. T. (1956b), *Nature* **177**, 39.

Ibrahim, R. K., Lawson, S. G. and Towers, G. H. N. (1961), *Can. J. Biochem. Physiol.* **39**, 873.

Imaseki, H., Wheeler, R. E. and Geissmann, T. A. (1965), *Tetrahedron Letters* **17**, 1785.

Ishikura, N. (1972), *Phytochemistry* **11**, 2555.

Jaffe, M. J. and Galston, A. W. (1967), *Pl. Physiol.* **42**, 848.

Janistyn, B., Barz, W. and Pohl, R. (1971), *Ztsch. Naturf.* **26b**, 973.

Jayasankar, N. P., Bandoni, R. J. and Towers, G. H. N. (1969), *Phytochemistry* **8**, 379.

Jeffrey, A. M., Jerina, D. M., Self, R. and Evans, W. C. (1972b), *Biochem. J.* **130**, 383.

Jeffrey, A. M., Knight, M. and Evans, W. C. (1972a), *Biochem. J.* **130**, 373.

Jurd, L. (1972). In *The Chemistry of Plant Pigments* (C. O. Chichester, ed.), p. 123. Academic Press, New York and London.

Jürgensmeyer, H. L. and Bopp, M. (1961), *Naturwissenschaften* **48**, 80.

Kindl, H. (1969a), *Eur. J. Biochem.* **7**, 340.

Kindl, H. (1969b), *Ztsch. physiol. Chem.* **350**, 1289.

Kindl, H. and Hoffmann-Ostenhof, O. (1966), *Phytochemistry* **5**, 1091.

Kiss, P. (1972), Diplomarbeit, University Freiburg/Br., Germany.

Klyne, W. and Wright, A. A. (1957), *Biochem. J.* **66**, 92.

Krishnamurty, H. G., Cheng, K.-J., Jones, G. A., Simpson, F. J. and Watkin, J. E. (1970), *Can. J. Microbiol.* **16**, 759.

Krishnamurty, H. G. and Simpson, F. J. (1970), *J. biol. Chem.* **245**, 1467.

Kuc, J. and Nelson, O. E. (1964), *Arch. Biochem. Biophys.* **105**, 103.

Lackmann, I. (1971), *Planta* **98**, 258.

Lewak, St. and Podstolski, A. (1966), *Bull. Acad. Polon. Sci.* **14**, 103.

Libbert, E., Schröder, R., Drawert, A. and Fischer, E. (1970), *Physiol. Plantarum* **23**, 287.

Magnus, V., Iskric, S. and Kveder, S. (1971), *Planta* **97**, 116.

Magnus, V., Iskric, S. and Kveder, S. (1973), *Planta* **110**, 57.

Mahler, H. R. and Cordes, E. H. (1966), *Biological Chemistry*, p. 404, 405. Harper and Row, Publishers, New York and London.

Maier, V. P. and Hasegawa, S. (1970), *Phytochemistry* **9**, 139.

Mansell, R. L. and Hagen, C. W., Jr. (1966), *Am. J. Bot.* **53**, 875.

Marrian, G. F. and Haslewood, G. A. D. (1932), *Biochem. J.* **26**, 1227.

McClure, J. W. and Wilson, K. G. (1970), *Phytochemistry* **9**, 763.

Meister, A. (1965), *Biochemistry of Amino Acids*, 2nd Ed., Vol. 1, pp. 294--369. Academic Press, New York.

Minamikawa, T., Jayasankar, N. P., Bohm, B. A., Taylor, J. E. P. and Towers, G. H. N. (1970), *Biochem. J.* **116**, 889.

Moritz, O. and Morf, H. (1956), *Arch. Pharm.* **289**, 632.

Mothes, K. (1969), *Experientia* **25**, 226.

Neish, A. C. (1965). In *Plant Biochemistry* (eds. J. Banner and J. E. Varner), p. 611. Academic Press, New York and London.

Neumann, J. and Avron, M. (1967), *Pl. Cell Physiol.* **8**, 241.

Noguchi, I. and Mori, S. (1969), *Arch. Biochem. Biophys.* **132**, 352.

Nöll, G. (1955), *Pharmazie* **10**, 609, 679.

Nonaka, F. (1967), *Saga Daigaku Nogaku Iho* **24**, 109.

Noveroske, R. L., Kuc, J. and Williams, E. B. (1964a), *Phytopathology* **54**, 92.

Noveroske, R. L., Williams, E. B. and Kuc, J. (1964b), *Phytopathology* **54**, 98.

Oka, T. and Simpson, F. J. (1971), *Biochem. biophys. Res. Comm.* **43**, 1.

Oka, T., Simpson, F. J. and Krishnamurty, H. G. (1972), *Can. J. Microbiol.* **18**, 493.

Ornston, L. N. (1971), *Bacteriol. Rev.* **35**, 87.

Padron, J., Grist, K. L., Clark, J. B. and Wender, S. H. (1960), *Biochem. biophys. Res. Comm.* **3**, 412.

Paris, R. D. and Jacquemin, H. (1970), *C.r. Acad. Sci. Ser.* D **270**, 3232.

Patschke, L., Barz, W. and Grisebach, H. (1964b), *Ztsch. Naturf.* **19b**, 1110.

Patschke, L., Hess, D. and Grisebach, H. (1964a), *Ztsch. Naturf.* **19b**, 1114.

Patschke, L., Hess, D. and Grisebach, H. (1964a), *Z. Naturf.* **19b**, 1114.

Peng, C. Y. and Markarkis, P. (1963), *Nature* **199**, 597.

Pickard, M. A. and Westlake, D. W. S. (1969), *Can. J. Microbiol.* **15**, 869.

Pickard, M. A. and Westlake, D. W. S. (1970), *Can. J. Biochem.* **48**, 1351.

Podstolski, A. and Lewak, St. (1970), *Phytochemistry* **9**, 289.

Powers. J. J. (1964), *Internatl. Symp. Fd. Microbiol.* 4th, Goteborg, Sweden, p. 59.

Procter, J. I. A. and Creasy, L. L. (1969), *Phytochemistry* **8**, 1401.

Puri, B. and Seshadri, T. R. (1955), *J. chem. Soc.* 1589.

Raa, J. and Overeen, J. C. (1968), *Phytochemistry* **7**, 721.

Ramwell, R. M. and Sherratt, H. S. A. (1964). In *Biochemistry of Phenolic Compounds* (ed. J. B. Harborne), p. 457. Academic Press, London and New York.

Rathmell, W. G. and Bendall, D. S. (1972), *Biochem. J.* **127**, 125.

Reuter, E.-W. (1970), Ph.D. dissertation, University Freiburg/Br., Germany.

Reznik, H. (1960), *Erg. Biol.* **23**, 14.

Ribbons, D. W. (1965), *Annual Rep. chem. Soc.* **62**, 445.

Rosa, N. (1966), Ph. D. thesis, Dalhousie Univ. Halifax, N.S.

Rossiter, R. C. (1967), *Aust. J. agric. Res.* **18**, 39.

Rossiter, R. C. (1970), *Aust. J. biol. Sci.* **23**, 469.

Rossiter, R. C. and Beck, A. B. (1966a), *Aust. J. agric. Res.* **17**, 29.

Rossiter, R. C. and Beck, A. B. (1966b), *Aust. J. agric. Res.* **17**, 447.

Rossiter, R. C. and Beck, A. B. (1967a), *Aust. J. agric. Res.* **18**, 23.

Rossiter, R. C. and Beck, A. B. (1967b), *Aust. J. agric. Res.* **18**, 561.

Sakamura, S., Watanabe, S. and Obata, I. (1965), *Agric. biol. Chem.* **29**, 181.

Sarapuu, L. P. (1965), *Fiziol. Rast.* **12**, 134.

Sarapuu, L. P. (1971), *Biokimia* **36**, 343.

Schultz, G. (1965), *Dt. tierärztl. Wschr.* **72**, 246.

Schultz, G. (1967), *Dt. tierärztl. Wschr.* **74**, 118.

Schultz, E. Engle, F. E. and Wood, J. M. (1974), *Biochemistry* **13**, 1768.

Schultz, E. and Wood, J. M. (1972), *Fed. Proc.* **31**, 477.

Schumacker, R. and Bastin, M. (1965), *Bull. Soc. Roy. Sci. Liege* **34**, 42.

Schwarze, P. (1958), *Handbuch der Pflanzenphysiologie*, Vol. 10, p. 523. Springer-Verlag, Berlin, Göttingen, Heidelberg.

Schwochau, M. E. (1968), private communication to H. Grisebach.

Sharma, H. K., Jamaluddin, M. and Vaidyanathan, C. S. (1972), *FEBS Letters* **28**, 41.

Shutt, D. A. and Braden, A. W. H. (1968), *Aust. J. agric. Res.* **19**, 545.

Simpson, F. J., Talbot, G. and Westlake, D. W. S. (1960), *Biochem. biophys. Res. Comm.* **2**, 15.

Stafford, H. A. (1969), *Phytochemistry* **8**, 743.

Staude, M. and Reznik, H. (1973), *Ztsch. Pflanzenphysiol.* **68**, 346.

Steiner, A. M. (1971), *Ztsch. Pflanzenphysiol.* **65**, 210.

Steiner, A. M. (1972), *Ztsch. Pflanzenphysiol.* **66**, 133.

Stickland, R. G. (1972), *Annuals Bot.* **36**, 459.

Stickland, R. G. and Sunderland, N. (1972a), *Annuals Bot.* **36**, 443.

Stickland, R. G. and Sunderland, N. (1972b), *Annuals Bot.* **36**, 671.

Street, H. E. (1973). In *Biosynthesis and its Control in Plants* (B. V. Milborrow, ed.), p. 93. Academic Press, London and New York.

Suzuki, H. (1962), *Arch. Biochem. Biophys.* **99**, 476.

Swain, T. (1962). In *The Chemistry of Flavonoid Compounds* (T. Geissmann ed.), p. 542. Pergamon Press, Oxford.

Tang, G. and Common, R. H. (1968), *Biochim. Biophys. Acta* **158**, 402.

Tateoka, T. N. (1970), *Bot. Mag. (Tokyo)* **83**, 49.

Taylor, A. O. and Zucker, M. (1966), *Pl. Physiol.* **41**, 1350.

Thomas, R. (1965), *Biogenesis of Antibiotic Substances,* p. 155. Publishing House of the Czechoslovak Acad. of Sciences, Prague.

Tissut, M. (1968), *Physiol. veg.* **6**, 351.

Tissut, M. and Egger, K. (1972), *Phytochemistry* **11**, 631.

Todt, D. (1962), *Ztsch. Bot.* **50**, 1.

Towers, G. H. N. (1964). In *Biochemistry of Phenolic Compounds* (J. B. Harborne ed.), p. 249. Academic Press, New York and London.

Towers, G. H. N. and Subba Rao (1972). In *Recent Advances in Phytochemistry* (V. C. Runeckles and J. E. Watkins eds.), Vol. 4, p. 1. Appleton-Century-Crofts, New York.

Uchiyama, Y. (1969), *Agric. biol. Chem.* **33**, 1342.

Udupa, S. R., Banerji, A. and Chadha, M. S. (1968), *Tetrahedron Letters* 4003.

Udupa, S. R., Banerji, A. and Chadha, M. S. (1969), *Tetrahedron* **25**, 5415.

Van Buren, J. P., Scheiner, D. M. and Wagenknecht, A. C. (1960), *Nature* **185**, 165.

Voirin, B. and Lebreton, P. (1972), *Phytochemistry* **11**, 3435.

Vollmer, K. O., Reisener, H. J. and Grisebach, H. (1965), *Biochem. biophys. Res. Comm.* **21**, 221.

Wehnert, H.-U. (1970), Ph.D. dissertation, University Hamburg, Germany.

Weissenböck, G. (1971), *Ztsch. Pflanzenphysiol.* **66**, 73.

Weissenböck, G. and Reznik, H. (1970), *Ztsch. Pflanzenphysiol.* **63**, 114.

Westlake, D. W. S. (1963), *Can. J. Microbiol.* **9**, 211.

Westlake, D. W. S., Roxburgh, J. M. and Talbot, G. (1961), *Nature* **189**, 510.

Westlake, D. W. S., Talbot, G., Blakley, E. R. and Simpson, F. J. (1959), *Can. J. Microbiol.* **5**, 621.

Whistance, G. R. and Threlfall, D. R. (1968a), *Biochem. J.* **109**, 482.

Whistance, G. R. and Threlfall, D. R. (1968b), *Biochem. J.* **109**, 577.

Wiermann, R. (1973), *Planta* **110**, 353.

Wong, E. (1966), *Phytochemistry* **5**, 463.

Wong, E. (1967), *Phytochemistry* **6**, 1227.

Yamamoto, I. (1969), *Residne Rev.* **25**, 161.

Zaprometov, M. N. (1959), *Dokl. Akad. Nauk SSSR* **125**, 1359.

Zaprometov, M. N. and Bukhlaeva, V. Ya. (1967), *Fiziol. Rast.* **14**, 804.

Zaprometov, M. N. and Bukhlaeva, V. Ya. (1968), *Fiziol. Rast.* **15**, 457.

Zaprometov, M. N. and Bukhlaeva, V. Ya. (1971), *Fiziol. Rast.* **18**, 787.

Zaprometov, M. N., Sarapuu, L. P. and Bukhlaeva, V. Ya. (1971), *Fiziol. Rast.* **18**, 23.

Zenk, M. H. (1966). In *Biosynthesis of Aromatic Compounds* (G. Billek ed.), p. 45. Pergamon Press, Oxford.

Zenk, M. H. (1967), *Ber. dt. bot. Ges.* **80**, 573.

Zenk, M. H. (1972), *Ztsch. physiol. Chem.* **353**, 123.

Zenk, M. H. and Müller, G. (1964), *Ztsch. Naturf.* **19b**, 398.

Zucker, M. (1972), *Annual Rev. Pl. Physiol.* 133.

Chapter 18

Physiology and Functions of Flavonoids

JERRY W. McCLURE

18.1 Introduction

This chapter is concerned with physiological factors that control levels of flavonoids accumulating in plants and with the biological implications of these accumulations both within the plant and in plant-animal (including man) interactions. Levels of flavonoids within plants are a reflection of the efficiency of biosynthesis (see Chapter 16) tempered by turnover and degradation (Chapter 17) during growth and development. These are all influenced by the external environment. The functions of flavonoids may be approached by a consideration of their general reactivity when introduced into biochemical systems, the many implications of their pigment characteristics, their possible mediation in plant growth and development, and their effects on animals and microorganisms.

Few generalizations on either the physiology or the functions of flavonoids are possible. Such generalizations would imply either that the many different flavonoid structures all work the same way in a given biological response, or that the same stimulus produces different flavonoids, each specific to the tissue being effected. The only generalizations that seem defensible are that the function of a specific flavonoid is a matter of its chemical reactivity within a specific subcellular compartment, and that each cell and tissue type has a highly regulated and responsive system for accumulating and/or coping with specific flavonoids.

Accordingly, much of the following is of necessity presented as a review of the literature; only in a few areas is it possible to make a critical evaluation of the numerous, and often conflicting, reports.

18.2 Physiology of flavonoid accumulation in plants

In 1799 Senebier observed that although most flowers develop anthocyanins only in the light, those of *Crocus* and *Tulipa* do so in darkness as well. By the time of Sachs (1863) two classes of anthocyanin-producing plants were well established; those that produced normal colouration without buds being previously exposed to light and those that developed normally coloured flowers only when the buds were exposed to light at least up to the time of flower opening. Physiological studies were refined by Mirande (1922) who showed that in scales detached from bulbs of *Lilium candidum*, blue and indigo lights were effective, red was slightly effective and green not effective, in promoting anthocyanin accumulation. The first quantitative studies on anthocyanins were apparently those of Kuilman (1930) who found that both light and temperature influenced anthocyanin levels in seedlings of *Fagopyrum esculentum*.

By 1925 Onslow, in the second edition of her book on anthocyanins, summarized the results of many workers who had studied the control of anthocyanin accumulation by light, temperature, oxygen pressure, exogenous sugars, drought, organic acid content of the tissues, and photosynthesis. Comparable work on the physiology of other flavonoids awaited the development of paper chromatography and sensitive techniques of isolation and quantitation.

18.2.1 Light and ionizing radiations

Light has been a favourite tool of plant physiologists interested in flavonoids. Light is one of the most important external environmental factors controlling plant growth and it can be '. . . introduced into an intact cell with a minimal perturbation of the biological object' (Siegelman, 1964). Action spectra for anthocyanins in seedlings (Hendricks and Borthwick, 1965) and callus cultures (Lackmann, 1971), and for C-glycosylflavones in seedlings (Carlin and McClure, 1973) have been reported. Photocontrol of precursor incorporation (Harper *et al.*, 1970) and flavonoid degradation (Steiner, 1971) is established for a few compounds.

In simple cases the action spectrum for a response can be fitted to the absorption spectrum of the light absorbing pigment (Allen, 1964), yet there is little similarity between reported action spectra for anthocyanins in several plants, as shown in Table 18.1. The inconsistencies in anthocyanin action spectra may often reflect the high levels of radiant energy required for anthocyanin accumulation. High intensity treatments would saturate many low energy level photoresponses and action spectra are usually meaningful only when the light intensity is such that light limits the reaction (Allen, 1964).

Reports of photocontrol of flavonoids in a range of plants are listed in Table 18.1. Although experimental techniques and energy levels are highly variable, most of the responses seem to be explainable on the basis of four classes of photoreactions: (1) the low-energy requiring red far-red reversible photochrome system; (2) a High Intensity Response that is satisfied with 1 J cm^{-2} or more of either far-red or blue light; (3) photosynthetic involvement for precursor or cofactor production, and (4) complex responses that work indirectly by such means as enhanced growth-hormone synthesis.

With our current knowledge of the enzymology of flavonoid biosynthesis one may ask where along the pathway from CO_2 to the completed glycoside does photocontrol of flavonoid accumulation operate? The answer appears to be that light works at many points. For example, Hahlbrock et al. (1971) have reported photocontrol of eight different enzymes involved in the synthesis of apiin, the 7-apiosylglucoside of apigenin in cell suspension cultures of Petroselinum crispum. The enzymes were rapidly increased in response to light and they could be separated into two groups. Group I enzymes (Table 18.2), which are not exclusively required for flavonoid synthesis, reach a maximum activity 15 hours after the onset of light and then decrease in activity throughout the 24-hour experimental period. Group II enzymes are involved more directly in the synthesis and conversion of intermediates into apiin and they continue to increase throughout the experimental period.

Although the photoreceptor characteristics have not yet been established for enzymes related to flavonoid synthesis in parsley cell suspension cultures, flavonoid glycosides are not found in dark-grown cultures (Hahlbrock et al., 1971) and the levels of these enzymes in the dark are, in most instances, less than 10% of that found under maximal light-induced conditions.

18.2.1.1 Phytochrome control of flavonoid accumulation

Phytochrome is a non-globular protein with a molecular weight of about 120 000 and it is composed of at least two subunits coupled with an open chain tetrapyrrole chromophore. This complex can undergo phototransformation to absorb maximally at either the red (P_r) or far-red (P_{fr}) regions of the spectrum (Fig. 18.1). Phytochrome is most prevalent in meristematic or recently meristematic tissues and appears to be a component of cellular membranes (Briggs and Rice, 1972).

Much work is done on the assumption that phytochrome undergoes the simple transformation:

$$P_r \underset{\text{far red light}}{\overset{\text{red light}}{\rightleftharpoons}} P_{fr}$$

Table 18.1   Some effects of light of selected wavelengths on the accumulation of flavonoids

Plant	Organ	Flavonoid(s)[1]	Photoresponse[2]	References
Brassica oleracea (Cruciferae)	seedling	anthocyanin	P_{fr} 690 nm	Ku and Mancinelli, 1972 Siegelman and Hendricks, 1957
B. rapa	seedling	anthocyanin	P_{fr} 750(620)450 nm HI-B, P_{fr} HI-FR	Ku and Mancinelli, 1972 Siegelman and Hendricks, 1957 Grill and Vince, 1965 Schneider and Stimson, 1971 Barz and Adamek, 1970
Cicer arietinum (Leguminosae)	seedling	formononetin biochanin A	P_{fr} (decr) P_{fr} (decr)	Scherf and Zenk, 1967a Scherf and Zenk, 1967b
Fagopyrum esculentum (Polygonaceae)	hypocotyl	cyanidin glycoside leucocyanidin rutin	P_{fr} (tr) HI-B, P_{fr} HI-B, P_{fr}	
Fragaria vesca (Rosaceae)	leaf disc	cyanidin 3-glucoside	P_{fr}	Creasy and Swain, 1966
Haplopappus gracilis (Compositae)	intact seedling wounded seedling callus FUB callus MSU callus A1	anthocyanin cyanidin 3-glucoside cyanidin 3-rutinoside	HI-B HI-B, –R (440)660 HI-B, 438(372) nm UV HI-B, –G, –R	Lackmann, 1971; Reinert, 1964 Lackmann, 1971 Groenwald and Zeevart, 1973 Strickland and Sunderland, 1972
Hordeum vulgare (Gramineae)	shoot	saponarin	P_{fr}, HI-B UV 620, 660 nm 580, 620, 660 nm	McClure and Wilson, 1970 McClure, unpublished Carlin and McClure, 1973
Lilium candidum (Liliaceae)	bulbs	lutonarin anthocyanin	HI-B	Mirande, 1922
Lycopersicon esculentum (Solanaceae)	seedling fruit fruit	anthocyanin flavonol (?) anthocyanin	P_{fr} 660 nm (600)650 nm	Ayers and Mancinelli, 1969 Pringer and Heinze, 1954 Siegelman and Hendricks, 1958
Malus pumila (Rosaceae)				
Nicotiana tabacum (Solanaceae)	leaf	rutin	UV	Lott, 1960

Species (Family)	Tissue	Pigment	Light	Reference
Perilla ocymoides (Labiatae)	seedling	anthocyanin	P_{fr}	Bulakh and Gordzinskii, 1970
Petroselinum crispum (Umbelliferae)	suspension culture	flavone glycoside	UV, P_{fr}	Wellmann, 1971
Petunia hybrida (Solanaceae)	petals	delphinidin 3-glucoside malvidin 3-glucoside petunidin 3-glucoside	P_{fr}	Steiner, 1972
Phaseolus vulgaris (Leguminosae)	seedling	anthocyanin	HI-FR	Withrow *et al.*, 1953
Pisum sativum (Leguminosae)	several	kaempferol 3-triglucoside quercetin 3-triglucoside (and *p*-coumaryl acylated derivatives of each)	P_{fr}, HI-B	(e.g. Galston, 1969)
	fruit	pisatin	UV	Hadwiger and Schwochau, 1971
Secale cereale (Gramineae)	seedling	anthocyanin	P_{fr}	Bulakh and Gordzinskii, 1970
Sinapis alba (Cruciferae)	seedling	cyanidin glycosides anthocyanin	HI-FR, P_{fr} P_{fr}	(e.g. Mohr, 1972) Ku and Mancinelli, 1972
Sorghum vulgare (Gramineae)	seedling	anthocyanin	HI-B, P_{fr} 470 nm	Downs and Siegelman, 1963
Spirodela intermedia (Lemnaceae)	frond	cyanidin 3-glucoside vitexin, orientin kaempferol and quercetin 3-glycosides	P_{fr}, HI-B	McClure, 1968
S. oligorhiza	frond	petunidin 3,5-diglucoside quercetin triglucosides	300, 705 nm	Ng *et al.*, 1964
Tropaeolum majus (Cruciferae)	seedling		P_{fr}, UV	Delaveau, 1967
Zea mays (Gramineae)	seedling	anthocyanin	HI-FR	Withrow *et al.*, 1953

[1] Many reports detail the accumulation of 'anthocyanin' without identifying the pigment or mixture of pigments. Increased accumulation is assumed. Decreases attributable to light are designated by (decr), trace amounts by (tr).

[2] Abbreviations as follows: HI = High Intensity responses to either blue (B), green (G), red (R) or far red (FR) light of 1 J cm^{-2} or more total irradiation. P_{fr} = low energy red, far-red reversible photoresponse. UV = ultraviolet light.

Table 18.2 *Differential photocontrol of enzymes involved in the synthesis of flavonoids in* Petroselinum crispum *suspension cultures*

Enzyme[1]	Photocontrol[2]
L-phenylalanine ammonia-lyase (PAL)	I
trans-cinnamic acid 4-hydroxylase	I
p-coumarate : CoA ligase	I
'chalcone synthetase'	requires light
chalcone-flavanone isomerase	II
UDP-glucose : apigenin 7-*O*-glucosyltransferase	II
UDP-apiose : 7-*O*-glucosylapigenin-apiotransferase	II
UDP-apiose synthetase	II
S-adenosylmethionine : luteolin 3'-*O*-methyltransferase	II

[1] Cultures were placed under continuous white fluorescent light for 24 hours and samples assayed at various times during the light period.
[2] Group I enzymes reach a maximum within 15 hours after transfer to the light and then decline in activity. Group II enzymes continue to increase throughout the 24-hour period. (*Data from* Hahlbrock *et al.*, 1971, and Kreuzaler and Hahlbrock, 1973.)

with P_{fr} slowly decaying back to P_r in the dark. The actual process seems far more complex. For example, physiological experiments indicate that the time taken for reversion in the dark in different species varies from less than 3 hours to more than 72 hours

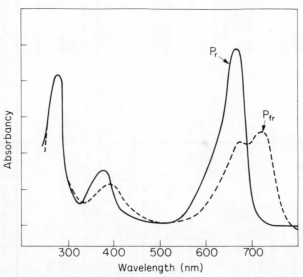

Figure 18.1 Absorption spectrum of purified *Avena sativa* phytochrome after 5 minutes of saturating far-red light (P_r) and 5 minutes of saturating red light (P_{fr}). (*Redrawn from* Briggs *et al.*, 1972).

(Cumming *et al.*, 1965; see also Kendrick and Hillman, 1971; Pike and Briggs, 1972). Klein and Edsall (1966) found that a variety of reducing agents could substitute for red light in promoting *Phaseolus vulgaris* leaf expansion while oxidants could substitute for far-red light in reversing the red-light-induced phenomenon. Although phytochrome transformation itself may not be altered by these agents, a redox reaction may be one of the early phytochrome-triggered reactions.

A part of the differences between action spectra for various flavonoids may be the screening of phytochrome by other pigments such as chlorophyll, protochlorophyll(ide) and carotenoids. For example, Grill (1972) found that *in vivo* phytochrome spectra were different for light- and dark-grown plants and that this was attributable primarily to chlorophyll. Action spectra and photoreversibility studies for saponarin (1) and lutonarin (2) accumulation in etiolated seedlings of *Hordeum vulgare* (Fig. 18.2) show phytochrome control for saponarin and implicate protochlorophyll(ide) screening of phytochrome in the action spectra of lutonarin (Carlin and McClure, 1973).

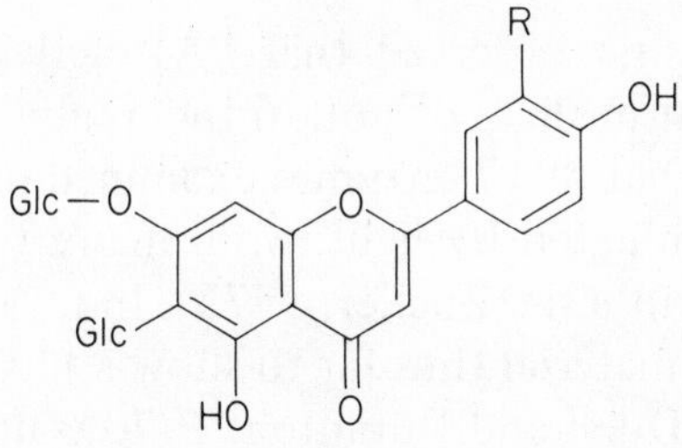

(1, Saponarin R = H)
(2, Lutonarin R = OH)

Early hypothesis about phytochrome control of flavonoids centred around its control of the incorporation of acetate (i.e. Hendricks and Borthwick, 1965). More recent ideas are that phytochrome acts in some way to affect membrane permeability, which could lead to many metabolic changes, including altered gene expression and enzyme induction (Briggs and Rice, 1972; Galston and Satter, 1972). With the realization that phenylalanine ammonia lyase (PAL) is involved in flavonoid synthesis, many workers turned their attention to photocontrol of this enzyme.

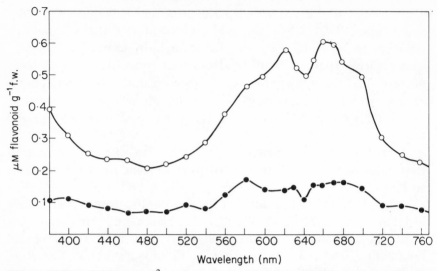

Figure 18.2 A 6.6 kerg cm^{-2} action spectrum for the increased accumulation of saponarin (○) and lutonarin (●) in 5 day old shoots of *Hordeum vulgare*. Previously etiolated plants were harvested 24 h after illumination. Linearity of response was determined within the range of 1 to 13 kerg cm^{-2}. (*Redrawn from* Carlin and McClure, 1973).

Zucker (1965) first observed that PAL activity is greatly stimulated by white light in slices of potato tuber, and Sacher *et al.* (1972) have since shown that of 17 enzymes examined in potato tuber discs, only PAL was stimulated by light. Still, many tissues show no light dependent PAL synthesis (Zucker, 1972). In fact, tuber slices are the only non-seedling material thus far to show a phytochrome-regulated synthesis of PAL (Durst and Duranton, 1970; Zucker, 1972).

Amrhein and Zenk (1971) studied PAL activity in acetone powders prepared from four- to seven-day-old etiolated seedlings from several families and found that in almost every instance light enhanced PAL activity (Table 18.3) and that striking differences are found between species. For example, over 200 times more PAL activity is found in the shoot of *Hordeum vulgare* than in the hypocotyl of *Cucumis sativus*. Furthermore, the enhancement of PAL by light varies from a factor of 0.96 (no effect) in *Zea mays* epicotyls to a factor of 6.75 in the epicotyl of *Pisum sativum*. If these are representative values it may explain why PAL may be limiting for flavonoid synthesis in some plants (where it is present in low amounts in the dark) and not limiting in others where it is in high levels without the mediatory effects of light.

Table 18.3   *The influence of 10 hours of white fluorescent light (25 000 lux) on PAL activity in previously etiolated seedlings*

Species	Organ	Specific activity of PAL[1] in dark grown tissues $m\mu M\ mg^{-1}\ min^{-1}$	Enhancement of PAL in response to light
Agrostemma githago (Caryophyllaceae)	hypocotyl	0.176	3.5
Amaranthus lividus (Amaranthaceae)	seedling	0.159	1.6
Cucumis sativus (Cucurbitaceae)	hypocotyl	0.047	5.1
Fagopyrum esculentum (Polygonaceae)	hypocotyl	1.2	6.0
Helianthus annuus (Compositae)	hypocotyl	0.052	3.7
Hordeum vulgare (Gramineae)	shoot	10.95	1.4
Pisum sativum (Leguminosae)	epicotyl	0.100	6.75
Zea mays (Gramineae)	root	15.8	1.0
	mesocotyl	4.65	1.0
	epicotyl	7.65	0.96

[1] PAL activity determined by incubating acetone powders with L-phenylalanine-U-^{14}C and determining the specific activity of the labelled cinnamic acid produced by the reaction mixture (*Data from* Amrhein and Zenk, 1971).

While purified phytochrome absorbs most strongly at around 660 nm, being converted to P_{fr} in the process, there is ample evidence for the photodestruction of phytochrome *in vitro* and *in vivo*. Various wavelengths of light given for protracted periods of time will lead to predictable steady-state concentrations of phytochrome within the tissues. Kendrick (1972) measured phytochrome decay in previously etiolated seedlings under conditions of constant illumination (Table 18.4) and found that in *Avena*, and perhaps in monocotyledons in general, decay rate is maximal at low P_{fr} concentrations and the decay curve is the same under continuous red, blue or incandescent light. In the dicotyledons *Mirabilis* and *Pisum*, decay depends on the P_{fr}/P_{total} maintained by each light source. *Amaranthus* is an exception to the pattern seen in the two other dicotyledons; under incandescent light *Amaranthus* phytochrome decay deviates from first-order decay kinetics. Both *Amaranthus* and *Mirabilis* (Centrospermae) lack reversion of phytochrome, yet they show no saturation of decay. All of these interactions may be significant in explaining phytochrome control of flavonoid accumulation, especially the lack of a uniform pattern in this phenomenon.

Table 18.4   *Decay of total phytochrome under continuous illumination with incandescent, red or blue light in* Amaranthus caudatus, Mirabilis jalapa, Avena sativa *and* Pisum sativum.

Light treatment	P_{fr}/P_{total} ratio maintained	Approximate percentage of total phytochrome remaining at the end of the experiment[1]			
		Amaranthus seedling	*Avena* coleoptile	*Mirabilis* hypocotyl	*Pisum* epicotyl
Incandescent	0.68	3 h 40%	3 h 10%	3 h < 20%	3 h 40%
Red	0.80	1 h 20%	3 h 10%	2 h < 10%	3 h 15%
Blue	0.22	3 h 20%	3 h 15%	3 h 40%	3 h 60%

(*Data adapted from* Kendrick, 1972).
[1] Total phytochrome determined by photoreversibility at 735 and 800 nm in a Ratiospect spectrophotometer.

18.2.1.2 The high intensity response in flavonoid accumulation

Plant responses potentiated by the photoreversible change of phytochrome are often modified by irradiation for a long time in the blue or far-red region. While the low-energy red, far-red reversible, responses are highly variable in their energy requirements in physiological systems (Furuya, 1968) these blue or far-red responses generally require 1 J cm^{-2} or more of energy and are called the high intensity response, HIR (Borthwick *et al.*, 1969; Mohr, 1972). In 1965, Hendricks and Borthwick summarized photocontrol of flavonoid levels and emphasized the distinctiveness between a HIR to produce acetyl groups and a low-energy phytochrome response to control the condensation of such groups into the A-ring of anthocyanins. Subsequent workers have carefully studied the HIR for anthocyanins (Mohr, 1972) and other flavonoid glycosides (Scherf and Zenk, 1967a).

The major question at this time is the nature of the photoreceptor(s) involved in this response. To date all attempts to identify a new pigment with absorption maxima corresponding to the action spectra for HIR responses have been unsuccessful. The major workers contend that either phytochrome (Mohr, 1972) or chlorophyll (Schneider and Stimson, 1972) is the pigment involved. Mohr (1972) has amassed an impressive amount of data to justify his contention that anthocyanin synthesis in *Sinapis alba*, controlled through the HIR, is due exclusively to maintaining a steady-state level of about 3% P_{fr}. In a recent review Mohr (1972) points out that most of the action spectra for the HIR do not coincide with the action spectra for photophosphorylation and that chlorophyll accumulation under continuous far-red light is very slow, much slower than the rate of anthocyanin accumulation. Furthermore, the amount of chlorophyll *a* in the cotyledons of *Sinapis* seedlings is only 2.2% of the amount found in the cotyledons of seedlings which are grown for the same period of time under low irradiance red light, a wavelength nearly ineffective as far as the HIR is concerned. Most tellingly, the extent of the HIR in anthocyanin accumulation in *Sinapis* is not related to the amount or rate of chlorophyll *a* synthesis.

Schneider and Stimson (1971, 1972) have investigated the role of chlorophyll in the HIR responses of anthocyanin synthesis in *Brassica rapa* seedlings. When *Brassica* seedlings are irradiated for 24 h at 720 nm, they synthesize chlorophyll *a* and an anthocyanin. Antimycin A and 2,4-dinitrophenol, which are inhibitors of cyclic photophosphorylation, also reduce anthocyanin synthesis. Non-cyclic photophosphorylation is inhibited by 3-(3,4-dichlorophenyl)-1,1-dimethylurea (DCMU) and by *o*-phenanthroline, compounds which promote both cyclic photophosphorylation and anthocyanin synthesis. On the basis of these findings they suggest that the HIR for anthocyanin synthesis in *Brassica* seedlings arises through photosynthetic activity. They report similar responses in *Sinapis alba* seedlings (Schneider and Stimson, 1972). The effects of red and far-red light on anthocyanin synthesis, and on ATP production by both cyclic and non-cyclic photophosphorylation, are compared in Fig. 18.3.

Blue light HIR are known for flavonoids (Table 18.1), but they have not been as extensively investigated as far red HIR. Mohr (1972) considers it likely that the blue light is absorbed predominantly by a flavoprotein. Schneider (personal communication) has found that compounds which uncouple photophosphorylation also block the blue HIR for anthocyanin accumulation in *Brassica.*

To the complexity of different patterns of P_{fr} activation in various plants (Table 18.4), and differential responses of enzymes of general secondary phenolic biosynthesis to light (Tables 18.2, 18.3), one must add the problem of differential photocontrol of different flavonoids in the same plant. Three examples will be given.

Steiner (1972) has investigated the influence of light on the accumulation of anthocyanins in isolated petals of *Petunia hybrida.* Working with the delphinidin phenotype of *Petunia,* the effects of

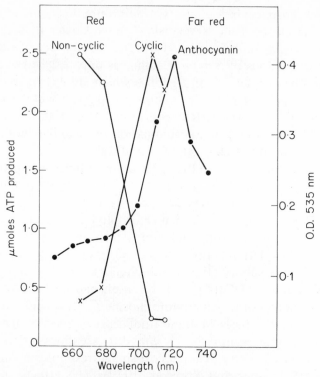

Figure 18.3 Effect of red and far-red spectral regions on anthocyanin accumulation and on ATP production by cyclic and non-cyclic photophosphorylation in *Brassica napus* seedlings. (*Redrawn from* Schneider and Stimson, 1971).

light on accumulation of the 3-glucosides of delphinidin, petunidin, and malvidin were measured. In the intensity range from 1900 to 19 000 lux of white light, there was an increase in the accumulation of all three anthocyanins with increasing light. In red light, no intensity dependency is observed. In either far red or blue, increasing light intensity decreases the accumulation of delphinidin 3-glucoside whereas the levels of petunidin and malvidin 3-glucoside remain about the same. This may be interpreted as red light saturation of the responses at low energies, white light (mixture of wavelengths) giving complex responses that cause a net increase, and blue or far red increases of methylation at the expense of delphinidin 3-glucoside.

Photocontrol of the relative accumulation of derivatives of kaempferol and quercetin in *Pisum sativum* has been extensively examined by Galston and his associates (see Galston, 1969). Peas accumulate quercetin 3-*p*-coumaroyltriglucoside (QGC) (4), quercetin 3-triglucoside (QG) and the corresponding kaempferol derivatives (KGC, 3 and KG). In the plumule brief red light (far red reversible) increases QGC predominately. In the internode tissue, on the other hand, the same light treatments cause an increase in KGC and KG, while quercetin derivatives cannot be detected. Thus the exact chemical consequences of the effect of phytochrome on B-ring hydroxylation of flavonoids of the pea plant depend on the nature of the tissues.

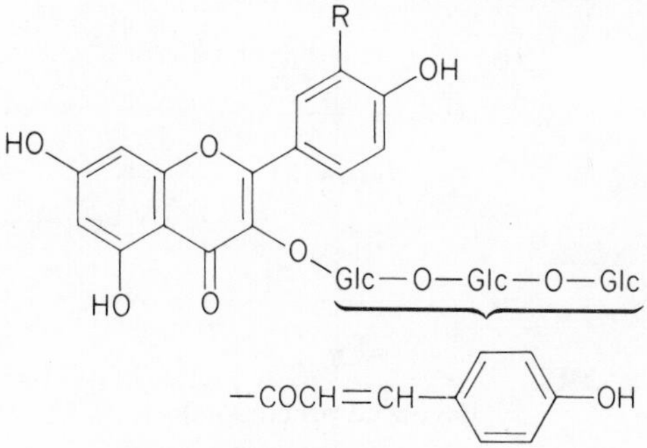

(3, Kaempferol 3−;
p− coumaryltriglucoside, R = H)

(4, Quercetin 3−;
p − coumaryltriglucoside, R = OH)

A final example of the photoregulation of different flavonoids within the same tissue is the case of the duckweed *Spirodela intermedia* which produces vitexin (5), orientin (6), and 3-glucosides of kaempferol, quercetin and cyanidin (McClure and Alston, 1966).

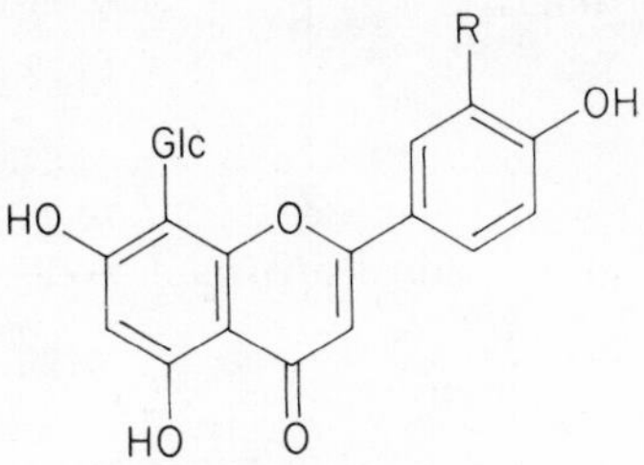

(5, Vitexin R = H)

(6, Orientin R = OH)

In total darkness only vitexin and kaempferol 3-glucosides accumulate. After 45 minutes of white or blue light daily, the levels of vitexin and kaempferol are increased and both orientin and quercetin 3-glucoside appear. Only in high intensity blue or white light for several days is cyanidin 3-glucoside detected (Fig. 18.4) (McClure, 1968).

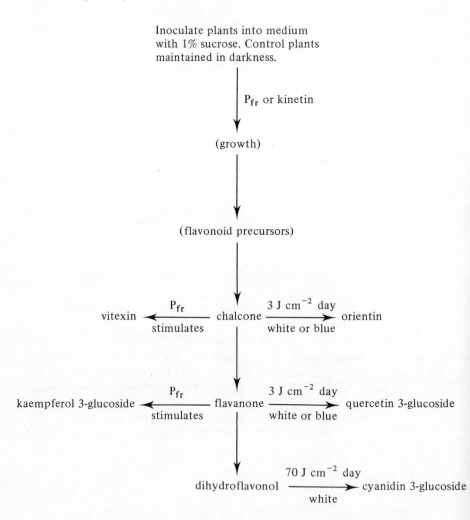

Figure 18.4 Hypothesis to explain photocontrol of *Spirodela intermedia* flavonoids at various biosynthetic stages. *Data from* McClure, 1968, 1970.

In short, there appears to be a general relationship between complexity of biosynthesis and complexity of photocontrol. It is unfortunate that the vast majority of the research on photocontrols of flavonoids has been done on the anthocyanins which are easy for the researcher to quantitate but more complex for the plant to synthesize.

18.2.1.3 Ultraviolet light and ionizing radiation

Many workers have considered that flavonoids have a function in screening plants from ultraviolet (UV) radiation since they all have considerable absorption in the 200 to 380 nm region (Chapter 2). It is also clear that in many cases these same wavelengths are very effective in inducing higher levels of flavonoids.

Irradiation of purified oat phytochrome with 365 nm light establishes about 73% of the phytochrome as P_{fr} in comparison to about 80% P_{fr} if illuminated with light within the 546 to 665 nm region (Butler *et al.*, 1964). This does not ensure that the same ratios would be established *in vivo*, nor does it take into account the considerable absorbance of UV light by the outer layers of cells. It is quite likely that at least a part of the UV responses of increased flavonoid levels (Table 18.1) are due to phytochrome. Under certain conditions the HIR may be satisfied by UV light since detailed HIR action spectra in *Lactuca sativa* show pronounced absorbance at around 635 nm and appreciable effectiveness down to 300 nm.

In a few cases UV light is the only effective wavelength for inducing flavonoids. Groenwald *et al.* (1967), in studying anthocyanin production in an achlorophyllous callus of *Haplopappus gracilis,* found it necessary to give the plants prolonged treatment with light of wavelengths less than 390 nm before any anthocyanin was formed. Hadwiger and Schwochau (1971) investigated UV induction of pisatin (7) and PAL in immature pods of *Pisum sativum.*

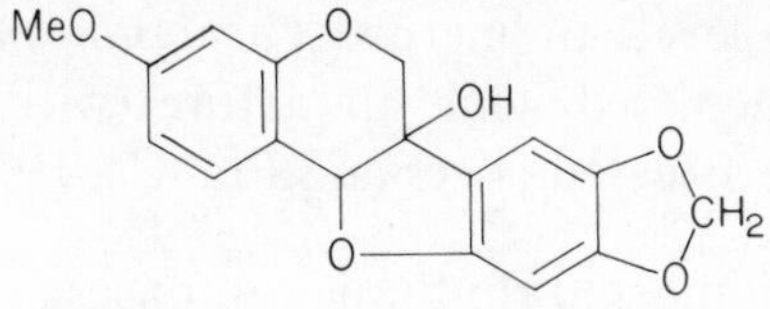

(7, Pisatin)

Thirty seconds of short wave light (254 nm) saturated the induction response yet pisatin was not detected in plants given 10 minutes of long wave light (366 nm) or in the control plants maintained in water. Blue light counteracts the effects of short wave UV light in pisatin induction (Hadwiger and Schwochau, 1971) which suggests that induction is due to conformational changes in the DNA.

Rutin (quercetin 3-rutinoside) content of tobacco plants is influenced by UV light and Lott (1960) found that when greenhouse grown plants were provided supplementary short wave UV light, their rutin content was increased 28%. Again, the total flavone glycoside content of cell suspension cultures of *Petroselinum crispum* is stimulated by UV light with maximal quantum efficiency found below 300 nm (Wellmann, 1971). This particular response is partly reversed by a subsequent irradiation with far red light and the far red effect is reversed by red. The red and far red effects on *Petroselinum* are ineffective without UV irradiation (Wellmann, 1971).

Ionizing radiation generally has an effect on flavonoid levels. Sparrow *et al.* (1968) measured the effects of X-rays and ^{60}Co gamma rays on anthocyanins in *Rumex crispus, R. hydrolapathum* and *R. sanguineus* and found increases up to twentyfold. The largest increases occurred at lethal or near-lethal doses. Anthocyanin content was increased in the leaves of 16 species irradiated chronically and 21 species irradiated acutely; it appears that enhanced anthocyanin formation is a common response of higher plants to ionizing radiation. Major differences were found between plants with respect to dosage responses. Dosage of ^{60}Co gamma irradiation necessary to yield observable increases in anthocyanin varied from 3.0 kR in *Acer saccharum* to 342 kR in *Luzula acuminata.* Lane and Constantine (1969) reported 600-fold increases in cyanidin 3-glucoside when seedlings of *Zea mays* were given 11 kR of ^{60}Co radiation. At levels from 150 to 300 kR, anthocyanin and flavonol glycoside levels of mature *Vaccinium macrocarpon* berries are significantly increased while immature (with little colouration) berries are scarcely affected (Lees and Francis, 1972).

18.2.1.4 Photoperiodic and rhythmic responses in flavonoid
         accumulation

Photoperiodicity is frequently controlled through the phytochrome system and several cases are known where photoperiod controls the level of flavonoids accumulating in various plant organs. Tso *et al.* (1970) examined the effects of photoperiod and end of daylight quality on rutin content in leaves of *Nicotiana tabacum* grown in long day (16 hours of light) or short day (8 hours of light) photoperiods which were, in some cases, terminated with 5 minutes of either red or far-red light. Plants grown under 16-hour photoperiods had a significantly higher concentration of rutin than those grown under 8 hours of light. Within each photoperiod the plants that received far-red light just before the dark had higher concentrations of rutin and other polyphenols than those receiving a terminal red light treatment. The reduced flavonoid level in plants given a terminal red light treatment, which should establish a higher level of P_{fr}, suggests a phytochrome control of turnover rates interacting with photocontrol of synthesis.

Taylor (1965) examined the effects of photoperiod on a number of polyphenols, including kaempferol and quercetin glycosides, in *Xanthium pennsylvanicum.* After 26 days of photoperiodic treatments, under six different conditions ranging from 8 to 20 hours of light during a 24-hour period, there was a good correlation between higher levels of flavonoids and longer treatments with light. When the dark periods were interrupted by 2 minutes of light – conditions which block flowering in these photoperiodically sensitive plants – levels of the quercetin glycosides were increased about 20%.

Phytochrome control of flowering and of flavonoid accumulation have been compared in *Fuchsia* sp. (Holland and Vince, 1968). The anthocyanin content of the leaves in different photoperiodic treatments varied independently of flowering. However, Neyland *et al.* (1963) found that anthocyanin accumulation and photoinduced flowering were inseparable in *Kalanchoe blossfeldiana.* Cyanidin 3- and 3,5-(di)glucosides were found only under short-day conditions that lead to flowering. In the same plants, leucocyanidin decreases as anthocyanins accumulate.

A frequent pattern of diurnal rhythms of flavonoid levels is that the level is higher at the end of dark period than at any other time during the 24-hour cycle. This pattern has been reported for apigenin 7-glucoside in *Silybum marianum* (Pahlich, 1969), for leucocyanidin in *Sedum album* (Combier and Lebreton, 1968) and for phloridzin (8) in young shoots of *Malus pumila* (Vardya and Sarapuu, 1967).

Endogenous circadian rhythms for leucoanthocyanins have been well characterized in *Sedum album* (d'Arcy, 1970). These cycles

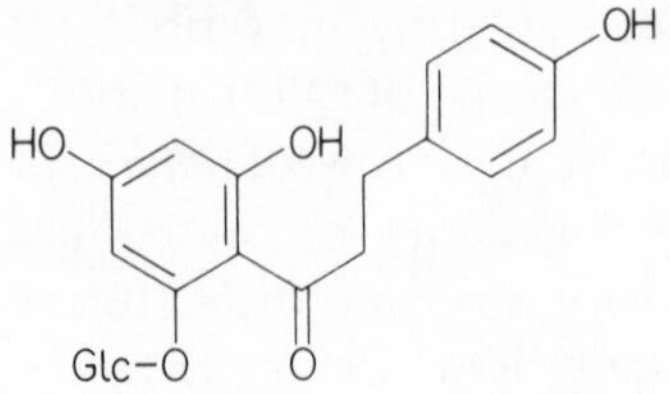

(8, Phloridzin)

show peaks at about 20 hours in continuous darkness and are temperature independent within a broad range. Saponarin content of dark grown barley seedlings also displays a rhythmic fluctuation under constant temperature conditions (McClure and Wilson, 1970).

18.2.2 Temperature effects

In temperate climates, it is a common observation that foliar anthocyanins increase with the onset of cold weather, although it is not usually possible to separate the effects of shorter photoperiods and lower temperatures. Overton (1899) reported that low temperatures favoured anthocyanin accumulation in *Hydrocharis morsus-ranae*. Similar reports have been made for anthocyanin increases due to cold treatments in *Impatiens balsamina* (Alston, 1959); *Malus pumila* (Creasy, 1968b); *Euphorbia pulcherrima* (Marousky, 1968); *Citrus sinensis* (Meredith and Young, 1968); and *Chrysanthemum morifolium* (Rutland, 1968). Anthocyanin levels are increased, while leucoanthocyanin levels are decreased, by cold treatment of the fruit of *Malus* (Creasy, 1968b) or of the fronds of the fern *Asplenium trichomanes* (Voirin and Lebreton, 1972).

Creasy and Swain (1966) studied flavan production in leaf discs from *Fragaria vesca* and found marked temperature effects on leucocyanidin and (+)-catechin. Low temperature (below 20°) did not favour flavan production. In fact, over four times as much total flavolan was produced at 30° than at 20°. Isoflavones may also be controlled by temperature. Rossiter and Beck (1966) grew *Trifolium subterraneum* at temperatures ranging from 9° to 36° and found six times as much formononetin (9), genistein (10), and biochanin A (11) in those grown at 15° compared to plants maintained at 36°.

Low temperature often causes an increase in PAL activity. Kozukue and Ogata (1972) studied chilling injury in *Capsicum*

(9, Formononetin R = H, R′ = Me)
(10, Genistein R = OH, R′ = H)
(11, Biochanin A R = OH, R′ = Me)

annuum fruits and found threefold higher levels in fruits stored at 6°
compared to 21°. Similarly Engelsma (1968, 1970) found that a
temperature below 10°, in either light or dark-grown *Cucumis sativus*
seedlings, caused a rise in PAL both during the course of the cold
treatment and after transfer to a higher temperature. At temperatures above 10°, the increase was apparently blocked by a PAL
inactivating system functioning best at these higher temperatures.
PAL is induced by light in the *Cucumis* hypocotyl but the duration
of this response is a function of temperature (Engelsma, 1970). The
increase and subsequent decline of PAL levels was completed within
4 h at 22° but required 20 h at 12.5°. At 10° the increase in PAL
was very slow but showed no indication of declining throughout a
24-hour experimental period.

In 1964 McClure and Alston reported that the pattern of
flavonoids in vegetative fronds of *Spirodela oligorhiza* and *S.
polyrhiza* were qualitatively stable to various changes in medium
composition, photoperiod, and high levels of plant growth regulators
or inhibitors. Reznik and Menschick (1969) examined flavonoids
from turions of *S. polyrhiza*, i.e. fronds produced primarily in
response to low temperature, and found additional compounds
tentatively identified as diglucosides of apigenin and luteolin. It is
likely that cold treatment relates to this qualitative change of
flavonoid glycosides.

18.2.3 Water and atmospheric changes

Aquatic plants are generally depauperate in secondary constituents
including flavonoids (McClure, 1970). Flavonoids levels of terrestrial
plants also respond to changes in water relations. For example,
Francis and Devitt (1969) measured the effects of 21 days of
waterlogging on the isoflavone content of *Trifolium subterraneum*

and found a general increase in formononetin, genistein and biochanin A in the leaves of 75 cultivars and 3 subspecies. Lee and Tukey (1972) studied the response of *Euonymus alatus* to intermittent mist and found an increase in PAL, various flavonols and flavans, but a decrease in the anthocyanins.

Bardzik *et al.* (1971) examined the effects of water stress on the activities of nitrate reductase, NADH oxidase and PAL in *Zea mays* leaves by withholding water for various periods starting 10 days after emergence of the seedling. NADH oxidase levels were essentially unaltered by various water deficits up to 40% and nitrate reductase levels decreased with increasing water stress, losing about half of the activity at a 20% water deficit. However, PAL was remarkably sensitive to water levels. As little as a 2% water deficit caused up to 40% reduction in PAL levels based on enzyme per unit protein. Saunders and McClure (1973) found that a brief dip in distilled water significantly increased the levels of both PAL (5 hours later) and saponarin (after 24 hours) in etiolated, guttating, seedlings of *Hordeum vulgare.*

Atmospheric changes may have profound effects on flavonoids. Thus, Goldstein *et al.* (1962), while studying the production of leucoanthocyanins in cambial cell cultures of *Acer pseudoplatanus*, found that a low O_2 level markedly inhibited accumulation. A change in the atmospheric composition may similarly effect the enzymes involved early in flavonoid biosynthesis. Russell (1971) studied properties of the cinnamic 4-hydroxylase of *Pisum sativum* seedlings and found that this enzyme was inhibited 70% in a 50% carbon monoxide atmosphere. The requirement of this enzyme for both NADPH and O_2 suggest dual roles for photosynthesis in promoting this aspect of the flavonoid pathway.

Koukol and Dugger (1967) studied the effects of ozone on *Rumex crispus* and found cyanidin glycosides only in plants exposed to ozone or to urban smog with a high ozone concentration. Increasing ozone levels up to 10 p.p.m. decreased anthocyanin content in *Petunia hybrida* and *Pelargonium hortense* but increased anthocyanins in *Euphorbia pulcherrima* (Craker and Feder, 1972). Hydrogen fluoride, another increasingly common air pollutant, leads to the loss of both cyanidin 3,5-diglucoside acylated with p-coumaric acid and dihydrokaempferol in the leaves of *Coleus blumei* (Powell and Lamprech, 1972).

18.2.4 Carbohydrates and flavonoid levels

Starting with Overton (1899) who showed that anthocyanin accumulation in the leaves of *Hydrocharis morsus-ranae* was markedly stimulated by several sugars, many workers have subsequently observed that leaf discs or thin pieces of tissue make more anthocyanin when floated on dilute sugar solutions. In the peel of apple (*Malus pumila*) various treatments known to increase the activity of the pentose phosphate pathway were correlated with increased anthocyanin accumulation (Faust, 1965). Detailed studies have been made of the effects of sugars on increasing anthocyanin accumulation in *Impatiens balsamina* (Arnold and Alston, 1961) and *Spirodela oligorhiza* (Thimann *et al.*, 1951). Sugars may exert their influence very early in the pathway to flavonoids. For example, Zucker (1968) showed that high rates of PAL activity in *Xanthium pennsylvanicum* leaf discs required both photosynthesis and exogenous sucrose. The light responses could be blocked by DCMU and presumably involved non-cyclic photophosphorylation which leads to increased carbohydrate synthesis. In *Allium cepa* leaf bases PAL is not light controlled but is strongly influenced by sugar levels and respiratory rates (Tissut, 1972).

Creasy and his associates have provided considerable insight into the relative roles of the phytochrome system, temperature, and carbohydrate status in the control of PAL and various flavonoids in leaf disc of *Fragaria vesca*. Creasy *et al.* (1965) found that light was necessary for anthocyanin synthesis in leaf discs floated on sucrose solutions. ATP stimulated anthocyanin accumulation in the light only when sucrose was omitted but stimulated this response in the dark only when sucrose was supplied, implicating photosynthetic production of both ATP and sugars in enhancing anthocyanin levels. Creasy (1968a) then examined PAL, anthocyanin, and leucoanthocyanin levels in illuminated discs and found that their increased production was dependent on a supply of CO_2. In the dark, CO_2 had no effect on the small rates of production. An external supply of sucrose could remove the CO_2 dependency and an external supply of CO_2 could remove the stimulation brought about by external sucrose. Both the uptake and metabolism of externally supplied sucrose, and the synthesis of flavonoids and PAL, were inhibited by DCMU, probably through its inhibition of CO_2 fixation or sucrose uptake. In 1966, Creasy and Swain found that phytochrome responses of flavan (catechins and leucoanthocyanins) accumulation in *Fragaria* leaf discs was influenced by the concentration of exogenous sugars. In 0.3 M or higher sucrose solutions a far-red light treatment given at the end of a 24-hour illumination with white fluorescent light had no effect on the subsequent dark synthesis of flavans or anthocyanins; but if lower (i.e. 0.05 M) concentrations of sucrose were used, a terminal far-red light treatment lowered the subsequent dark synthesis of these flavonoids. Thus high sugar concentrations may mask the phytochrome response in flavonoid synthesis.

Whole seedlings may respond to exogenous sugars. Margna *et al.* (1972) grew *Fagopyrum esculentum* seedlings on 1% (w/v) solutions of glucose, fructose, sucrose or mannitol under various conditions of light and darkness. When compared to control plants, anthocyanin levels were not promoted but the rutin content was significantly increased. The pattern was changed when excised hypocotyls were floated on the solutions; both anthocyanins and rutin were stimulated by feeding sugars to detached hypocotyls.

18.2.5 Mineral nutrition

As mineral nutrition controls the metabolism of the total plant, it is not surprising that an imbalance of minerals would change the balance of flavonoids. Many workers have suggested that a deficiency in certain elements may cause an increase in soluble sugars which are then diverted into flavonoids; unfortunately, such suggestions have rarely been substantiated by data from sugar analyses.

Several workers have determined flavonoid levels in response to fertilization with the macronutrients nitrogen, phosphorus and potassium. Asen *et al.* (1959) found that while varying ratios of N, P, and K^+ had no significant effect on the level of delphinidin 3-glucoside in sepals of *Hydrangea macrophylla,* increased nitrogen levels were inhibitory to the accumulation of kaempferol glycosides. In this plant the ratio of NO_3^- to NH_4^+ is important in controlling sepal colour. *Hydrangea* bracts display varying amounts of blue colouration depending on the amount of aluminium available to form a copigmentation complex with the anthocyanin, and NH_4^+ is antagonistic to aluminium uptake.

When Butler *et al.* (1967) determined the effect of N, P, and K^+

on isoflavone levels in *Trifolium pratense* they found that phosphorus deficiency markedly increased isoflavones while the other elements had no specific effects. In *Lycopersicon esculentum* rutin (quercetin 3-rutinoside) levels in the leaf are increased by fertilization with N, P, and K^+, with P and K^+ having the greatest effect (Jablonowski and Jedrych, 1971). Lawanson *et al.* (1972), in following anthocyanin accumulation in *Zea mays* seedlings, found that a deficiency of potassium caused anthocyanin to appear within five days while ten and fifteen days respectively were required for anthocyanin accumulation in response to phosphorus and nitrogen deficiency.

In *Trifolium subterraneum*, both nitrogen and phosphorus deficiency double the level of formononetin and genistein but they have little effect on biochanin A (Rossiter, 1969; Rossiter and Beck, 1966). When *Trifolium* plants are deprived of zinc, the content of all three isoflavones is increased for the first 29 days but subsequently decreases below control values as the characteristic 'little leaf' symptom is developed (Rossiter, 1967). Sulphur deficiency almost doubles the levels of all of these isoflavones (Rossiter and Barrow, 1972).

Nitrogen deficiency causes phloridzin accumulation in roots of *Malus pumla* (Hutchinson *et al.*, 1959), increased rutin in *Fagopyrum esculentum* leaves (Krause and Reznik, 1972); precocious rutin production in *Nicotiana tabacum* seedlings (Sheen, personal communication), and increased anthocyanin levels in embryos isolated from *Brassica oleracea* (Szweykowska, 1959). In contrast high levels of nitrogen lead to higher levels of rutin and total phenols in mature *Nicotiana tabacum* plants (Tso *et al.*, 1967).

Phosphorus deficiency, in severe cases, almost always leads to an increase in flavonoids. For example, the rutin content of *Lycopersicon esculentum* leaves is increased thirteenfold by withholding phosphorus (Ulrychova and Sosnova, 1970) and marked increases in anthocyanins, rutin and C-glycoflavones are found in phosphorus deficient *Fagopyrum esculentum* plants (Krause and Reznik, 1972).

The effect of trace elements, especially boron and copper, on flavonoids has been studied in several plants, often with confusing results (Table 18.5). Boron has not yet been demonstrated to be required by algae or other plants that do not accumulate lignin or flavonoids. Boron deficiency in higher plants normally causes an

Table 18.5   *Effects on flavonoid accumulation of trace element deficiency or excess*

Plant	Organ	Flavonoid	Element withheld (−) or applied (+)	Effect of treatment	References
Fagopyrum esculentum (Polygonaceae)	seedling	rutin	− cobalt	slight increase	Grinkevich et al., 1967
Pisum sativum (Leguminosae)	fruit	cyanidin pisatin	+ copper	marked increase induces synthesis	Perrin and Cruickshank, 1965
Malus pumila (Rosaceae)	fruit peel whole fruit	anthocyanins anthocyanins	+ cobalt + cobalt	marked decrease marked increase	Faust, 1965
Spirodela oligorrhiza (Lemnaceae)	frond	petunidin 3,5-diglucoside	− boron − copper − molybdenum −zinc	inhibits inhibits inhibits inhibits	Thimann and Edmondson, 1949

accumulation of simple phenols that may serve as lignin and flavonoid precursors (Lee and Aronoff, 1967; Coke and Whittington, 1968; Dugger, 1973). Emphasis has been placed on the role of boron in inhibiting catechol oxidase activity in plants (Racusen, 1970), and King (1971) reports that when enzymes from cotton leaves are extracted with borate, far better results are obtained than with PVP or other additives designed to reduce the level of reactive phenolics. When phenolic production is measured in response to boron deficiency, increases of 220-fold have been reported for simple cinnamic acid derivatives in *Helianthus annuus* leaves (Watanabe *et al.*, 1964) and the level of rutin and leucoanthocyanins is increased threefold by withholding boron from buckwheat (Shkolnik and Abysheva, 1971). In marked contrast, Rajaratnam *et al.* (1971) report that boron-deprived oil palms showed a total absence of leucoanthocyanins before morphological deficiency symptoms were evident. However, there often appears to be an inverse relationship between leucoanthocyanin levels and the levels of anthocyanins and flavonol glycosides (Sections 18.2.1–18.2.4).

Edmondson and Thimann (1950) report that phenylthiocarbamate (PTC) and other copper-complexing agents in concentrations too low to inhibit growth, markedly inhibited anthocyanin accumulation in *Spirodela oligorhiza*. When PTC and excess copper were added to the media, anthocyanin content was increased. They concluded that anthocyanin accumulation required a copper-containing enzyme. This requirement is probably specific for the anthocyanins of this plant since various copper levels in the medium have no discernible effect on the production of ten other flavone and flavonol glycosides in the same clone (McClure and Alston, 1964, 1966). Oka and Simpson (1971) have described a quercetinase which is a copper-containing dioxygenase and this may relate to copper involvements in determining flavonoid levels in plants.

18.2.6 Intact versus isolated plant systems

Much use has been made of callus or suspension cultures for studying flavonoid synthesis in plants. Many advantages accrue from thus using rapidly growing cell lines in axenic culture. Microbial contamination is eliminated and relatively large amounts of material of apparently similar developmental stages may be obtained from the same inoculum. By this technique one may also uncover biochemical pathways not normally expressed in the intact plant. For example, Strickland and Sunderland (1972) grew callus cultures of *Haplo-pappus gracilis* and found cyanidin 3-glucoside and cyanidin 3-rutino-side, anthocyanins not found in the intact mature plants. Also, pisatin, usually found in *Pisum sativum* plants only in response to infection, is produced in high yields in *Pisum* callus cultures (Bailey, 1970). In fact, the production of flavonoids not produced in the parent plant may be a common characteristic of callus and cell suspension cultures. This alteration in flavonoid chemistry has been carefully documented in cell suspension cultures of *Rosa rugosa* (Davies, 1972), in callus cultures of *Camellia sinensis* (Forrest, 1969), and in callus cultures of *Citrus paradisi* (Thorpe *et al.*, 1971).

Excised apices, hypocotyls, leaf discs, petals, and the like are also favoured material for studying the accumulation of flavonoids. In some instances the excised organ responds to external stimuli in the same manner as the attached organ. Troyer (1964) measured anthocyanin in both intact and excised *Fagopyrum esculentum* hypocotyls and found similar patterns in both, yet the degree of response was much less for the excised hypocotyl. Still, excised organs do not always respond in the same manner as intact ones. For example, Pjon and Furuya (1967) showed that although a low energy red (far-red reversible) treatment inhibited elongation of intact *Avena sativa* or *Oryza sativa* coleoptiles, the same treatment promoted elongation of these organs when they were excised and floated on 0.02 M pH 7.0 phosphate buffer. *Hordeum vulgare* shoots respond to light by increased saponarin synthesis, but excised shoots are completely unresponsive to light in a variety of media (McClure and Wilson, 1970; McClure, unpublished). Klein and Hagen (1961) found that detached petals of *Impatiens balsamina* made cyanidin and quercetin glycosides, while these compounds could not be detected in intact petals from the same genotype.

The interaction of organs in flavonoid accumulation, especially in seedlings, has been examined in several laboratories. Dittes *et al.* (1971) examined PAL levels in cotyledons and hypocotyls of *Sinapis alba* in response to continuous far red light. In the dark the cotyledons had a very low level of PAL. Light initiated an increase in PAL that persisted for about 24 hours. In hypocotyls from the same plants, a relatively high level of PAL is found in the dark and light has less effect on this enzyme. Kinnersley and Davies (1972) investigated anthocyanin accumulation in *Sinapis* and found that levels in both the cotyledon and the hypocotyl were dependent on illumination of the cotyledon. This suggests that the cotyledon provides either substrate or a stimulus for anthocyanin accumulation in the hypocotyl, perhaps through phytochrome control of PAL. In excised *Brassica napus* hypocotyls a combination of blue light with exogenous glucose, phenylalanine and sodium acetate is able to substitute for illumination of attached cotyledons (Grill, 1969).

18.2.7 Substrate induction and similar regulatory mechanisms

It is difficult to carry out meaningful experiments with flavonoids added to biological systems as they or their oxidation products may rapidly combine with, and deactivate, many enzyme systems in cellular compartments which they do not contact under normal conditions.

With this critical limitation in mind, it still appears that there is some evidence from feeding experiments for substrate induction and feedback repression in the control of flavonoid accumulation. For example, ferulic acid has been reported to control the accumulation of peonidin in *Petunia hybrida* petals through substrate induction (Hess, 1967). When exogenous ferulic acid and ^{14}C-acetate are supplied to *Petunia* petals and the radioactivity of the peonidin and cyanidin determined after several hours of incubation, the presence of ferulic acid doubles the level of labelled peonidin while labelled cyanidin levels are unchanged (Hess, 1967). Addition of chloramphenicol at low levels essentially blocks the stimulatory effect of ferulic acid (Hess, 1967), and when added at a critical stage, 2-thiouracil can block peonidin 3-glucoside without changing the rate of accumulation of cyanidin 3-glucoside (Hess, 1964).

Induction must be coupled with repression and patterns suggesting feedback inhibition has been demonstrated with very low levels of *p*-coumaric acid added to purified preparations of cinnamic acid 4-hydroxylase from *Pisum sativum* (Russell, 1971) and tyrosine ammonia-lyase (TAL) from *Hordeum vulgare* (Kindl, 1970).

Chorismate mutase is another early enzyme involved in flavonoid biosynthesis and two forms can be detected in *Phaseolus aureus*. One of the purified *Phaseolus* enzymes, but not the other, was inhibited by low levels of either L-phenylalanine or L-tyrosine (Gilchrist *et al.*, 1972). Chu and Widholm (1972) determined

chorismate mutase activity in tissue cultures of *Nicotiana tabacum,*
Oryza sativa, Daucus carota and *Lycopersicon esculentum* when the
cultures were fed either L-phenylalanine or L-tyrosine. The pool sizes
of these amino acids increased within the tissues from four- to
twentyfold but did not alter the level of chorismate mutase. From
these results they concluded that there was no repression of this
enzyme *in vivo.*

An important contribution to our understanding of the regulation
of flavonoids has been made with the garden pea (*Pisum sativum*), and
Attridge *et al.* (1971) present evidence for end product inhibition of
PAL by *Pisum* flavonoids. They found that the kaempferol and
quercetin 3-triglucosides were inhibitory to *Pisum* PAL at concentrations as low as 0.2 mM. This inhibition was apparently not due to
non-specific enzyme deactivation, as preincubation of PAL from
Pisum with quercetin aglycone for 90 min caused no significant loss
of activity. Their interpretation was that the flavonoid glycosides
operate by inhibiting PAL at a non-catalytic (allosteric) site of the
enzyme. As PAL appears to be an important enzyme in the
biosynthesis of flavonoids, PAL inhibition by flavonoids would be an
effective feedback regulatory mechanism.

Hadwiger and Schwochau (1969) have presented a model for
pisatin induction in *Pisum sativum* pods based on the induction of
specific mRNAs. They propose that the pea genome contains an
operon with an adjacent operator site and a polycistronic structural
gene where enzymes for pisatin production are coded. The operon is
assumed to be under the control of a regulator gene which
continuously produces a specific repressor substance. This repressor
substance is thought to occupy the operator site in healthy pods and
prevent the synthesis of pisatin. If repressor synthesis is inhibited,
then the genes for pisatin production are activated and this
isoflavonoid accumulates at the site of infection. This model may
explain the induction of pisatin in non-infected tissue by application
of such diverse materials as polylysine, actinomycin D, bovine
pancreatic ribonuclease, or heavy metals (Hadwiger and Schwochau,
1970).

18.2.8 Mechanical damage and pathogenic attack

Pathogens may alter the flavonoid composition of afflicted plants
either by mechanical damage (wounding) or by processes which

depend on the presence of a pathogenic metabolite to change flavonoid accumulation. Hadwiger *et al.* (1970) showed that PAL levels of immature pods from *Phaseolus vulgaris* and *Pisum sativum* increase within 8 hours of inoculation with several pathogenic or nonpathogenic microorganisms. This response was also brought about by fungal spore suspensions and such compounds as $CuCl_2$, and Actinomycin D. It is significant that these treatments did not stimulate PAL levels in seedlings of either *Triticum vulgare, Zea mays* or *Linum usitatissimum.*

Keen *et al.* (1972) fed ^{14}C-L-phenylalanine and ^{14}C-isoliqui-ritigenin (12) to *Glycine max* hypocotyls inoculated with the pathogenic fungus *Phytophthora megasperma.* After 16 hours, isoflavones appeared and after 48 hours hydroxyphaseollin (13) accounted for over half of the label in the ethanol soluble fractions. Malvidin and several flavone glycosides were present in the plants throughout the experimental period but none of these was quanti-tatively changed by infection. Similar results have been reported in *Trifolium repens* (Wong and Latch, 1971).

Olah and Sherwood (1971) infected *Medicago sativa* with the fungus *Ascochyta imperfecta* and noted that both healthy and infected leaves had glycosides of formononetin, daidzein, 7,4'-dihydroxyflavone, 7,3',4'-trihydroxyflavone and tricin (14). As

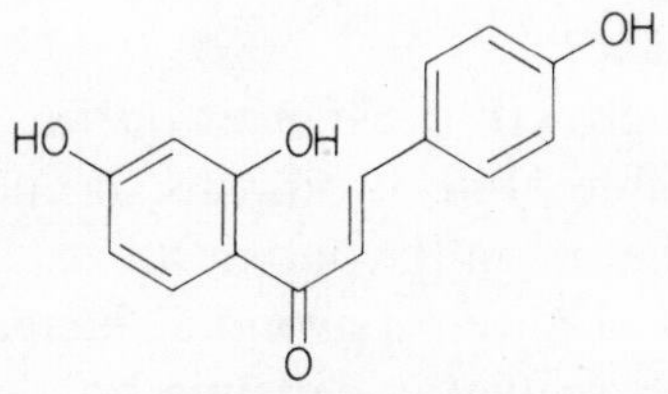

(12, Isoliquiritigenin)

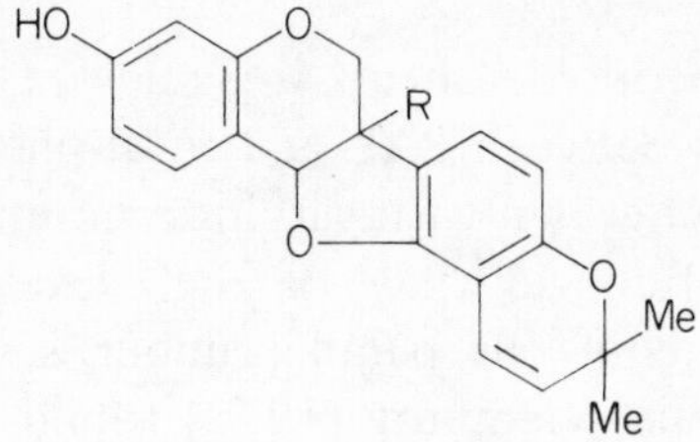

(13, Hydroxyphaseollin R = OH)
(17, Phaseollin R = H)

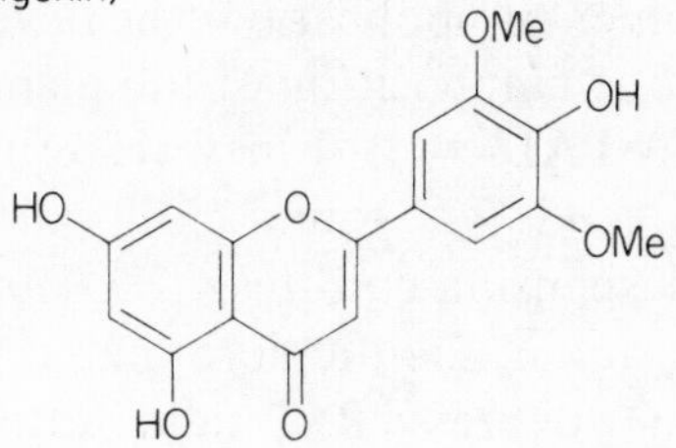

(14, Tricin)

the leaf spots characteristic of this disease developed, flavonoid levels increased and considerable amounts of aglycones were detected. By the eleventh day of infection, 17 flavonoids not found in healthy leaves were detected in infected plants. When *Medicago* is infected with a broad range of fungi, pathogens induce the formation of coumestrol (15) while nonpathogens have no effect. Viral pathogens seem generally less specific in controlling flavonoids. Kaempferol and quercetin glycoside levels are raised significantly in viral infected leaves of *Prunus persica* and *Prunus avium* (Geissman, 1956). Viral infection also increases the hesperitin (16) content of *Citrus sinensis* (Hanks and Feldman, 1969) and anthocyanins in *Nicotiana tabacum* (Sosnova and Ulrychova, 1972). In *Matthiola incana* virus infection appears to block accumulation of an acylated pelargonidin glucoside while stimulating kaempferol derivatives (Feenstra *et al.*, 1963).

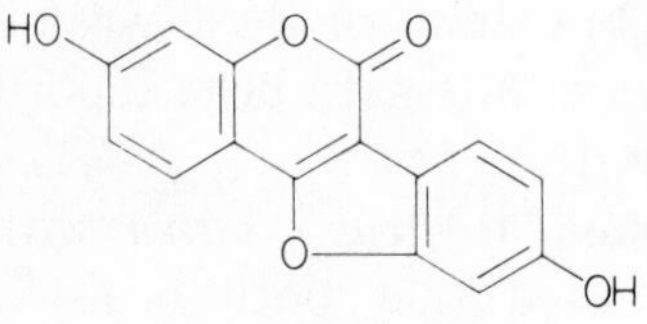

(15, Coumestrol) 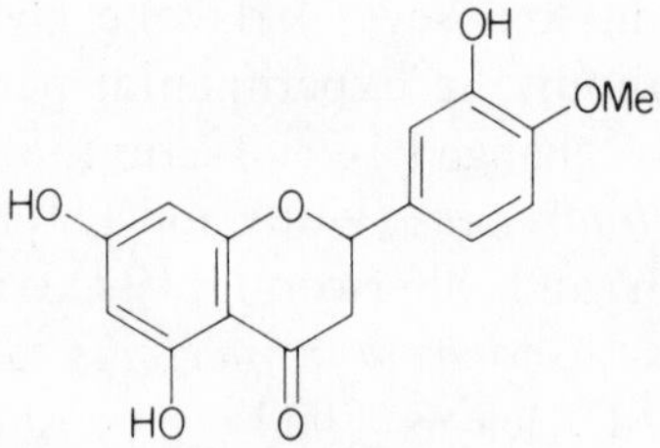   (16, Hesperitin)

18.2.9 Plant growth substances

The effects of added growth substances on flavonoid levels in plants are summarized in Table 18.6. From this, it appears that the effects of plant growth regulators on flavonoids are as diverse as their effects on growth. Whether a given flavonoid increases or decreases in response to a growth regulator depends on the concentration and manner of application, the developmental stage of the tissue, and the genetic composition of the plant.

As PAL control may relate to flavonoid control, it is relevant that Davies (1972) found close parallels between PAL and total phenols in suspension cultures of *Rosa rugosa* grown with varying amounts of the auxin 2,4-dichlorophenoxyacetic acid (2,4-D). High levels of 2,4-D depress PAL and total phenols in parallel under several experimental conditions. Leonova and Gamburg (1972) found that naphthaleneacetic acid, another auxin, markedly decreased PAL in tissue cultures of *Nicotiana tabacum*. In seedlings of *Zea mays*,

Lycopersicon esculentum, and *Phaseolus vulgaris*, gibberellic acid promotes PAL (Reid and Marsh, 1969).

Kang and Burg (1973) have recently shown that phytochrome induced anthocyanin in etiolated *Brassica oleracea* seedlings is reduced by ethylene at 10 p.p.b. Etiolated seedlings of this species produce sufficient ethylene to influence their anthocyanin accumulation, and reduction of endogenous ethylene in the tissues by hypobaric treatment markedly accelerates anthocyanin levels. Phytochrome activation may initiate ethylene evolution from a tissue and allow anthocyanin synthesis to occur.

In many physiological responses CO_2 blocks ethylene effects, and it reverses the effects of ethylene on anthocyanin accumulation in *Euphorbia pulcherrima*, but not in *Sorghum vulgare, Brassica oleracea,* or *Vaccinium macrocarpon* (Craker and Wetherbee, 1973). In fact, both ethylene and CO_2 increase anthocyanin levels in *Sorghum vulgare* seedlings, both decrease anthocyanin in *Brassica oleracea* seedlings, while in *Vaccinium macrocarpon* fruits, ethylene increases anthocyanin levels while CO_2 has no effect (Craker and Wetherbee, 1973).

Ethylene markedly increases PAL in tissues as diverse as *Pisum sativum* seedlings (Hyodo and Yang, 1971), *Citrus paradisi* flavedo (Riov *et al.*, 1969), and root tissue from *Brassica napobrassica* and *Pastinaca sativa* (Rhodes and Wooltorton, 1971). In the *Citrus* flavedo, light has no effect on PAL and in *Brassica* and *Pastinaca*, ethylene is without effect on several other enzymes such as TAL, shikimate dehydrogenase and the transaminase controlling the formation of phenylalanine. However, both PAL and cinnamic 4-hydroxylase levels are markedly increased by ethylene in *Pisum* epicotyls maintained in the dark (Hyodo and Yang, 1971).

18.2.10 Inhibitor studies

Flavonoid accumulation and the balance between synthesis and degradation must involve enzyme synthesis. For this reason, it is likely that any inhibitor of enzyme synthesis will have an eventual effect on the level of flavonoids present at a given time. However, by judiciously applying antibiotics or analogues of metabolites involved in protein synthesis, one may determine some features about the point at which an environmental stimulus may control a biochemical pathway.

Table 18.6 *Effects of plant growth substances on flavonoid accumulation in plants*

Plant	Organ	Flavonoid[1]	Growth substance[2]	Response[3]	References
Brassica oleracea (Cruciferae)	seedling	anthocyanin	ethylene, IAA	inhibits	Kang and Burg, 1973
Convolvulus sepium (Convolvulaceae)	stems	rutin	2,4-D	inhibits	Tronchet, 1966
Cymbidium hybrids (Orchidaceae)	flower	anthocyanin	GA, NAA	promotes	Arditti et al., 1971
Daucus carota (Umbelliferae)	callus	anthocyanin	GA	inhibits	Schmitz and Seitz, 1972
			IAA, NAA or 2,4-D	required	Alfermann and Reinhard, 1971
Dimorphotheca sp. (Compositae)	callus	anthocyanin	kinetin and IAA	required	Ibrahim et al., 1971
			benzyl adenine	inhibits	Ball and Arditti, 1972
Gladiolus sp. (Iridaceae)	flower	anthocyanin	kinetin or 2,4-D	promotes	Halevy and Shilo, 1970
			CCC	promotes	
Glycine max (Leguminosae)	callus	daidzein	IAA, 2,4-D	required	Miller, 1969
Haplopappus gracilis (Compositae)	callus	anthocyanin	NAA	inhibits (complex)	Blakely and Steward, 1961
	suspension culture	anthocyanin	NAA, IAA, 2,4-D or kinetin	promotes	Gamborg et al., 1970 Constabel et al., 1971
	callus	cyanidin 3-glucoside and 3-rutinoside	2,4-D	inhibits	Strickland and Sunderland, 1972
Helianthus tuberosus (Compositae)	callus	anthocyanin	kinetin and IAA	required	Ibrahim et al., 1971
Hordeum vulgare (Gramineae)	shoot	saponarin	acetylcholine	inhibits in light only	Saunders and McClure, 1973
Impatiens balsamina (Balsaminaceae)	hypocotyl	anthocyanin	NAA	promotes	Arnold and Albert, 1964
	petals	anthocyanin	kinetin	promotes	Klein and Hagen, 1961
Linum usitatissimum (Linaceae)	callus	anthocyanin	kinetin and IAA	required	Ibrahim et al., 1971

Species (Family)	Tissue	Pigment[1]	Growth regulator[2]	Effect[3]	Reference
Lycopersicon esculentum (Solanaceae)	fruits	rutin	CCC	promotes	Jablonowski, 1972
	leaves	rutin	2,4-D	inhibits	Van Bragt *et al.*, 1965
Malus pumila (Rosaceae)	callus	anthocyanin	kinetin and IAA	required	Ibrahim *et al.*, 1971
Olea europaea (Oleaceae)	fruit	anthocyanin	kinetin or zeatin	promotes	Shulman and Lavee, 1972
Phaseolus aureus (Leguminosae)	suspension culture	coumestrol	ethylene or NAA	no effect	Berlin and Barz, 1971
			IAA, NAA, or kinetin	promotes	
Pisum sativum (Leguminosae)	fruit	pisatin	ethylene	induces	Chalutz and Stahmann, 1969
	leaves	quercetin	GA	inhibits	Moore and Pecket, 1972
	internodes	kaempferol 3-triglucoside	GA	inhibits	Russell and Galston, 1968
Rosa multiflora (Rosaceae)	callus	anthocyanin	kinetin and IAA	required	Ibrahim *et al.*, 1971
Sorghum vulgare (Gramineae)	seedling	anthocyanin	ethylene	complex	Craker *et al.*, 1971
	internodes	anthocyanin	IAA, 2,4-D or GA	inhibits	Vince, 1968
Spirodela intermedia (Lemnaceae)	frond	vitexin, orientin and cyanidin 3-glucoside	GA	inhibits	McClure, 1970
S. oligorhiza	frond	petunidin 3,5-diglucoside	GA	inhibits	Furuya and Thimann, 1964
S. polyrhiza	frond	cyanidin 3-glucoside	GA	inhibits	
Trifolium pratense (Leguminosae)	seedling	biochanin A formononetin anthocyanin	kinetin	promotes	Dedio and Clark, 1971
T. subterranean	leaves	formononetin	2,4-D	inhibits	Rossiter *et al.*, 1973
Vaccinium macrocarpon (Ericaceae)	fruit	anthocyanin	ethylene	promotes	Craker, 1971

[1] Many reports detail the accumulation of 'anthocyanin' without identifying the pigment or mixture of pigments.

[2] Abbreviations: CCC = chloroethyltriethylammonium chloride; GA = gibberellic acid; IAA = indoleacetic acid; NAA = naphthaleneacetic acid; 2,4-D = 2,4-dichlorophenoxyacetic acid.

[3] Responses are invariably complex and dose-dependent. The overall effects are summarized in this column.

Coupled induction of pisatin and PAL in *Pisum satavum* (Had-wiger and Schwochau, 1971) and of phaseollin (17) and PAL in *Phaseolus vulgaris* (Hess and Hadwiger, 1971) is accomplished by 9-aminoacridine and a broad range of similar compounds which react with DNA. These compounds have the potential of changing the conformation of DNA and Hadwiger and Schwochau (1971) suggest that they operate by dissociation of a repressor compound which allows the DNA to assume a more desirable conformation for transcription.

The requirement for *de novo* enzyme synthesis, as judged by the effects of applying various inhibitors of reactions between DNA and the completed polypeptide, have been demonstrated for flavonoid accumulation in many laboratories (e.g. Bibb and Hagen, 1972; Creasy *et al.*, 1965; Hartmann, 1970; Hess, 1966; Longevialle, 1969; Margna *et al.*, 1969; Mohr and Bienger, 1967; Mohr and Senf, 1966; Rossiter and Barrett, 1970; Straus, 1960). Inhibitors of protein synthesis may block the accumulation of one flavonoid, but not another, in the same tissue. For example, Stafford (1966) found that light-independent apigeninidin (18) and luteolinidin (19) synthesis in *Sorghum vulgare* internodes was inhibited by chloramphenicol and L-ethionine but not by actinomycin D or 8-azaguanine. In the same tissue, light dependent synthesis of cyanidin was blocked by all four of these inhibitors.

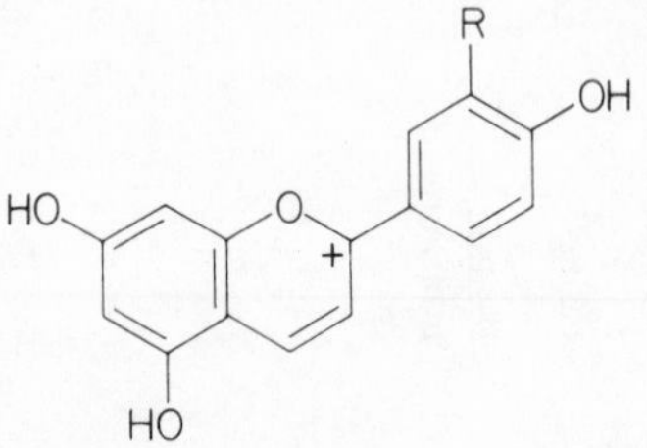

(18, Apigeninidin R = H)
(19, Luteolinidin R = OH)

The time of inhibitor treatment may also be critical. Lange and Mohr (1965) found that phytochrome-induced synthesis of antho-cyanin in *Sinapis alba* was completely blocked by 10 mg ml^{-1} of actinomycin D if given before or at the onset of light, but only partly blocked if given at the end of the lag phase before anthocyanin accumulation occurred. They conclude that actinomycin D, at a suitable concentration, will block the formation of new kinds of mRNA while still permitting the continued synthesis of mRNAs already in production at the time the antibiotic is added. In other work on *Sinapis alba*, Wagner *et al.* (1967) found that chloramphenicol within the range of 20 to 40 mg ml^{-1} increases phytochrome-mediated anthocyanin synthesis; they surmise that this antibiotic blocks protein synthesis in the plastids more strongly than in the cytoplasm and increases the pool of phenylalanine in the cytoplasm, leading to increased levels of anthocyanin synthesis.

The method of application may critically determine the response to inhibitors. Faust (1965) found that dipping whole apple *(Malus pumila)* fruits into solutions of chloramphenicol, streptomycin, cycloheximide or ethionine increased anthocyanin accumulation. When discs of fruit peel were floated on solutions of these inhibitors, they all inhibited anthocyanin accumulation.

18.3 Localization of flavonoids

Most workers have assumed that flavonoids are synthesized in the vacuole, that they remain there, and that they are not in contact with the cytoplasmic enzymes and organelles. This assumption has apparently been made solely on the basis of visual detection of anthocyanins in vacuoles. Recent investigations suggest that this view is not correct and that flavonoids may be synthesized within the cytoplasm, specifically within the plastids, and some fractions eventually transported to the vacuole.

18.3.1 Translocation of flavonoids

Only water, inorganic ions, a few sugars and vitamins appear to be moved appreciable distances in intact growing plants (Milthorpe and Moorby, 1969). Grafting experiments have shown that the translocation of flavonoids across grafts is unlikely (Delaveau, 1964; Schultz, 1969) while flavans and flavanones do not pass mycorrhizal barriers in *Pinus radiata* and *Pseudotsuga menziessii* (Hillis and Ishikura, 1969).

Macleod and Pridham (1966) investigated the rates of translocation of phenolic compounds introduced into the apical leaves of
Vicia faba. Exogenous kaempferol and quercetin moved at rates of
18 and 12 cm h^{-1} respectively. In an attempt to detect phenols
naturally present as phloem constituents, they placed aphids (*Macro-
siphum pisi*) on the stems of *V. faba* plants and analysed the
honeydew for phenolic compounds. Tyrosine was conclusively
identified and two possible cinnamic derivatives, but no flavonoids
were present. Thus there is no evidence supporting translocation of
flavonoids within the vascular system of plants. Flavonoids appear to
be formed within the tissues, probably within the cells, in which they
accumulate.

18.3.2 Localization within specific organs, tissues and cells

Flavonoids are excreted in considerable amounts by certain plants.
Barz *et al.* (1970) report that *Cicer arietinum* excretes isoflavones
through roots into the culture medium. Egger *et al.* (1970) have
found large amounts of free aglycones of rhamnocitrin (7-*O*-methyl
kaempferol) and kaempferol in the resinous oil of *Aesculus hippo-
castanum,* while oil from the buds of *Populus nigra* have been shown
to contain a range of mostly methylated flavonoid aglycones
(Wollenweber and Egger, 1971). Harborne (1968) reports that the
farinas of 35 species of the Primulaceae have flavone as a major
constituent. Flavonoid aglycones are also excreted from the fronds
of the ferns *Pityrogramma tartarea* and *P. calomelanos* (Star and
Mabry, 1971). The presence of these flavonoids as aglycones
probably relates to their solubility in the waxy cuticle. While most
flavonoids in the leaf of plants are normally found as glycosides (e.g.
Geissman, 1962; Harborne, 1967), aglycones may well be present, in
trace amounts, in the leaf cutins of many species.

Restriction of flavonoids to specific tissues within an organ is
probably the rule in plants. Anthocyanins in foliage leaves, for
example, are frequently restricted to either the epidermis or the
mesophyll (Parkin, 1903) (see Table 18.7). Anthocyanin in seedlings
is most frequently present only in the outermost layers of cells. A
survey for tissue sites of red (presumably anthocyanin) pigmentation
showed that in 42 species, the pigment was in a single layer of cells in
the epidermis, in 19 species, in the two outermost layers of cells, and
in only 5 species, in several layers of cells (Nozzolillo, 1972).

Table 18.7   *Tissue localization of anthocyanins in foliage leaves*

Period of anthocyanin accumulation[1]	Number of species examined	Site of anthocyanin accumulation		
		Epidermis only	Mesophyll only	Both epidermis and mesophyll
Transitory anthocyanins	235	20%	64%	16%
Autumnal anthocyanins	81	11%	78%	11%
Permanent anthocyanins	54	70%	17%	13%
Accidental anthocyanins	30	–	most	–

(*Data from* Parkin, 1903.)
[1] Transitory anthocyanins may be found, e.g. only in young growth. Accidental anthocyanins are those accumulating in response to injury, disease, cold, etc.

Fruits present still another pattern. Tronchet (1972) studied epidermal peels, seeds, and pulp of 70 fruits; in all cases, the superficial layer had the same flavonoids as were present in the epidermis of other aerial parts of the plant. Flavonoids however were conspicuously absent from the pulp of most fruits but quite frequently present in the seeds. Variegated leaves provide yet another pattern of flavonoid distribution. Delaveau and Paris (1968) studied the flavones of variegated leaves in *Acer negundo, A. pseudoplatanus, Euonymus japonicus, Ligustrum ovalifolium* and *Cornus alba.* In all but *C. alba,* higher concentrations of quercetin and kaempferol glycosides were found in the yellow areas than in the green areas.

Restriction of flavonoids to certain cells within a tissue is common. Esau (1965) reports that in tissues forming anthocyanins, a high level within one cell appears to have no carryover or gradient effect on the presence or absence of anthocyanins in adjoining cells. Striking patterns are often produced by this disjunction of anthocyanins. The lurid colors of some garden forms of *Delphinium* are attributable to cells containing red anthocyanins side-by-side with cells containing blue anthocyanins, and in *Crocus aureus*, the outside of the perianth shows green stripes which are due not to chlorophyll but to a combination of blue anthocyanins overlaying cells with yellow soluble pigments (Bidgood, 1905). Patterns of cell-specific flavonoid accumulation are apparently determined very early in the development of the tissue. Hess and Endress (1972) report that protoplasts isolated from *Nemesia strumosa* petals during the early bud stage will develop into a mixture of either red or red-blue cells containing variable amounts of cyanidin 3-xylosylglucoside.

18.3.3 Subcellular localization of flavonoids

It is still not entirely clear where, within the cell, flavonoids are synthesized. Poltis (1959) describes 'cyanoplast' forming within the vacuoles of several of the Solanaceae at the onset of anthocyanin accumulation. These cyanoplasts are small spherical structures in which intense anthocyanin pigmentation is evident. As the cell matures, the anthocyanins appear to leak from these cyanoplasts and give rise to the characteristic diffuse pattern of anthocyanin within vacuoles. These small intensely pigmented bodies are easily seen in the epidermal cells of many young seedlings. This suggests that the flavonoids are being synthesized in the vacuole; however, it does not rule out their synthesis in other compartments of the cytoplasm and their transference into the vacuole.

In some apparently rare instances, anthocyanins may even accumulate in plant cell walls or in intracellular spaces. Gertz (1906) reported anthocyanin in cell walls of *Sorghum vulgare* and *Pontederia crassipes* and in the red phloem fibres of *Oxalis ortgiesii.* There are even reports of blue starch grains from some lines of *Zea mays* (Gertz, 1906). Wiermann (1970), in studying the accumulation of flavonoids in pollen grains, concludes that the final steps of synthesis of flavonol glycosides and anthocyanins may occur in the outer layers of the pollen wall in *Tulipa hybrida.*

One approach to determining the site of flavonoid biosynthesis is to isolate subcellular fractions from plants, and screen them for related enzymes or for the flavonoid itself. However, so far this procedure has given equivocal results. It is true that the enzyme PAL has been found in respectable levels in cytoplasmic microsomes or glyoxysomes isolated from either *Quercus pedunculata* roots (Alibert et al., 1972) or from *Ricinus communis* endosperm (Kindl and Ruis, 1971). However, Camm and Towers (1972), in studying the intracellular distribution of enzymes of phenolic metabolism in tubers from *Solanum tuberosum*, found that shikimate dehydrogenase, prephenate dehydrogenase, and an *O*-methyltransferase acting on caffeic acid, were not associated with any particulate fraction. Again, PAL was found in all particulate fractions, yet the majority of the PAL activity was not particulate.

An alternative approach is to prepare highly purified, hopefully intact and functional, organelles and determine the ability of the isolated system to accomplish the biological reactions. Several recent experiments along these lines suggest that the chloroplast may be involved in flavonoid synthesis. For example, a number of enzymes relating to phenolic metabolism have been found in isolated chloroplasts and many appear to be phenolases which oxidize a range of phenolic substrates (Arnon, 1949; Bartlett *et al.*, 1972; Bokuchava *et al.*, 1970; Mayer, 1966; Neumann and Avron, 1967; Parish, 1972; Sato, 1966; Stafford, 1969).

Indirect evidence for specific types of PAL activity in plastids was found in certain hybrids of *Oenothera berteriana* x *Oe. odorata* (Hachtel and Schwemmle, 1972). Cytoplasmic inheritance has been intensively studied in *Oenothera* and the type of cytoplasm in these crosses determined the degree to which PAL levels were influenced by light. More directly, both PAL and TAL have been found in highly purified chloroplast and etioplast preparations from *Hordeum vulgare* (Saunders and McClure, 1972).

Oettmeier and Heupel (1972) isolated *Spinacia oleracea* chloroplasts and found p-coumaryl-meso-tartaric acid and several flavonoid glycosides, one yielding 3-methoxy-6,7-methylenedioxy-5,3',4'-trihydroxyflavone on hydrolysis. Similarly, saponarin is recovered in high yield from chloroplasts and etioplasts isolated from *Hordeum vulgare* by several aqueous and non-aqueous techniques of differential and gradient centrifugation (Saunders and McClure, 1972). When the isolating media for the *Hordeum* plastids was contaminated at various stages of isolation with near saturating levels of various anthocyanins, or glycosides of flavones and flavonols, only the flavonoids found in whole leaf methanolic extracts of *Hordeum* were recovered from the plastids. This suggests that flavonoids recovered from *Hordeum* plastids are localized within the plastids and that they are not artifacts or contaminants from a cytoplasmic pool that adheres to the plastids during isolation and subsequent purification. Similar results have been obtained with chloroplast from *Lycopersicon esculentum* and *Daucus carota;* when purified *Daucus* chloroplast are incubated with UDP-^{3}H-glucose and ^{14}C-luteolin, they synthesize and accumulate ^{14}C-luteolin 7-^{3}H-glucoside (Saunders *et al.*, 1973).

Chloroplasts from *Impatiens balsamina*, isolated by various techniques, contain flavonoids up to 2% of their dry weight and in a pattern typical of the flavonoid composition of whole leaf extracts

(Weissenboeck *et al.*, 1971). The level of flavonoids in *Impatiens* chloroplast is markedly increased by illumination (Weissenboeck, 1972). When these plastids are disrupted and thylakoid fractions obtained, the ratio of kaempferol to quercetin derivatives is 7 : 1 while whole plastids have these flavonoids at a ratio of 3 : 1. This suggests further compartmentalization of specific flavonoids within the chloroplast (Weissenboeck, personal communication).

When etioplast from two varieties of *Avena sativa* were examined, each had three different *C*-glycosylflavones and one produced a kaempferol glycoside while the other contained isovitexin (6-*C*-glucosylapigenin). These last two flavonoids were in much higher relative concentration in the etioplast than in whole leaf extracts; furthermore, different patterns of flavonoid glycosides are found when the plastids and cytoplasmic preparations are compared (Weissenboeck, 1973). These results all suggest that flavonoids are synthesized within the plastids and eventually certain fractions may be transported to the central vacuole.

Gifford and Stewart (1968) detected material in *Kalanchoe* apices which had many characteristics of flavonoids. This material accumulated in the cisternae of plastid membranes and was transferred from these plastids to the central vacuole as the cells matured. In root tissues of *Reaumuria palastenia* tannins are detected in the cisternae of small membrane-covered bodies which inflate, coalesce, and form into the central vacuole (Ginzburg, 1967).

If plastids are the site of flavonoid synthesis, then it is curious that there are no unambiguous reports of cytoplasmic inheritance in flavonoids. It may well be that early steps in flavonoid biosynthesis are regulated, in part, by the genome of the plastid (Hachtel and Schwemmle, 1972) while the later stages are controlled by nuclear genes. Regulation of plastid functions by nuclear genes is well known for several aspects of non-phenolic biosynthesis (Givan and Leech, 1971). The reports of different patterns of photoregulation of phenylpropanoid enzymes and enzymes more directly involved with the later stages of flavonoid biosynthesis (Table 18.2) may relate to such a proposed scheme.

18.3.4 State of flavonoids within the cell

One must finally consider the chemical associations of flavonoids within various cellular compartments. The cell vacuolar solubility of flavonoids is generally due to the sugars attached to them. Other types of covalent association, e.g. acylation with malonic acid (Kreuzaler and Hahlbrock, 1973) or linkage with bisulphate, may also increase solubility and allow flavonoids to accumulate in the cell vacuole in relatively high concentration. In flower tissue, the anthocyanins are closely associated, probably via hydrogen bonding, with flavones in loose co-pigment complexes and such associations may also aid pigment solubility.

18.4 Functions of flavonoids

18.4.1 Reactivity in biological systems

The most reactive biological compounds are usually nitrogenous; these compounds are bases and have strong affinities for the acidic groupings of reactive sites. In contrast, phenolic rings can react only weakly with specific receptor groups, primarily by hydrogen bonding. However, phenolic substituents usually improve the solubility characteristics of compounds and the phenols are in general easily moved across biological membranes (Parke, 1968). Thus one may predict that flavonoids have relatively weak influences, but on a broad range of biological .phenomena, since altered membrane characteristics appear to be a major means by which organsims control their biochemistry (Oaks and Bidwell, 1970).

18.4.1.1 Flavonoids as antioxidants

Ascorbic acid is a universal component of plant cells, especially abundant in cells of high metabolic activity (Mapson, 1970). It is a good reductant, and the ease with which ascorbate may be oxidized and its oxidation products reduced, fulfil many of the requirements of an electron carrier. The oxidation of ascorbate occurs in the presence of O_2 and is catalysed by metals or by enzymes with metallic prosthetic groups. Effective enzymes are ascorbic acid oxidase, phenolase, cytochrome oxidase and peroxidase (Mapson, 1970). Several flavonoids serve as antioxidants for ascorbic acid, apparently by chelating metals from the reaction mixture. This chelation is dependent primarily on the 3-hydroxyl-4-carbonyl and the 3'4'-dihydroxyl groupings (Clements and Anderson, 1961; Samorodova-Bianka, 1965). The sparing effect of flavonoids on ascorbate oxidation may explain many of the interactions of flavonoids and ascorbate in the voluminous, but often contradictory, literature on flavonoids and vitamin C.

Ascorbate may, in turn, protect flavonoids from oxidation. Grommeck and Markakis (1964) found that purified cyanidin 3-gentiobioside, cyanidin 3-rhamnoside and pelargonidin 3-glucoside are decolorized by low levels of H_2O_2 and horseradish peroxidase. Ascorbate added to this system inhibited decolorization of the anthocyanins to one tenth the rate of the control, apparently by reducing an early oxidation product of anthocyanin breakdown. This scheme has implications in flavonoid metabolism (Chapter 17).

Pratt (1965) investigated the antioxidant effects of several flavonoids on beef slices and in carotene-lard solutions. Both quercetin and myricetin, and their 3-glucosides, were significantly effective antioxidants at concentrations ranging from 0.4 to 2.5×10^{-5} M. They attributed the antioxidant effects to chelation of metallic ions and accepting free radicals (Pratt and Watts, 1964). Quercetin is an effective antioxidant for fatty acid methyl esters of sunflower oil or linseed oil (Letan, 1966). Methylation of quercetin at the 3- or 5-positions slightly reduced its effectiveness while methylation at either the 3'- or 4'-positions drastically reduced antioxidant abilities. On the basis of histological examinations and coincidence of occurrence within the same tissue, Van Fleet (1969) suggests that a major function of flavonoids is to serve as antioxidants for lipids and polyacetylenes in plant tissues.

18.4.1.2 Flavonoids as enzyme inhibitors

Flavonoids are a nuisance in that they must be removed from homogenates of plant material before certain enzyme activities can be demonstrated. Loomis (1969) and Loomis and Battaile (1966) have reviewed techniques for reducing the level of flavonoids and other polyphenols in enzyme preparations and these reactivities bear on the function of flavonoids. Effective treatments for removal of flavonoids include addition of large amounts of insoluble polyvinyl-pyrrolidone, addition of excess reducing agents, or using borate buffers (Loomis, 1969). Some success in blocking the reactivity of flavonoids and proteins has been obtained by including a small amount of Tween 80 in the reaction mixture (Firenzuoli et al., 1969).

Phenolases (o-diphenyl: O_2 oxidoreductases) catalyse the reaction o-diphenol + ½ O_2 ⇌ o-quinone + H_2O (catecholase activity) and many have the additional ability to hydroxylate phenols to o-diphenols (cresolase activity). Ascorbic acid oxidase and catalase will also oxidize flavonoids, at least *in vitro* (Brown, 1967). Challice and Williams (1970) studied phenolase specificity in a number of plants and found that dihydrochalcone glycosides, such as phloridzin, were much better substrates than were various flavones, flavonols, or dihydroflavonols. Van Buren (1960) reported a similar phenomenon in the enzymatic decolourization of anthocyanins in sour cherries, since it was necessary to add catechol for activity.

If flavonoids react with phenolase, the products may inhibit enzymes by non-specific binding to the enzyme, by competing or reacting with the substrate, by oxidation of sulphydryl groups controlling tertiary enzyme structure, or through complexing metallic prosthetic groups (Rich, 1969). If phenolases are localized primarily in the plastids, and perhaps in mitochondria, of healthy tissues (Parish, 1972), then as the tissues age or are wounded or attacked by pathogens, the phenolases may be released into the milieu of the cell and form quinones from flavonoids or other polyphenols.

Several workers have added flavonoids to isolated enzyme systems. At levels of 1 to 4 μM, malvidin 3-glucoside inhibits malate dehydrogenase and glutamate decarboxylase isolated from the bacterium *Salmonella enteritidis*. Malate dehydrogenase was inhibited non-competitively while glutamate decarboxylase showed a competitive or non-competitive inhibition depending on a number of experimental factors (Wheeler *et al.*, 1967). Carpenter *et al.* (1967) compared a number of anthocyanins from *Vitis vinifera* and found that they were not inhibitory to α-glucan phosphorylase or glutamate decarboxylase, but were to glycerol dehydrogenase, malate dehydrogenase and hexokinase.

DeSwardt *et al.* (1967) discovered an interesting correlation between the level of low molecular weight leucoanthocyanins and the activity of pectin methylesterase in preclimacteric fruits of *Musa paradisiaca*. They suggested that in the green fruits the monomers inhibit this enzyme and the fruit remains firm, while polymerization of the flavonoids releases the enzyme from inhibition and the fruit softens.

Flavones and flavonols with 3'-, 4'- and 7-hydroxyls are potent inhibitors of bovine pancreatic ribonuclease (Mori and Noguchi, 1970). Methoxylation at the 6- or 8-position decreases inhibitory activity and no inhibition was found for various flavanones and flavanonols.

Protein synthesis may be inhibited by flavonoids, at least *in vitro*. Parups (1967) found that hesperitin, naringenin, quercetin, and their glycosides, plus phloridzin, inhibited incorporation of ^{14}C-leucine into tuber slices of *Solanum tuberosum* or cell free extracts of *Escherichia coli*. In all cases, the aglycones were much more effective inhibitors than their glycosides.

Substituted cinnamic acids are often considered to be more reactive in isolated enzyme systems than are the flavonoids (Pridham, 1963). However, Van Sumere *et al.* (1972) found that of a series of phenolic acids and related compounds, only caffeic acid (at 10^{-3} M) decreased ribonuclease activity by more than 25% while quercetin inhibited this enzyme 100% at 10^{-3} M and 11% at 10^{-5} M. They also found that the amino acid activating system of *Hordeum vulgare* embryos was inhibited by flavonoids, but only those that could be converted into quinones. For example, at 5×10^{-4} M quercetin showed a 30% inhibition in this reaction while similar levels of kaempferol were ineffective.

While measuring ATP production in plant mitochondria by the firefly luciferase method, Stenlid (1970) found that some of the flavonoids, especially the isoflavones, were inhibitory to the luciferin-luciferase light generating system at concentrations as low as 2×10^{-7} M.

18.4.1.3 Flavonoids as precursors of toxic substances

When exogenous flavonoids are applied to biological systems it is usually not possible to differentiate between the direct effects of the flavonoids and effects that might be more appropriately attributed to their degradation products. Phloridzin is a powerful inhibitor of respiration in animal tissues (Stenlid, 1968) yet in several chromatographic solvents it decomposes to *p*-coumaric acid and phloroglucinol β-glucoside which may themselves be inhibitors (Grochowska, 1966). Hamdy *et al.* (1961) studied the effects of an anthocyanin and its degradation products on bacterial growth by a sensitive disc bioassay. *Escherichia coli*, *Lactobacillus casei* and *Staphylococcus aureus* were grown on media containing either pelargonidin 3-glucoside or its heat degradation products. The heated and unheated pigments were equally inhibitory to the growth of both *E. coli* and *S. aureus,* while *L. casei* was markedly inhibited by the unheated anthocyanin and not affected by the heated products.

Rice and Pancholy (1973) studied changes in tannins and nitrification occurring between 15 and 60 cm depths in the soil during successional changes in Oklahoma. Soils from early successional stages at three different sites had high nitrates and low total tannins. As the vegetation reached mature stages, nitrates decreased as tannins increased. Foliage of predominant species were found to have considerable amounts of condensed tannins. When these condensed tannins were isolated and added to soil suspension cultures of *Nitrosomonas* or *Nitrobacter* sp. they completely inhibited nitrification at a level of only a few parts per million in the soil. Rice (personal communication) has found that quercetin and myricetin, present as glycosides in several of the plants, were inhibitory to nitrification in *Nitrosomonas* at 10^{-6} M in soil suspension cultures. Various flavonoids, accumulating in the soil over a period of time, could play a prominent role in ecological succession.

Symbiotic nitrogen fixation is an alternative to relying completely on nitrogen fixation by free-living bacteria, and it is interesting that exogenously applied quercetin will increase the number of nodules containing nitrogen fixing bacteria in the root system of *Lotus corniculatus* (Molina and Alexander, 1967).

18.4.2 Flavonoids as pigments

All flavonoids have a high absorbance (log ϵ 4.0 to 4.5) in the 250 to 270 nm range where proteins and nucleic acids have high absorptivity. Flavones and flavonols absorb significantly in the range from 330 to 350 where NAD and NADP cofactors absorb strongly, and anthocyanins strongly absorb in the range from about 520 to 560 nm where the mammalian eye has its highest sensitivity and where plant photoreceptors such as chlorophyll, carotenoids and phytochrome have minimal absorptivity. It is probably in the area of visual perception and light absorbance that the strongest case can be made for unambiguous functions of flavonoids (e.g. Harborne, 1972).

18.4.2.1 Flavonoids and visible colour

It is well known that a wide range of insects have visual systems which allow them to be particularly sensitive to flavone and flavonol glycosides absorbing near 350 nm and to the yellow flavonoids such as chalcones, aurones, 3-deoxyanthocyanins, or flavonols with extra hydroxyl or methoxyl substituents in the 6- or 8-position (Harborne, 1972). On the basis of colour vision and from field observations, it is also clear that bees generally prefer blue and yellow, butterflies pink or white, birds red, and moths white (Harborne, 1972). Most correlations between flower colour and the types of insects that visit them can be attributed, at least in part, to the flavonoids. Kevan (1972) made extensive field studies in the Canadian Arctic and found that white and yellow flowers are predominant. Here the high UV reflectance from the flowers gives them a greater luminance and they stand out from their backgrounds in both colour and brightness to the insect. Horovitz (1969) traced the relationship between flower colour and bees in blue or white flowering forms of the California annual *Lupinus nanas*. Reflectance of the flowers in the blue region (400 to 470 nm) was closely associated with high female outcrossing, while in the white flowers there was a very low incidence of outcrossing.

Some of the most complex and interesting relationships between insects and pollination are found in flowers of the Orchidaceae. Anthocyanins, with possibly other flavonoids, produce elaborate patterns that, to the insect, promises food, deceives it by producing structures with the form and colour of copulation partners, or are repulsive. Other patterns evoke the insects' anger, attack, and eventual pollination. Post-pollination responses in the orchids are often rapid and the flavonoid-determined pattern may be obscured in a matter of a few hours. It is likely that these changes discourage the pollinator from again visiting the flower and increase the likelihood that it will visit an unpollinated member of the same species (Van Der Pijl and Dodson, 1966).

Ultraviolet patterns in flowers have been examined by televising images through a visible-absorbing, ultraviolet-transmitting filter (Eisner *et al.*, 1969). The images show that flowers pollinated by birds or bats generally lack UV nectar guides or other special markings. By contrast, flowers pollinated by insects have characteristic markings and are presumably distinctive. When Horovitz and

Cohen (1972) photographed UV reflectance patterns of flowers of 26 eastern Mediterranean wild species and three cultivated species of crucifers, they found striking and complex patterns associated with the veins near the base of the petals. Patuletin (20) and quercetagetin (21) have been identified as the UV absorbing and reflecting compounds in the nectar guides of *Rudbeckia hirta* (Thompson *et al.*, 1972). Bloom and Vickery (1973) found that pattern-partitioning in yellow flowered *Mimulus luteus* was due to mixtures of carotenoids and flavonoids. The carotenoids were uniformly present, while cyanidin 3-glucoside and quercetin 3-glucoside were localized within the red spots. The yellow flavonol herbacitrin (8-hydroxykaempferol 7-glucoside) and the anthocyanin were mutually exclusive.

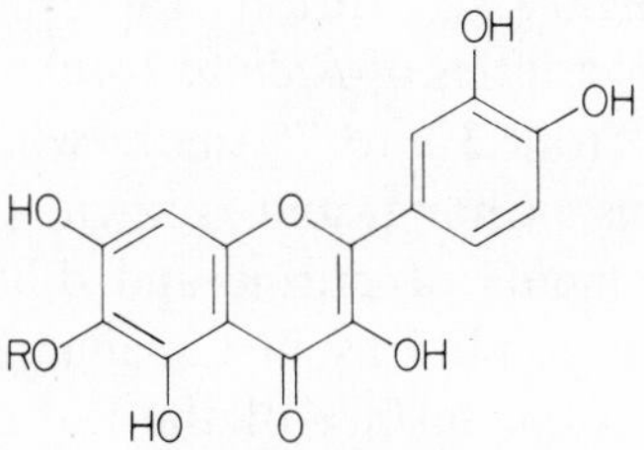

(20, Patuletin R = Me)
(21, Quercetagetin R = H)

Birds have powerful vision but a very poor sense of smell and most bird-flowers are odourless and pigmented with anthocyanin in shades ranging from orange to scarlet. Plants which have flowers with sufficient calorific rewards to attract birds consistently as pollinators must limit the availability of their nectar to effective pollinators. Red and orange are not conspicuous to insects, excepting a few butterflies, and to the insect these flowers usually blend into the background foliage. High levels of anthocyanins uniformly distributed throughout the tissues would also dampen the visibility of the flavone and flavonol glycosides in the ultraviolet range. Thus, although birds have no intrinsic preference to red, this is the only colour that is at once inconspicuous to most insects and an excellent visual signal to birds (Raven, 1972; Grant and Grant, 1968).

18.4.2.2 Flavonoids as light screens

Flavonoids are quite stable to the visible and ultraviolet regions of the spectrum. Kaneta and Sugiyama (1971) irradiated quercetin and luteolin with a 200 W high pressure mercury lamp while bubbling air through the solutions. Quercetin was not noticeably altered until after 8 hours of these drastic conditions and luteolin was stable for the 67 hours of the experiment.

Caldwell (1971), in reviewing the possible roles of flavonoids in absorbing solar ultraviolet irradiation, especially at high elevations, found that there was no simple correlation between elevation and epidermal flavonoid content. However, when ethanolic extracts of epidermal peels from many plants were examined, he noted that most of the ultraviolet absorbance was attributable to flavonoids.

Several workers have suggested that anthocyanins and other flavonoids may play a role in alpine and northern regions by absorbing light and warming the tissues. When Nagornaya and Kotsur (1970) measured light absorption by leaves of anthocyanin-containing and green varieties of *Ricinus communis, Brassica oleracea*, and *Perilla* sp., only from 3% to 7% more white light was absorbed by the red varieties. Chong and Brawn (1969) made careful temperature measurements of purple and dilute sun red strains of *Zea mays* under field conditions in Canada. The purple lines were about 1.2° warmer on the surface of the leaf or at a depth of 1 cm within the stalk. No differences in yield were found between these two strains and the role of anthocyanins in significantly increasing the temperature of plant tissues is questionable.

Flavonoids may play subtle roles as light filters in plant growth. Zenk (1967) has suggested that there is a relationship between *C*-glycosylflavones and phototropism of *Avena sativa* coleoptiles. Light of 450 nm is most effective in this response and when the soluble fractions of the tissues are examined, 84% of the absorbance at 450 nm is due to these flavonoids. At 370 nm, still a highly effective wavelength for phototropism, 98.3% of the light is absorbed by the flavone fraction while only 1% is absorbed by carotenoids and less than one percent by various flavins. Thus the flavonoids of the primary leaves create a transverse light gradient throughout the coleoptile and shadow the photoreceptor(s) for phototropism in an ideal manner.

18.4.2.3 Photosensitizing and energy transferring compounds

Nishie *et al.* (1968) studied the effects of 31 flavonoids, mostly flavones and flavonols, on growth and survival of the protozoan *Tetrahymena pyriformis*. These flavonoids had no effect in the dark and were not cytotoxic at the levels employed. When the cultures were exposed to a sunlamp for 20 min the methoxylated flavonoids were markedly photodynamic. For example, at 4×10^{-6} M kaempferol 7,3′,4′-trimethyl ether gave a 100% kill within 10 min. The fully hydroxylated flavonol in contrast was photodynamically ineffective.

The accumulation of flavonoid aglycones on the outer layer of cuticles may play a role in the photodecomposition of either natural or synthetic toxins falling on these surfaces. Ivie and Casida (1971) report that quercetin and related substituted 4-chromones accelerate the photoalteration of chlorinated cyclodiene insecticides under field conditions.

18.4.3 Flavonoids in control of plant growth and development

The chemical reactivity and solubility of flavonoid glycosides and their proclivity to complex with proteins ensures that many will cause metabolic changes, even if in non-specific ways, when applied to plants. They may also interact more specifically as cofactors for enzyme reactions requiring hydroxylated phenols as electron acceptors.

18.4.3.1 Flavonoids in concert with plant growth hormones

In reports of flavonoid promotion or inhibition of physiological responses normally attributable to specific plant growth hormones, it is not always possible to separate promotive or inhibitory effects of the flavonoids themselves from those of the hormone. This separation has however been attempted for interactions between flavonoids, indoleacetic acid (IAA), and enzymes that can oxidize indoleacetic acid.

Tang and Bonner (1947) found an enzyme from etiolated *Pisum sativum* that destroyed IAA in the presence of O_2. A few years later, Galston and Baker (1953) noted that while the concentration of IAA optimal for elongation of etiolated pea epicotyls was about 10^{-7} M, even momentary exposure to red light shifted this optimal auxin level to about 10^{-5} M. They also noted that the ability of the tissue to cause the disappearance of IAA from solution was sharply altered by prior irradiation of the tissue with red light. This led to the conclusion that indoleacetic acid oxidase (IAA oxidase) was photoregulated.

This enzyme is a peroxidase requiring O_2, Mn^{2+}, and a mono-hydroxyl phenolic cofactor (Hare, 1964). Hillman and Galston (1957) found that the decrease in IAA oxidase activity following illumination was due to the production of dialysable inhibitors that Furuya et al. (1962) identified as kaempferol and quercetin 3-tri-glucosides (KG and QG) and their p-coumaric acid esters (KGC and QGC). Significantly, monohydroxyl phenols such as KGC and KG are cofactors for IAA oxidase while dihydroxyl phenols such as QGC and QG are inhibitors, and the levels of these kaempferol and quercetin derivatives are under close phytochrome control in Pisum sativum (Galston, 1969).

In Pisum P_{fr} controls flavonoid hydroxylation patterns in a tissue-specific pattern. In the plumule, light causes an increase of QGC but not KGC, while the same treatment of internode tissue increases KGC but not QGC. Galston (1969) has concluded that red light promotes the growth of buds and simultaneously promotes the synthesis of the auxin-sparing quercetin glycosides. On the contrary, P_{fr} in stem tissues causes preferential accumulation of kaempferol glycosides, cofactors for IAA destruction, and growth is inhibited. Flavonoid regulation of IAA oxidase is not limited to Pisum since monohydroxyl flavonoids are cofactors, and the dihydroxyl flavonoids inhibitors, in such tissues as flower buds of Prunus persica (Ritzert et al., 1972), leaves of Lupinus luteus (Kornelyuk and Volynets, 1971), and seedlings of Triticum vulgare (Plotnikova et al., 1968).

Stenlid and Saddik (1963) evaluated the effects of 20 flavonoids, (flavones, flavonols, isoflavones, flavanones, chalcones and antho-cyanidins) at 10^{-6} to 3×10^{-5} M on IAA oxidase preparations from Pisum sativum roots. In every instance, a single 4'-hydroxyl substituent increased enzyme activity while the 3'4'-dihydroxylated flavonoids were inhibitory. A free hydroxyl group at the 7-position improved the cofactor nature of the flavonoids; this may explain why flavonol 3-glucosides (with free 7-hydroxyl substituents) are better cofactors than flavone glycosides which are commonly present as 7-glycosides.

Many factors complicate the scheme of flavonoid control of IAA-oxidase through B-ring hydroxylation patterns. When Pisum buds are excised prior to irradiation, neither growth, IAA oxidase activity, nor flavonoid content is altered and this is clearly a whole plant response (Galston, 1969). Photocontrols of *Pisum* flavonoids are complex and Smith and Harper (1970) have shown both phytochrome-mediated rapid responses of low magnitude and longer term responses of greater magnitude dependent on continuous irradiation. More importantly, none of the photocontrolled changes in flavonoid levels within the tissue is closely correlated with photocontrolled changes in the growth rate of the organ (Smith and Harper, 1970).

Psenak *et al.* (1970) report that simple phenolic glycosides influence IAA oxidase activity only if β-glucosidase is present. The reports that flavonoid aglycones are very effective cofactors or inhibitors of IAA oxidase (Stenlid and Saddik, 1963) support this implication that the flavonoid aglycones may be the effective mediators and that glycosylation or acylation may primarily determine the solubility and availability of the flavonoid to the site of IAA oxidase localization within the cell.

Sano (1971) has purified a highly active peroxidase from etiolated *Pisum sativum* seedlings which oxidizes IAA and is markedly inhibited by quercetin at 10^{-5} M. Both quercetin and IAA are non-competitively destroyed by this peroxidase, and each modifies the kinetic behaviour in the destruction of the other. In the absence of IAA, quercetin was altered by this enzyme first into a short-lived purple compound with maximal absorbance at 530 nm and then, within ten seconds, into a stable compound with maximal absorbance at 490 nm. When IAA was added to the reaction mixture shortly after quercetin, oxygen consumption and IAA oxidation were strikingly reduced. Thus IAA and quercetin have a mutual sparing effect in the presence of this peroxidase.

When exogenous gibberellic acid is applied to light grown plants of genetically dwarfed *Pisum sativum* they develop into tall plants. Corcoran *et al.* (1972) applied gibberellic acid alone and with various tannin preparations to dwarf and normal strains of *Pisum*. All tannins, including purified Carob condensed tannin, functioned as gibberellin antagonists even at a ratio of tannin to gibberellin of one to a thousand.

Mashtakov *et al.* (1971) determined combined effects of flavonoids and abscisic acid on growth of several plants. Rutin at 10^{-5} M generally relieved the inhibitory effects of abscisic acid, whereas apigenin had little effect or even enhanced the inhibition.

18.4.3.2 Flavonoids as plant growth regulators

Until abscisic acid was isolated and identified as a powerful growth inhibitor of higher plants, it was generally concluded that compounds with broad inhibitory actions were probably not naturally involved in control processes (Addicott and Lyon, 1969). However, it is now clear that flavonoids could be involved in control processes directly, although conclusive proof that this is so has yet to appear.

For example, exogenously applied naringenin at 10^{-4} M or lower concentrations stops budding in explants of Cichorium intybus (Bagni and Fracassini, 1966), markedly inhibits the elongation of Triticum vulgare coleoptiles, blocks the gibberellic acid controlled release of sugars in Hordeum vulgare endosperm, and induces seed dormancy in some light sensitive varieties of Lactuca sativa (Phillips, 1961; White et al., 1972). Naringenin 7-glucoside is found at levels of approximately 150 μg per bud in Prunus persica during dormancy of the bud and this level falls markedly just prior to bud opening (Corgan, 1965). When various flavanones, including naringenin and hesperidin, are applied to Citrus sinensis seedlings just before vigorous spring growth is initiated, they prolong dormancy by as much as seven weeks (Feldman et al., 1966). Thus, the conclusion might be drawn that naringenin functions as an endogenous regulator of dormancy. More recently, Corgan and Peyton (1970) dissected over 1000 dormant Prunus persica buds and found that while the bud scales have high levels of flavanones, sensitive analytical techniques could not detect naringenin in the pooled floral cup which is the tissue involved in bud opening. Abscisic acid was detected in the floral cup and it, not naringenin, appears to be the inducer of dormancy in Prunus buds. Bagni (personal communication) has found that naringenin in Prunus buds is restricted to glands of the bud scales.

Phloridzin is a potent inhibitor of sugar uptake in animals and will serve as a cofactor for IAA oxidase in plants (Stenlid, 1968). Large amounts of phloridzin are produced in leaves, twigs and roots of most species of the genus Malus (Challice and Williams, 1970). At very low concentrations, phloridzin is inhibitory to the growth of Malus pumila seedlings in water culture (Börner, 1959). Phloridzin will block sugar utilization in Avena sativa coleoptile segments and at high (10^{-3} M) concentrations it is inhibitory to sugar metabolism in

Saccharum hybrids (Bieleski, 1960). Podstolski and Lewak (1970) have reported a relatively specific β-glucosidase for phloridzin that appears during the last stages of stratification in *M. pumila*, suggesting that the conversion of phloridzin to phloretin plays a role in controlling germination.

Kefeli and Turetskaya (1967) compared the inhibitory effects of phloridzin and isosalipurposide (22) with those of a large number of apparently non-specific inhibitors such as carbon tetrachloride, 2,4-dinitrophenol, oxytetracycline and C_5 to C_{16} aliphatic alcohols. These were evaluated on the budding and rooting of *Salix* cuttings and the elongation of *Triticum vulgare* coleoptile segments. Their general conclusions were that since it required from 50 to 100 times higher concentrations of the flavonoids than the other inhibitors to block the responses, the flavonoids were not specific antihormones and simply acted non-specifically on growth processes.

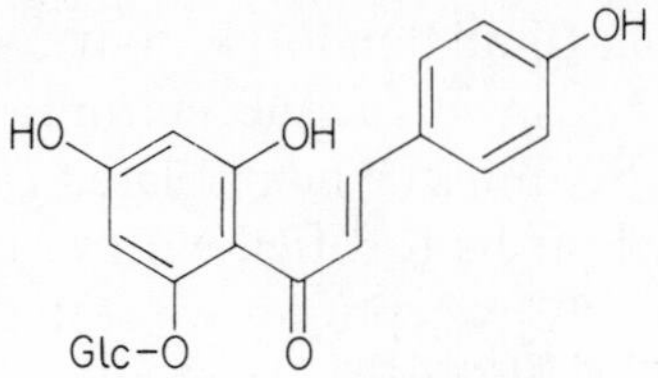

(22, Isosalipurposide)

Stenlid (1968) attempted to evaluate the overall physiological effects of phloridzin and twelve related flavonoids in the range of 10^{-4} to 10^{-5} M. He measured their effects on IAA oxidase from *Pisum sativum* roots, oxidative phosphorylation in *Cucumis sativus* hypocotyls, and the elongation and uptake of sugars and auxins in *Triticum vulgare* roots. As expected, the 4'-hydroxyl flavonoids were stimulators of IAA oxidase and this activity apparently explained many of the responses. Other effects seem explainable on the basis of uncoupling oxidative phosphorylation. The glycosides were generally less effective uncouplers than were the aglycones.

Control of the IAA oxidase system through flavonoids does not discount the possibility that flavonoids may also interact in other responses normally attributed to auxins. For example, Durmishidze *et al.* (1970) compared the effect of substituents at C-3 of the flavonoid ring on the growth of isolated *Parthenocissus tricuspidata* and *Daucus carota* tissue, and the elongation of axenic *Zea mays* and *Phaseolus vulgaris* seedlings; (+)-catechin stimulated, quercetin inhibited, and rutin had no effect on these responses. Again, Zinsmeister (1964) applied condensed tannins from *Rhus semialata* to *Avena sativa* coleoptile segments and found only slight inhibition, about 8%, after incubation in tannin concentrations up to 0.1% (w/v) for 20 h or more. This tissue is very sensitive to IAA, and if tannin, at levels down to 0.001, was added to the IAA then elongation was blocked. The most likely explanation is that the IAA-tannin complex (Leopold and Plummer, 1961) which was formed, reduced the concentration of exogenous, but not endogenous, IAA.

Callus cultures provide convenient material for the evaluation of plant growth regulators and both (+)-catechin and rutin stimulate growth of *Daucus carota* tissue explants (Naef, 1971). The most pronounced effects of flavonoids on tissue cultures have been reported by Steward (1968). He suggests that many of the promotive effects of coconut milk (liquid endosperm of *Cocos nucifera*) and the endosperm of *Aesculus* sp. on tissue cultures are due to their high content of leucoanthocyanins. When isolated *Daucus carota* explants were cultured on basal media fortified with various flavonoids, highly variable but almost always promotive effects were found. For example, pelargonidin 3-glucoside (level not reported) increased the fresh weight of the *Daucus* tissue up to 240% and the cell number by 428%. Quercetin increased fresh weight and cell number by 53% and 25%, respectively, while cyanidin decreased these responses by 6% and 11% (Steward, 1968).

Ion uptake in *Triticum vulgare* roots is significantly blocked by various flavones, flavonols, isoflavones and flavanones in the range of 10^{-5} to 10^{-6} M (Stenlid, 1961) or by a large number of anthocyanidins (Stenlid, 1962). Almost all of these flavonoids that blocked ion uptake were effective in alleviating the inhibition of root growth caused by auxins or by inhibitory sugars such as mannose or 2-deoxy-D-glucose. Many of these responses seem attributable to uncoupling oxidative phosphorylation (Stenlid, 1962).

18.4.3.3 Flavonoids as phytoalexins

Phytoalexins are antimicrobial metabolites that are absent, or present in low amounts, in healthy plants and that accumulate in high concentrations in or around cells damaged by many different stimuli, particularly infection by pathogenic fungi and bacteria (Deverall, 1972; Kuc, 1972a). This area has received intensive attention from plant pathologists in recent years and excellent reviews are those of Deverall (1972), Hare (1966), Ingham (1972), Kuc (1972a, 1972b), and Kosuge (1969).

To date over 20 phytoalexins have been chemically characterized. All are low molecular weight compounds and many are flavonoids. The induction of these flavonoids by microbial infection has been considered (Section 17.2.8). Most of the flavonoid phytoalexins are pterocarpans (see Chapter 14) and the best known are pisatin from *Pisum sativum* and phaseollin from *Phaseolus vulgaris*. The mode of action of these compounds is under active study. Phaseollin, for example, may be responsible for the cessation of growth of fungal germ-tubes in hypersensitive host cells. At 15 mg l^{-1} (within the probable range of accumulation within *Phaseolus* tissues) phaseollin decreases the ability of *Rhizoctonia solani* to take up glucose $-U-^{14}C$ and blocks cytoplasmic streaming within the hyphae (Van Etten and Bateman, 1971). These responses suggest alteration of membranes, and Van Etten and Bateman (1971) found that phaseollin causes a complete lysis of sheep erythrocytes at 23 mg l^{-1}. From these observations, they suggest that phaseollin either acts on the plasma membrane of the fungus or affects some process needed for membrane function.

The role of phytoalexins in disease resistance is a subtle one and one example of the complexity of fungus-flavonoid-host relationships can be seen in *Medicago sativa*. The leaves of *Medicago* normally contain glycosides of formononetin, daidzein, tricin and several other flavonoids. In plants infected with *Ascochyta imperfecti*, leaf spots develop but the concentration of flavonoid glycosides is unchanged. However, the corresponding aglycones are released in the leaves three to four days after inoculation and increase in concentration through the seventh day (Olah and Sherwood, 1971). This change is correlated with increasing β-glucosidase activity. The pH optima of the enzyme from healthy plants is 6.5 while that from the infected leaves or from pure cultures of *A. imperfecti* is 5.0. In disc gel electrophoresis, the enzyme from diseased leaves corresponds to the glycosidase from the fungus (Olah and Sherwood, 1973). This suggests that the fungus invades the plant, produces a β-glucosidase which degrades the flavonoid glycosides, and the aglycones in turn inhibit the further spread of the fungus. A similar pattern operates in the resistance of certain varieties of *Malus pumula* to the fungus *Venturia inaequalis.* Phloridzin, a major flavonoid of most *Malus* cultivars, stimulates the growth of the fungus on defined media. However, its aglycone phloretin is markedly inhibitory to this organism (Barnes and Williams, 1961) and β-glucosidase accumulates in *Malus* leaves in response to infection (Noveroske *et al.,* 1964).

Induction of phenolase as well as of β-glucosidase may take place during microbial invasion of the plant and examples of increases in levels of β-glucosidase and/or phenolase in response to pathogenic attack are numerous (i.e. Fric, 1968; Gagnon, 1967; Hildebrand and Sands, 1966; Nye and Hampton, 1966; Raa and Overeem, 1968; Seevers and Daly, 1970).

It should not be expected that all plants respond to fungal attack in the same manner since there are susceptible and resistant varieties, and there must be many ingenious biochemical devices to ensure that a fungus can, or cannot, successfully invade and grow within a certain plant tissue. Many other unrelated types of secondary constituent (e.g. acetylenic esters) have been implicated as phyto-alexins in plants (Deverall, 1972; Kuc, 1972a; Ingham, 1972). In many cases where flavonoids are involved, they may simply be precursors on pathways parallel to the active constituents. For example, Walker and Stahmann (1955) show that *Coletotrichum circinans,* a soil fungus that lives on the dead outer scales of *Allium cepa* and parasitizes the living tissues, is resistant to the anthocyanins and flavonols of the outer scales but is inhibited by catechol and protocatechuic acid produced from these flavonoids by these scales.

18.4.3.4 Flavonoids in respiration and in photosynthesis

If mitochondria are isolated from plant tissues without taking precautions to reduce the levels of polyphenols, uncoupled and non-reactive preparations are obtained (Hulme and Jones, 1963; Stokes *et al.,* 1968). The classic inhibitor of mitochondrial phosphorylation is 2,4-dinitrophenol but Stenlid (1970), using several methods including the firefly luciferase method for direct measurement of ATP, has found that a few flavonoid aglycones are just as effective. He examined the effects of 43 flavonoids, natural and synthetic, on ATP production in mitochondria isolated from *Zea mays* coleoptiles, *Cucumus sativa* roots, or *Pisum sativum* roots and found that flavonoid aglycones with free 3- and 4'-positions were most effective. While there are apparently no reports of flavonoids naturally present in plant mitochondria, and the likelihood of their reaching the mitochondria in healthy tissues cannot yet be evaluated, it is quite likely that flavonoids and mitochondria would interact in parasitized or wounded tissues, especially if glycosidase activity were present.

One of the most effective inhibitors of oxidative phosphorylation in animal mitochondria is the flavonoid rotenone (23). In fact, Ernster *et al.* (1963) calculate that rotenone completely inhibits DPN-flavin linked electron transport at levels as low as 24 nmoles per

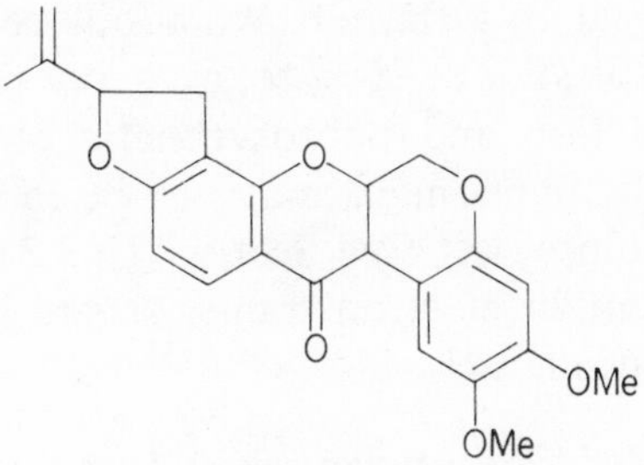

(23, Rotenone)

gram of mitochondrial protein. Rotenone is also a powerful inhibitor of the reduction of DPN and at 110 nmoles per gram of mitochondrial protein it inhibits this reaction about 87% (Löw and Vallin, 1963). The NADH-coenzyme Q reductase system and NADH-cytochrome *c* reductase in beef heart mitochondria are strongly inhibited by similar levels of rotenone (Merola *et al.,* 1963).

Wawra and Webb (1942) isolated a chalcone-protein complex from *Citrus limon* that could serve as an electron transporter in mammalian tissues and which they believed might function in a cyclic electron transport system to bypass cytochromes. While a cyclic system of flavonoid oxidation and reduction explains much of the complicated change that occurs during such commercial processes as, e.g. tea processing (Sanderson, 1972), it must be remembered that these changes occur during processing when the tissues are macerated and undergoing disintegration of subcellular organelles. Flavonoids will effectively alter levels of reduced cofactors and O_2 in cell free systems by uncoupling mitochondria

(Stenlid, 1970) or by serving as phenolase substrates (Challice and Williams, 1970). Thus oxygen consumption and cofactor oxidation in cell free systems is not evidence for a terminal oxidase system based on an o-diphenol $\rightleftharpoons$ o-quinone type of reaction. Hanson *et al.* (1967) concluded that there was little evidence to assign a major respiratory role to such a scheme, and no strongly supportive evidence has appeared since that time.

With the discovery of photosynthetic phosphorylation it was recognized that chloroplast isolated from *Spinaca oleracea* required a water-soluble constituent for a vigorous rate of ATP synthesis (Whatley *et al.*, 1955). Krogmann and Stiller (1962) examined these aqueous extracts and found an unidentified compound with spectral characteristics similar to a flavone. When quercetin or its glycosides were added to the plastids at 10^{-5} M, in lieu of the aqueous fraction, both the Hill reaction and photosynthetic phosphorylation were enhanced. While the phenomena may not be related, the occurrence of flavonoids in chloroplast (see Section 18.3.3) coupled with their possible involvement in electron transport, are interesting problems for future investigation.

18.4.3.5 Flavonoids in morphogenesis and sex determination in
          plants

If flavonoids initiate, interact with, or inhibit the growth of plants then morphological changes will be observed. These changes may, in turn, indirectly alter the time or sequence of reproductive development.

When Bastin (1966) examined rooting of hypocotyls excised from *Impatiens balsamina* seedlings he found a linear correlation between the number of roots produced, the level of endogenous auxin, and the level of anthocyanin. He suggested that wounding enhanced the synthesis of anthocyanin which, in turn, inhibited IAA oxidase and allowed a fuller expression of auxin in root initiation. Poapst *et al.* (1970) took stem sections of *Phaseolus vulgaris*, placed them in solutions of various flavonoids at 1 to 3 mg l^{-1}, and followed lateral root initiation. Phloridzin, its aglycone phloretin, and quercetin, were mildly effective initiators. However, oxidation of the flavonoids by bubbling O_2 through the test solutions for 5 to 30 min or by adding mushroom tyrosinase, yielded much more effective products. The authors conclude that oxidation products of the flavonoids controlled the IAA oxidase system and that auxin levels controlled rooting. However, these data are subject to other interpretations. For example, Stafford (1968) examined the relationship between adventitious roots and the levels of cyanidin, luteolinidin and apigeninidin accumulating in isolated first internodes of *Sorghum vulgare*. She found that root initiation was not always correlated with the accumulation of anthocyanins, and that subsequent growth of the roots was frequently inversely correlated with anthocyanin content. She also points out two critical anatomical limitations. First, the anthocyanins are produced initially in the outer part of the cortex while the adventitious roots arise from cells just inside the endodermal layer surrounding the central vascular stele. Secondly, analysis of the phenolic compounds of whole tissues fails to distinguish between those accumulated in the vacuole, and the (apparently) small amounts that may be in compartments within the cytoplasm where regulation and interactions may be expected to occur (Stafford, 1968).

Tendrils of *Pisum sativum* contain high concentrations of flavonoids, mainly quercetin 3-triglucosyl *p*-coumarate (QGC) and this compound is most abundant near the highly responsive apex of the tendril (Jaffe and Galston, 1968). After mechanical stimulation and during coiling of the tendril, the QGC concentration drops, the kinetics of QGC disappearance being correlated with the kinetics of coiling. Either aqueous extracts of unstimulated *Pisum* tendrils or 10 μM solutions of QGC inhibit contact coiling of excised tendrils while extracts of coiled tendrils do not. Jaffe and Galston (1968) suggest that flavonoids may intervene at at least two places in contact coiling: (1) An actomysin-like contractile ATPase is involved in coiling and QGC may control the activity of this enzyme; (2) since QGC is an effective inhibitor of IAA oxidase in *Pisum* tissues, altered flavonoid levels would change levels of auxin. However, co-ordinated growth responses of this sort may also depend on other experimental factors. At least one other attempt to correlate endogenous or exogenous QGC levels with coiling in *Pisum* tendrils has been unsuccessful (T. Swain, personal communication).

In 1944 Kuhn *et al.* reported that isorhamnetin (quercetin 3'-methyl ether) was an extraordinarily effective sex hormone in the unicellular green alga *Chlamydomonas eugametos*. Reports of flavonoids as sex hormones were also made for peonidin in *Chlamydo-*

monas (Moewus, 1951) and for rutin in both *Chlamydomonas* and *Forsythia intermedia* (Moewus, 1950a). Working with 26 mutant clones of *Chlamydomonas,* an elaborate biosynthetic scheme for quercetin was published (Birch *et al.,* 1953).

The work on flavonoids in *Chlamydomonas* is no longer generally accepted. Subsequent workers have neither found flavonoids in this alga nor have they been able to confirm their effectiveness as sex hormones. Ryan (1955) invited Moewus to work in his laboratory but could not obtain substantial confirmation of the original claims. The other claims are also probably bogus. When Hartshorne (1958) examined anthocyanins from both male and female flowers of several plants he could find no significant differences. Druzhkov (1972) reported that there was no relationship (cf. Moewus, 1950b) between rutin levels and short or long styled flowers in *Fagopyrum sagittatum* or *F. emarginatum.*

18.4.4 Pharmacological and clinical applications

The vast majority of flavonoids are completely innocuous in the diet (Swain and Bate-Smith, 1962). Physiological activities of flavonoids, and of plant phenolics in general, appear to arise most commonly if natural barriers or detoxification mechanisms are overloaded by amount, circumvented by the manner of administration, or foiled by uncommon compounds such as methylenedioxy ethers or isoprenoid structures (Singleton and Kratzer, 1969). Singleton and Esau (1969), in summarizing a vast amount of information on physiological activities of polyphenols on animals, conclude that while flavonoids may have a number of generally beneficial physiological effects on animals, they are not equally physiologically active. Furthermore, flavonoids in effective doses generally lack any deleterious effects.

There are, however, two major questions. Are flavonoids valuable dietary constituents? Is the presence of flavonoids in plants of significance in plant-animal interactions? The involvement of flavonoids in the pollination phenomena seem clear (Section 18.4.2). Their role in the diet of man and insects has generated many reports and much misunderstanding.

18.4.4.1 Taste and allied sensations

Bate-Smith (1972), in considering attractants and repellents in higher animals, reports that taste preferences seem to be quite similar in man and a number of animal tasters. For example, hedgehogs, rats and humans detect bitterness to an equal degree and it is equally repellent to all species. However, a taste for tannin such as the condensed tannins of coffee and tea, seems to be a particularly human trait not shared by grazing animals or insects (Morton, 1972). In fact, bird resistance in *Sorghum vulgare* is apparently due to astringent leucoanthocyanins of the aleurone (Harris *et al.*, 1970). Relations between structure and taste of flavonoids are also considered in Chapter 11.

In the area of plant-insect relationships, the palatibility of flavonoids is quite important. Fraenkel (1969) contends that host selection through secondary plant substances by insects is the key to agricultural entomology. In a review on the co-evolutionary patterns between butterflies and plants, Ehrlich and Raven (1964) conclude that flavonoids and other secondary substances play the leading role in determining the kind of plants upon which the larvae feed. Since insects are the most numerous of all animals and few plants are free from attack by some insect, it is not surprising that co-evolutionary patterns have developed which make a given flavonoid attractive to one insect, repugnant to another, toxic to a third and completely uninteresting to the rest (Table 18.8).

Several non-visual methods by which insects detect flavonoids have been reported. The caterpillar of *Pieris brassicae*, like all lepidopterous larvae, has on each maxilla two chemoreceptive hairs.

Table 18.8   *Selected examples of flavonoids in insect diets that serve as feeding stimulants, feeding deterrents or toxins*

| Insect | Dietary flavonoid | | | |
	Morin	Quercitrin	Rutin	Isoquercetin
Anthonomus grandis (Col.)[1]	deterrent	stimulant	deterrent	stimulant
Bombyx mori (Lep.)	stimulant	deterrent	deterrent	stimulant
Heliothis virescens (Lep.)	toxic	toxic/stimulant[2]	toxic/stimulant	toxic/stimulant
Hypantria cunea (Lep.)	stimulant	–	–	–
Pectinophora gossypiella (Lep.)	toxic	toxic	toxic	toxic

(*Data from* Schoonhoven, 1972; original references cited therein.)
[1] Col., Coleoptera; Lep., Lepidoptera.
[2] Both toxicity and stimulatory activity to feeding have been reported.

With two receptor cells on these hairs it detects mustard oil glycosides and a third cell is stimulated by flavonoids such as pelargonidin, malvidin or cyanidin (Schoonhoven, 1972). The silkworm *Bombyx mori* has a very sensitive and discriminating receptor cell for flavonoids; quercetin 3-rhamnoside is stimulating to silkworm feeding while quercetin 3-glucoside is inactive (Ishikawa and Hirao, 1965).

Once consumed by the insect, flavonoids may be detoxified since a phenolase appears to be an invariable component of the saliva of all phytophagous bugs as well as some of the carnivorous ones (Miles, 1968). Such dietary habits of the larvae often leave telltale traces in the adult insect. Flavonoids have been identified from the wings of a number of butterflies (Feltwell and Valadon, 1970; Morris and Thomson, 1963) and moths (Ford, 1941), from the cocoon of the silkworm (Fujimoto and Hayashia, 1960) and even in the hairs of the bumblebee (Stein, 1961).

18.4.4.2 Antibiotic effects

Powers (1964) studied the effects of 24 anthocyanins, leucoanthocyanins and phenolic acids on *Salmonella typhosa, S. enteritidis, Shigella paradysenteriae, Staphylococcus aureus, Aerobacter aerogenes, Proteus vulgaris, Escherichia coli, Lactobacillus casei, L. acidophilus,* and a putrefactive anaerobe. Most of the anthocyanins and leucoanthocyanins inhibited respiration and reproduction at the level of one or two μmoles if glucose were present in the medium. In the absence of glucose the flavonoids were metabolized. Of more than 20 flavonoids investigated, no compound was devoid of inhibitory activity toward one or more of the ten bacteria studied. Antihelmitic activities have also been reported for flavonoids. Laliberte *et al.* (1967) evaluated 116 chalcones and analogues on pinworms in mice. The chalcones were generally effective, especially those with few hydroxyl substituents.

Flavonoids of many types have antiviral effects in animals. In human cell lines (HeLa cells) *Herpesvirus hominis* is inhibited by quercetin at levels of 300 μg ml^{-1} (Pusztai *et al.*, 1966) but not by rutin (Beladi *et al.*, 1965) or dihydroquercetin (Bakay *et al.*, 1968).

When quercetin was added to cultures of several viruses associated with human maladies, viruses with an envelope were inhibited while those lacking such an envelope were moderately, or completely, resistant (Puszatai *et al.*, 1966). Anthocyanins at high concentrations may also confer resistance to the tobacco mosaic virus in *Lycopersicon esculentum* (Ulrychova *et al.*, 1973) and in *Nicotiana tabacum* (Ulrychova and Brcak, 1967).

18.4.4.3 Flavonoids and malignancy

With the extensive screening programmes of plant products for anti-cancer drugs, it is not surprising that claims have been made that flavonoids may contribute to, or be effective in combatting, certain types of cancer. For example, Morton (1972) presents circumstantial evidence to associate human oesophogeal cancer to the consumption of excessive amounts of condensed tannins characteristic of astringent beverages such as tea or coffee.

Several flavonoids are moderately effective against laboratory cultures of malignant cells. Eupatin (24) and eupatoretin (25) (Kupchan *et al.*, 1969) and either centaureidin (26) or 6-demethoxy-centaureidin (Kupchan and Bauerschmidt, 1971) are all moderately effective against a carcinoma from the nasopharynx.

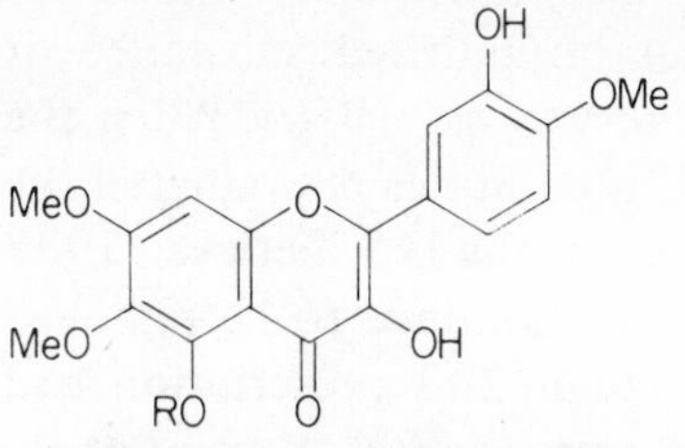

(24, Eupatin R = H)
(25, Eupatoretin R = Me)

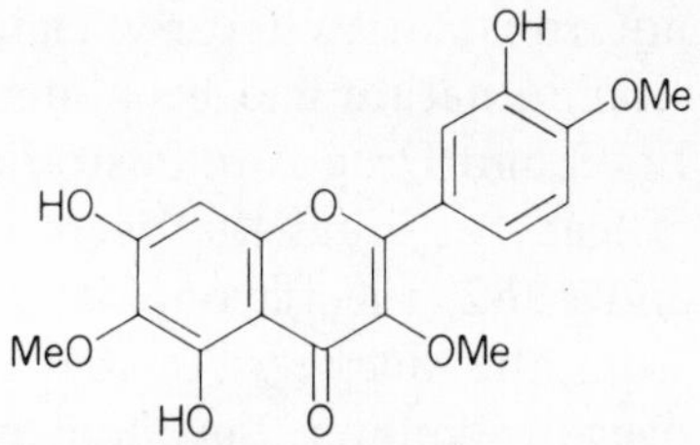

(26, Centaureidin)

Dittman *et al.* (1972), using manometric techniques to measure glycolysis and anaerobic respiration of human brain tumour slices, noted an increase in glycolysis as malignancy progressed, and quercetin and its glycosides were weakly inhibitory to these tissues. Wattenberg and Leong (1970) found that quercetin pentamethyl ether and rutin were strongly and moderately effective inhibitors respectively of benzo(a)pyrene induced pulmonary adenoma in mice. They suggested that the flavonoids induced a benzy(a)pyrene hydroxylase which detoxified the carcinogen.

18.4.4.4 Vitamin P and bioflavonoids

In 1936 Rusznyak and Szent-Gyorgyi provided a sample of impure ascorbic acid to a physician for administration to a patient suffering from subcutaneous capillary bleeding. The patient was cured. Later, they provided a more purified preparation which had no effect on other patients suffering from the same malady. Returning to their impure fractions, they found contaminants with the characteristics of flavones in preparations from the fruit of both *Capsicum annuum* and *Citrus limon*. Laboratory experiments showed that periodic intravenous injections of these 'bioflavonoids' restored normal capillary resistance in a fortnight. They called this preparation Vitamin P (Rusznyak and Szent-Gyorgyi, 1936) and claimed that it reduced haemorrhages, extended the effects of ascorbic acid, and reduced vascular purpurea. Subsequent research has emphasized the therapeutic effects of several flavonoids, primarily rutin, its methylated derivatives, and the flavanones from *Citrus* fruits (Kefford and Chandler, 1970; Griffith *et al.,* 1955).

American science did not take kindly to Vitamin P and the name was eventually dropped from the roles of effective preparations by the American Society of Biological Chemists and by the American Institute of Nutrition on the grounds that subsequent studies had not substantiated early claims and that no purified substance of vitamin nature had been identified (Vickery *et al.,* 1950). When the Food and Drug Administration of the United States re-evaluated the efficacy of drugs that were first marketed in the U.S. between 1938 and 1962, the flavonoids were the first compounds to be removed from the market. At that time, more than 200 prescription and over-the-counter flavonoid preparations were in use (Meyers *et al.,* 1972).

DeEds (1968), in attempting to summarize almost a thousand research reports of the physiological effects of flavonoids, concluded that flavonoids with free hydroxyls at the $3'4'$-positions exert beneficial physiological effects on the capillaries through: (1) chelating metals and thus sparing ascorbate from oxidation; (2) prolonging epinephrine action by the inhibition of O-methyltransferase; and (3) stimulating the pituitary-adrenal axis. Robbins (1973) and Srinivasan *et al.* (1971) have presented evidence that flavonoids may play another important role in the circulatory system by acting on the aggregation of erythrocytes.

These therapeutic effects may involve more than the capillary wall. Vogel (1971) points out that although capillary permeability may be altered by the administration of various flavonoids, the test most commonly applied is that of capillary resistance. This test is normally made by applying a slight vacuum to an epidermal region and noting colour changes of the skin as blood leaves the capillaries. Vogel (1971) points out that changes in capillary resistance are more likely a property of the total tissue in which the capillaries are embedded. Thus the lymphatic system and the movement of solutes from the surrounding tissues must be considered in the overall response. Flavonoids reported effective in increasing capillary resistance include flavanones, flavonols, isoflavones, catechin, flavandiols, and chalcones (Gabor, 1971). Hendrickson and Kesterson (1964) tabulate 52 diseased states that respond to flavonoids that have the property of increasing capillary resistance. Martin (1954) provides an attractive explanation for the complexity of these responses by suggesting that there is no diseased state which will not benefit by assuring proper capillary strength.

In an attempt to reconcile claims for the beneficial effects of flavonoids in the human diet, Szent-Gyorgyi (1972) points to contrasting dietary habits in different countries and Gabor (1972) contends that the beneficial effects of flavonoids are often apparent only after prolonged treatments of a month or more. Gabor (1972) also warns that if flavonoids are used indiscriminately in attempts to correct many types of vascular damage without taking into consideration the cause of the disorder, contradictory data are inevitable.

Some work has been done with animals raised on artificial diets free of, or very deficient in, flavonoids. Benko et al. (1970) report severe brain edema and subpleural haemorrhages in rats kept on a diet deficient in flavonoids. When rats on the flavonoid-deficient diet were given 50 mg kg^{-1} of rutin derivatives or hesperidin every other day, the maladies were significantly reversed.

In order for dietary flavonoids to have a biological effect they must survive the digestive enzymes, perhaps the intestinal flora, and be transported to active sites. Shutt et al. (1967) have found both aglycones and glucuronides of the isoflavones formononetin, genistein and biochanin A at levels up to 40 μg l^{-1} in the blood serum of sheep feeding on oestrogenic clover. Unfortunately, similar measurements have apparently not been reported for man.

18.4.4.5 Toxic flavonoids

There are apparently no reports of flavonoid toxicity in man, although Ogiso et al. (1972) have recently reported isolation of the macrocyclic flavonoid glycosides poriolide (27) and isoporiolide (28)

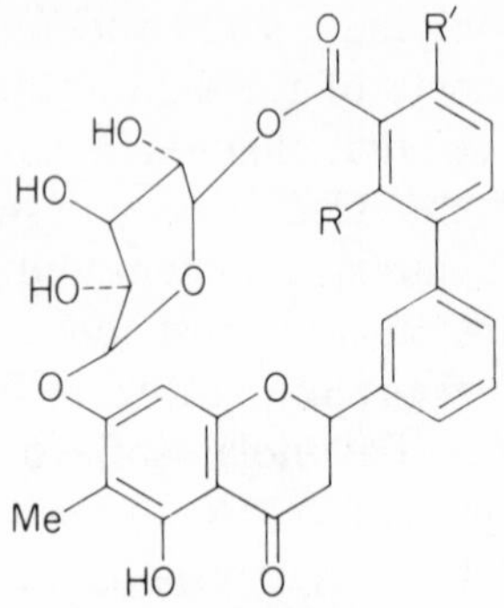

(27, Poriolide R = H, R′= OH)
(28, Isoporiolide R = OH, R′= H)

from *Leucothoe keiskei* which have a LD_{50} of 1 mg kg^{-1} when injected intravenously into mice.

The rotenoids have a low toxicity to mammals (i.e. LD_{50} of 3 g kg^{-1} for rotenone in rabbits), but are extremely potent inhibitors of mitochondrial oxidation in insects and in fish at levels in the range of 1 μg (Fukami and Nakajima, 1971). As rotenone degradation in mice and in houseflies proceeds at least in part through the same pathways *in vitro*, the reasons for the selective toxicity of rotenone are not yet understood (Fukami *et al.*, 1967).

Flavones and isoflavones are considered by Anjanexulu and Ramachandrarow (1964) to be inherently toxic to fish, and Chari and Seshadri (1948) found that chrysin (5,7-dihydroxy-) and galangin (3,5,7-trihydroxy-flavone) are considerably toxic to fish while flavones having more hydroxyl groups are only slightly toxic. Methylation of a hydroxyl group increases the toxicity, presumably through increasing lipid solubility. Thus 7-methoxyflavone is about one tenth as toxic to fish as is rotenone and 5,6,7-trimethoxyflavone is toxic to the rice fish *Oryzias latipes* when added to the water at a concentration of 5 ppm (Chari and Seshadri, 1948).

18.4.4.6 Flavonoid oestrogens

For over 40 years it has been realized that certain forage plants may contain substances whose effects on animals are characteristic of animal oestrogens; i.e. initiating oestrus in immature animals or interfering with normal reproduction (Bickoff, 1968a). Breeding abnormalities in sheep became a major problem in Australia in the years following 1940. The major cause of this was traced to *Trifolium subterraneum* by Bennetts *et al.* (1946). Bradbury and White (1951) isolated the isoflavone genistein, in levels up to 0.7%, from leaves of the Dwalganup variety of *T. subterraneum*. In 1954 Biggers and Curnow showed that genistein was oestrogenic in mice, and the infertility syndrome in sheep which became known as the 'clover disease' was related to this compound (Bickoff, 1968a).

Oestrogenic activity in *Medicago sativa* and *Trifolium repens* had been noted in the United States and Bickoff *et al.* (1957) isolated the potent oestrogen coumestrol from these plants and showed that it was from 30 to 100 times more oestrogenic than the isoflavones from *T. subterraneum*. In the discussion of a paper by Biggers (1959), Whalley suggests that the stilbene-like structure of these compounds might explain their oestrogenic activity and that they are analogous to such potent synthetic oestrogens as diethylstilboestrol (29).

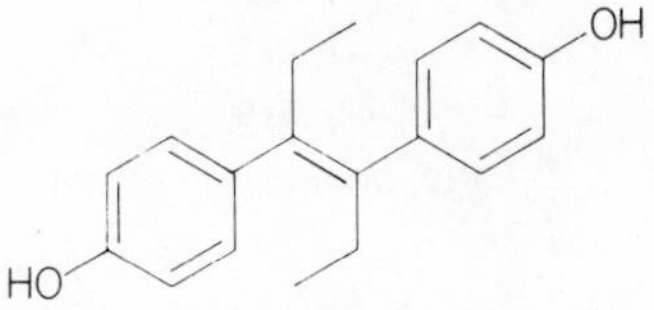

(29, Diethylstilboestrol)

There appears to be an isoflavone tolerance mechanism in sheep since Morley *et al.* (1969) report that ewes can become non-responsive to the effects of genistein or biochanin A but not to formononetin, after prolonged feeding on *T. subterraneum*. Also, when the effects of these flavonoids on the oestrogenic cycle is overlooked, they generally have no ill effects on livestock and in fact improve the commercial carcass quality of lambs (Johnston *et al.*, 1965).

Pretorius *et al.* (1958) studied the oestrogenic activity of a number of other classes of flavonoids and found that quercetin and kaempferol 3-rhamnosylgalactoside-7-rhamnoside had a small amount of activity. They calculated that 50 mg of these compounds had the equivalent activity of 0.034 μg of diethylstilboestrol. Bickoff (1968b) reports that tricin has slight oestrogenic activity and that a compound formed by the alkaline hydrolysis of quercetin, but not quercetin itself (Pretorius *et al.*, 1958), is weakly effective.

In contrast, Yamashita (1965) examined several flavonoids in a bioassay for oestrogenic activity based on carbonic anhydrase levels in an isolated rabbit uterus challenged with progesterone. The flavonoids alone had no effect on the response and were considered non-oestrogenic. However, when the flavonoids were administered with progesterone, hesperetin and morin (30) in 1 to 10 mg levels blocked the progesterone effects. When these flavonoids were given by stomach tube they had no effect on the assay. However, one must recognize that these flavonoids may be non-specific enzyme inhibitors in such experimental systems and that assays with isolated organs from laboratory animals may bear no relation to the response of a ruminant (Bickoff, 1968a).

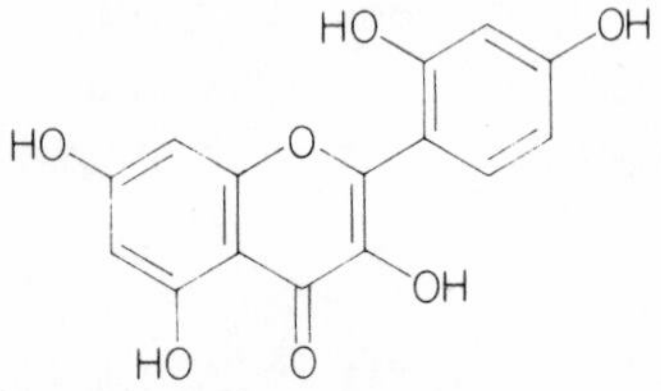

(30, Morin)

18.4.5 Conclusions regarding the biological functions of flavonoids

What would be the implications of the loss of the ability to make a certain class of functional flavonoid in a large group of plants? The answer seems to be provided in the case of the ten families of the order Centrospermae, where the anthocyanins have been replaced by betacyanins. A representative example of a betacyanin is betanidin (31), a red pigment from *Beta vulgaris* (Mabry *et al.,* 1972). The flavone and flavonol glycosides of the betalain-containing plants are the same as those in any other plant group (Mabry *et al.,* 1972). Any function such as pollination attractiveness that is attributable to anthocyanins in other families appears to have been specifically replaced by betalains in these plants. The physiological factors that control betacyanin accumulation in the Centrospermae are very similar to those controlling anthocyanins in other families. This similarity of accumulation in response to stimuli, the distinctiveness of their biochemical pathways and their apparently complete exclusiveness of occurrence within the plant kingdom, all suggest that there is a functional role for a red, water soluble pigment that

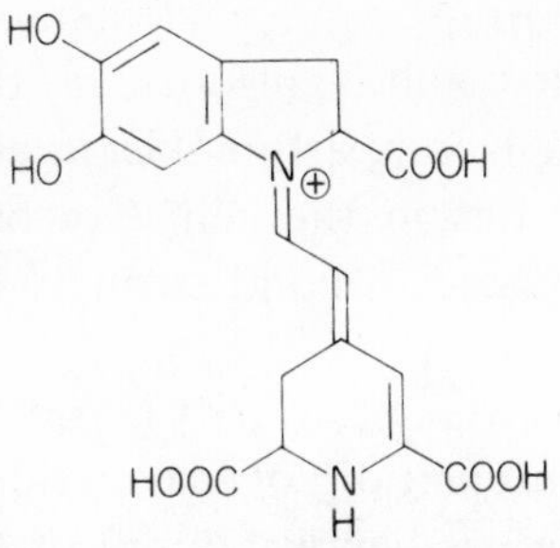

(31, Betanidin)

may be satisfied by anthocyanins in some families and betacyanins in others (Birch, 1973).

It is useful to consider the potential biological reactivity of flavonoids with reference to their chemistry; their solubility, absorptivity, relative resonance contributions, ionization constants and reactivity with heavy metals. With this in mind, it is obvious that flavonoids differing only slightly in structure and substitution may to varying degrees react, fail to react, or interfere with a biological receptor site such as a membrane, a particular enzyme, or another low molecular weight compound. When these chemical variables are placed in perspective with the biological variables, the difficulty of trying to fit all of the flavonoids, or even certain classes, into any rigorous functional role is obvious.

Rather than attempt the onerous, and presumptive, task of summarizing the biological functions of flavonoids, the author contacted several active researchers, who were asked to provide vignettes summarizing their understanding and insights in the area of research that they best understood. These summaries, with a few introductory comments, are presented below.

The role of flavonoids in plant growth and development at the subcellular level, at least as determined by the application of exogenous flavonoids, focuses on the interactions between inhibition and promotion. These, in concert, imply regulatory roles. Goran Stenlid (Royal Agricultural College, Uppsala) writes: 'Flavonoids may be important regulators of respiration as many types act as uncouplers and inhibitors of ATP formation. They may be oxidized by various enzymes, and quinonoid substances formed can e.g. influence the level of ascorbic acid. Their chelating properties and interaction with SH groups can affect membranes and enzymes and indirectly permeability and transport in the cell. A crucial and unsettled point is their compartmentation and to what extent they really reach various sites in the cell. Glycosylation, hydroxylation pattern and type of C-2—C-3 bond strongly affect the physiological activities'.

Arthur Galston (Yale University), one of the more active spokesmen on the interaction of flavonoids and auxins, considers them in the following perspective: 'In etiolated peas, red light (P_{fr}) promotes synthesis of all flavonoids, especially quercetin, through induction of PAL and cinnamic hydroxylases. Gibberellins inhibit P_{fr} action in this system. Since kaempferol is a cofactor of IAA oxidase and quercetin an inhibitor, red light greatly decreases apparent IAA oxidase activity. It is not known whether the altered flavonoid or IAA oxidase patterns are related to morphogenetic changes initiated by red light'.

At the level of whole plant responses, a complex pattern has been determined for specific tissue interactions in response to light and mechanical stimulation. On this matter, Josette Tronchet (University of Besançon) has this to say: 'The level of flavonoids accumulating in epidermal cells is controlled by light and they, in turn, provide a screen against ultraviolet light. The same flavonoid glycosides are found in all epidermal layers of organs exposed to the light but they are often different in other tissues. These epidermal flavonoids are resistant to inclement weather but responsive to shock; quercetin derivatives often disappear while kaempferol derivatives are unaltered or even accumulate. Under certain conditions corollas may accumulate flavonoids characteristic of epidermal layers. Pollen and stigmas normally show a total disparity of flavonoids, perhaps in a regulatory fashion'.

The most apparent role of flavonoids is one of high visibility and attractiveness to bird and insect pollinators. However, they appear to function in more involved responses than simply announcing the availability of food. As Joseph Arditti (University of California, Irvine) puts it: 'Bird-pollinated flowers are predominantly red, whereas insect-pollinated ones are more often blue or may reflect ultraviolet light. Both are normally coloured by flavonoids. In addition to making flowers conspicuous by solid colours, flavonoids often form intricate patterns such as nectar guides or images resembling other organisms (e.g., prey, enemies, copulation partners, etc.) which attract pollinating vectors. Distribution patterns or content of flavonoids within floral tissues may change after pollination and alter or obliterate the visual clues. This may render the flowers no longer attractive to a pollinator. In plants which depend on very specific pollination vectors and/or where the number of pollinated flowers may be small, prevention of a second, essentially "wasted," visit may have survival value. Thus, flavonoids probably serve to attract and repel pollinators'.

Great strides have been made in understanding the biochemistry and physiology of infectious plant diseases. In several instances, especially in fungal diseases of the Leguminosae, the role of flavonoids seems assured. I. A. M. Cruickshank (Division of Plant Industry, Canberra) writes: 'Phenolic compounds, including many flavonoids which occur as normal metabolites in uninfected plant tissues, accumulate at the site of infection. Correlations between varietal resistance and flavonoid content have been established in some instances. Recent studies, however, have emphasized that post-infectionally formed antifungal compounds, phytoalexins, which are abnormal metabolites of healthy tissues, are of primary significance in immunity and probably of major importance in varietal resistance. The isoflavonoids pisatin, phaseollin, 6-hydroxyphaseollin, medi-carpin and maackiain are included in this group'.

Beneficial effects of flavonoids in the human diet have been bitterly argued for at least two decades. In the United States the flavonoids were removed from the list of effective drugs on the grounds that no purified substances of vitamin nature had been identified (see p. 1034). The controversy is far from resolved, but recent discoveries may at least explain many of the early inconsistencies. R. C. Robbins (University of Florida, Gainsville) summarizes the present situation in this field. 'Flavonoids that exhibited beneficial effects on abnormal capillary permeability and fragility were once known as vitamin P. After failure to establish these compounds as vitamins, they were termed bioflavonoids, and subsequent research showed that they exerted beneficial but inconsistent effects in more than 50 diseases. Recent research has revealed that flavonoids act on blood cell aggregation, a phenomenon that generally accompanies illness and injury. Action of flavonoids on aggregation is consistent with beneficial effects on capillaries and in disease, since aggregation impairs the microcirculation and in itself enhances symptoms of disease and induces pathology. Flavonoids evidently reduce aggregation by membrane surface effects and there is evidence of specificities in action, which appears to explain the inconsistent effects. Recent research shows that flavonoids with multiple methoxyl or ethoxyl groups are effective inhibitors of blood cell aggregation. Present evidence indicates medicinal use of flavonoids under conditions where aggregated erythrocytes or platelets occur and promote a variety of pathological effects'.

Acknowledgments

The author is extremely grateful to the many researchers who provided reprints, prepublication copies of their manuscripts, personal communications, or vignettes for inclusion in the summary section.

References

Addicott, F. T. and Lyon, J. L. (1969), *A. Rev. Pl. Physiol.* **20**, 139.

Alfermann, W. and Reinhard, E. (1971), *Experientia* **27**, 353.

Alibert, G., Ranjeva, R. and Boudet, A. (1972), *Biochem. biophys. Acta* **279**, 282.

Allen, M. B. (1964), In *Photophysiology* (A. C. Giese, ed.), Vol. 1, p. 83–110. Academic Press, New York and London.

Alston, R. E. (1959), *Genetica* **30**, 261.

Amrhein, N. and Zenk, M. H. (1971), *Z. Pflanzenphysiol.* **64**, 145.

Anjanexulu, A. S. R. and Ramachandrarow, L. (1964), *Symp. Syn. Heterocyclic Comp. Physiol. Interest* (Hyderabad, India) p. 47, [*Chem. Abstr.* **68**, 77167].

Arditti, J., Flick, B. and Jeffrey, D. (1971), *New Phytol.* **70**, 333.

Arnold, A. W. and Albert, L. S. (1964), *Pl. Physiol. Lancaster* **39**, 307.

Arnold, A. W. and Alston, R. E. (1961), *Pl. Physiol. Lancaster* **36**, 650.

Arnon, D. I. (1949), *Pl. Physiol. Lancaster* **24**, 1.

Asen, S., Stuart, N. W. and Siegelman, H. W. (1959), *Proc. Am. Soc. Hort. Sci.* **73**, 495.

Attridge, T. H., Stewart, G. R. and Smith, H. (1971), *FEBS Lett.* **17**, 84.

Ayers, J. and Mancinelli, A. L. (1969), *Pl. Physiol. Lancaster* **44**(S), 19.

Bagni, N. and Fracassini, D. S. (1966), *Experientia* **22**, 1.

Bailey, J. A. (1970), *J. Gen. Microbiol.* **61**, 409.

Bakay, M., Mussi, I., Beladi, I. and Gabor, M. (1968), *Acta. Microbiol. Acad. Sci. Hung.* **15**, 223.

Ball, E. A. and Arditti, J. (1972), *Am. J. Bot.* **59**(S), 671.

Bardzik, J. M., Marsh, H. V. Jr. and Havis, J. R. (1971), *Pl. Physiol. Lancaster* **47**, 828.

Barnes, E. H. and Williams, E. B. (1961), *Can. J. Microbiol.* **7**, 525.

Bartlett, D. J., Poulton, J. E. and Butt, V. S. (1972), *FEBS Lett.* **23**, 265.

Barz, W. and Adamek, C. (1970), *Planta* **90**, 191.

Barz, W., Adamek, C. and Berlin, J. (1970), *Phytochemistry* **9**, 1735.

Bastin, M. (1966), *Photochem. Photobiol.* **5**, 423.

Bate-Smith, E. C. (1972), In *Phytochemical Ecology* (J. B. Harborne, ed.), pp. 45–56, Academic Press, London and New York.

Beladi, I., Pusztai, R. and Bakai, M. (1965), *Naturwissenschaften* **13**, 402.

Benko, S., Gabor, M., Varkonyi, T., Antal, A. and Foldi, M. (1970), *Physiol. Chem. Phys.* **2**, 110.

Bennetts, H. W., Underwood, E. J. and Shier, F. L. (1946), *Aust. Vet. J.* **22**, 2.

Berlin, J. and Barz, W. (1971), *Planta* **98**, 300.

Bibb, P. C. and Hagen, C. W. Jr. (1972), *Am. J. Bot.* **59**, 305.

Bickoff, E. M. (1968a), *Oestrogenic Constituents of Forage Plants* Commonwealth Agr. Bur., Berkshire, England.

Bickoff, E. M. (1968b), *Am. Perfum. Cosmet.* **83**, 59.

Bickoff, E. M., Booth, A. N., Lyman, R. L. and others (1957), *Science N.Y.* **126**, 969.

Bidgood, J. (1905), *J. R. Hort. Soc. London.* **29**, 463.

Bieleski, R. L. (1960), *Aust. J. Biol. Sci.* **13**, 221.

Biggers, J. D. (1959), In *Pharmacology of Plant Phenolics* (J. W. Fairbair, ed.) pp. 51–70. Academic Press, London and New York.

Biggers, J. D. and Curnow, D. H. (1954), *Biochem. J.* **58**, 278.

Birch, A. J. (1973), *Pure Appl. Chem.* **33**, 17.

Birch, A. J., Donovan, F. W. and Moewus, F. C. (1953), *Nature* **172**, 902.

Blakely, L. M. and Steward, F. C. (1961), *Am. J. Bot.* **48**, 351.

Bloom, M. and Vickery, R. K. (1973), *Phytochemistry* **12**, 165.

Bokuchava, M. A., Shalamberidze, T. Kh. and Soboleva, G. A. (1970), *Dokl. SSSR* **192**, 1374.

Börner, H. (1959), *Contrib. Boyce Thompson Inst.* **20**, 39.

Borthwick, H. A., Hendricks, S. B., Schneider, M. J., Taylorson, R. B. and Toole, V. K. (1969), *Proc. Natl. Acad. Sci. U.S.A.* **64**, 479.

Bradbury, R. B. and White, D. E. (1951), *J. chem. Soc.* 3447.

Briggs, W. R. and Rice, H. V. (1972), *A. Rev. Pl. Physiol.* **23**, 293.

Briggs, W. R., Rice, H. V., Gardner, G. and Pike, C. S. (1972). In *Recent Advances in Phytochemistry* (eds. V. C. Runeckles and T. C. Tso), Vol. 5, pp. 35–50. Academic Press, New York and London.

Brown, B. R. (1967), In *Oxidative Coupling of Phenols* (W. I. Taylor and A. R. Battersby, eds.) pp. 167–202. Marcel Dekker, Inc., New York.

Bulakh, A. A. and Gordzinskii, D. M. (1970), *Fiziol. Rast.* **2**, 21.

Butler, G. W., Steemers, M. A. and Wong, E. (1967), *N.Z. J. Agric. Res.* **10**, 312.

Butler, L. W., Hendricks, S. B. and Siegelman, H. W. (1964), *Photochem. Photobiol.* **3**, 521.

Caldwell, M. M. (1971), In *Photophysiology* (A. C. Giese, ed.) Vol. 6, p. 131–177. Academic Press, New York and London.

Camm, E. L. and Towers, G. H. N. (1972), Book of abstracts, Phytochemical Society of North American meetings (Syracuse, New York) p. 12.

Carlin, R. M. and McClure, J. W. (1973), *Phytochemistry* **12**, 1009.

Carpenter, J. A., Wang, Y. P. and Powers, J. J. (1967), *Proc. S. Exp. Biol. Med.* **124**, 702.

Challice, J. S. and Williams, A. H. (1970), *Phytochemistry* **9**, 1261.

Chalutz, E. and Stahmann, M. A. (1969), *Photopathology* **59**, 1973.

Chari, N. N. and Seshadri, T. R. (1948), *Proc. Indian Acad. Sci. A***27**, 128.

Chong, C. and Brawn, R. I. (1969), *Can. J. Pl. Sci.* **49**, 513.

Chu, M. and Widholm, J. M. (1972), *Physiol. Plant.* **26**, 24.

Clements, C. A. B. and Anderson, L. (1961), *Ann. N.Y. Acad. Sci.* **136**, 339.

Coke, L. and Whittington, W. J. (1968), *J. Exp. Bot.* **19**, 295.

Combier, H. and Lebreton, P. (1968), *C. r. hebd. Seanc. Acad. Sci., Paris* **267**, 421.

Constabel, F., Shyluk, J. P. and Gamborg, O. L. (1971), *Planta* **96**, 306.

Corcoran, M. R., Geissman, T. A. and Phinney, B. O. (1972), *Pl. Physiol. Lancaster* **49**, 323.

Corgan, J. N. (1965), *Proc. Amer. Soc. Hort. Sci.* **86**, 129.

Corgan, J. N. and Peyton, C. (1970), *J. Amer. Soc. Hort. Sci.* **95**, 770.

Craker, L. E. (1971), *Hort. Sci.* **6**, 137.

Craker, L. E. and Feder, W. A. (1972), *Hort. Sci.* **7**, 59.

Craker, L. E., Standley, L. A. and Starbuck, M. J. (1971), *Pl. Physiol. Lancaster* **48**, 349.

Craker, L. E. and Wetherbee, P. J. (1973), *Pl. Physiol. Lancaster* **51**, 436.

Creasy, L. L. (1968a), *Phytochemistry* **7**, 1743.

Creasy, L. L. (1968b), *Proc. Am. Soc. hort. Sci.* **93**, 716.

Creasy, L. L., Maxie, E. C. and Chichester, C. O. (1965), *Phytochemistry* **4**, 517.

Creasy, L. L. and Swain, T. (1966), *Phytochemistry* **5**, 501.

Cruickshank, I. A. M. and Perrin, D. R. (1965), *Aust. J. Biol. Sci.* **18**, 817.

Cumming, B. G., Hendricks, S. B. and Borthwick, H. A. (1965), *Can. J. Bot.* **43**, 825.

d'Arcy, A. (1970), These, Universite de Lyon, 124 pp.

Darwin, C. (1876), *The Movements and Habits of Climbing Plants,* 2nd ed., revised, Appleton and Co., New York.

Davies, M. E. (1972), *Planta* **104**, 50.

Dedio, W. and Clark, K. W. (1971), *Pestic. Sci.* **2**, 65.

DeEds, F. (1968), In *Comprehensive Biochemistry* (M. Florkin and E. H. Stotz eds.) Vol. 20, p. 127–171. Elsevier Publishing Co., Amsterdam.

Delaveau, P. (1964), *C. r. hebd. Seanc. Acad. Sci., Paris* **258**, 318.

Delaveau, P. (1967), *Physiol. veg.* **5**, 357.

Delaveau, P. and Paris, R. R. (1968), *C. r. hebd. Seanc. Acad. Sci., Paris* **267**, 317.

DeSwardt, G. H., Maxie, E. C. and Singleton, V. L. (1967), *S. Afr. J. agric. Sci.* **10**, 641.

Deverall, B. J. (1972), *Proc. R. Soc. London B.* **181**, 233.

Dittes, L., Rissland, I. and Mohr, H. (1971), *Z. Naturf.* **26b**, 1175.

Dittman, J. H., Herrmann, D. and Palleske, H. (1972), *Arzneim. Forsch.* **21**, 1999.

Downs, R. J. and Siegelman, H. W. (1963), *Pl. Physiol. Lancaster* **38**, 25.

Druzhkov, A. A. (1972), *Izv. Akad. Nauk Turkm. SSR, Ser. Biol. Nauk* **1**, 9. [*Chem. Abstr.* **77**, 2862].

Dugger, W. M. (1973), *Adv. in Chem.* 1973 **123**, 112.

Durmishidze, S. V., Meski, A. B. and Gotsiridze, A. A. (1970), *Nauk Gruz. SSR* **58**, 445 [*Chem. Abstr.* **73**, 65248].

Durst, F. and Duranton, H. (1970), *C. r. hebd. Seanc. Acad. Sci., Paris* **270**, 2940.

Edmondson, Y. H. and Thimann, K. V. (1950), *Archs. Biochem.* **25**, 79.

Egger, K., Wollenweber, E. and Tissut, M. (1970), *Z. Pflanzenphysiol.* **62**, 464.

Ehrlich, P. R. and Raven, P. H. (1964), *Evolution* **18**, 856.

Eisner, T., Silberglied, R. E., Aneshansley, D., Carrel, J. E. and Howland, H. C. (1969), *Science N.Y.* **66**, 1172.

Engelsma, G. (1968), *Acta Bot. Neer.* **17**, 499.

Engelsma, G. (1970), *Planta* **91**, 246.

Ernster, L., Dallner, G. and Azzone, G. F. (1963), *J. Biol. Chem.* **238**, 1124.

Esau, K. (1965), *Plant Anatomy,* John Wiley and Sons, Inc., New York and London.

Faust, M. (1965), *Proc. Am. Soc. hort. Sci.* **87**, 1; 10.

Feenstra, W. J., Johnson, B. L., Ribereau-Gayon, P., and Geissman, T. A. (1963), *Phytochemistry* **2**, 273.

Feldman, A. W., Hanks, R. W. and Collins, R. J. (1966), *Phytopathology* **56**, 1312.

Feltwell, J. and Valadon, L. R. G. (1970), *Nature* **225**, 969.

Firenzuoli, A. M., Vanni, P. and Mastronuzzi, E. (1969), *Phytochemistry* **8**, 61.

Ford, E. B. (1941), *Roy. Ent. Soc.* (London) Proc. **16**, 65.

Forrest, G. I. (1969), *Biochem. J.* **113**, 765.

Fraenkel, G. (1969), *Ent. Exp. Appl.* **12**, 473.

Francis, C. M. and Devitt, A. C. (1969), *Aust. J. agr. Res.* **20**, 819.

Fric, F. (1968), *Biologia* (Bratislava) **24**, 54.

Fujimoto, N. and Hayashiya, K. (1960), *Nippon Sanshigaku Zasshi* **29**, 495 [Chem. Abstr. 61, 9810].

Fukami, H. and Nakajima, M. (1971), In *Naturally Occurring Insecticides* (M. Jacobson and D. G. Crosby, eds.), pp. 71–98. Marcel Dekker, Inc., New York.

Fukami, J., Yamamoto, I. and Casida, J. E. (1967), *Science N.Y.* **155**, 713.

Furuya, M. (1968), In *Progress in Phytochemistry*, Vol. 1. (L. Reinhold and Y. Liwschitz, eds.), pp. 347–406. Interscience, London, New York and Sydney.

Furuya, M., Galston, A. W. and Stowe, B. B. (1962), *Nature* **193**, 456.

Furuya, M. and Thimann, K. W. (1964), *Arch. Biochem. Biophys.* **108**, 109.

Gabor, M. (1971), Intl. Flavonoid Symposium (Nyon-Friborg Switzerland) Plenary Lecture.

Gabor, M. (1972), *The Anti-inflammatory Action of Flavonoids,* Akademiai Kiado, Budapest.

Gagnon, C. (1967), *Can. J. Bot.* **45**, 2119.

Galston, A. W. (1969), In *Perspectives in Phytochemistry* (J. B. Harborne and T. Swain, eds.), pp. 193–204. Academic Press, London and New York.

Galston, A. W. and Baker, R. S. (1953), *Am. J. Bot.* **40**, 512.

Galston, A. W. and Satter, R. L. (1972), In *Recent Advances in Phytochemistry,* Vol. 5. (V. C. Runeckles and T. C. Tso, eds.), pp. 51–80. Academic Press, New York and London.

Gamborg, O. L., Constabel, F., LaRue, T. A. G., Miller, R. A. and Steck, W. (1970), *Colloques internationaux Can. natl. Res. Council* **193**, 335.

Geissman, T. A. (1956), *Archs. Biochem. Biophys.* **60**, 21.

Geissman, T. A. (1962), *The Chemistry of Flavonoid Compounds.* Macmillan, New York.

Gertz, O. (1906), *Studier of ver Anthocyan.* Lund, Sweden.

Gifford, A. M., Jr. and Stewart, K. D. (1968), *Am. J. Bot.* **55**, 269.

Gilchrist, D. G., Woodin, T. S., Johnson, M. L. and Kosuge, T. (1972), *Pl. Physiol. Lancaster* **49**, 52.

Ginzburg, C. (1967), *Bot. Gaz.* **128**, 1.

Givan, C. V. and Leech, R. M. (1971), *Biol. Rev.* **46**, 409.

Goldstein, J., Swain, T. and Tjhio, K. H. (1962), *Archs. Biochem. Biophys.* **98**, 176.

Grant, K. A. and Grant, V. (1968), *Hummingbirds and their Flowers.* Columbia University Press, New York.

Griffith, J. Q., Jr., Krewson, C. F. and Naghski, J. (1955), *Rutin and Related Flavonoids: Chemistry, Pharmacology, Clinical Applications,* Mack Publishing Co., Easton, Pennsylvania.

Grill, R. (1969), *Planta* **85**, 42.

Grill, R. (1972), *Planta* **108**, 185.

Grill, R. and Vince, D. (1965), *Planta* **67**, 122.

Grinkevich, N. I., Kovalskii, U. V. and Gribovskaya, I. F. (1967), *Chem. Abstr.* **74,** 84086.

Grochowska, M. J. (1966), *Pl. Physiol. Lancaster* **41**, 432.

Groenwald, E. G., Lee, P. and Zeevart, J. A. D. (1967), In *Plant Research, 1967. MSU/AEC Plant Research Laboratory*, pp. 25—26. Michigan State University Press, East Lansing, Michigan.

Groenwald, E. G. and Zeevart, J. A. D. (1973), unpublished results.

Grommeck, R. and Markakis, P. (1964), *J. Food Sci.* **29**, 53.

Hachtel, W. and Schwemmle, B. (1972), *Z. Pflanzenphysiol.* **68**, 127.

Hadwiger, L. A., Hess, S. L. and Broembsen, S. (1970), *Phytopathology* **60**, 332.

Hadwiger, L. A. and Schwochau, M. E. (1969), *Phytopathology* **59**, 223.

Hadwiger, L. A. and Schwochau, M. E. (1970), *Biochem. Biophys. Res. Comm.* **38**, 683.

Hadwiger, L. A. and Schwochau, M. E. (1971), *Pl. Physiol. Lancaster* **47**, 346; 588.

Hahlbrock, K., Ebel, J., Ortmann, R., Sutter, A., Wellmann, E. and Grisebach, H. (1971), *Biochem. Biophys. Acta* **244**, 7.

Halevy, A. H. and Shilo, R. (1970), *Physiol. Plantarum* **23**, 820.

Hamdy, M. K., Pratt, D. E., Powers, J. J. and Somaatmadja, D. (1961), *J. Fd. Sci.* **26**, 457.

Hanks, R. W. and Feldman, A. W. (1969), *Soil and Crop Sci. Soc. of Florida Proc.* **29**, 306.

Hanson, K. R., Zucker, M. and Sondheimer, E. (1967), In *Phenolic Compounds and Metabolic Regulation,* (B. J. Finkle and V. C. Runeckles, eds.), pp. 69—93. Appleton-Century-Crofts, New York.

Harborne, J. B. (1967), *Comparative Biochemistry of the Flavonoids,* Academic Press, London and New York.

Harborne, J. B. (1968), *Phytochemistry* **7**, 1215.

Harborne, J. B. (1972), In *Recent Advances in Phytochemistry* (V. C. Runeckles and J. E. Watkin, eds.), Vol. 4, pp. 107—141. Appleton-Century-Crofts, New York.

Hare, R. C. (1966), *Bot. Rev.* **32**, 95.

Hare, T. C. (1964), *Bot. Rev.* **30**, 129.

Harper, D. B., Austin, D. J. and Smith, H. (1970), *Phytochemistry* **9**, 497.

Harris, H. B., Cummins, D. G. and Burns, R. E. (1970), *Agron. J.* **62**, 633.

Hartmann, W. (1970), *Biochem. Physiol. Pflanzen* **161**, 472.

Hartshorne, J. N. (1958), *Nature* **182**, 1382.

1048 THE FLAVONOIDS

Hathway, D. E. (1959), *Biochem. J.* **71**, 533.

Hendricks, S. B. and Borthwick, H. A. (1965), In *Chemistry and Biochemistry of Plant Pigments* (T. W. Goodwin, ed.), pp. 405–439. Academic Press, London and New York.

Hendrickson, R. and Kesterson, J. W. (1964), *Technical Bulletin* 684, Florida Agric. Exp. Sta. Gainesville, Florida.

Hess, D. (1964), *Planta* **61**, 73.

Hess, D. (1966), *Z. Pflanzenphysiol.* **54**, 356.

Hess, D. (1967), *Naturwissenschaften* **54**, 289.

Hess, D. and Endress, R. (1972), *Z. Pflanzenphysiol.* **68**, 441.

Hess, S. and Hadwiger, L. A. (1971), *Pl. Physiol. Lancaster* **48**, 197.

Hildebrand, D. C. and Sands, D. C. (1966), *Phytopathology* **56**, 881.

Hillis, W. E. and Ishikura, N. (1969), *Aust. J. Biol. Sci.* **22**, 1425.

Hillman, W. S. and Galston, A. W. (1957), *Pl. Physiol. Lancaster* **32**, 129.

Holland, R. W. K. and Vince, D. (1968), *Nature,* **219**, 511.

Horovitz, A. (1969), *Diss. Abstr.* **70**, 21889.

Horovitz, A. and Cohen, Y. (1972), *Am. J. Bot.* **59**, 706.

Hulme, A. C. and Jones, J. D. (1963), In *Enzyme Chemistry of Phenolic Compounds* (J. B. Pridham, ed.), pp. 97–120. Pergamon Press, Oxford.

Hutchinson, A., Taper, C. D. and Towers, G. H. N. (1959), *Can. J. Biochem. Physiol.* **37**, 901.

Hyodo, H. and Yang, S. F. (1971), *Archs. Biochem. Biophys.* **143**, 338.

Ibrahim, R. K., Thakur, M. L. and Permannan, B. (1971), *Lloydia* **34**, 175.

Ingham, J. L. (1972), *Bot. Rev.* **38**, 343.

Ishikawa, S. and Hirao, T. (1965), *Bull. Sericulture Exp. Sta., Tokyo* **20**, 21.

Ivie, G. W. and Casida, J. E. (1971), *J. Agr. Fd. Chem.* **19**, 410.

Jablonowski, W. (1972), *Bromatol. Chem. Toksykol.* **5**, 27; (*Chem. Abstr.* **77**, 110428).

Jablonowski, W. and Jedrych, D. (1971), *Bromatol. Chem. Toksykol.* **4**, 161. [*Chem. Abstr.* **75**, 139552].

Jaffe, M. J. and Galston, A. W. (1968), *A. Rev. Pl. Physiol.* **19**, 417.

Johnston, W. K., Jr., Anglemier, A. F., Fox, C. W., Oldfield, J. E. and Sather, L. (1965), *J. Animal Sci.* **24**, 718.

Kaneta, M. and Sugiyama, N. (1971), *B. Chem. S. Japan* **44**, 3211.

Kang, B. G. and Burg, S. P. (1973), *Planta* **110**, 227.

Keen, N. T., Zaki, A. I. and Sims, J. J. (1972), *Phytochemistry* **11**, 1031.

Kefeli, V. I. and Turetskaya, R. Kh. (1967), *Wissen. Z. Univ. Rostock* **16**, 675.

Kefford, J. F. and Chandler, B. V. (1970), *The Chemical Constituents of Citrus Fruits,* Academic Press, New York and London.

Kendrick, R. E. (1972), *Planta* **102**, 286.

Kendrick, R. E. and Hillman, W. S. (1971), *Am. J. Bot.* **58**, 424.

Kevan, P. G. (1972), *Can. J. Bot.* **50**, 2289.

Kindl, H. (1970), *H. S. Zeit. Physiol. Chem.* **351**, 792.

Kindl, H. and Ruis, H. (1971), *Z. Naturf.* B **26**, 1379.

King, E. E. (1971), *Phytochemistry* **10**, 2337.

Kinnersley, A. M. and Davies, P. J. (1972), *Pl. Physiol. Lancaster* **49**, 63.

Klein, A. O. and Hagen, C. W., Jr. (1961), *Pl. Physiol. Lancaster* **36**, 1.

Klein, R. M. and Edsall, P. C. (1966), *Pl. Physiol. Lancaster* **41**, 949.

Kornelyuk, V. N. and Volynets, A. P. (1971). *DOKL Akad. Nauk. Beloruss.* SSR **15**, 949. [Chem. Abstr. 76, 55189].

Kosuge, T. (1969), *Ann. Rev. Phytopath.* **7**, 195.

Koukol, J. and Dugger, W. M., Jr. (1967), *Pl. Physiol. Lancaster* **42**, 1023.

Kozukue, N. and Ogata, K. (1972), *J. Fd. Sci.* **37**, 708.

Krause, J. and Reznik, H. (1972), *Z. Pflanzenphysiol.* **68**, 134.

Kreuzaler, F. and Hahlbrock, K. (1972), *FEBS Letters* **28**, 69.

Kreuzaler, F. and Hahlbrock, K. (1973), *Phytochemistry* **12**, 1149.

Krogmann, D. W. and Stiller, M. L. (1962), *Biochem. Biophys. Res. Comm.* **7**, 46.

Ku, P. K. and Mancinelli, A. L. (1972), *Pl. Physiol. Lancaster* **49**, 212.

Kuc, J. (1972a), *Ann. Rev. Phytopathology* **10**, 207.

Kuc, J. (1972b), *Microb. Toxins* **8**, 211.

Kuhn, R. and Löw, I. (1944), *Ber. Dt. Chem. Ges.* **77**, 202.

Kuhn, R., Moewus, F. and Löw, I. (1944), *Ber. Dt. Chem. Ges.* **77**, 219.

Kuilman, L. (1930), *Rec. Trav. Bot. Neerl.* **27**, 287.

Kupchan, S. M. and Bauerschmidt, E. (1971), *Phytochemistry* **10**, 664.

Kupchan, S. M., Siegel, C. W., Knox, J. R. and Udayamurthy, M. S. (1969), *J. Org. Chem.* **34**, 1460.

Lackmann, I. (1971), *Planta* **98**, 258.

Laliberte, R., Campbell, D. and Bruderlein, F. (1967), *Can. J. Pharm. Sci.* **2**, 37.

Lane, F. E. and Constantine, M. J. (1969), *Pl. Physiol. Lancaster* **44**(S), 16.

Lange, H. and Mohr, H. (1965), *Planta* **67**, 107.

Lawson, A. O., Akindele, B. B., Fasalojo, P. B. and Akpe, B. L. (1972), *Z. Pflanzenphysiol.* **66**, 251.

Lee, C. I. and Tukey, H. B. (1972), *J. Am. Soc. Hort. Sci.* **97**, 97.

Lee, S. and Aronoff, S. (1967), *Science N.Y.* **158**, 798.

Lees, D. H. and Francis, F. J. (1972), *J. Am. Soc. Hort. Sci.* **97**, 128.

Leonova, L. A. and Gamburg, K. Z. (1972), *Fiziol. Rast.* **19**, 71.

Leopold, A. C. and Plummer, T. H. (1961), *Pl. Physiol. Lancaster* **36**, 589.

Letan, A. (1966), *J. Fd. Sci.* **31**, 518.

Longevialle, M. (1969), *Bull. Soc. Bot. Fr.* **116**, 399.

Loomis, W. D. (1969), *Methods in Enzymology* **13**, 555.

Loomis, W. D. and Battaile, J. (1966), *Phytochemistry* **5**, 423.

Lott, H. V. (1960), *Planta* **55**, 480.

Löw, H. and Vallin, I. (1963), *Biochem. biophys, Acta* **69**, 361.

Mabry, T. J., Kimler, L. and Chang, C. (1972), In *Recent Advances in Phytochemistry,* Vol. 5 (V. C. Runeckles and T. C. Tso, eds.), pp. 105–134. Academic Press, New York and London.

Macleod, N. J. and Pridham, J. B. (1966), *Phytochemistry* **5**, 777.

Mapson, L. W. (1970). In *The Biochemistry of Fruits and their Products,* Vol. 1 (A. C. Hulme, ed.), pp. 369–384. Academic Press, New York and London.

Margna, U., Margna, E. and Otter, M. (1969), *Eesti NSV Tead. Akad. Toim., Biol.* **18**, 291.

Margna, U., Vainjarv, T. and Margna, E. (1972), *Eesti NSV Tead. Akad. Toim.*, *Biol.* 21, 141.

Marousky, F. J. (1968), *Proc. Am. Soc. Hort. Sci.* 92, 678.

Martin, G. (1954), *Exp. Med. Surg.* 12, 535.

Mashtakov, S. M., Volynets, A. P. and Kornelyuk, V. N. (1971), *Fiziol. Rast.* 18, 802.

Mayer, A. M. (1966), *Phytochemistry* 5, 1297.

McClure, J. W. (1968), *Pl. Physiol. Lancaster* 43, 193.

McClure, J. W. (1970), In *Phytochemical Phylogeny*, (J. B. Harborne, ed.), pp. 233–268. Academic Press, London and New York.

McClure, J. W. and Alston, R. E. (1964), *Nature* 201, 311.

McClure, J. W. and Alston, R. E. (1966), *Am. J. Bot.* 53, 849.

McClure, J. W. and Wilson, K. G. (1970), *Phytochemistry* 9, 763.

Meeuse, B. J. D. (1961), *The Story of Pollination*, Ronald Press, New York.

Meredith, F. I. and Young, R. H. (1968), *Proc. First Intl. Citrus Symp.* 1, 271. [*Chem. Abstr.* 74, 28849].

Merola, A. J., Coleman, R. and Hansen, R. (1963), *Biochem. Biophys. Acta* 73, 638.

Meyers, F. H., Jawetz, E. and Goldfien, A. (1972), *Review of Medical Pharmacology*, 3rd edition. Lange Medical Publications, Los Altos, California.

Miles, P. W. (1968), *A. Rev. Phytopath.* 6, 137.

Miller, C. O. (1969), *Planta* 87, 26.

Milthorpe, F. L. and Moorby, J. (1969), *A. Rev. Pl. Physiol.* 20, 117.

Mirande, M. (1922), *C. r. hebd. Seanc. Acad. Sci., Paris* 175, 496.

Moewus, F. (1950a), *Angew. Chem.* 62, 496.

Moewus, F. (1950b), *Biol. Zentr.* 69, 181.

Moewus, F. (1951), *Erg. Enzymforsch.* 12, 173.

Mohr, H. (1972), *Lectures on Photomorphogenesis*, Springer-Verlag, New York, Heidelberg, Berlin.

Mohr, H. and Bienger, I. (1967), *Planta* 75, 180.

Mohr, H. and Senf, R. (1966), *Planta* 71, 195.

Molina, J. A. E. and Alexander, M. (1967), *Can. J. Microbiol.* 13, 819.

Moore, K. and Pecket, R. C. (1972), *Annals Bot.* 36, 109.

Mori, S. and Noguchi, I. (1970), *Archs. Biochem. Biophys.* 139, 444.

Morley, F. H. W., Bennett, O. and Axelsen, A. (1969), *Aust. J. Exp. Agr. Anim. Husb.* 9, 569.

Morris, S. J. and Thomson, R. H. (1963), *J. Insect Physiol.* 10, 337; 391.

Morton, J. F. (1972), *Quart. J. Crude Drug Res.* 12, 1829.

Naef, J. (1971), *C. r. hebd. Seanc. Acad. Sci., Paris* 272, 1089.

Nagornaya, R. V. and Kotsur, N. V. (1970), *Nauk. Pr., Ukr. Silskogospod. Akad.* 31, 103. [Chem. Abstr. 75, 85372].

Neumann, J. and Avron, M. (1967), *Pl. Cell Physiol.* 8, 241.

Neyland, M., Ng, Y. L. and Thimann, K. V. (1963), *Pl. Physiol. Lancaster* 38, 447.

Ng, Y. L., Thimann, K. V. and Gordon, S. A. (1964), *Arch. Biochem. Biophys.* 107, 550.

Nishie, K., Waiss, A. C. and Keyl, A. C. (1968), *Photochem. Photobiol.* **8**, 223.
Noveroske, R. L., Williams, E. B. and Kuc, J. (1964), *Phytopathology* **54**, 98.
Nozzolillo, C. (1972), *Can. J. Bot.* **50**, 29.
Nye, T. G. and Hampton, R. E. (1966), *Phytochemistry* **5**, 1187.
Oaks, A. and Bidwell, R. G. S. (1970), *A. Rev. Pl. Physiol.* **21**, 43.
Oettmeier, W. and Heupel, A. (1972), *Z. Naturf.* **B 27**, 177.
Ogiso, A., Sato, A., Sato, S. and Tamura, C. (1972), *Tetrahedron Letters* **30**, 3071.
Oka, T. and Simpson, F. J. (1971), *Biochem. Biophys. Res. Commun.* **43**, 1.
Olah, A. F. and Sherwood, R. T. (1971), *Phytopathology* **61**, 65.
Olah, A. F. and Sherwood, R. T. (1973), *Phytopathology* **63**, 739.
Onslow, M. W. (1925), *The Anthocyanin Pigments of Plants*, 2nd edition. Cambridge University Press.
Overton, E. (1899), *Jahrb. Wiss. Bot. (Leipzig)* **33**, 171.
Pahlich, E. (1969), *Flora* (Jena) **158 B**, 443.
Parish, R. W. (1972), *Z. Pflanzenphysiol.* **66**, 176.
Parke, D. V. (1968), *The Biochemistry of Foreign Compounds*, Pergamon Press, Oxford.
Parkin, J. (1903), *Rep, Brit. Assn. London* **1903**, 862.
Parups, E. V. (1967), *Can. J. Biochem.* **45**, 427.
Perrin, D. R. and Cruickshank, I. A. M. (1965), *Aust. J. Biol. Sci.* **18**, 803.
Phillips, I. D. J. (1961), *Nature,* **192**, 240.
Pike, C. S. and Briggs, W. R. (1972), *Pl. Physiol. Lancaster* **49**, 514.
Pjon, C. J. and Furuya, M. (1967), *Pl. Cell Physiol.* **8**, 709.
Plotnikova, I. V., Runkova, L. V. and Ugolik, N. A. (1968), *Byull. Gl. Bot. Sada* **68**, 57. [Chem. Abstr. **69**, 75773].
Poapst, P. A., Durkee, A. B. and Johnston, F. B. (1970), *J. Hort. Sci.* **45**, 69.
Podstolski, A. and Lewak, S. (1970), *Phytochemistry* **9**, 289.
Poltis, J. (1959), *Bull. Torrey Bot. Club* **86**, 387.
Powell, R. D. and Lamprech, W. O. (1972), *Pl. Physiol. Lancaster* **49**, 16.
Powers, J. J. (1964), Proc. Fourth International Symposium on Food Microbiology (Goteborg, Sweden), pp. 59–75.
Pratt, D. E. (1965), *J. Fd. Sci.* **30**, 737.
Pratt, D. E. and Watts, B. M. (1964), *J. Fd. Sci.* **29**, 27.
Pretorius, P. J., Pieterse, P. J. and Homersma, P. J. (1958), *S. Afr. J. Lab. Clin. Med.* **4**, 289.
Pridham, J. B. (1963). (ed.), *Enzyme Chemistry of Phenolic Compounds*, Pergamon Press, Oxford.
Pringer, A. A. and Heinze, P. H. (1954), *Pl. Physiol. Lancaster* **29**, 467.
Psenak, M., Jindra, A. and Kovacs, P. (1970), *Biol. Plantarum* **12**, 241.
Pusztai, R., Beladi, I., Bakai, M., Musci, I. and Kukan, E. (1966), *Acta Micro. Acad. Sci. hung.* **13**, 113.
Raa, J. and Overeem, J. C. (1968), *Phytochemistry* **7**, 721.
Racusen, D. (1970), *Can. J. Bot.* **48**, 1029.
Rajaratnam, J. A., Lowry, J. B., Avadhani, P. N. and Corley, R. H. V. (1971), *Science* N.Y. **172**, 1142.

Raven, P. H. (1972), *Evolution* **26**, 674.

Reid, P. D. and Marsh, H. V., Jr. (1969), *Z. Pflanzenphysiol.* **61**, 170.

Reinert, J. H. C. (1964), *Naturwissenschaften* **51**, 87.

Reznik, H. and Menschick, R. (1969), *Z. Pflanzenphysiol.* **61**, 348.

Rice, E. L. and Pancholy, S. K. (1973), *Am. J. Bot.* **60**, 691.

Rich, S. (1969), In *Fungicides – An Advanced Treatise*, Vol. 2 (D. C. Torgeson, ed.), pp. 447–476. Academic Press, New York and London.

Riov, J., Monselise, S. P. and Kahan, R. S. (1969), *Pl. Physiol. Lancaster* **44**, 631.

Ritzert, R. W., Hemphill, D. D., Sell, H. M. and Bukovac, M. J. (1972), *J. Amer. Soc. Hort. Sci.* **97**, 48.

Robbins, R. C. (1973), *J. Clin. Pharm.* (in press).

Rodes, M. J. C. and Wooltorton, L. S. C. (1971), *Phytochemistry* **10**, 1989.

Rossiter, R. C. (1967), *Aust. J. Agric. Res.* **18**, 39.

Rossiter, R. C. (1969), *Aust. J. Agric. Res.* **20**, 1043.

Rossiter, R. C. and Barrett, D. W. (1970), *Aust. J. Exp. Agr. Anim. Husb.* **10**, 729.

Rossiter, R. C., Barrett, D. W. and Klein, L. (1973), *Aust. J. agric. Res.*, in press.

Rossiter, R. C. and Barrow, N. J. (1972), *Aust. J. Agric. Res.* **23**, 411.

Rossiter, R. C. and Beck, A. B. (1966), *Aust. J. Agric. Res.* **17**, 29; 447.

Russell, D. W. (1971), *J. Biol. Chem.* **246**, 3870.

Russell, D. W. and Galston, A W. (1968), *Pl. Physiol. Lancaster* **43**(S), 15.

Rusznyak, I. and Szent-Gyorgyi, A. (1936), *Nature* **138**, 27.

Rutland, R. B. (1968), *Proc. Am. Soc. Hort. Sci.* **93**, 576.

Ryan, F. J. (1955), *Science N. Y.* **122**, 470.

Sacher, J. A., Towers, G. H. N. and Davies, D. D. (1972), *Phytochemistry* **11**, 2383.

Sachs, J. (1863), *Beilage zu Bot. Leipzig* **21**, 30.

Samorodova-Bianka, G. B. (1965), *Biokhimiya* **30**, 213.

Sanderson, G. W. (1972), In *Recent Advances in Phytochemistry,* Vol. 5, (V. C. Runeckles and T. C. Tso, eds.), pp. 247–316. Academic Press, New York and London.

Sano, H. (1971), *Biochem. Biophys. Acta* **227**, 565.

Sato, M. (1966), *Phytochemistry* **5**, 385.

Saunders, J. A. and McClure, J. W. (1972), *Am. J. Bot.* **59**(S), 673.

Saunders, J. A. and McClure, J. W. (1973), *Pl. Physiol. Lancaster* *51*, 407.

Saunders, J. A., McClure, J. W. and Wallace, J. W. (1973), *Am. J. Bot.* **60**(S) (in press)

Scherf, H. and Zenk, M. H. (1967a), *Z. Pflanzenphysiol.* **56**, 203.

Scherf, H. and Zenk, M. H. (1967b), *Ztsch. Pflanzenphysiol.* **56**, 401.

Schmitz, M. and Seitz, U. (1972), *Ztsch. Pflanzenphysiol.* **68**, 259.

Schneider, M. J. and Stimson, W. (1971), *Pl. Physiol. Lancaster* **48**, 312.

Schneider, M. J. and Stimson, W. (1972), *Proc. Natl. Acad. Sci.* U.S.A. **69**, 2150.

Schoonhoven, L. M. (1972), In *Recent Advances in Phytochemistry,* Vol. 5, (V. C. Runeckles and T. C. Tso, eds.), pp. 197–224. Academic Press, New York and London.

Schultz, G. (1969), *Z. Pflanzenphysiol.* **61**, 29.

Schwochau, M. and Hadwiger, L. (1968), *Arch. Biochem. Biophys.* **126**, 731.

Seevers, P. M. and Daly, J. M. (1970), *Phytopathology* **60**, 1642.

Senebier, J. (1799), *Physiol. veg. (Genève)* **5**, 53.

Shkolnik, M. Y. and Abysheva, L. N. (1971), *Bot. Zh.* **56**, 543.

Shulman, Y. and Lavee, S. (1972), *J. Am. Soc. Hort. Sci.*, in press.

Shutt, D. A., Axelsen, A. and Lindner, H. R. (1967), *Aust. J. Agr. Res.* **18**, 647.

Siegelman, H. W. (1964), In *Biochemistry of Phenolic Compounds,* (J. B. Harborne, ed.), pp. 437–456. Academic Press, London and New York.

Siegelman, H. W. and Hendricks, S. B. (1957), *Pl. Physiol. Lancaster* **32**, 393.

Siegelman, H. W. and Hendricks, S. B. (1958), *Pl. Physiol. Lancaster* **33**, 185.

Singleton, V. L. and Esau, P. (1969), *Phenolic Substances in Grapes and Wine, and Their Significance,* Academic Press, New York and London.

Singleton, V. L. and Kratzer, F. H. (1969), *Agr. Fd. Chem.* **17**, 497.

Smith, H. and Harper, D. (1970), *Phytochemistry* **9**, 477.

Sosnova, V. and Ulrychova, M. (1972), *Biol. Plantarum* **14**, 133.

Sparrow, A. H., Furuya, M. and Schwemmer, S. S. (1968), *Radiation Bot.* **8**, 7.

Srinivasan, S., Lucas, T., Burrowes, C. B., Wanderman, N. A., Redner, A., Bernstein, S. and Sawyer, P. N. (1971), *European Conf. Microcirculation* **6**, 394. [*Chem. Abstr.* **76**, 121485].

Stafford, H. A. (1966), *Pl. Physiol. Lancaster* **41**, 953.

Stafford, H. A. (1968), *Pl. Physiol. Lancaster* **43**, 318.

Stafford, H. A. (1969), *Phytochemistry* **8**, 743.

Star, A. R. and Mabry, T. J. (1971), *Phytochemistry* **10**, 2817.

Stein, G. (1961), *Z. Naturf.* **10b**, 129.

Steiner, A. M. (1971), *Z. Pflanzenphysiol.* **66**, 133.

Steiner, A. M. (1972), *Z. Pflanzenphysiol.* **68**, 266.

Stenlid, G. (1961), *Phys. Plantarum* **14**, 659.

Stenlid, G. (1962), *Phys. Plantarum* **15**, 598.

Stenlid, G. (1968), *Phys. Plantarum* **21**, 882.

Stenlid, G. (1970), *Phytochemistry* **9**, 2251.

Stenlid, G. and Saddik, K. (1963), *Phys. Plantarum* **16**, 110.

Steward, F. C. (1968), *Growth and Organization in Plants,* Addison-Wesley, Reading, Massachusetts.

Stokes, D. M., Anderson, J. W. and Rowan, K. S. (1968), *Phytochemistry* **7**, 1509.

Straus, J. (1960), *Pl. Physiol. Lancaster* **35**, 645.

Strickland, R. G. and Sunderland, N. (1972), *Ann. Bot.* **36**, 443; 671.

Swain, T. and Bate-Smith, E. C. (1962), In *Comparative Biochemistry,* Vol. 3 (M. Florkin and H. S. Mason, eds.), pp. 755–809. Academic Press, New York and London.

Szent-Gyorgi, A. (1972). Preface to *The Anti-inflammatory Action of Flavonoids* (M. Gabor author). Akademiai Kiado, Budapest.

Szweykowska, A. (1959), *Acta Soc. Bot. Polon.* **28**, 539. [*Chem. Abstr.* **54**, 13273].

Tang, Y. W. and Bonner, J. (1947), *Arch. Biochem.* **13**, 11.

Taylor, A. O. (1965), *Pl. Physiol. Lancaster* **40**, 273.

Thimann, K. V. and Edmondson, Y. H. (1949), *Arch. Biochem.* **22**, 33.

Thimann, K. V., Edmondson, Y. H. and Radner, B. S. (1951), *Arch. Biochem.* **34**, 305.

Thompson, W. R. Meinwald, J., Aneshansley, D. and Eisner, T. (1972), *Science* New York **177**, 528.

Thorpe, T. A., Maier, V. P. and Hasegawa, S. (1971), *Phytochemistry* **10**, 711.

Tissut, M. (1972), *Physiol. Veg.* **10**, 381.

Tronchet, J. (1966), *Ann. Sci. Univ. de Besancon, Bot.* **3**, 60.

Tronchet, J. (1972), *Bull. Soc. Bot. France* **119**, 555.

Troyer, J. R. (1964), *Pl. Physiol. Lancaster* **39**, 907.

Tso, T. C. Kasperbauer, M. J. and Sorokin, T. P. (1970), *Pl. Physiol. Lancaster* **45**, 330.

Tso, T. C., Sorokin, T. P., Engelhaupt, M. E., Anderson, C. E., Bortner, C. E., Chaplin, J. F., Miles, J. D., Nichols, B. C., Shaw, L. and Street, O. E. (1967), *Tobacco Sci.* **165**, 30.

Ulrychova, M. and Brcak, J. (1967), *Phytopath. Z.* **58**, 87.

Ulrychova, M. and Sosnova, V. (1970), *Biol. Plantarum* **12**, 231.

Ulrychova, M., Sosnova, V. and Cech, M. (1973), *Proc. 7th Conference of Czechslovak Plant Virologists,* (Novy Smokovec) (in press).

Van Bragt, J., Rohrbaugh, L. M. and Wender, S. H. (1965), *Phytochemistry* **4**, 963.

Van Buren, J. P. (1960). *Nature* **185**, 165.

Van Der Pijl, L. and Dodson, C. H. (1966), *Orchid Flowers, Their Pollination and Evolution,* Univ. of Miami Press, Coral Gables, Florida.

Van Etten, H. D. and Bateman, D. F. (1971), *Phytopathology* **61**, 1363.

Van Fleet, D. S. (1969), In *Advancing Frontiers of Plant Science,* Vol. 23, (L. Chandra, ed.), p. 65–89. Impex, New Delhi.

Van Sumere, C. F., Dedonder, A. and Pe, I. (1972), Book of Abstracts, Phytochemical Society of North America meetings (Syracuse, New York) p. 17.

Vardya T. and Sarapuu, S. (1967), *Fiziol. Rast.* **14**, 551.

Vickery, H. B., Nelson, E. M., Almquist, H. J. and Elvehjem, A. C. (1950), *Chem. and Engr. News* **33**, 2827.

Vince, D. W. (1968), *Planta* **82**, 261.

Vogel, G. (1971), In *Pharmacognosy and Phytochemistry* (H. Wagner and L. Horhammer eds.), p. 370–386. Springer-Verlag, Berlin, Heidelberg and New York.

Voirin, B. and Lebreton, P. (1972), *Phytochemistty* **12**, 3435.

Wagner, E., Bienger, I. and Mohr, H. (1967), *Planta* **75**, 1.

Walker, J. C. and Stahmann, M. A. (1955), *A. Rev. Pl. Physiol.* **6**, 351.

Watanabe, R., Stok, J., Chorney, W. and Wender, S. H. (1964), *Phytochemistry* **3**, 391.

Wattenberg, L. W. and Leong, J. L. (1970), *Cancer Res.* **30**, 1922.

Wawra, C. W. and Webb, J. L. (1942), *Science* N.Y. **96**, 302.

Weissenboeck, G. (1972), *H-S Z. Physiol. Chem.* **353**, 136.

Weissenboeck, G. (1973), *Ber. Dt. Bot. Ges.* **86**, 351.

Weissenboeck, G., Tevini, M. and Reznik, H. (1971), *Z. Pflangenphysiol.* **64**, 274.

Wellmann, E. (1971), *Planta* **101**, 283.

Whatley, F. R., Allen, M. B. and Arnon, D. I. (1955), *Biochem. Biophys. Acta* **16**, 605.

Wheeler, O. R., Carpenter, J. A., Powers, J. J. and Hamdy, M. K. (1967), *Proc. Soc. Exp. Biol. Med.* **125**, 651.

White, J. C., Hillman, J. and Phillips, I. D. J. (1972), *J. Exp. Bot.* **23**, 987.

Wiermann, R. (1970), *Planta* **95**, 133.

Withrow, R. B., Klein, W. H., Price, L. and Elstad, V. (1953), *Pl. Physiol. Lancaster* **28**, 1.

Wollenweber, E. and Egger, K. (1971), *Phytochemistry* **10**, 225.

Wong, E. and Latch, G. C. M. (1971), *N. Z. J. Agr. Res.* **14**, 633.

Yamashita, K. (1965), *J. Endrocrin.* **32**, 259.

Zenk, M. H. (1967), *Z, Pflanzenphysiol.* **56**, 122.

Zinsmeister, H. D. (1964), *Planta* **61**, 130.

Zucker, M. (1965), *Pl. Physiol. Lancaster* **40**, 779.

Zucker, M. (1968), *Pl. Physiol. Lancaster* **43** (S), 26.

Zucker, M. (1972), *Annual Rev. Pl. Physiol.* **23**, 133.

Chapter 19

The Biochemical Systematics of Flavonoids

J. B. HARBORNE

19.1 Introduction

Modern plant biochemical systematics represents an interdisciplinary approach to the many problems of plant classification and can be dated from 1962, the year in which the first international conference on this topic was held. From the proceedings of this meeting, which were published in the following year (Swain, 1963), it is clear that flavonoids were already regarded as potentially important taxonomic markers since no less than a third of the 18 chapters which made up the volume dealt *inter alia* with these constituents. In addition, one of the very first examples of a significant correlation between chemistry and taxonomy was the discovery of a relationship between sectional classification and flavonoid heartwood constituents in the genus *Pinus* (Erdtman, 1963).

Since 1962, most plant surveys in taxonomically interesting groups have included flavonoid studies and indeed these pigments now occupy a pre-eminent position as the most favoured of all plant constituents as taxonomic markers. The reasons why flavonoids are preferred to the terpenoids or alkaloids in systematic studies are not far to seek. They have particular advantages over most other low molecular weight constituents in that they are universally distributed in vascular plants, they show considerable structural diversity, they are so chemically stable that they can be detected in herbarium tissue and, finally, they are easily and rapidly identified. This last factor is by no means the least important, since it is responsible for their popularity among experimental taxonomists, who normally have only limited facilities for biochemical work. Flavonoids are more readily separated and detected on two-dimensional chromatograms than most other plant compounds. Furthermore, the commonly occurring pigments can be identified by relatively simple methods, the only expensive instrument required being a spectrophotometer.

In spite of the fact that a considerable body of data on flavonoid distribution patterns has accumulated in the last ten years, the impact of flavonoid studies on formal plant systematics is slight (cf. Heywood, 1973). Inevitably, it will take many years before flavonoid data can be absorbed into all systems of classification. Nevertheless, much potentially valuable data have been revealed for taxonomic purposes and it remains ultimately for the taxonomist to assimilate these results, together with all the other available chemical and biological information, into his classifications.

The fact that flavonoids occur universally in all vascular plants means that a representative flavonoid pattern can be elicited from all major plant phyla. This can be used for exploring relationships at the higher levels of classification and, since different flavonoids can be placed in a biosynthetic sequence, the results have a phylogenetic content. The evolution of flavonoids, in fact, is discussed separately in the chapter following this one, so that it will not be mentioned further here.

The emphasis in this chapter is on the possible utility of flavonoids at levels of classification below that of the plant order, i.e. at the family, genus or species level. A number of examples will be mentioned where flavonoid patterns have been found to correlate with existing classifications, e.g. at the sectional level in the genus *Pinus* (Erdtman, 1963) or at the tribal level in the family Plumbaginaceae (Harborne, 1967a). It is an important preliminary to establish such correlations before flavonoids can be utilized with confidence in the classification of difficult plant groups. Several cases have indeed now been recorded where flavonoids have helped to support a revision of the existing classification, e.g. at the subfamily level in the family Gesneriaceae (Harborne, 1966b, 1967b). More important still, the flavonoid data may throw new light on a plant classification which cannot be obtained from any other source. For example, the fact that the Caryophyllaceae contains purple-red anthocyanins while the other families in the same order, the Centrospermae, have the mutually exclusive betacyanin pigments, has provided a completely new facet on the complex inter-familial relationships within this order and has even led to the suggestion that the Caryophyllaceae be removed from the order (see e.g. Mabry, 1966).

Much attention has been paid recently to exploring flavonoid patterns at the plant population level. Such studies have significantly aided in the identification of hybrids between closely similar species when growing sympatrically. They have also revealed interesting correlations between flavonoid patterns and plant geography. For example, a plant species originating from the Northern temperate region of the world may well have evolved a different flavonoid pattern during its southern migration and establishment in Southern temperate regions (Moore *et al.*, 1970). Correlations between flavonoids and plant geography lead directly to the question of flavonoid function; that this may lie in the field of plant-animal interactions has often been suggested (see Chapter 18). However, joint chemical, systematic and ecological studies are still required to fully explain the many variations in flavonoid distribution patterns that have been observed.

19.2 Flavonoids as taxonomic markers

19.2.1 Presence/absence characters

In most systematic studies of flavonoids, ascertainment is based on determining the presence or absence of a particular character. Therefore, characters which are easily detectable, even at very low concentrations, are definitely preferable, since it is usually equally obvious when they are detectably absent from a taxon. For this reason, the most useful flavonoid characters from the practical viewpoint are those which are visually coloured or those which exhibit fluorescence in UV light. Coloured flavonoids include the orange, red and purple anthocyanins and the various yellow pigments, i.e. yellow flavonols, chalcones and aurones. Leucoanthocyanidins and proanthocyanidins (see Chapter 10) may also be included here because, although not coloured themselves, they give anthocyanidins when treated with acid. Their presence or absence can be determined very simply by heating plant tissue directly in 2M HCl for 30 min at 100° and noting whether a red colour develops. It is important to confirm the identification by chromatographing the concentrated extract against standard markers of pelargonidin, cyanidin and delphinidin, since certain other classes of plant constituent also give colours when leaf tissue is heated with mineral acid. In the case of the anthocyanins themselves, their presence or absence is usually very readily scored for by visual means. They are occasionally masked *in vivo* by other types of plant pigment, but in such cases, extraction with methanolic acid and chromatography will separate them from interfering substances.

Of those flavonoids which are yellow in colour, the chalcones and aurones are the most readily detected, since they usually give a characteristic visual 'anthochlor' colour change to orange or red in the presence of ammonia. Their detection and natural distribution is discussed at length in Chapter 9, so that further mention here is unnecessary. The two other major groups of yellow flavonoids are the flavonols and flavones with extra hydroxyl substituents either in the 6- or 8-positions. These can be recognized by their visual yellow colour and by their appearance on paper in UV light as dark absorbing spots which are not affected by ammonia. Flavonols with 8-hydroxylation such as herbacetin (1) and gossypetin (2) have distinctly different distribution patterns (Table 19.1) from those with hydroxylation in the 6- position (Table 19.2). Gossypetin, which is by far the commonest 8-hydroxyflavonol, has very similar chromatographic properties to the isomeric 6-hydroxyquercetin, namely quercetagetin (3), but can be distinguished from it during systematic surveys by the fact it gives a blue colour on paper with alcoholic sodium acetate, while quercetagetin is unaffected (Harborne, 1968). As can be seen from Table 19.1, gossypetin is an interesting taxonomic marker at the family level, occurring as it does in several pairs of related families (e.g. Malvaceae and Sterculiaceae, Ericaceae and Empetraceae, Ranunculaceae and Papaveraceae). While 6-hydroxyflavones scutellarein (4), 6-hydroxyluteolin (5) and their methyl ethers) can occur very occasionally in the same families (e.g.

Table 19.1   *Distribution of gossypetin in the angiosperms*

Family	Genera (and species)	Frequency and form*
Ranunculaceae	*Ranunculus* spp.	in flowers, as a glycoside
Papaveraceae	*Meconopsis* and *Papaver* spp.	in flowers, as glycosides
Leguminosae	*Acacia constricta*	in leaf
	Lotus corniculatus	in flowers, with the 7- and 8-methyl ethers
Crassulaceae	*Sedum album, S. acre*	in flowers, herbacetin 8-methyl ether
Malvaceae	*Abutilon, Gossypium, Hibiscus* spp.	frequent in flowers; also in leaves
Sterculiaceae	*Chiranthodendron pentadactylon*	in leaves
	Fremontia californica	in flowers and leaves
Rutaceae	*Xanthoxylum acanthopodium*	in seeds, as 8-4'-dimethyl ether
Ericaceae	*Rhododendron* and 15 other genera	occasionally in flowers, often in leaves
Empetraceae	*Empetrum, Corema,* and *Ceratiola* spp.	in leaves of all known taxa
Primulaceae	*Primula, Dionysia* and *Douglasia*	as flower pigment, sometimes with herbacetin
Compositae	*Chrysanthemum segetum*	as flower pigment
Restionaceae	*Calorophus lateriflorus*	in stems and inflorescence
	Restio spp.	

*For references, see Harborne, 1969a.

Table 19.2   *Distribution of 6-hydroxyflavonols and*
6-hydroxyflavones in the angiosperms

Family	Genera (and species)	Frequency and form*
Betulaceae	*Alnus glutinosa* and *A. sieboldiana*	bud excretion as methylated derivatives
Amaranthaceae	*Alternanthera phylloxeroides*	6-methoxyluteolin, patuletin
Chenopodiaceae	*Spinacia oleracea*	and quercetagetin 6,3'-DME
Polygonaceae	*Polygonum* and *Atraphaxis* spp.	scutellarein and 5-methyl-6-hydroxykaempferol
Saxifragaceae	*Chrysosplenium* spp.	methylated flavonols
Rosaceae	*Kerria japonica* and *Sorbaria* spp.	scutellarein, 6-methoxyluteolin
Leguminosae	*Distemonanthus, Prosopis* and *Tephrosia*	Patuletin and other methylated derivatives
Polemoniaceae	*Ipomopsis* and 8 other genera	Patuletin, patuletin 7- methyl ether and 6-hydroxykaempferol 6, 7-dimethyl ether
Verbenaceae	*Callicarpa, Cyanostegia, Lippia* and *Vitex*	methylated flavonols and 5,6,7-trimethoxyflavone
Labiatae	many genera (e.g. *Scutellaria*)	6-Hydroxyluteolin, scutellarein and derivatives
Plantaginaceae	*Plantago*	6-Hydroxyluteolin in 10/26 spp.
Buddleiaceae	*Buddleia* and *Chilianthus*	6-Hydroxyluteolin
Scrophulariaceae	*Digitalis, Linaria Scrophularia, Veronica* spp.	6-hydroxyluteolin and methylated derivs.
Myoporaceae	*Eremophila fraseri*	6-hydroxymyricetin 3,6,7,4'-tetramethyl ether
Globulariaceae	*Globularia, Lytanthus* spp.	6-Hydroxyluteolin in all spp. examined
Bignoniaceae	*Catalpa, Jacaranda, Millingtonia, Stereosperum* and *Tecoma* spp.	Scutellarein and 6-hydroxyluteolin frequent
Pedaliaceae	*Pedalium* and *Sesamum* spp.	6-methoxyluteolin
Rubiaceae	*Gardenia lucida*	highly methylated 6-hydroxyflavones
Valerianaceae	*Valerianella*	6-hydroxyluteolin in 6 of 40 spp.
Compositae	*Chrysanthemum, Coreopsis, Cirsium, Lasthenia* and many others	6-hydroxyflavonols and 6-hydroxyflavones as such, or methylated, frequent

*Taxonomic arrangement follows Cronquist (1968).

the Leguminosae) together with 8-hydroxyflavonols, they occur predominantly in herbaceous plants, in families of the Tubiflorae and in the Compositae (Table 19.2). In some cases, they are very consistent family markers (in the Globulariaceae, present in all taxa examined) and in other cases (e.g. the Labiatae) they are confined to particular tribes or subfamilies.

Of the flavonoids which are easily detectable in UV light by their intense fluorescence, two groups are of particular interest from the systematic viewpoint, namely flavonols and flavones with either methylation or glycosylation in the 5-position. 5-Methylated flavonols include azaleatin (5-methylquercetin) (6) which has an

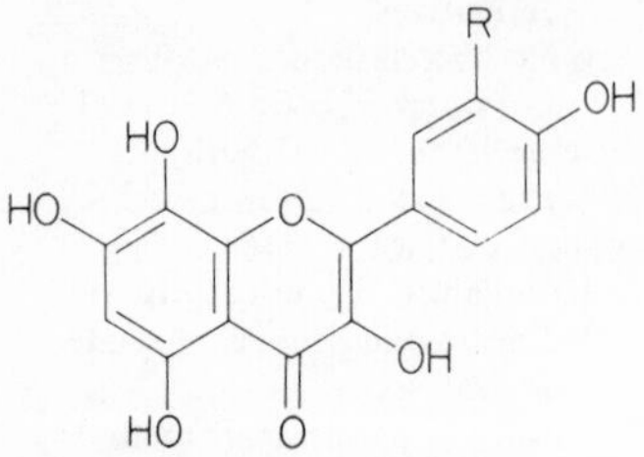

(1, Herbacetin R = H)
(2, Gossypetin R = OH)

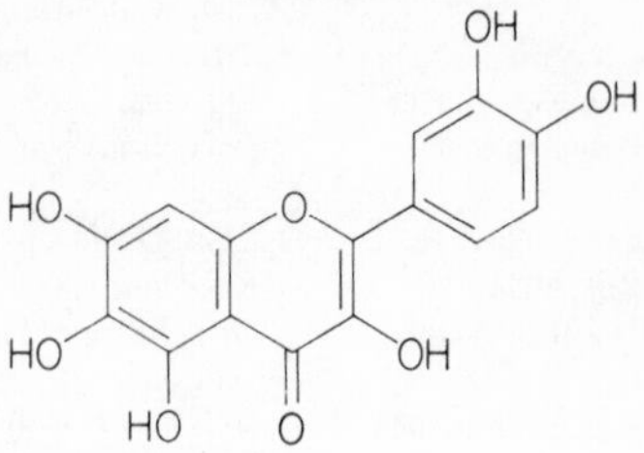

(3, Quercetagetin)

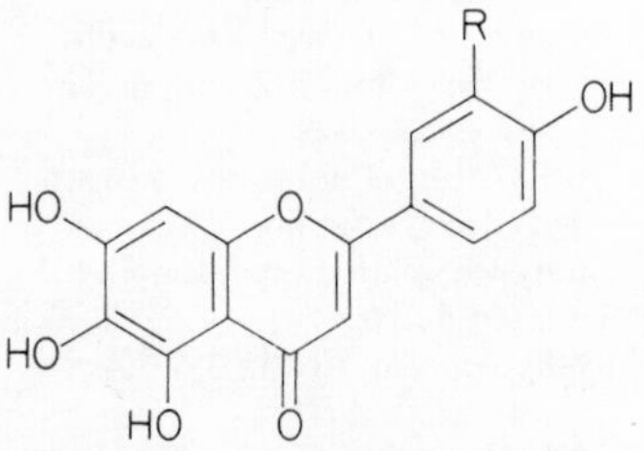

(4, Scutellarein R = H)
(5, 6—Hydroxyluteolin R = OH)

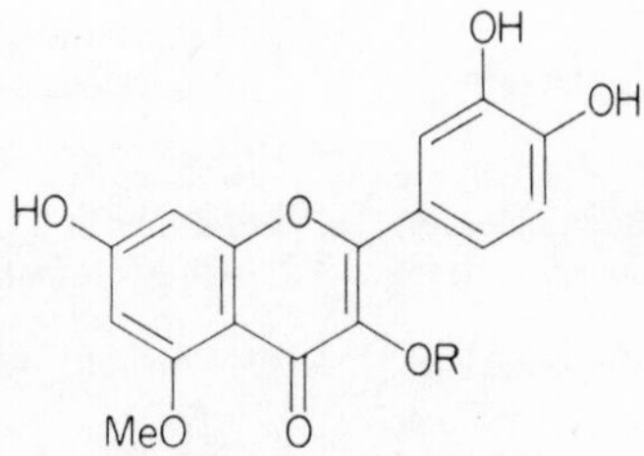

(6, Azaleatin R = H)
(7, Caryatin R = Me)

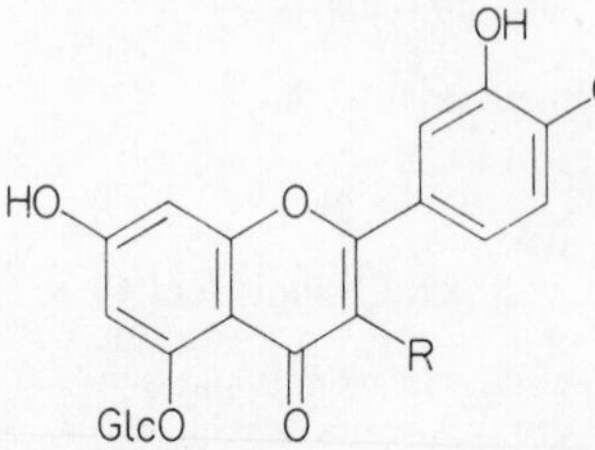

(8, Quercetin 5—glucoside R = OH)
(9, Luteolin 5—glucoside R = H)

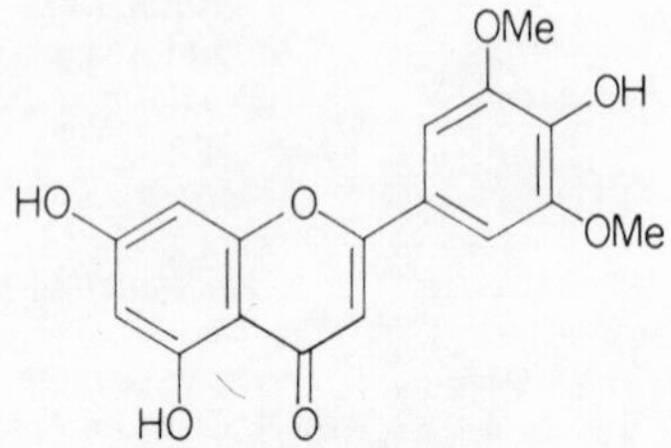

(10, Tricin)

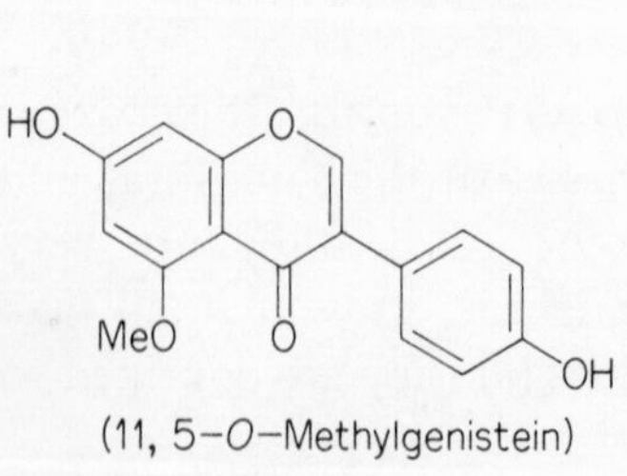

(11, 5—O—Methylgenistein)

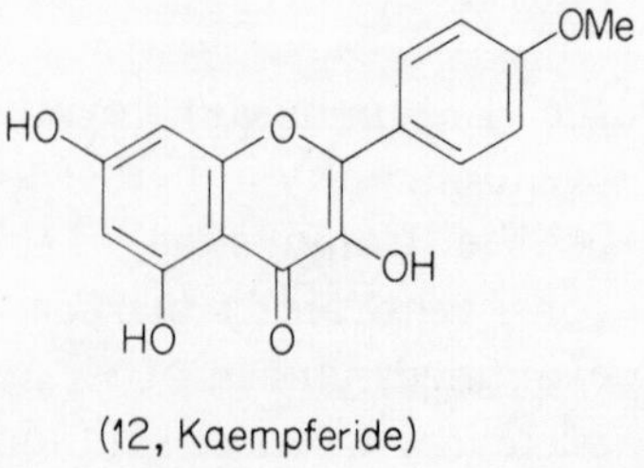

(12, Kaempferide)

intense yellow fluorescence and caryatin (7) (the corresponding 3,5-dimethyl ether) which has an intense blue fluorescence. 5-Glycosylated flavonoids of taxonomic interest include the 5-glucosides of quercetin (8), kaempferol, luteolin (9), and tricin; these all exhibit the same intense yellow fluorescences as the flavonol 5-methyl ethers, but they can be distinguished by the fact that fluorescence disappears on mild acid treatment whereas the 5-methyl ethers remain unaffected. In any case, the two classes of fluorescent compound have very different distribution patterns (Table 19.3) and have not yet been detected together in the same plant.

Fluorescence can be induced in the majority of flavonoids by forming the aluminium chelates, e.g. by treating chromatograms with alcoholic aluminium chloride when flavonoids appear as intense yellow green fluorescent spots. Such a procedure increases several fold the level of detection of most flavonoids. However, it is of little use for distinguishing different compounds, since most substances give the same colour.

The commonly occurring flavones and flavonols normally lack both visible colour and intense UV fluorescence. They are detected

Table 19.3 *Plant families containing fluorescent flavonol derivatives*

Plant family	Typical species containing compounds
Flavonol 5-methyl ethers	
Eucryphiaceae	*Eucryphia glutinosa* (Poepp. et Endl) Baill.
Juglandiaceae	*Carya pecan* (Massh.) Eugl. et Graebn.
Lauraceae	*Bielschmiedia miersii* (Gay) Kosterin.
Dilleniaceae	*Tetracera portobellensis* Beurl.
Ericaceae	*Rhododendron calostrotum* Balf and Ward.
Plumbaginaceae	*Plumbago capensis* Thunb.
Flavonol and flavone 5-glucosides	
Rosaceae	*Docyniopsis tschonoski* Koidzumi
Leguminosae	*Galega officinalis* L.
Umbelliferae	*Torilis arvensis* (Hudson) Link.
Labiatae	*Lamium album* L.
Compositae	*Anthemis cotula* L.
Caprifoliaceae	*Leycesteria formosa* Wall.
Gramineae	*Triticum monococcum* L.
Cyperaceae	*Carex riparia* Curtis
Palmae	*Chamaerops humilis* L.

Data from Harborne (1969b), Glennie and Harborne (1971) and references therein.

on chromatograms in UV light, when they appear as either bright yellow to brown colours or darkly absorbing, changing to brighter colours in the presence of ammonia vapour. O-Methylation (other than at the 5-position, which has already been mentioned, see above) of flavones and flavonols is often of taxonomic interest but such substitution is sometimes difficult to detect or to distinguish from lack of methylation. For example, isorhamnetin (quercetin 3'-methyl ether) can be confused with quercetin or kaempferol in some solvents, while 7-O-methylmyricetin has the same R_f and colour as quercetin in the Forestal solvent, although it does separate from it in phenol-water (Harborne, 1967a). Related methyl ethers may also not be easily distinguished: the 7- and 4'-methyl ethers of kaempferol have identical R_fs in all solvents that have been tried and can only be distinguished by their different spectral characteristics (see Bate-Smith and Harborne, 1971).

One flavone methyl ether of especial taxonomic interest is tricin (10), which is easily distinguished on paper chromatograms by its high mobility in phenol and medium mobility in butanol-acetic acid-water. Its natural distribution (Table 19.4) is rather typical of many flavonoid markers in that it occurs in one group of related plants (in grasses, sedges and palms) where it has taxonomic significance and yet it also has a number of apparently unrelated occurrences. There is, perhaps, some sense in its distribution in the monocotyledons, since as it is present in the Liliaceae, which has a

Table 19.4   *Natural distribution of the flavone tricin*

Family	Frequency*
Gramineae	in 91% of 118 spp. surveyed
Palmae	in 51% of 125 spp. surveyed
Cyperaceae	in 47% of 62 spp. surveyed
Liliaceae	rare; in *Uvularia sessiliflora* and 3 *Colchicum* spp.
Iridaceae	rare; in *Crocus cambessedesii* and possibly two other *Crocus* spp.
Leguminosae	rare; in *Medicago sativa* only
Orobanchaceae	rare; in *Orobanche ramosa* and *O. arenaria* but not in 8 other spp. tested

*As a leaf constituent, except in the Orobanchaceae where it occurs in the seed. For references, see Harborne (1967b); Bate-Smith (1968); Harborne (1971c); Williams *et al.* (1971, 1973); records in the Liliaceae from unpublished data of C. A. Williams.

central position in most systems describing the phylogeny of the monocots, it could be expected to occur in such derived families as the grasses, sedges and palms and also the Iridaceae, as is indeed the case.

Some classes of flavonoid (e.g. isoflavones, certain flavanones) are difficult to detect in plant extracts, especially when present in low amounts. In such cases, it may be more useful to consider them as *either* accumulating in the tissue (more than 0.5% dry wt) *or* occurring in trace amount. This applies, for example, to the isoflavones in leaf tissue of herbaceous Leguminosae. While all species of the Leguminosae – Lotoideae probably have the genetic potential for synthesizing isoflavones, these compounds only occur in large amount in relatively few taxa. In the genus *Trifolium,* for instance, fourteen species can be distinguished by a high isoflavone content (0.8–5% dry wt) from 86 others with only trace amounts (< 0.25%) (Francis *et al.*, 1967). Again, the tribe Genisteae is characterized by having species which accumulate isoflavone (particularly 5-*O*-methylgenistein (11)), while genera in neighbouring tribes (e.g. *Lupinus*) are poor in leaf isoflavone (Harborne, 1969c).

All classes of flavonoid, except anthocyanins, can be detected in herbarium leaf tissue and many plant surveys have been usefully extended by carrying out analyses on the relatively small samples that can be spared from herbarium specimens, usually with excellent results. It is important to remember that changes may occur during the drying or storage processes, so that data from herbarium analyses should be critically evaluated and related whenever possible to results on fresh tissue, either of the same or of a related species. Even the anthocyanins may survive the drying process, but they cannot be detected with reliance in dried floral tissue because of their relative instability.

Significant qualitative differences in flavonoids between fresh and herbarium tissue are probably the exception rather than the rule. One such exception was noted in *Dillenia,* by Bate-Smith and Harborne (1971) who found that while fresh tissue showed the presence of dihydrokaempferide glycosides, herbarium tissue showed only the presence of the related flavonol methyl ether, kaempferide (12). Storage obviously causes oxidation of the dihydroflavonol to flavonol, the rate of oxidation probably being enhanced in this case by the presence of a methoxyl in the 4'-position. The same authors also noted that leucoanthocyanidin content was much higher in fresh tissue of several *Dillenia* spp. than that recorded earlier by Kubitski (1968) in herbarium material of the same plants. While qualitative variation between fresh and herbarium material is uncommon, quantitative variation is more likely and should be allowed for whenever systematic comparisons are made.

19.2.2 Replacement characters

A problem with flavonoid characters, as with most other types of chemical marker used in systematic studies, is the difficulty of scoring for the recessive condition, i.e. for absence. Presence is clear enough, but the distinction between presence in trace amount and complete absence is sometimes difficult to ascertain. This difficulty is accentuated, of course, when only limited amounts of plant tissue are available for analysis, which is often the case when critical species are being examined. It can, however, be circumvented by scoring for chemical characters which replace each other, so that no 'negative' scoring as such is involved. Situations where flavonoids act as replacement characters are not common, but a number of examples may be mentioned.

An unrelated biosynthetic replacement occurs in the case of the anthocyanin and betacyanin pigments, two classes of compound which are mutually exclusive in their occurrence in plants. In this instance, the usual purple to red anthocyanins of higher plants are replaced by the alkaloidal betacyanins in all but one of the families of the order Centrospermae. The one exception is the Caryophyllaceae, a family which retains the anthocyanin character. This difference also applies to the yellow flower pigments, since the Centrospermae have betaxanthins, pigments which are closely related in structure to the betacyanins, instead of the more usual yellow carotenoids. The Caryophyllaceae is again exceptional since, where the yellow pigments have been identified, they have been found to belong to the flavonoid class. Thus, a yellow chalcone provides the colour in the yellow carnation *Dianthus caryophyllus* (Harborne, 1966a) and a 6-hydroxyflavonol gives the yellow colouration in *D. knappii* (Harborne, 1964).

Yellow flavonol pigment may also appear to replace carotenoids in other plant families. In the Primulaceae, gossypetin or herbacetin are the major yellow corolla pigments in the sections Vernales and

Sikkimenses of the genus *Primula* (and also in *Dionysia* and *Douglasia*) replacing carotenoid which is the major type of flower pigment in other sections of *Primula* and in most other genera of the family (e.g. *Lysimachia*) (Harborne, 1968). In Gesneriaceae, there is a similar situation in that several species of one subfamily (Cyrtandroideae) have yellow flower colour based on chalcones and aurones, while an equal number of species of the other subfamily, the Gesnerioideae, have carotenoids in the flowers (Harborne, 1966b, 1967b).

Probably the most systematically useful flavonoids which act as replacement characters are the flavonols and the flavones. A general evolutionary trend in leaf tissue of many plant groups is the gradual replacement of flavonol by flavone and it is significant, since it is a changeover often correlated with morphology or cytology. Flavonols and flavones, unlike anthocyanins and betacyanins, are not mutually exclusive in their occurrence and types can be detected which have both flavonol and flavone.

Perhaps the most extensive study of this phenomenon has been in plants of the family Umbelliferae, of which over 200 species have been surveyed (Table 19.5). Here, there is a clear changeover in

Table 19.5   *Distribution of flavones and flavonols in the Umbelliferae*

Subfamily and tribe	No. of species with Flavonol	Flavone*	Generic ascertainment
Hydrocotyloideae			
1. Hydrocotyleae	16	0 ⎫	13/34
2. Mulineae	7	0 ⎭	
Saniculoideae			
1. Saniculeae	26	0 ⎫	7/9
2. Lagoecieae	2	0 ⎭	
Apioideae			
1. Echinophoreae	2	0	2/5
2. Scandiceae			
(a) Scandicinae	1	16 ⎫	17/21
(b) Caucalinae	2	19 ⎭	
3. Coriandreae	3	0	2/5
4. Smyrnieae	21	5	22/29
5. Apieae	79	20	52/85
6. Peucedaneae	46	1	23/41
7. Laserpitieae	7	11	4/8
8. Dauceae	5	7	3/4

*Results based on a leaf survey (Crowden *et al.*, 1969), but similar data have also been obtained from a survey of fruits (Harborne and Williams, 1972).

pattern such that the twelve tribes that have been analysed can be classified according to whether the majority of species have (1) only flavonol, (2) mainly flavonol or (3) mainly flavone. It is correlated with evolutionary trends within the family in that the tribes with mainly flavone (Dauceae, Caucalideae, Laserpitieae) are generally regarded as being highly specialized in their floral and fruit morphology. The same progression can be observed within individual tribes. For example, in the Caucalideae, there are two genera (*Astrodaucus, Artedia*) with only flavonol, three (*Daucus, Orlaya, Pseudorlaya*) with flavonol plus flavone and four (*Caucalis, Torilis, Chaetosciadium, Turgenia*) with flavone alone. With only one exception, this replacement of flavonol by flavone is correlated with increasing basic chromosome number, from 8 to 16 (Harborne, 1971b).

Another example of flavonol replacing flavone is in the Caprifoliaceae, where Bohm and Glennie (1971) surveyed 52 taxa representing 11 genera for leaf flavonoids. They found that kaempferol and quercetin were confined mainly to *Viburnum, Sambucus* and *Weigelia,* whereas the flavones luteolin and apigenin predominated in *Symphoricarpos, Triostemma* and *Lonicera.* There were some intermediate species: thus of 24 flavonol-containing species, six had flavones as well and, conversely, six of 20 flavone-containing species had flavonols as well. In a further example, in *Valerianella,* 34 species from five sections of the genus were found to have flavonols whereas all six species from a sixth section had flavones instead (Greger and Ernet, 1973). Yet again, in *Carex,* a group of five related 'primitive' species were discovered to have the flavonols kaempferol and quercetin, whereas 38 of the 39 other species examined had tricin, luteolin and glycoflavone instead (Harborne, 1971c). Similar examples could be quoted from flavonoid studies of other plant groups and there is little doubt that many more instances of this type of discontinuity will be discovered in the future.

19.2.3 Spot pattern characters

Flavonoids lend themselves to easy detection in alcoholic plant extracts (together with related phenolics) following one- or two-dimensional chromatography on paper or thin layers of cellulose. By examination in UV light with and without ammonia vapour, a series of spots with a range of colour reactions and R_fs can be observed. Such data, when measured on a group of related taxa, can be used directly in taxonomic studies without further chemical identification. Indeed, there are many examples in the literature where this type of procedure has been employed and the data obtained used directly for computer analysis, with or without the inclusion of other types of taxonomic character.

Parups *et al.* (1966), for example, examined the flavonoid profiles of 11 species of *Trifolium* using one- and two-dimensional TLC. They recorded the presence/absence of some 140–190 characters and used the data in a cluster analysis to determine the relationships of the various species; three groupings of the species were produced by this procedure. In a similar study of *Aquilegia* (Ranunculaceae), Taylor and Campbell (1969) analysed flavonoid patterns in extracts of leaves and flowers of 18 species, recorded presence/absence of 87 spots, (22 being anthocyanin) and obtained five chemical groupings as a result. Again, Iiyama and Grant (1972) examined spot data for 15 species of *Avena* (Gramineae) but found no correlation with nuclear DNA content of the same plants. A final example may be taken from the work of Asker and Frost (1970) who detected phenolic patterns in seven *Potentilla* species (Rosaceae) and used the results (over 100 spots) to note differences between the species; variations in pattern between juvenile and mature plants were also noted.

Probably, the main danger of using spot data without identification of individual constituents is the inevitable over-estimation of the number of components or characters really present and hence of the taxonomic distance between taxa. There are two main reasons for this. Artifacts may be formed during extraction, concentration or chromatography giving rise to 'false' spots. For example, a labile flavonol di- or tri-glycoside might easily be partly broken down to intermediate glycosides and even to aglycone during this procedure, so that one component in the original plant could give three or more spots on a 2D chromatogram. Secondly, the same flavonoid character may be erroneously counted more than once. Thus, if four flavonols occur in the same plant each combined as the 3-rhamnoside and 3-glucoside, the actual number of chemical characters present is six (one for each aglycone and one for each glycosidic type). However, on 2D chromatograms, twelve spots will appear (i.e. eight glycoside spots and four aglycone spots) so that each character could effectively be counted twice.

Other problems in using spot data in taxonomic studies include the difficulty, except when comparing chromatograms of very closely related taxa, of being sure that the same position on a chromatogram is always occupied by the same compound. It is also an oversimplification to assume all spots with the colour reactions of flavonoids are really flavonoids; occasionally, other classes of plant constituent (e.g. phenolic alkaloids) give very similar colours.

While it may not be possible, because of limitations of plant material, to identify all flavonoids present on 2D chromatograms of a given group of related plants, there is much to be said for identifying at least the major or more variable constituents, before applying the data to systematic studies. Even where components cannot be identified, their homogeneity should be tested by eluting them individually and re-chromatographing them in several solvent systems. This procedure will also indicate if any are labile compounds, breaking down during handling and chromatography to give other components apparent on the original 2D chromatogram. Only thus can errors of overestimation be kept to a minimum in such work.

19.3 Correlations with existing classifications

Before flavonoids can be considered as useful systematic markers, it must be demonstrated that they are correlated in their distribution patterns with more conventional taxonomic criteria derived from morphology and anatomy. In other words, the occurrences of flavonoids in plants should generally show some agreement with existing classifications. Indeed, there is no doubt that this is so. The predictive value of flavonoid patterns is considerable. Only rarely does one fail to find an unusual flavonoid, reported in one particular species, to occur also in some of the related species, i.e. in the same genus or in related genera. The same is true at the family level. The Leguminosae, for example, are known to have a rich and distinctive flavonoid pattern (for review, see Harborne, 1971a) and one can confidently predict that a previously unexplored species or genus in the family will yield on phytochemical analysis one or more characteristic legume flavonoids, e.g. a chalcone, a 5-desoxyflavonol or an isoflavone.

Where deliberate surveys have been carried out in restricted plant groups, correlations between flavonoids and existing classifications have frequently been uncovered. In the very first comprehensive examination of flavonoid patterns in a genus, namely in *Pinus,* a very good correlation was discovered between the presence of flavones and flavanones in the heartwood and the classification of the genus into subgenera and sections based on conventional morphological characters (Erdtman, 1963). This and a few other more recent examples are collected together in Table 19.6, to illustrate the point that such correlations have been observed in widely differing plant groups and with different flavonoid markers.

Such correlations are often incomplete and from the purely taxonomic viewpoint, flavonoids rarely as yet provide completely satisfactory markers for separating different taxa. In any case, complete ascertainment of flavonoid characters within a group is still relatively rare. However, as Bate-Smith (1962) has pointed out, it is the incomplete correlations which are often the most interesting. They do at least suggest that the affinities of the plants in question may not be all that they seem. To take just one illustration, there is the case of gossypetin occurring in the leaves of *Galax aphylla,* Diapensiaceae. This is an exceptional occurrence, since surveys indicated that gossypetin, while being a regular constituent of the Ericaceae and Empetraceae, is generally absent from the related Epacridaceae, Clethraceae and Diapensiaceae (Harborne and Williams, 1973). Flavonoid chemistry at least suggests that *Galax*, a monotypic genus, may be misplaced in the Diapensiaceae; it is interesting that the affinities of precisely the same plant have been questioned for quite other reasons by Hutchinson (1974) during a revision of the taxonomy of this family.

Before leaving the question of flavonoid correlations, it is worth considering whether the discovery of such correlations has any value to systematics in general. If chemistry broadly agrees with existing classification, it could be argued that it does no more than reinforce the view that plants can be quite satisfactorily classified using biological characters alone. However, if we take the case of the Plumbaginaceae, where flavonoid chemistry has been demonstrated to correlate closely with tribal divisions (Table 19.7) it can be seen that there are some benefits to be gained. For example, one taxonomist on viewing the *combination* of biological and chemical differences between the two tribes, has suggested that they are such that each tribe deserves to be raised to familial level (Linczevski,

Table 19.6 *Some flavonoid correlations with existing classifications*

Plant group	Flavonoid characters	Number of species surveyed	Correlation	Reference
Centrospermae	Anthocyanin in leaf and flower	c. 200	Absent from all but one family (Caryophyllaceae)	Mabry, 1966
Gesneriaceae	Anthocyanin and other pigments	81	3-Deoxyanthocyanin in New World, isocotylous subfamily	Harborne, 1966b, 1967b; Lowry, 1972
Plumbaginaceae	Flavonoids in root, leaf and flower	55	with tribal classification, e.g. azealatin in the Plumbagineae, etc.	Harborne, 1967a
Primulaceae	Flavonoids of leaf and petal	100	with generic and subfamily classification e.g. 3′,4′-dihydroxyflavone in *Primula* and 3 other genera in the Primuleae	Harborne, 1968
Genisteae	Flavonoids of leaf	128	isoflavone accumulation as a tribal character	Harborne, 1969c
Acacia	Flavonoids in bark	61	7,8,4′-tri OH and 7,8,3′,4′-tetra OH flavonoids in sections Plurinerves and Juliflorae	Tindale and Roux, 1969
Eucalyptus	Flavonoids of leaf	80% of known spp.	Myricetin, leucocyanidin and rutin with series and sectional classification	Hillis, 1967
Ilex	Anthocyanins of fruit	24	Cyanidin and pelargonidin 3-glucosides and 3-xyloglucosides and sectional classification	Santamour, 1973
Iris	Flavonoids of leaf	25	Flavonols, glycoflavones and leucoanthocyanidins and sectional classification	Bate-Smith, 1968
Malus and *Pyrus*	Flavonoids and other leaf phenolics	all known spp.	Dihydrochalcones in *Malus*, replaced by arbutin in *Pyrus*	Challice and Williams, 1968
Pinus	Flavones in heartwood	52	Present in subgenus Haploxylon but absent from subgenus Diploxylon	Erdtman, 1963
Rhododendron	Flavonoids of leaf and petal	206	Presence of caryatin in subgenus Hymenanthes; absence of gossypetin from subgenus Pentanthera	Harborne and Williams, 1971
Valerianella	Flavonoids of leaf	40	6-Hydroxyflavones only in section Siphonocoela	Greger and Ernet, 1973

Table 19.7 *Chemical and biological differences at the tribal level in the family Plumbaginaceae*

	Plumbagineae	Staticeae
No. genera studied	4/4	6/7
No. species studied	11/30	44/286
	Chemical characters	
Quinones in root	Plumbagin uniformly present	Plumbagin uniformly absent
Flavonols in leaf and/or flower	5-O-methylation frequent, 7-O-methylation in 2 spp.	Myricetin characteristically in leaf, 5-O-methylation rare (1 sp.), 3-O-methylation in 1 sp.
Anthocyanins in flowers	A-ring methylation common	A-ring methylation absent
	Glycosidic patterns: 3-glucoside, 3-galactoside or 3-rhamnoside	Glycosidic patterns: 3-rhamnoside-5-glucoside or 3,5-diglucoside
Leucoanthocyanidins	Rare in root, uncommon in leaf or flower	Frequent in root, leaf and flower
	Biological characters	
Basic chromosome number	$X=6$ or 7	$X=8$ or 9
Pollen type	Uniformly monomorphic	mainly dimorphic
Key systematic character	Stamens free from the corolla	Stamens attached to the corolla

From Harborne (1967a).

1968). Perhaps, a more interesting question to ask is why these two groups, while retaining many morphological characters in common, should have diverged so greatly in their flavonoid chemistry. The answer perhaps lies in their different geographical origins and their need to adapt to different insect pollinators and/or predators; certainly it suggests that an ecological study of the situation would be rewarding. Finally, the flavonoid data may be of value in determining evolutionary trends within the family. It is worth noting that the Australasian shrub *Aegialitis annulata,* regarded as a primitive relict species, has a different and simpler flavonoid pattern than other members of the Plumbaginaceae. In terms of herbaceous habit, geography, pollen morphology and chromosome number, the Staticeae seem to be more advanced than the Plumbagineae (Baker, 1948). Some chemistry supports this, i.e. the loss of flavonol 5-O-methylation and of plumbagin synthesis (primitive woody characters) in the Staticeae, with the gain in anthocyanidin 5-glycosylation. On the other hand, the frequency of leucoanthocyanidin and of myricetin in the Staticeae is unexpected and does not fit in with the above scheme. Clearly, the evolutionary relationships between the two groups are complex and deserve further analysis from other points of view.

19.4 Utility of flavonoids in revising plant classifications

No present system of plant classification is anywhere near perfect and one of the major activities of plant taxonomists today is to revise and improve existing classifications. Many areas of the systems of Engler and Prantl and of Bentham and Hooker, for example, have not been revised since the turn of the century and much work needs to be done just to bring them up-to-date. Revision is necessary partly because new taxa are regularly being discovered but also because new data (e.g. from cytology or scanning electron microscopy) become available on the known species. Better methods of handling taxonomic data (e.g. numerical taxonomy) and fresh ideas on evolutionary relationships in floral morphology may also provide the spur for revision. In all this activity chemistry, and particularly flavonoid chemistry, can play a part.

Firstly, chemical data can be obtained from a group of plants to see if they fit in better with the revised than with the original system, i.e. whether they can provide support for a suggested reclassification. Secondly, in a more decisive way, flavonoid chemistry can be employed to provide a decision between two or more competing re-classifications, which are approximately equally satisfactory from the biological viewpoint. Thirdly, and preferably, flavonoid and other chemistry can be used as an integral part of a taxonomic revision, based on a re-examination of all possible characters present in a given group of plants.

There are now several examples in the literature where flavonoid chemistry can be said to have served a useful purpose in confirming the correctness of a taxonomic decision to re-classify or rearrange a particular plant group. Two will be mentioned, regarding reclassification within the families Gesneriaceae and Ericaceae. The Gesneriaceae were at one time separated into two subfamilies on the basis of whether the ovary of the inflorescence was superior or inferior (Fritsch, 1893–4). Burtt (1962), after an exhaustive study of the biology of the family, decided that a more satisfactory separation into subfamilies was produced by using geographical origin ⟨New

World versus Old World) and an unusual seedling character, aniso-cotyly (development from one instead of from both cotyledons). Chemical examination of flavonoid pigments of representative taxa provided excellent support for the re-classification, the data cutting across the older system of Fritsch (Harborne, 1966b, 1967b; Lowry, 1972). Burtt's isocotylous New World Gesnerioideae regularly contain a rare group of 3-desoxyanthocyanins (in 29 of 36 spp. surveyed) as orange red flower and leaf pigments, and yellow corolla colour is provided in these plants by carotenoid. By contrast, the anisocotylous Old World Cyrtrandoideae lack desoxyanthocyanin (absent from all of 45 spp. surveyed) and when yellow pigments are present, they are based on chalcones or aurones (in all of 12 yellow flowering species examined). Both subfamilies have ordinary antho-cyanins, but again glycosidic patterns are different, 3-rutinosides occurring characteristically in the Gesnerioideae, 3-sambubiosides and 3-arabinosylglucosides in the Cyrtandroideae. These differences in floral pigmentation are probably related to geographical factors, i.e. New World species have different pollinating vectors (e.g. humming birds) from the Old World taxa.

In the Ericaceae, the difficulty of classification stems from the extremely variable vegetative habit. The earlier and widely used system of Drude (1897—8) was recognized to be unsatisfactory by Watson et al. (1967) and a much needed re-classification, based on the analysis of 60 mainly morphological characters, was recently proposed by Stevens (1971). From the flavonoid viewpoint, the family is rich in unusual constituents and there is little difficulty in scoring plants for at least ten flavonoid characters. Indeed, a survey of 344 species (Harborne and Williams, 1973) provided useful support for many of Steven's decisions on tribal and subfamilial boundaries. The subfamily Rhododendroideae, which contains the ornamentally important genus *Rhododendron,* was found to be particularly rich in rare flavonols (e.g. gossypetin (2), azaleatin (6) and caryatin (7); for formulae, see p. 1062). Here, the chemical data are such as to suggest that the tribes Rhodoreae and Phyllodoceae might be placed even nearer to each other than indicated by Stevens. Gossypetin, for example, is a particularly consistent character in these two tribes (see Table 19.8), but is quite absent from all the other five tribes in the subfamily.

Turning now to examples where flavonoid data may provide a

Table 19.8   *The distribution of flavonoid aglycones in leaves of the genera of the subfamily Rhododendroideae* (classification according to P. F. Stevens)

Taxa	No. of species examined/no. of known species	Gossypetin	5-O-methyl-flavonols	Caryatin	Dihydro-flavonols
Bejarieae					
Bejaria	4/30	–	+	+	+
Rhodoreae					
Rhododendron	206/500	+	+	+	+
Therorhodion	1/3	+	+	–	–
Tsusiophyllum	1/1	–	–	–	+
Menziesia	3/7	+	–	–	–
Ledum	3/10	+	–	–	–
Cladothamneae					
Cladothamnus	1/1	–	–	–	–
Elliottia	1/1	–	–	–	–
Epigaeae					
Epigaea	2/3	–	+	–	+
Phyllodoceae					
Kalmia	2/8	+	–	–	–
Phyllodoce	4/6	+	+	+	+
Kalmiopsis	1/1	–	+	–	+
Rhodothamnus	1/2	+	+	–	+
Bryanthus	1/1	+	–	–	+
Ledothamnus	1/5	+	–	–	–
Leiophyllum	1/1	+	–	–	–
Loiseleuria	1/1	+	–	–	–
Daboecieae					
Daboecia	2/2	–	+	–	+
Diplarcheae					
Diplarche	1/1	–	–	–	–

Quercetin is common to all species, kaempferol and myricetin are variably present. From Harborne and Williams (1973).

decision between alternative classifications, let us consider the *Chrysanthemum* complex in the tribe Anthemideae of the Compositae. The decision here is one familiar to taxonomists of lumping all taxa together into the same genus, as proposed by Hoffmann (1894) or splitting them into a number of smaller genera, as suggested by Briquet and Cavallier (1916–7) and supported by Heywood (1959). Cytological data have not resolved the problem, since the base number ($x = 9$) is uniform throughout the group. Chemistry could clearly help; if homogenous, it would favour Hoffmann and if heterogenous, it would favour Briquet. In fact, flavonoid chemistry is variable, at least within a sample of 21 representative species (Harborne *et al.*, 1970). Although flavones are regularly present, there are variations in the methylation and glycosylation patterns and in whether flavonols and 6-hydroxy-flavonols occur (Table 19.9). Supporting evidence for heterogeneity within the group has come from a study of the polyacetylenes of root and leaf, since a considerable number of different compounds have been found to occur variously within these plants (Bohlmann *et al.*, 1964). There seems little doubt therefore that splitting the original *Chrysanthemum* complex into a number of smaller genera is supported by the chemical findings.

A second example of where flavonoid chemistry provides some weighting in favour of one classification over another is the case of the tribal boundaries of the Genisteae, in the Leguminosae. Roth-maler (1944) and Tutin *et al.* (1968) include some 18 genera in the tribe, with the centre around *Genista* and *Cytisus*. By contrast, Hutchinson (1964) adds into the same group several genera which are distinct on morphological and distributional grounds (Gibbs, 1966) and then divides them arbitrarily into no less than four tribes. *Genista* and *Cytisus*, which are morphologically and chemically difficult to separate, are placed in different tribes by Hutchinson. Chemical examination (Harborne, 1969c) revealed a range of flavonoids in the leaf. However, only two characters—accumulation of isoflavone and presence of leucoanthocyanidin—had a bearing on

Table 19.9   *Chemical variation in the chrysanthemum complex*

Genera sensu Briquet	Luteolin and Apigenin	Chrysoeriol	Quercetin	Quercetagetin	Patuletin	Flavone glucuronides	Flavone α-glucosides
Anthemis	+	−	−	−	+	+	−
Chrysanthemum	+	+	+	+	+	+	−
Dendranthema	+	−	+	−	−	+	−
Leucanthemum	+	−	−	−	−	+	−
Matricaria	+	−	−	−	−	−	+
Tripleurospermum	+	−	−	−	−	−	−
Tanacetum	+	+	+	−	−	+	−

Data from Harborne *et al.* (1970); similar results on the aglycones have been obtained independently by Greger (1969).

the taxonomic problem. Briefly, all taxa included in the Genisteae by Rothmaler characteristically accumulated isoflavone, and especially 5-methylgenistein which is a marker for the tribe, and lacked leucoanthocyanidin. By contrast, taxa excluded by Rothmaler but included by Hutchinson (such as *Hypocalyptus, Loddigesia, Argyrolobium* and *Lupinus*) lacked isoflavone and most of them synthesized lencoanthocyanidin. Thus Rothmaler's circumscription of the tribe is the preferred one, from the chemical viewpoint.

Finally, with regard to the employment of flavonoid chemistry as an integral part of programmes of taxonomic revision, it is perhaps too early to comment in detail. One programme on *Baptisia* has been in progress since the early 1960s (Alston and Turner, 1963); although the flavonoid data have been published (Markham *et al.*, 1970), the results have not yet been related in detail to the systematics of the group. Another programme on the tribe Caucalideae of the Umbelliferae was commenced in 1965 and flavonoid results on leaf and fruit have appeared (Crowden *et al.*, 1969; Harborne and Williams, 1972). Again, the synthesis of these data with those from biological characters has yet to be carried out, although a preliminary computer analysis, using some eighty biological and three flavonoid characters, has been described (McNeill *et al.*, 1969).

Obviously, the combined biological-chemical approach is highly desirable, since it means that precisely the same plant populations that are being scored for biological characters are being studied chemically. This is both economical and avoids errors in plant identification, a frequent source of confusion in the past. In the future, it is to be hoped that chemical studies of systematically difficult plant taxa will be initiated at the same time as the biological studies. Certainly, the flavonoids are among the most promising group of chemical markers to look for in such surveys.

19.5 Flavonoids at the species level

19.5.1 Intraspecific variation

Can one identify a plant species from its flavonoid pattern? The answer to this question is a qualified yes, but it very much depends on the plant group being studied and the degree of structural variation in the flavonoids present therein. Markham *et al.* (1970), in describing the identification of 62 flavonoids in 17 species of the genus *Baptisia*, reported that each species had a characteristic flavonoid pattern. However, Dement and Mabry (1972), in examining 44 populations of 13 species in the closely related legume genus *Thermopsis,* found that all the species had practically the same pattern, containing 12 flavones and 19 isoflavones. Thus, there was apparently no differentiation in flavonoid complement at the species level. These two are extreme examples and most genera probably fall somewhere in between, i.e. the majority of species will have individual patterns but a few may be identical, even in spite of morphological differences.

There are now a number of examples available of genera where sampling has been sufficiently detailed to show that species do have different flavonoid patterns. All five known species of *Eucryphia* (Eucryphiaceae) have been examined (Bate-Smith *et al.,* 1967) and each species has a characteristic pattern of leaf constituents. In addition, each species has at least one specific flavonoid present: *E. cordifolia,* azaleatin 3-arabinosylgalactoside; *E. glutinosa,* azaleatin 3-galactoside; *E. moorei,* quercetin 3-triglycoside; *E. milliganii,* dihydroquercetin 3-glycoside; and *E. lucida,* kaempferol 3,7-dimethyl ether. In *Plumbago* (Plumbaginaceae), seven of the ten known species have been surveyed and each has a distinctive pattern of flavonoids in leaf and petal (Harborne, 1967a). In *Daucus* (10 spp.) and *Torilis* (5 spp.) (Umbelliferae), distinction at the species level is provided by the flavonoids present in the fruits (Harborne and Williams, 1972). In *Pyrus* (Rosaceae) a survey of leaf of 17 species showed that there was a basic pattern common to all species and a number of other flavonoids with selective distribution. As the authors, Challice and Williams (1968), put it: 'every single species has a different combination of these rarer constituents'. In *Primula* (Primulaceae), a genus which is perhaps exceptionally rich in flavonoids, species can be distinguished on the basis of petal and leaf flavonoids and of different simple flavones in the farina (Harborne, 1968).

There are, of course, other genera where surveys have revealed only limited variation at the species level. *Impatiens aurantiaca* (Balsaminaceae) has a unique anthocyanidin, aurantinidin or 6-hydroxypelargonidin as the flower pigment (Jurd and Harborne,

1968). A survey of 18 species, including other orange flowered species, failed to reveal any other source, so this appears to be the only unusual species, in terms of anthocyanidin production, in the genus (Clevenger, 1971). Again, in *Briza,* a survey of 16 species showed that differences in flavonoid pattern were present but these were more closely related to geographical origin and ploidy than to speciation within the genus (Williams and Murray, 1972). Finally, another genus similar to *Thermopsis* has been discovered, namely *Thelesperma* (Compositae), in which 11 species have the same flavonoid profile. Variation was detected once in *T. simplicifolium* but this was related to ploidy and not to speciation (Melchert, 1966).

From these examples, it is clear that, although the chances are reasonable, it is impossible to predict whether flavonoids will provide species-specific markers for taxonomic purposes. When looking for differences at the species level, it should be remembered that they are more common in the glycosidic than in the aglycone pattern; also they may be present in one tissue or at one stage of growth and not another. How far differences in flavonoid patterns between species can be utilized in biosystematic studies will be discussed briefly in a later section (Section 19.5.3).

19.5.2 Infraspecific variation

In the early days of biochemical systematics, little attempt was made to screen more than one sample of a given species for flavonoids. However, with the increasing awareness of chemical variation in essential oils and other chemicals at the infraspecific level and with the closer collaboration between flavonoid chemists and systematists, more has been done to survey plant populations during recent years. Indeed, in recent studies, an effort has nearly always been made to check for variation below the species level. Also, an increasing number of papers have appeared concerned with sampling populations of species complexes or hybrid swarms, using flavonoid data in conjunction with cytological and morphological approaches.

The degree of flavonoid variation among populations of the same species clearly depends to some extent on the particular type of plant being examined. One might expect morphologically variable taxa such as the wild carrot, *Daucus carota* L., to show chemical variation, as indeed they do, but this is not necessarily so. Conversely, one might predict that a morphologically stable species would be uniform in its flavonoid pattern, but this is by no means always true; even among such taxa, chemical races may be present on occasions.

The problem of assessing infraspecific variation, apart from the labour of collecting representative samples and analysing them all by two-dimensional chromatography, is that of fluctuations in flavonoid levels. This applies particularly to the analysis of leaf tissue, which is the most readily available material for such work. Ideally, samples from different populations should be collected as seed material and then grown under identical environmental conditions. Unless this is done, variation may occur such that a major constituent in one sample becomes a minor one in a second or is undetectable in a third. Bearing this factor in mind, it appears from the results from many laboratories that there is an almost continuous range of variation, from species which are constant in their flavonoids to those in which well defined chemical races appear. In between, there are the majority in which the major components are generally uniform but which vary in some of the minor compounds.

Chemical stability at the species level is exemplified in the Lemnaceae, where 22 species from four genera showed predictive and distinctive flavonoid patterns, even after the sampling of 186 clones (McClure and Alston, 1966). Only one of the 22 species showed variation, *Lemna perpusilla,* and this was only in one flavonoid of seven present. A family where minor infraspecific variation is apparent is the Empetraceae. Data obtained from surveying 92 samples of six species are shown in Table 19.10 (Moore *et al.,* 1970). It is a fairly typical situation, with major constituents being uniformly present and minor components being irregular. The degree of variation present is probably that to be expected, considering the amount of heterozygosity present in most plant populations.

Where the variation becomes more clearcut in major flavonoids, it is possible to recognize chemical races. Some examples are given in Table 19.11, but this phenomenon does not appear to be common. It is presumably due to the fact that the compounds concerned are controlled by dominant-recessive, rather than by the more usual additive, inheritance in the leaves. In flowers, by contrast, dominant-recessive inheritance is the rule rather than the exception. Indeed, variation in flower pigmentation is quite common, and there are

Table 19.10   *Percentage frequency of distribution of flavonoids in leaves of six species of the Empetraceae*

Species	No. samples analysed	F1	F2	F3	F4	F5	F6	F7	F8	F9	F10	F11	F12	F13	F14	F15	F16	F17
Empetrum rubrum Vahl ex Willd.	18	83	100	17	100	6			83						6			
Empetrum nigrum L.	24	20	100	96	88	4	8	4	32	84	4				4	4		
Empetrum eamesii Fern. and Wieg.	30	77	100	93	93	19	13	3	3	22	10			6	3	6	13	
Corema album(L.) D.Don	8		100	100					100	100			13					
Corema conradii (Torrey) Torrey ex Loud.	4		75	100	100	50				100	100	50						
Ceratiola ericoides Michx.	8		100	50	100				50	50	100	25						75

From Moore *et al.*, 1970.

Key: F1, quercetin 3-arabinoside; F2, quercetin 3-galactoside; F3, quercetin 3-rutinoside; F4, gossypetin 3-galactoside; F9, methylated flavonol; others are unidentified.

Table 19.11   *Examples of 'chemical races' in flavonoid content*

Species	Races recognized	Reference
Eucalyptus sideroxylon A. Cunn. ex Wools Myrtaceae	Several, varying in presence/absence of five flavonol glycosides, dihydro-kaempferol 3-rhamnoside, catechins and *C*-methylflavones	Hillis and Isoi, 1965
Daucus carota L. Umbelliferae	in leaf, presence/absence of chrysoeriol 7-glucoside, etc. in fruit, presence/absence of luteolin 4'-glucoside	Harborne, 1971a
Foeniculum vulgare Miller Umbelliferae	Quercetin and kaempferol 3-glucuronides present/absent (3-arabinosides uniform) in leaf	Harborne and Saleh, 1971
Galium mollugo Rubiaceae	Hesperidin present/absent	Hegnauer, 1959
Pityrogramma triangularis (Kaulf) Maxon, Pityrogrammaceae	1. Ceroptene 2. Kaempferol derivative A 3. Kaempferol derivative B 4. Ceroptene + kaempferol derivatives	Smith *et al.,* 1971
Pterocarpus indicus Leguminosae	1. Pterocarpin, homopterocarpin 2. Angolensin 3. Homopterocarpin 4. Pterocarpin, angolensin, isoliquiritigenin, formononetin	Cooke and Rae, 1964
Smilax glycyphylla Liliaceae	1. Phloretin 2'-rhamnoside 2. Mangiferin (a xanthone) 3. Phenolics absent	Williams, 1964
Solanum chacoense and *S. stoloniferum* Solanaceae	in leaf, presence/absence of quercetin 3-(2^G-glucosylrutinoside), quercetin 3-sophoroside and luteolin 7-glucoside	Harborne, 1962

many species where presence of (anthocyanin) colour is a dominant and variable character among wild populations.

Knowledge of infraspecific variation in flavonoids has been utilized for identifying different cultivars in a range of ornamental and crop plants. While is it rarely possible to completely identify a particular cultivar from its leaf flavonoid pattern, cultivars can frequently be separated into groups according to the constituents that are present or absent.

19.5.3 Chemistry of hybrid plants

Chromatographic studies of leaf flavonoids in both natural and artificial hybrids have shown that inheritance is normally additive, although occasionally either some parental constituents are missing or some additional 'hybrid' compounds are present. The utility of flavonoid data for determining the parental origins of natural hybrids has been most extensively explored in one particular legume genus *Baptisia* (Alston and Turner, 1963; Brehm and Alston, 1964; Alston and Hempel, 1964). These chromatographic profiles have provided the only unambiguous method of identifying the parentage of some hybrids in natural *Baptisia* populations, cytological and morphological characters being insufficient for this purpose. This important early contribution to the subject of plant chemosystematics was carried out using mainly unidentified flavonoid spots, although most of the leaf flavonoids of *Baptisia* have been fully characterized subsequently (Markham *et al.,* 1970). The success of the *Baptisia* work has encouraged other systematists to utilize spot pattern data, together with other biosystematic approaches, for analysing natural hybrid swarms. Because of space limitations, only two of the many papers published on this topic can be mentioned here.

In one typical case, Hunter (1967) surveyed 36 clones, from seven different geographical sites, of a natural hybrid between *Vernonia lindheimeri* and *V. interior* (Compositae). Sixteen flavonoids or other phenols were regularly present. Hybridity was established by the fact that eight compounds were derived from one parent, three from the other, with five being common to both species involved. In another example, Quinn and Rattenbury (1972) analysed leaf twigs of a putative hybrid, *Dacrydium laxifolium* x *D. intermedium* growing naturally in the Tongariro National Park in New Zealand. Using flavonoid profiles and cytology, they were able to prove the plant was a true hybrid, making this the first documented case of hybridization in this gymnosperm genus. One parent *D. laxifolium* had five species-specific chromatographic components, four of which appeared in the hybrid, while the other parent had ten species-specific components, nine of which came through in the hybrid. The two spots missing from chromatograms of the hybrid were only trace constituents in the parents so their absence is not surprising.

It could be argued that in such studies, chemical identification of the individual flavonoids is not crucial to solving the biological problem present. This is true, but nevertheless it would be useful to know the chemical background in a reasonable number of such situations. Certainly, chemistry is essential if hybrid substances are to be identified and their origins traced. For example, a new hybrid substance was detected by Alston et al. (1965) in the leaves of *Baptisia leucantha* x *B. sphaerocarpa* and *B. alba* x *B. tincoria.* It was identified as quercetin 3-rutinoside-7-glucoside. Information on the flavonoids in leaf and flower of all the species involved was necessary before it could be established that the ability to synthesize quercetin 3-rutinoside came from one pair of parental species and the ability to synthesize quercetin 7-glucoside came from the other two parents. A remarkable facet of this investigation was the discovery that the latter ability is restricted in the parents to the floral tissue; thus it is only in the hybrid that the regulatory mechanism has been broken down to allow the synthesis of flavonol 7-glucoside in the leaf.

One of the few instances where the chemical analysis of parents and hybrids is relatively complete is with plants of the Appalachian *Asplenium* complex. The original chromatographic analyses of these ferns were carried out in 1963 by Smith and Levin, who showed that it was possible to identify hybrids derived from two and three parental species using spot pattern data and genome analysis. The clarity of the chromatographic data in elucidating the parental origin of the various natural hybrids is such that this has been quoted as one of the classic examples of additive inheritance of chemical characters in plants (e.g. Briggs and Walters, 1969). More recent chemical work (Smith and Harborne, 1971; Harborne et al., 1973) has confirmed the correctness of the original deductions in all cases. All three of the parental species involved are capable of synthesizing kaempferol and adding glucose to its 3- and 7-hydroxyl groups. One species, *Asplenium platyneuron,* has the added ability to methylate kaempferol in the 3- and 4'-positions and to acylate it with an aliphatic acid. A second species, *A. rhizophyllum,* has the capability of synthesizing kaempferol 3-sophoroside and 4'-glucoside and to acylate these glycosides with caffeic acid. The third species, *A. montanum,* lacks all these special enzymes but instead synthesizes four C-glycosylxanthones, including mangiferin. All these compounds appear, in appropriate combinations, in the various hybrid ferns; in some cases, new 'hybrid' components are also present, but, because of the scarcity of plant material, these have yet to be identified.

Natural hybridization sometimes involves changes in ploidy level and when this happens, some alteration in flavonoid synthesis can be expected. Thus it has been shown that doubling up the chromosomes within the same species can sometimes affect flavonoid patterns. The simple change from diploid to autotetraploid in the grass *Briza media* has been found to alter flavonoid synthesis by increasing the degree of hydroxylation in the B-ring. Luteolin derivatives accumulate at the expense of apigenin derivatives in both natural and artificially produced autotetraploids of this plant (Murray and Williams, 1973). In *Phlox* (Polemoniaceae), hybrid allotetraploids derived from several diploid species, including *P. drummondii,* have been analysed and been found to produce novel glycoflavones not present in any of the diploids. Here, the change is in degree of glycosylation and while diploids have di- and tri-*C*-glycosyl-luteolins, the tetraploids have mainly mono-*C*-glycosyl-luteolins. The authors, Levy and Levin (1971) suggest that these latter compounds are accumulated or modified precursors of the parental glycosylflavones, the change being due in this case to gene regression.

From the practical viewpoint, chemical analysis of hybrids can be useful for tracing the origin of cultivars in crop or ornamental plants. In *Citrus,* for example, fourteen flavanone glycosides such as naringin and neohesperidin occur variously in the fruits of lemons, oranges and grapefruits. Albach and Redman (1969) analysed by one-dimensional TLC fruit from 49 hybrids of 18 crosses and found that the patterns were useful in evaluating the origin of established *Citrus* varieties. Again, in *Ulmus,* Santamour (1972) has used leaf flavonoids to determine the origin and identity of cultivated elms. Quercetin 3, 7-diglucoside, for example, is a useful marker substance since it occurs in the European white elm *Ulmus laevis* but is absent from the morphologically similar American elm *U. americana.* Furthermore, it is inherited in an additive fashion, since it could be detected in the hybrid *U. laevis* x *U. parvifolia.*

19.6 Flavonoids and plant geography

The idea that flavonoid distribution in plants may be related to climatic factors dates back many years (see e.g. Onslow, 1925). One of the conclusions of an early comprehensive survey of flower tissues for anthocyanin pigment (Beale *et al.,* 1941) was that delphinidin types predominated in alpine and temperate floras whereas pelargonidin types were more frequent in tropical climates. Subsequent work (Forsyth and Simmonds, 1954) showed that this was derived from a biased sample and that there is no simple relationship between the extent of anthocyanidin hydroxylation and geography. More recently, however, significant correlations have been noted at the family level. Examples have been given earlier in this chapter where anthocyanidin-type is related to habitat in the families Gesneriaceae and Plumbaginaceae. Here, chemical differences are clearly related to the function of anthocyanins as floral pigments and the effect on flower colour of natural selection by animal pollinators.

In the case of flavonoids in leaves and other plant parts, correlations between chemistry and plant geography have only emerged during the last five years. Some nine examples are listed in Table 19.12. A variety of flavonoid types are involved and there is no obvious relationship with the synthesis of anthocyanin in the flowers of the same plants. These variations must have some other cause and may be connected with a role for leaf flavonoids in providing protection from either microbial attack or animal predators. A change in habitat of a plant species may produce a change in the predatory fauna, so that the pressure to produce a particular pattern of flavonoids in the leaf may ease and a different set of compounds could be produced. If changes occur in leaf flavonoids with geographical migration, then the study of these constituents might help solve some of the outstanding problems in phytogeography. This is indeed the philosophy behind two recently developed research programmes, one sponsored by the International Biological Programme and concerned with xerophytic plants in North and South America and the other, at the University of Reading, concerned with the relationships between the Northern and Southern hemisphere temperate floras.

The place of flavonoids in the IBP Programme set up to solve the problem of American amphitropical desert disjunctions has been discussed by Solbrig (1972) and some of the first results have been reported by Mabry (1973). Research has been concentrated on three xerophytic genera, *Larrea* (Zygophyllaceae), *Prosopis* and *Cercidium* (both Leguminosae), all of which show conspicuous disjunctions in their distributions in the two subcontinents. In *Larrea,* there are

Table 19.12 *Correlations between flavonoid patterns and phytogeography*

Genus	Flavonoid(s)	Geographical correlation	Reference
Briza	Glycoflavones and others in leaf	Vitexin derivatives in 6 Eurasian spp., replaced by orientin derivatives in 10 S. American spp.	Williams and Murray, 1972
Carex	Flavones in leaf	Luteolin 7-glucoside in N. temperate plants, replaced by tricin 5-glucoside in S. temperate plants.	Aye, 1969
Dalbergia	Isoflavonoids in heartwood	Isoflavones in Asian and African spp; isoflavans and pterocarpans in S. American spp.	Braga de Oliveira *et al.,* 1971
Empetrum	Fruit anthocyanins and leaf flavonoids	Delphinidin derivs. and rutin in N. temperate sp.; cyanidin derivs. and infrequent rutin in S. temperate sp.	Moore *et al.,* 1970
Eucryphia	Flavonol 5-methyl ethers in leaf	Azaleatin and caryatin in S. American spp., replaced by other flavonoids in Australasian spp.	Bate-Smith *et al.,* 1967
*Pyrus**	Flavones and flavonols in leaf	Luteolin and apigenin in eastern Asian spp., kaempferol and quercetin in W. Asian and European spp.	Challice and Williams, 1968
Rhododendron	8-Hydroxyflavonols and flavonol 5-methyl ethers in leaf	Gossypetin and azealatin widespread in Asian spp. but absent from only known Australian sp. *R. lochae*	Harborne and Williams, 1971
Senecio	Glycoflavones and flavonols in leaf	Quercetin 3-glucoside in spp. from Madagascar, Kenya and Canaries; glycoflavones and rutin in S. and SW. African spp.	Glennie *et al.,* 1971
Valerianella	Flavonols and 6-hydroxyflavones in leaf	Kaempferol and quercetin in Asiatic spp., replaced by 6-hydroxyluteolin in Mediterranean and N. American spp.	Greger and Ernet, 1973

*A similar progression has been noted in *Malus* (Williams, 1972) but the details have not yet been published.

certainly differences in leaf flavonoid patterns of plants collected in Arizona, New Mexico, Utah and in N.W. Argentina, but these plants have yet to be analysed in detail (Hunziker *et al.*, 1972). In *Prosopis*, it has been found that two South American species, separated by several thousands of miles, i.e. *P. alba* from the Argentine and *P. chilensis* from Chile, have essentially the same eleven flavonol glucosides and glycoflavones in the leaves. Such stability is not unusual but what is much more remarkable is that two disjunct varieties of *Prosopis reptans* separated by a vast geographical distance (one from South Texas and the other from central Argentina) have nearly identical flavonoid patterns. As Mabry (1973) suggests: 'ecotypic variation may not occur if the disjunct populations occupy ecosystems which do not require a physiologically different organism'. Certainly in this case, the desert niches occupied by the plant in the two Americas are remarkably similar in climate and vegetational structure.

While there is a land bridge between desert regions in North and South America, much greater barriers separate the Northern temperate floras of Scandinavia and Great Britain from the Southern temperate floras of Tierra del Fuego, South Chile and the Falkland Islands. Only long range dispersal can explain how a variety of temperate plant species have managed to establish themselves in such widely separated habitats. A biosystematic study of relationships of plants found growing in both North and South temperate regions has been in progress for several years (Moore, 1972). Among genera that have been examined are *Acaena, Carex, Empetrum* and *Plantago*. Significant differences have been observed in the flavonoid profiles of plant populations in the two habitats.

The most detailed study has been of *Empetrum*, in which genus 31 flavonoids and other phenolics have been found in the leaves (Moore *et al.*, 1970). *E. nigrum* of Northern temperate areas is represented in the Southern hemisphere by *E. rubrum* and the two species differ modally in their flavonoids (see Table 19.10 for details). In particular, quercetin 3-arabinoside occurs in 83% of samples of *E. nigrum* but only in 20% of samples of *E. rubrum*. On the other hand, quercetin 3-rutinoside is much less frequent in *E. nigrum*, but is common in *E. rubrum* (17 and 96% respectively). There is also a striking difference in fruit colour, which involves the replacement of delphinidin derivatives in the black berries of *E. nigrum* by cyanidin derivatives in the red berries of *E. rubrum*. The two taxa are practically identical in their biological attributes and they are probably best regarded as forms of the same species. The chemical differences have presumably been established as a result of the migration of the species from the Northern to Southern hemisphere. Within *E. rubrum*, it is interesting to find a progressive change in flavonoid profile in populations collected respectively from central Chile, Southern Chile, Tierra del Fuego and Gough Island. This latter population is depauperate in flavonoids (six constituents; compared to up to 11 in other samples), suggesting that a loss in flavonoids accompanies isolation on an island, remote from the mainland.

In other genera and species, smaller chemical differences have been noted between disjunct populations. The sea plantain *Plantago maritima* has 14 flavonoids in the leaf, but the only major amphitropical discontinuity is in the occurrence of 6-hydroxyluteolin 7-glucoside. This compound is present in all but four of 27 European samples, and in five of six North American populations, but is quite absent from 10 South American samples. The distribution pattern of the structurally related 6-hydroxyluteolin 7-glucuronide shows the same trend, i.e. it is present in 45% of Northern samples, compared to 30% in Southern plants (Moore *et al.*, 1972).

A limited survey of *Carex* populations growing in North and South temperate floras (Aye, 1969) showed that some species showed amphitropical discontinuities while others were constant in leaf flavonoid pattern. Thus in *Carex microglochin* (2N and 4S populations examined) and *C. macloviana* (10N, 5S), luteolin 7-glucoside is confined to Northern populations; it is either absent from or is replaced by tricin 5-glucoside in Southern populations. On the other hand, in *Carex curta* (4N/9S) and *C. magellanica* (5N/5S) there were no differences between the geographically separated populations.

Correlations between flavonoids and phytogeography have been detected along other routes of plant migration. One well established route is from South-East Asia, through the Middle East to the European Mediterranean area and then onto North America. Both the rosaceous genera *Malus* and *Pyrus* follow this route in their speciation and significant correlations with flavonoid chemistry have been noted (see Table 19.12; also Challice and Westwood, 1973). Similar but less pronounced differences have also been noted by Bate-Smith (1968) for *Crocus* and *Iris*. Some plants, particularly those of central Eurasian origin, appear to have evolved phyto-geographically by radiation in several directions. In *Dillenia, Geranium* and *Ulmus*, this advancement and dispersal appears to be correlated with an impoverishment in flavonoids, although the degree of impoverishment is less marked in some taxa than in others (Bate-Smith and Richens, 1973).

One final example may be taken from the phytogeography of the Southern hemisphere, where there is a disjunction between the Australasian and South American flora. These floras are clearly related, having some 50–60 genera in common, and much evidence indicated that the direction of migration is generally from East (Australia) to West (South America). There are, however, two possible exceptions, one *Schizeilema,* where the chromosomal data indicate a flow in the opposite direction (Moore, 1972) and the other is *Eucryphia,* where the chemical results also suggest a reverse of the normal migratory route (Bate-Smith *et al.,* 1967). In *Eucryphia,* the two South American taxa have the primitive woody angiosperm markers, azaleatin and caryatin, whereas in the three Australasian species, they are replaced by more diverse flavonoid types. Clearly, other plants need to be examined before the possibility of a West-East migratory route can be established, but this example does indicate that flavonoid data may help in both confirming the fact of migration and also in determining its direction.

The use of flavonoids in phytogeographical studies has only just begun. Their use must obviously be closely linked with biological research programmes so that all other aspects of varying plant populations are studied. While other secondary constituents and also proteins and enzyme patterns have been shown to vary with geography, there is little doubt that flavonoids offer a number of advantages over other chemical approaches to solving biological problems of a phytogeographical nature. This is a rapidly expanding area of research and one can confidently predict many interesting developments in the future.

Acknowledgements

The author is grateful to Dr T. Swain for his perceptive comments on this manuscript. He thanks his colleagues at Reading for their assistance in the flavonoid studies in progress at this University.

Financial assistance is acknowledged from the Science Research Council and the Centre National de la Recherche Scientifique (programme RCP 286).

References

Albach, R. F. and Redman, G. H. (1969), *Phytochemistry* **8**, 127.

Alston, R. E. and Hempel, K. (1964), *J. Heredity* **55**, 267.

Alston, R. E., Rosler, H., Naifeh, K. and Mabry, T. J. (1965), *Proc. Natl. Acad. Sci.* **54**, 1458.

Alston, R. E. and Turner, B. L. (1963), *Biochemical Systematics,* Prentice Hall, New Jersey.

Asker, S. and Frost, S. (1970), *Hereditas* **65**, 241.

Aye, T. T. (1969), M.Sc. Thesis, University of Reading.

Baker, H. G. (1948), *Ann. Bot. NS.* **12**, 207.

Bate-Smith, E. C. (1962), *J. Linn. Soc. Bot.* **58**, 39.

Bate-Smith, E. C. (1968), *J. Linn. Soc. Bot.* **60**, 325.

Bate-Smith, E. C. and Harborne, J. B. (1971), *Phytochemistry* **10**, 1055.

Bate-Smith, E. C. and Richens, R. H. (1973), *Biochemical Systematics* **1**, 141.

Bate-Smith, E. C., Davenport, S. M. and Harborne, J. B. (1967), *Phytochemistry* **6**, 1407.

Beale, G. H., Price, J. R. and Sturgess, V. C. (1941), *Proc. Roy. Soc. Lond.* **130B**, 113.

Bohlmann, F., Arndt, C., Bornowski, H., Kleine, K. M. and Herbst, P. (1964), *Chem. Ber.* **97**, 1179.

Bohm, B. A. and Glennie, C. W. (1971), *Canad. J. Bot.* **49**, 1799.

Braga de Oliveira, A., Gottlieb, O. R., Ollis, W. D. and Rizzini, C. T. (1971), *Phytochemistry* **10**, 1863.

Brehm, B. G. and Alston, R. E. (1964), In *Taxonomy, biochemistry and serology* (C. A. Leone, ed.), pp. 265–272. Ronald Press, New York.

Briggs, D. and Walters, S. M. (1969), *Plant variation and Evolution* p. 229. Weidenfeld and Nicolson, London.

Briquet, J. and Cavallier, F. (1916–7), In *Flore des Alpes maritimes* (E. Burnat, ed.), vol. 6, pp. 1–344, Lyon.

Burtt, B. L. (1962), *Notes Roy. Botan. Garden Edinb.* **24**, 205.

Challice, J. S. and Westwood, M. N. (1973), *Bot. J. Linn. Soc.* **67**, 121.

Challice, J. S. and Williams, A. H. (1968), *Phytochemistry* **7**, 1781.

Clevenger, S. (1971), *Evolution* **25**, 669.

Cooke, R. G. and Rae, J. D. (1964), *Australian J. Chem.* **17**, 379.

Cronquist, A. (1968), *The Evolution and Classification of Flowering Plants.* Thomas Nelson, London.

Crowden, R. K., Harborne, J. B. and Heywood, V. H. (1969), *Phytochemistry* **8**, 1963.

Dement, W. A. and Mabry, T. J. (1972), *Phytochemistry* **11**, 1089.

Drude, O. (1897–98), In *Die Naturlichen Pflanzenfamilien* (A. Engler and K. Prantl, eds.), **3**, (8), 63.

Erdtman, H. (1963), In *Chemical Plant Taxonomy* (T. Swain, ed.), pp. 89-126. Academic Press, London.

Forsyth, W. G. C. and Simmonds, N. W. (1954), *Proc. Roy. Soc. Lond.* **142**, 549.

Francis, C. M., Millington, A. J. and Bailey, E. T. (1967), *Australian J. Agr. Res.* **18**, 47.

Fritsch, K. (1893–4), In *Die Naturliche Pflanzenfamilien* (A. Engler and K. Prantl, eds.), **4**(3B), 133.

Gibbs, P. E. (1966), *Notes R. Botan. Gdn. Edinb.* **27**, 11.

Glennie, C. W. and Harborne, J. B. (1971), *Phytochemistry* **10**, 1325.

Glennie, C. W., Harborne, J. B., Rowley, G. D. and Marchant, C. J. (1971), *Phytochemistry* **10**, 2413.

Greger, H. (1969), *Naturwissenschaften* **56**, 467.

Greger, H. and Ernet, D. (1973), *Phytochemistry* **12**, 1693.

Harborne, J. B. (1962), *Biochem. J.* **84**, 100.

Harborne, J. B. (1964), *Phytochemistry* **4**, 647.

Harborne, J. B. (1966a), *Phytochemistry* **5**, 111.

Harborne, J. B. (1966b), *Phytochemistry* **5**, 589.

Harborne, J. B. (1967a), *Phytochemistry* **6**, 1415.

Harborne, J. B. (1967b), *Phytochemistry* **6**, 1643.

Harborne, J. B. (1968), *Phytochemistry* **7**, 1215.

Harborne, J. B. (1969a), *Phytochemistry* **8**, 177.

Harborne, J. B. (1969b), *Phytochemistry* **8**, 419.

Harborne, J. B. (1969c), *Phytochemistry* **8**, 1449.

Harborne, J. B. (1971a), In *Chemotaxonomy of the Leguminosae* (J. B. Harborne, D. Boulter and B. L. Turner, eds.), pp. 31–72. Academic Press, London.

Harborne, J. B. (1971b), In *The Biology and Chemistry of the Umbelliferae* (V. H. Heywood, ed.), pp. 293–314. Academic Press, London.

Harborne, J. B. (1971c), *Phytochemistry* **10**, 1569.

Harborne, J. B. and Saleh, N. A. M. (1971), *Phytochemistry* **10**, 399.

Harborne, J. B. and Williams, C. A. (1971), *Phytochemistry* **10**, 2727.

Harborne, J. B. and Williams, C. A. (1972), *Phytochemistry* **11**, 1741.

Harborne, J. B. and Williams, C. A. (1973), *Bot. J. Linn. Soc.* **66**, 37.

Harborne, J. B., Heywood, V. H. and Saleh, N. A. M. (1970), *Phytochemistry* **9**, 2011.

Harborne, J. B., Williams, C. A. and Smith, D. M. (1973), *Biochemical Systematics* **1**, 51.

Hegnauer, R. (1959), *Pharm. Ztg. ver. Apotheker-Ztg.* **104**, 382.

Heywood, V. H. (1959), *Proc. Bot. Soc. British Isles* **3**, 177.

Heywood, V. H. (1973), In *Chemistry In Evolution and Systematics* (T. Swain, ed.), pp. 355–376. Butterworth's, London.

Hillis, W. E. (1967), *Phytochemistry* **6**, 845.

Hillis, W. E. and Isoi, K. (1965), *Phytochemistry* **4**, 541.

Hoffmann, O. (1894), *Naturl. Pflfam,* **4**(5) 267.

Hunter, G. E. (1967), *Am. J. Bot.* **54**, 473.

Hunziker, J., Palacios, R. A., Valesi, A. G. and Puggio, L. (1972), *Proc. 1st Latin-American Symp. Bot.* 265.

Hutchinson, J. (1964), *The Genera of Flowering Plants* Vol. 1. Oxford University Press, London.

Hutchinson, J. B. (1974), *Genera of Flowering Plants* Vol. 3, in press. Oxford, Clarendon Press.

Iiyama, K. and Grant, W. F. (1972), *Canad. J. Bot.* **50**, 1529.

Jurd, L. and Harborne, J. B. (1968), *Phytochemistry* **7**, 1209.

Kubitski, K. (1968), *Ber. Deut. Botan. Ges.* **81**, 238.

Levy, M. and Levin, D. A. (1971), *Proc. Nat. Acad. Sci. U.S.A.* **68**, 1627.

Linczevski, I. (1968), In *Novitates Systematicae Planterum Vascularium* (S. Czerepanov, ed.), pp. 171–177. Nauka Sons, Leningrad.

Lowry, J. B. (1972), *Phytochemistry* **11**, 3267.

McClure, J. W. and Alston, R. E. (1966), *Amer. J. Bot.* **53**, 849.

Mabry, T. J. (1966), In *Comparative Phytochemistry* (T. Swain, ed.), pp. 231–244. Academic Press, London.

Mabry, T. J. (1973), In *Chemistry in Evolution and Systematics* (T. Swain, ed.), pp.377–400. Butterworths, London.

McNeill, J., Parker, P. F. and Heywood, V. H. (1969), In *Numerical Taxonomy* (A. J. Cole, ed.). Academic Press, London.

Markham, K. R., Mabry, T. J. and Swift, W. T. (1970), *Phytochemistry* **9**, 2359.

Melchert, T. E. (1966), *Amer. J. Bot.* **53**, 1015.

Moore, D. M. (1972), In *Taxonomy Phytogeography and Evolution* (D. H. Valentine, ed.), pp. 115–138. Academic Press, London.

Moore, D. M., Harborne, J. B. and Williams, C. A. (1970), *Bot. J. Linn. Soc.* **63**, 277.

Moore, D. M., Williams, C. A. and Yates, B. (1972), *Bot. Notiser* **125**, 261.

Murray, B. G. and Williams, C. A. (1973), *Nature, Lond.* **243**, 87.

Onslow, M. W. (1925), *The Anthocyanin Pigments of Plants.* 2nd Edn, University Press, Cambridge.

Parups, E. V., Proctor, J. R., Meredith, B. and Gillett, J. M. (1966), *Canad. J. Bot.* **44**, 1177.

Quinn, C. J. and Rattenbury, J. A. (1972), *New Zealand J. Bot.* **10**, 427.

Rothmaler, W. (1944), *Fedde Repert.* **53**, 137.

Santamour, F. S. (1972), *Bull. Torrey. Bot. Club* **99**, 127.

Santamour, F. S. (1973), *Phytochemistry* **12**, 611.

Smith, D. M. and Harborne, J. B. (1971), *Phytochemistry* **10**, 2117.

Smith, D. M. and Levin, D. A. (1963), *Amer. J. Bot.* **50**, 952.

Smith, D. M., Craig, S. P. and Santarosa, J. (1971), *Am. J. Bot.* **58**, 292.

Solbrig, O. T. (1972), In *Taxonomy Phytogeography and Evolution* (D. H. Valentine, ed.), pp. 85–100. Academic Press, London.

Stevens, P. F. (1971), *Bot. J. Linn. Soc.* **64**, 1.

Swain, T. (ed.). (1963), *Chemical Plant Taxonomy* Academic Press, London.

Taylor, R. J. and Campbell, D. (1969), *Evolution* 23, 153.

Tindale, M. D. and Roux, D. G. (1969), *Phytochemistry* 8, 1713.

Tutin, T. G., Heywood, V. H. (ed.). (1968), *Flora Europea* Vol. II, pp. 86–104. Cambridge Univ. Press, Cambridge.

Watson, L., Wiliams, W. T. and Lance, G. N. (1967), *Proc. Linn. Soc. Lond.* 178, 25.

Williams, A. H. (1964), *Nature Lond.* 202, 824.

Williams, A. H. (1972), *Phytochemistry* 11, 874.

Williams, C. A. and Murray, B. G. (1972), *Phytochemistry* 11, 2507.

Williams, C. A., Harborne, J. B. and Clifford, H. T. (1971), *Phytochemistry* 10, 1059.

Williams, C. A., Harborne, J. B. and Clifford, H. T. (1973), *Phytochemistry* 12, 2417.

Chapter 20

Evolution of Flavonoid Compounds

T. SWAIN

20.1 Introduction

The world around us has been shaped by evolution. In social terms, evolutionary processes seem fast. The changes which have taken place in the organization of human societies over the last 1000 or even 100 years are indeed enormous and have had, and will continue to have, important ecological and evolutionary consequences for ourselves and other living organisms. But in biological terms change is very slow. It has taken two million years for man to evolve from the tool-using *Australopithecus,* and placental mammals had been evolving for well on 100 million years before even the first of the higher primates arose nearly 35 million years ago (Stebbins, 1966; McAlester, 1968). And when it is remembered that the biochemical differences between ourselves and modern day representatives of these early primates are extremely small, e.g. one or two amino acid differences in the structure of a few proteins (Dayhoff, 1969), or a few minor changes in excretory patterns (Smith, 1968), it is obvious that biochemical evolution is even slower still. It seems likely, therefore, that we will only be able to determine striking changes in biochemical evolution by looking at the differences between many major diverse taxa. Unfortunately, such data are scarce. We know a lot about the biochemistry of a very few organisms: man himself, his domestic animals and cultivated plants, and a few selected pathogens which prey on him or on them.

However, over the past two decades, information has become increasingly available about the distribution of various types of discrete chemical compounds in a wide variety of plants, and a great deal has been uncovered about their biosynthetic interrelationships. Admittedly, 90% of the information is about angiosperms, but nevertheless certain trends in biochemical evolution in the Plant Kingdom as a whole are becoming apparent. An examination of such trends in the evolution of flavonoids and their importance in plants forms the subject of this chapter. We start with a short review of the evolution of the Plant Kingdom itself and a discussion of the underlying problems of determining evolutionary processes in biochemical terms, and then relate these concepts to the biochemical elaboration of the flavonoids.

20.2 Evolution of plants

Life on earth began about 3.2 billion (10^9) years ago in what is called the Pre-Cambrian era (Ponnamperuma, 1972). How it began or even where it began cannot be and probably never will be completely answered. It seems likely that it arose through the molecular aggregation of complex polymers which can demonstrably be formed *in vitro* from simple precursors by the application of electrical or other sources of energy (Fox, 1973). But so far, prebiotic bio-chemistry has only been able to explain the way in which biological molecules are formed, their possible modes of replication and combination, and it has not been able to determine how the complex interplay of controlled biochemical reactions which determine the activity of the simplest living organism or even cellular organelle ever arose.

The evidence that anuclear (procaryotic) organisms were present in the early Pre-Cambrian Era rests on two separate sets of observations: the direct examination with light and electron micro-scopes of structurally preserved forms resembling many of today's microbiota in ancient sedimentary rocks, and the detection of accompanying chemical substances of presumed biological origin (Schopf, 1970; Echlin, 1970). Even though there is some contention about the very earliest reported microfossils, the most conservative estimates agree that living forms like present bacteria and blue-green algae are preserved in the rocks of the Transvaal Supergroup and Canadian Gunflint Iron Formations which are around two billion years old (Echlin, 1970).

The evolution of life between its first detection and the first abundant macrofossil record of simple worm-like animals which is found in the Australian Edicara Hills Formation of about 600 million years ago is a complete mystery (McAlester, 1968). Obvi-ously all these macrofossils of soft bodied metazoans must have been remains of eucaryotic organisms, that is organisms in which the nucleus is enclosed in a membrane, since no present-day multicellular procaryotes, which have no such membrane, are known to exist. There is evidence of nuclei in spheroidal green-algal-like microfossils from the Bitter Springs Formation which is 1000 million years old (Schopf, 1970). But the change from unicellular form to the differentiated multicellular complex organization of an early annelid is vast and there are, so far, no intermediates known.

We are faced, therefore, with a major difficulty in discussing the evolution of any group of naturally-occurring chemical compounds. Over two and a half billion years of increasing biological complexity must have been brought about by changes in various physical parameters affecting the earth's surface. For example the percentage of oxygen in the atmosphere, the diminution in volcanic activity and radioactivity, the decrease in the amount of ultraviolet light reaching the earth's surface and so on (Gass *et al.*, 1971). Such changes over this length of time must have had numerous biochemical consequences, but none of these can be charted against a known increase in morphological complexity. Instead, we find in the Plant Kingdom a scanty fossil record of single-celled organisms for about 1.5 to 2.5 billion years of Pre-Cambrian time, and then a sudden upsurge of multicellular green algae in the late Cambrian (*c.* 520 million years ago) with an increasing complexity of forms and probably the introduction of red and brown algae into the Ordovician, Silurian and early Devonian Periods (see Table 20.1). And then, there is the clear appearance of more complex organization in the earliest land plants. From then on, about 400 million years ago, the fossil record is sufficient to pinpoint the major morphological and anatomical changes that have taken place in the land flora (Banks, 1970). Many of these appear to have occurred in the first 50 million years or so of land plant evolution (Table 20.1). Indeed, only the gymnosperms and angiosperms have arisen since that time (*c.* 350 million years ago) (Table 20.1, Fig. 20.1) (Banks, 1970).

Because we can equate the structure of the fossils of ancient land plants to modern forms, it is relatively easy to suggest probable phylogenetic affinities in today's flora. This does not mean, however, that we can assign accurate ancestral relationships between the major groups. Indeed, the fossil record contains many examples which cannot be easily assigned to any present day taxon, and although some are regarded as representing intermediate forms, others have more doubtful affinities. Based on the observed morphology, especially of the reproductive organs of present and past plants, we can delineate the general overall affinities of higher plants and these are as shown in Fig. 20.1 (Banks, 1970; Bell and Woodcock, 1972; Sporne, 1965, 1970).

It is generally assumed that the first land plants arose from some class of advanced heterotrichous green algae, perhaps ancestors of today's Chaetophorales (Chlorophyta) (Banks, 1970). Although the fossil record points to primitive vascular plants (tracheophytes) having evolved before the bryophytes (Banks, 1970), this seems

Table 20.1   *Classification and age of Plant Kingdom*

Division* (number of species)	Subdivision	Class	Common name	Age of oldest known fossil (millions of years)
Procaryotes				
Schizophyta (1200)		Schizomycetes	Bacteria	3200?
Cyanophyta (1200)		Cyanophyceae	Blue-green algae	2800
Eucaryotes				
Chlorophyta (6500)		Chlorophyceae	Green algae	1000
		Charophyceae	Stoneworts	450
Phaeophyta (1500)		Phaeophyceae	Brown algae	400–600
Rhodophyta (3000)		Bangiophyceae Florideophyceae	Red algae	600–900
Fungi (3200)			Fungi	460
Bryophyta (23000)		Hepaticae (9000)	Liverworts	380
		Anthocerotae (30)	Hornworts	?
		Musci (14000)	Mosses	280
Tracheophyta (297000)	Psilopsida (3)		Psilophytes	400
	Lycopsida (1300)		Lycopods	380
	Sphenopsida (25)		Horsetails	380
	Pteropsida (295600)	Filicinae (9300)	Ferns	375
		Gymnospermae (640)	Gymnosperms	350
		Angiospermae (286000)	Angiosperms	135

*Several algal divisions are omitted.

unlikely in view of the close similarity of the protonemal phase in the bryophytes and the algae. However, the bryophytes have never been a major component of the world's vegetation and many of today's forms, which are epiphytic on angiosperms, perhaps represent a result of a later burst of evolution (Watson, 1967).

The primitive vascular plants give rise to groups leading more or less directly to plants akin to modern *Psilotum*, lycopods, (Lycopsida), horsetails (Sphenopsida) and ferns (Filicinae), whereas the evolution of angiosperms and the various groups of gymnosperms

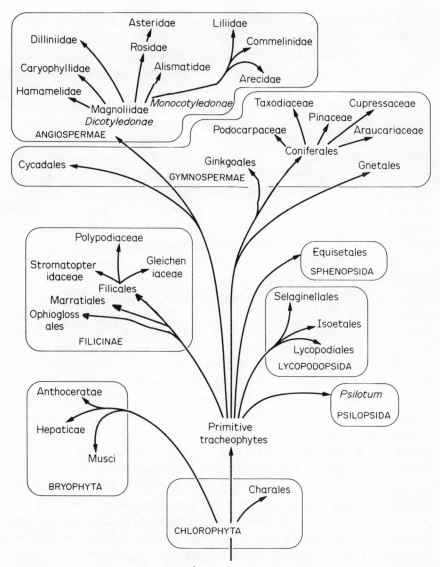

Figure 20.1 Evolution of higher plants.

probably arose through some now extinct intermediate forms such as the seed ferns (Pteridosperms) and the Cordaitales. The relationship of a number of other extinct forms, Bennettitales and Caytoniales, is less clear but their origin was probably in the seed ferns which they closely resemble (Banks, 1970; Sporne, 1965).

It is believed that the ferns arose from early groups of psilophytes in middle Devonian times. Three distinct modern orders are recognized (Copeland, 1947): Ophioglossales, Marratiales and Filicales. The latter, which contains the majority of fern species, is sometimes subdivided into three (Emberger, 1960) or four (Pichi-Sermolli, 1959) extra orders. Although the Ophioglossales are regarded as the most primitive group they have no well-established fossil record and are not regarded as truly ancestral to other groups (Sporne, 1970; Bell and Woodcock, 1972). The Marratiales are regarded as an intermediate group perhaps ancestral to the Filicales (Cronquist, 1971) among which certain genera of the Polypodiaceae are regarded as one of the most advanced. The most primitive members of the Filicales are the Osmundaceae, Stromatopteridaceae and Gleicheniaceae (Sporne, 1970). These last two families are regarded by Bierhorst (1971) as being very closely related to *Psilotum*.

The modern gymnosperms are usually divided into four orders; Cycadales, Coniferales, which contain the bulk of gymnosperm species, Ginkgoales, and Gnetales (Bell and Woodcock, 1972). Some authorities place the Taxaceae as a distinct order (Sporne, 1965). The Cycads are quite distinct from the rest and are usually treated as being closer to the line leading to the angiosperms. The Ginkgoales is today a monotypic order whose fossil record certainly does not extend back more than 250 million years (Banks, 1970; Sporne, 1965) and perhaps less. It has no direct affinities with the other two orders of gymnosperms with which it is grouped, but probably constitutes a parallel evolutionary line which separated early on. The Gnetales are a relatively diverse group which appear to be much more recent, the fossil record in the form of pollen only going back 60 million years or so (Sporne, 1965). Although at one time it was believed that this group showed features which indicated that they were intermediate between gymnosperms and angiosperms, this is no longer held to be true and they are regarded as a specialized offshoot of the conifer line (Bell and Woodcock, 1972). The fossil record of the Coniferales extends well back into the Carboniferous era. Of the seven modern families, the Podocarpaceae, Taxaceae and Araucariaceae are regarded as being overall the most primitive and the Taxodiaceae and Cupressaceae being the most advanced (Sporne, 1965).

All modern authorities divide the angiosperms into two major divisions, monocotyledons and dicotyledons and it is often implied that the former were derived from the latter early on in angiosperm evolution (Takhtajan, 1969; Cronquist, 1968). There are a number of reports of angiosperm fossils in pre-Cretaceous eras, but these are usually regarded as questionable. Fossil pollen, leaves, fruit and other remains, all clearly attributable to the angiosperms, however, have been recognized in Cretaceous strata over 100 million years old but they do not become a major constituent of fossil floras until upper Cretaceous times. However, most modern angiosperm families are not represented in the fossil record until the early Tertiary times (60 million years ago). On the basis of several morphological characters, including the possession of uniaperturate pollen, several modern authorities propose that the Magnoliidae are the most primitive group of modern flowering plants and ancestral to all other subclasses (Fig. 20.1: Cronquist, 1968; Takhtajan, 1969). Other authorities, however, believe on the basis of the fossil record, on chemical grounds, and by different interpretation of the morphological evidence that the Hamamelidae, Dilleniidae and Rosidae are equally as ancient as the Magnoliidae (Bate-Smith, 1972; Kubitski, 1969; Barnard, 1973). Barnard (1973) has further suggested that the monocots, especially the Arecidae, must have arisen from the basic angiosperm stock (assuming a monophyletic origin) early on in the lower Cretacean expansion of the division. An arrangement of the angiosperm subclasses based on Barnard's data, which takes into account the fossil record but is otherwise arranged according to Cronquist (1968), is shown in Fig. 20.2.

20.3 Biochemical evolution

Evolution implies change, change which imparts to an organism a better fitness for survival. Since the form and function of organisms is ultimately determined by the biochemical information enclosed in their genes, it is obvious that changes in their overall fitness must be reflected by changes in their biochemistry. But what sort of changes will these be? Will they involve increasing complexity of biochemical pathways or, perhaps, just the reverse? The answer, as we shall see, is that both types of change occur, and this is shown as much by the flavonoids as by any other group of plant products.

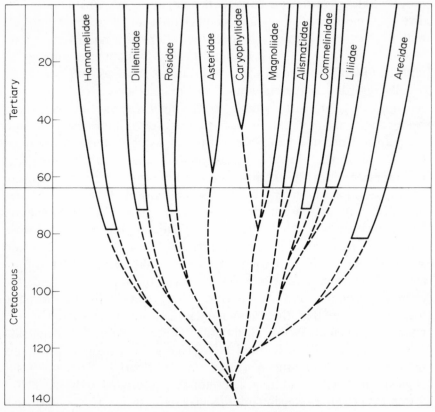

Figure 20.2 Relationship and evolution of the angiosperms.

It should be noted as a corollary to the argument presented above
that no biochemical mechanism is likely to survive in a given
population unless it leads to a product which confers some definite
advantage, however marginal. In plants, it is often presumed that the
so-called secondary products which they contain are of little use
because they have no obvious function in primary metabolism
(Weinberg, 1970). One could, of course, dismiss animal behaviour
patterns on the same score, but plainly this is nonsense. We now have
sufficient knowledge about the important ecological roles played by
plant secondary compounds (Harborne, 1972a; Whittaker and Feeny,
1971) to conclude that the very diversity of such products which a
plant contains determines its fitness to exist in a given environment
(Swain, 1974). In other words, every natural product has, or had, a

purpose in the evolutionary strategy of the taxon concerned. It is interesting to note that the overwhelming majority of known naturally occurring organic compounds are of plant origin, and plants can thus be regarded as being more sophisticated biochemically than either animals or protists. The increased size of the genome which marks evolutionary advancement in both plants and animals is thus devoted to different purposes: the elaboration of diverse compounds for many differing roles on the one hand, and the development of increasingly complex behaviour patterns on the other. This can be summed up in the aphorism 'plants produce, animals act'.

There is roughly 1000-fold more DNA in the nucleus of a higher organism than in that of the average bacterium. Much of this extra genetic information will, of course, be used in controlling the intricate processes of development of the multicellular forms, a good deal will be used for the biosynthesis of secondary products or the ordering of behaviour patterns, and yet more may, under normal circumstances, be redundant and act as a pool for future evolutionary change (Swain, 1974). Much, however, will be involved in more exact biochemical control mechanisms, many of which have only been uncovered in recent years (Milborrow, 1973). However, in the case of almost all secondary metabolites, we are at present still in the process of examining the nature of individual steps in the overall biosynthetic pathways and have not, so far, looked closely at the underlying biochemical control which determines, for example, the genetically defined mixtures of flavonoids responsible for different petal colours (Harborne, 1967a). Consideration of this one important area of flavonoid evolution, therefore, is denied to us. Another related subject about which we have little or no information, is the variation in the structure and relative activity of enzymes involved in the biosynthetic pathways of secondary metabolism: for example, knowledge of the amino acid sequence of an enzyme such as chalkone-flavanone isomerase (see Chapter 16) from different organisms would be expected to yield much useful information about the overall evolution of flavonoid biosynthesis.

It should also be noted that we still know too little about the role of flavonoids *per se* in plants and, more especially, the value to the plant of forming two or even several different derivatives of the same aglycone (Harborne 1967a, 1972b). This should not deter us from believing that the answers to such questions will eventually be found.

Even then, however, we will not be completely certain that the answers are meaningful in evolutionary terms. We have to rely exclusively, or nearly so, for biochemical information on examination of extant forms. Many of these may have partially or completely changed many of their biochemical parameters in comparison to the cognate organisms which existed at the time when the phyla first evolved. For example, while it is true that, as judged by the fossil record, the morphological and anatomical changes have been minimal between members of the Charales in Devonian times (400 million years ago) and those of the present day, we have no means of knowing whether the ability to synthesize flavonoids was present in the ancient forms. There is good evidence, however, that many biochemical features are just as conservative as morphological characters and often even more so since there is usually only one or two solutions to a given biochemical operation. One may note, for example, the extreme conservation of such enzymic cofactors as NAD, ATP, coenzyme A. Even among secondary compounds, there is a relative stability in their distribution as shown by the flavonoids in present day major divisions of the Plant Kingdom discussed here, and this is paralleled by other groups of compounds such as the terpenoids and alkaloids (Swain, 1974).

20.4 Distribution of flavonoids in the Plant Kingdom

Flavonoids are ubiquitous in higher plants, paralleling the distribution of the hydroxycinnamic acids (1a–c) and of lignin. They are absent from all algae, including the blue-greens, except for two of the higher members of the Charophyceae (Chlorophyta) (see below). Reports of catechin-like tannins in the Phaeophyta (Haug and Larsen, 1958) require further investigation. Although isoflavones were reported in cell cultures of the bacterium *Mycobacterium phlei* (Hudson and Bentley 1969) they were later found to be due to contamination (Grisebach, 1972). The presence of flavonoids in a chemical sense in fungi, however is well authenticated. Chloroflavonin (2) and dihydrochalkone (3a) have been isolated from *Aspergillus candidus* (Richards *et al.*, 1969) and *Phallus impudicus* (List and Freund, 1968) respectively. These two compounds have not been found in any higher plant and cannot be regarded as typical flavonoids, but this also rules out contamination. It is not unexpected to find flavonoids in fungi, since many members of this plant division possess the ability to synthesize cinnamic acid intermediates (1a and b) from phenylalanine and have a wide and varied polyketide metabolism. Furthermore, they can form mixed shikimate-polyketide metabolites such as hispidin (4) from caffeic acid and two acetate units (Bu'Lock and Smith, 1961), congeners of which occur in higher plants (Klohs, 1967) and which have been described as 'failed flavonoids'. So far, however, no flavonoids other than (2) and (3a) have been reported. Indeed, it has recently been reported (Marchelli and Vining, 1973) that the B-ring of chloro-flavonin (2) is indeed synthesized from the shikimic acid route. It should be noted that many compounds in fungi, e.g. fulvic acid (5) have congenors with chlorine substitution. It is, however, possible that dihydrochalkone (3a) in *Phallus impudicus* arises by an atypical pathway. For example, 4,4'-dihydroxychalkone (3b) and an unknown trihydroxychalkone have recently been found in decomposed

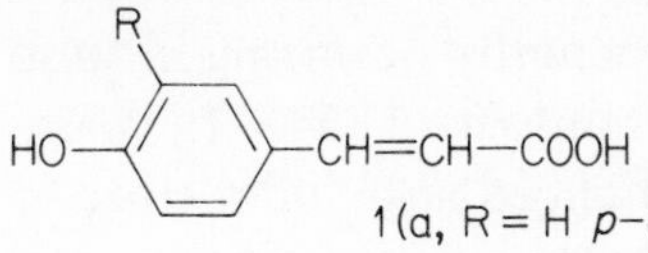

HO—⟨⟩—CH=CH—COOH

1(a, R = H *p*–coumaric acid)
 (b, R = OH Caffeic acid)
 (c, R = OMe Ferulic acid)

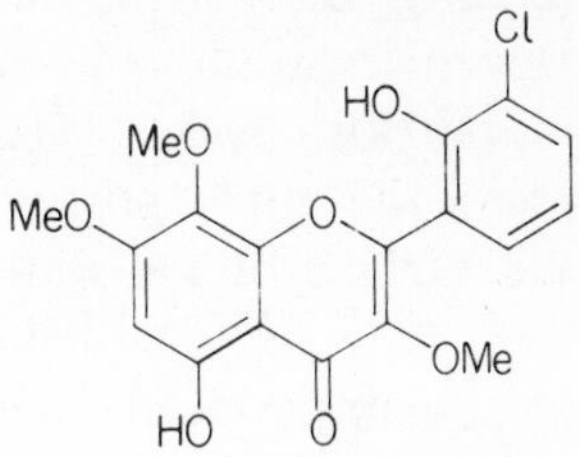

(2, Chloroflavonin)

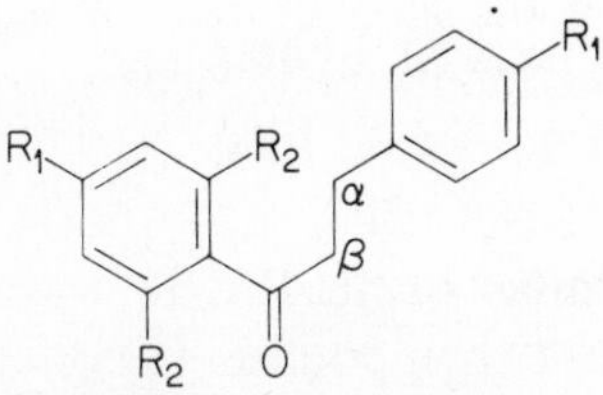

3 (a, R_1R_2 = H $\alpha\beta$ = dihydro
   Dihydrochalkone)
 (b, R_1 = OH R_2 = H $\alpha\beta$ = double
   bond
   4, 4'–dihydroxychalkone)
 (c, R_1 = OMe R_2 = OH $\alpha\beta$ = double
   bond
   2',6'–dihydroxy–4,
   4'–dimethoxychalkone)
 (d, R_1 = OMe R_2 = OH $\alpha\beta$ = dihydro)

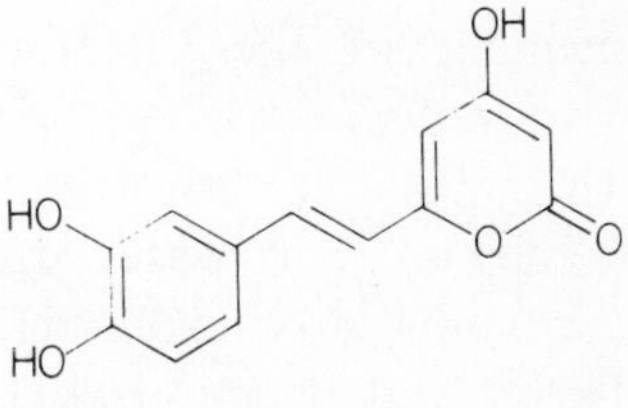

(4, Hispidin)

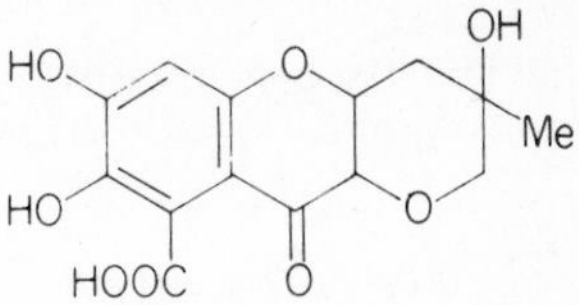

(5, Fulvic acid)

human liver (Fox *et al.*, 1973) where they are perhaps formed by anaerobic microbial degradation of tyrosine (or dopa) to give the corresponding benzaldehyde which then condenses with *p*-hydroxy-phenylpyruvic acid. Similar reactions may occur in *P. impudicus*. All other reports of flavonoids in microorganisms should be treated with suspicion, since careful reinvestigation of several reports showed that they were false (Harborne, 1972b). In many other cases, close scrutiny shows that the evidence presented is insufficient for any firm conclusions to be drawn as to the type of compounds present.

As mentioned above, the most primitive group of plants in which true flavonoids have been found are the members of the green algal family, the Charophyceae. Glycoflavones, similar to vicenin (6a) and lucenin (6b) were reported from both *Chara* species (T. J. Mabry, unpublished results) and *Nitella hookeri* (Markham and Porter, 1969). The Charophyceae (stoneworts), which are regarded as advanced members of the Chlorophyta (Bell and Woodcock 1972), (cf. Fig. 20.1) made their first appearance in the fossil record in early Devonian strata (380 million years ago) apparently occurring in fresh or brackish water, as they do today (Grambast, 1964). In many cases, the upper portions of these heterotrichous algae float close to the surface of the water and in spite of their primitive organisation they may be regarded as prototypes of 'amphibious' plants which presumably preceded true land plants.

No flavonoids were detected in another heterotrichous species, *Draparnaldia,* which grows in damp mud on the lake shores and is a member of the Chaetophorales which, as mentioned earlier, are favoured as ancestors of land plants (Fritsch, 1945; Banks, 1970). This alga did, however, contain bound caffeic acid (1b). Another species of fresh water alga, *Lemanea,* a member of the Rhodophyta, contained no flavonoids or related phenolic compounds (Cooper-Driver and Swain, 1974). However, a compound closely related to the flavonoids has recently been found in all classes of algae, including the blue-greens. This is the dihydrostilbene lunularic acid (7) (Pryce, 1972). This compound, which was first isolated from liverworts (see below) where it acts as a growth hormone, indicates that mixed polyketide-shikimic acid pathways are very ancient (Pryce, 1971).

It should be noted that lignin has not been detected in any fungal or algal phyla. The absence of lignin in fungi, which contain most of

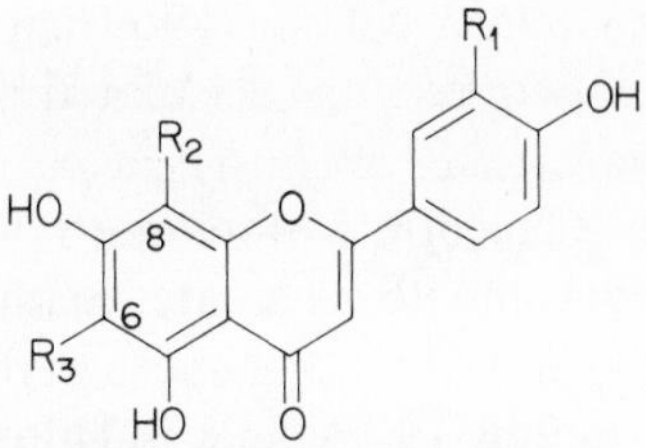

6 (a, R_1 = H R_2 R_3 = Glc Vicenin)
  (b, R_1 = OH R_2 R_3 = Glc Lucenin)
  (c, R_1 R_2 R_3 = H Apigenin)
  (d, R_1 = OH R_2 = OMe R_3 = H
      8—methoxyluteolin)
  (e, R'_1 = OH R_2 R_3 = H Luteolin)

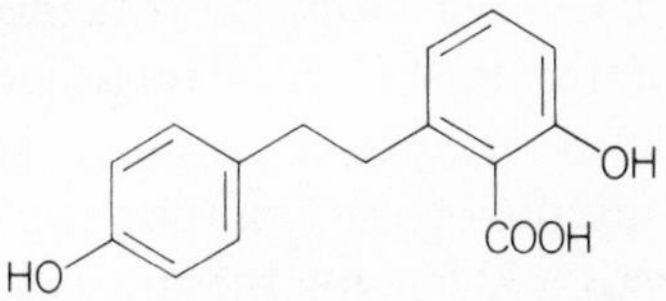

(7, Lunularic acid)

the enzymes required for its elaboration, may be due to variations in the phenolic methylation patterns which lead to precursors having no ability to polymerize by a free-radical mechanism (Swain, 1970). Alternatively, the absence of a cellulosic cell well in all true fungi (Swain, 1974) may be important, for it is possible that acylation of cellulose may precede lignification. In the algae, as with primitive Bryophyta and even some aquatic angiosperms, the absence of lignin again may be due to lack of an acylated cell wall. Although cellulose is present, it is apparently encrusted with silica in the Characeae and this may prevent the proper orientation of lignin precursors.

The most primitive land plants, the Bryophyta, like *Chara,* have also been shown to contain glycoflavones. These compounds have been detected in several species of moss (Musci) drawn from each subclass (Melchert and Alston, 1965; McClure and Miller, 1967). Three species of the moss genus *Bryum* have been shown to contain glycosides of the 3-deoxyanthocyanidins, luteolinidin (8a) (Bendz *et al.,* 1962; Nilsson, 1967) a rare class of compounds in higher plants (Harborne, 1967b). More recently, Bendz and her co-workers have reported other unexpected compounds in bryophytes including 5′,8″-biluteolin from *Dicranium scoparium* (Lindberg *et al.,* 1974) and several complex apigenin-*O*-glucosides and a scutellarein glucoside from *Bryum* (Nilsson and Bendz, 1974). The latter is a relatively advanced flavonoid.

The liverworts (Hepaticae) also contain glycoflavones (Markham *et al.,* 1969). Two vicenin-like compounds (6a) were isolated from *Hymenophytum flabellatum* and similar compounds were found in a number of other species (Markham *et al.*, 1969; Nilsson 1969 and 1973). More recently Markham and his co-workers have isolated a number of hitherto-regarded 'advanced flavonoids' from liverworts. Thus, acacetin-7-*O*-rhamnosylgalacturonide (acacetin is 4'-*O*-methyl-apigenin 6c) was obtained from *Reboulia hemispherica* (Markham *et al.*, 1972), a polysaccharide derivative (M.W. 3200) of 8-methoxy-luteolin (6a) from *Monoclea forsteri* (Markham, 1972) and 7-*O*-glu-curonides and rhamnosylglucuronides of apigenin, chrysoeriol (9a) and tricin (9b) from *Marchantia folicea* (Markham and Porter, 1973). As mentioned earlier, most bryophytes are epiphytic and it is not entirely surprising to find they generally lack lignin (Siegel, 1968). However, several workers (Bland *et al.*, 1968; Siegel, 1969) have shown that lignin is present in mosses. For example, the axes of the giant (0.5 m) New Zealand mosses *Dawsonia* and *Dendroligotrichum* contain up to 10% of the polymer although it was lacking from five other mosses and three liverworts. There is evidence, however, that the related oxidative polymerases may be present (Siegel, 1968).

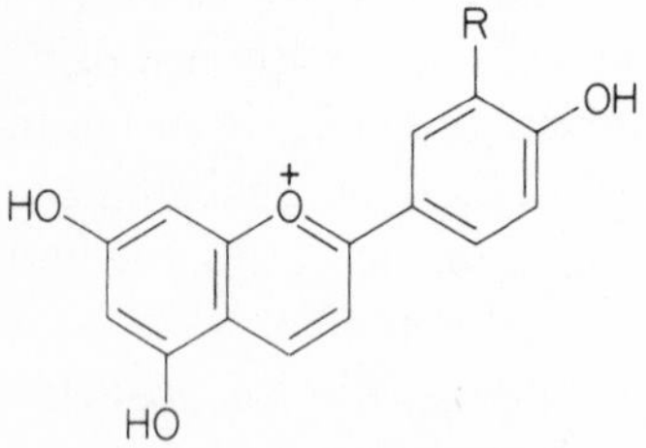

8 (a, R = OH Luteolinidin)
  (b, R = H Apigeninidin)

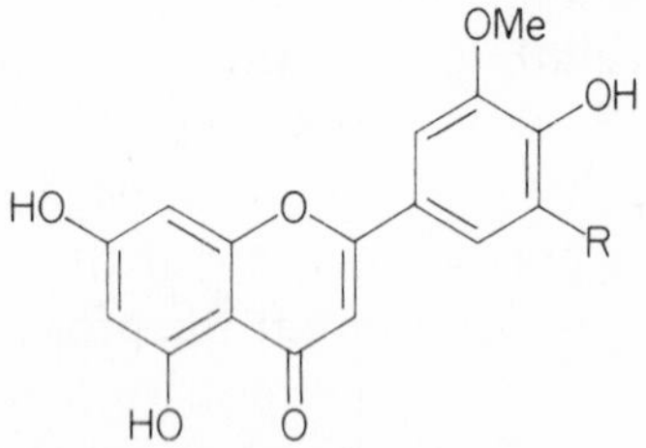

9 (a, R = H Chrysoeriol)
  (b, R = OMe Tricin)

It should be noted that the growth hormone lunularic acid (7) which as mentioned above is present in all classes of the algae and liverworts, is totally absent from the mosses and all higher plants. Here, its function is taken over by the terpenoid constituent abscisic acid (10) to which it has a formal structural analogy (Pryce, 1972). Other stilbenes are found mainly in heartwoods of gymnosperms and a few angiosperms.

One other report of flavonoids in liverworts is the unexpected finding of kaempferol (11a) and quercetin (11b) in *Corsinia coriandrina* (Reznik and Wierman, 1966). This genus is regarded by some as being one of the most advanced members of the March-antiales (Parihar, 1962), but the occurrence of flavonoids is very surprising (see below).

The most primitive tracheophyte, *Psilotum* (Psilopsida), contains none of the flavonoids mentioned above. Instead, *P. triquetrum* contains the biflavones, amentoflavone (12a) and perhaps hinoki-flavone (13) (Voirin and Lebreton, 1971; Voirin, 1970). The other psilophyte genus, *Tmesipteris*, was found to contain an unknown flavonoid (Voirin, 1970).

A more interesting distribution of flavonoids is found in the

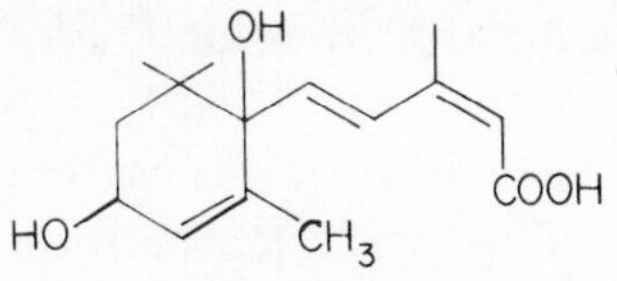

(10, Abscisic acid)

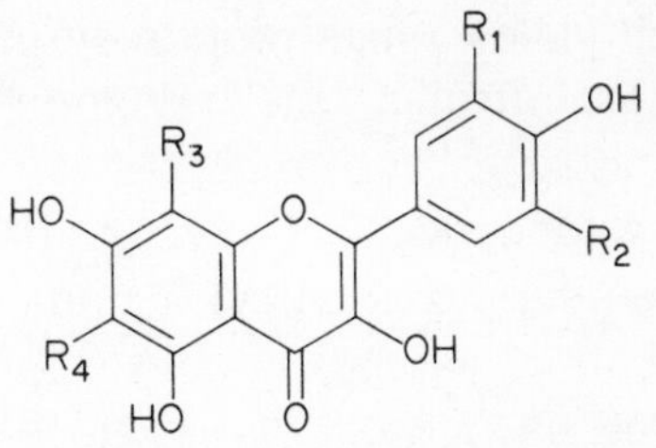

11 (a, $R_1 R_2 R_3 R_4$ = H   Kaempferol)
   (b, R_1 = OH $R_2 R_3 R_4$ = H   Quercetin)
   (c, R_3 = OH $R_1 R_2 R_4$ = H   Herbacetin)
   (d, $R_1 R_3$ = OH $R_2 R_4$ = H   Gossypetin)
   (e, $R_1 R_2$ = OH $R_3 R_4$ = H   Myricetin)
   (f, $R_1 R_4$ = OH $R_2 R_3$ = H   Quercetagetin)

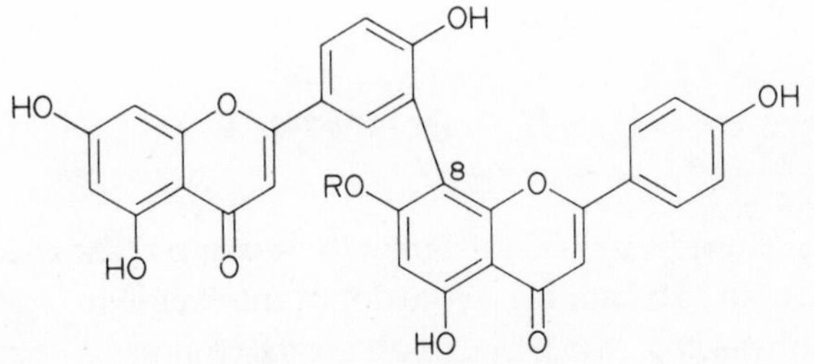

12 (a, R = H   Amentoflavone)
   (b, R = Me   Sotetsuflavone)

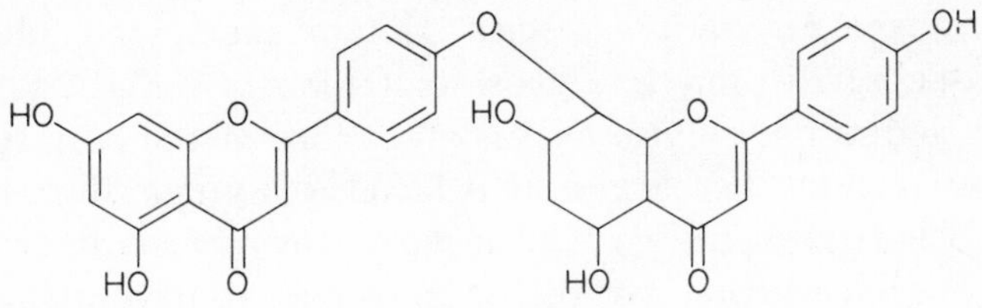

(13, Hinokiflavone)

Lycopsida (Voirin, 1972): the Isoetales (*Isoetes*) contain apigenin (6c) and luteolin (6e) *O*-glycosides; Lycopodiales (*Lycopodium, Lepidotis, Huperzia* and *Diphasium*) consistently contain, besides these compounds, an unknown glycoside, possibly of genkwanin (7-*O*-methylapigenin) (Voirin, 1970); Selaginellales, (*Selaginella*) on the other hand, contains only amentoflavone (12a) and what is probably its 7″-methyl ether, sotetsuflavone (12b). It should be noted that *Selaginella* differs from the Lycopsidales by having lignin which yields a high ratio of syringaldehyde (14a) to vanillin (14b) on alkaline nitrobenzene oxidation (White and Towers, 1967). *Isoetes* had earlier been reported as also containing lignin which gave syringaldehyde. The Lycopodiales, on the other hand, although containing no syringyl lignin, can be divided into those genera containing alcohol-soluble compounds yielding syringic acid on hydrolysis (*Lycopodium* and *Diphasium*) and those which do not (*Huperzia* and *Lepidotis*) (Towers and Maass, 1965). It should be noted that syringyl lignin is regarded as an advanced angiosperm character and is more or less totally absent from the gymnosperms and ferns (Towers and Gibbs, 1953).

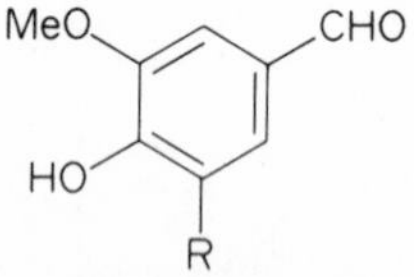

14 (a, R = OMe Syringaldehyde)
    (b, R = H   Vanillin)

The monogeneric class, Equisetales (horsetails), show four further advancements in flavonoid evolution; the regular formation of flavonoids containing the 3-hydroxyl group (but see *Corsinia* above); the introduction of an hydroxy group at C-8; the production of a trihydroxy B-ring; and the synthesis of C-4–C-8(6) linked polymers; the proanthocyanidins or leucoanthocyanins, (see Chapter 10). The majority of the *Equisetum* species contain kaempferol (11a) and quercetin (11b) di- and tri-glycosides (Saleh *et al.*, 1972; Voirin, 1970). In addition *E. fluviatile* has glycosides of the corresponding 8-hydroxyflavonols, herbacetin (11c) and gossypetin (11a) (Saleh *et al.*, 1972). Herbacetin glycosides were also present in two other *Equisetum* species; one of these, *E. sylvaticum*, along with *E.*

telmateja, contained proanthocyanidins, the former having leuco-delphinidin (15a) as well as leucocyanidin (15b) (Voirin, 1970). Leucodelphinidin was also found in traces in *E. palustre.*

The ferns (Filicinae) show little or no advancement on the Equisetales. However, no biflavones have been reported and glyco-flavones and flavone *O*-glycosides are rare (Ueno *et al.,* 1963; Voirin, 1970). The Ophioglossales are marked by containing mainly glyco-sides of kaempferol and quercetin and the 3-*O*-methyl ether of the latter, (Voirin, 1970). This methyl ether is relatively uncommon in the Plant Kingdom, and in ferns is found only in two species of *Polystichum,* in the rather advanced Aspidiaceae. Neither *Ophio-glossum* nor *Botrychium* contains any proanthocyanidins, which characterise the ferns generally (Voirin, 1970; Cooper-Driver and Swain, 1973). Originally Voirin (Voirin, 1970; Voirin and Lebreton, 1971) reported the absence of proanthocyanidins in the Marratiales, but this was shown later to be untrue (Cooper-Driver and Swain, 1973), even for the rather primitive *Marratia* itself. We can thus conclude that the Marratiales contain a full complement of 'regular' flavonoids (*O*-glycosides of kaempferol and quercetin, leucocyanidin and leucodelphinidin) which abound in almost all families of the Filicales. Thus, no distinction can be properly made between the eusporanginate (Marratiales) and the sporanginate (Filicales) mem-bers of the Filicinae. The Ophioglossales, however, do stand apart.

One other evolutionary relationship which has been proposed is that between the Psilophyta and the primitive fern families Stro-matopteridaceae and Schizaeaceae (Bierhorst, 1971), but this is not borne out by their flavonoid chemistry. The two fern families do not contain any biflavones which are present in *Psilotum* but have the 'normal' complement of flavonoids found in other ferns: flavonols and proanthocyanidins (Cooper-Driver and Swain, 1974).

The only other flavonoid compounds whose occurrence in ferns should be noted because of their rarity are 4'-*O*-methylkaempferol, 5-*O*-methylquercetin, myricetin (11e) and leucopelargonidin (15c), all of which have scattered occurrences in single species (Voirin, 1970; Akabori and Hasegawa, 1970). In addition, the red colouration of the juvenile fronds of several diverse species has been shown to be due to the presence of 3-deoxyanthocyanins (8a,b) (Harborne 1966). Two new classes of flavonoids in the ferns are the chalkones (3c) and dihydrochalkones (3d) from *Pityrogramma chrysophylla* and the

C-methyl derivatives, 6-methylchrysin (16a), pinocembrin (17a) the flavanone corresponding to chrysin (16b), matteucinol (17b) and farrerol (17c) (both 6,8-di-*C*-methyl derivatives of flavanones). The latter group show a link between the ferns and the gymnosperms which contain many such compounds (Harborne, 1967a)

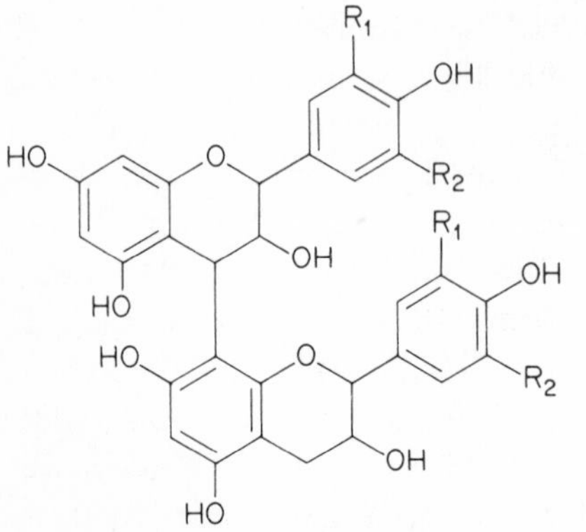

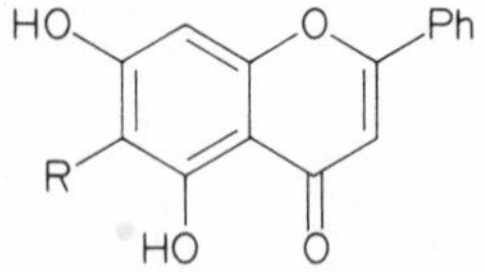

16 (a, R = Me 6-methylchrysin)
   (b, R = H Chrysin

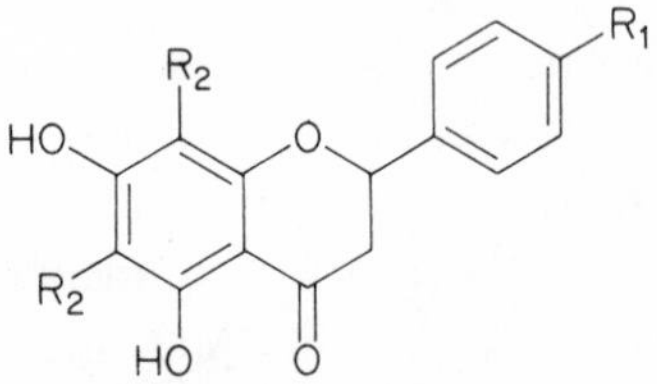

15 (a, R₁ R₂ = OH Leucodelphinidin)
   (b, R₁ = OH R₂ = H Leucocyanidin)
   (c, R₁ R₂ = H Leucopelargonidin)
   (dimeric forms shown)

17 (a, R₁ R₂ = H Pinocembrin)
   (b, R₁ = OMe R₂ = Me Matteucinol)
   (c, R₁ = OH R₂ = Me Farrerol)

The gymnosperms themselves contain very few new types of flavonoids. The characteristic flavonoid compounds which occur in the majority of gymnosperm orders are primarily the biflavones, which are characteristic constituents, along with flavones, flavanones, flavonols and flavanonols. As mentioned above, many of the last named four groups are found as 6- or 8-*C*-methyl derivatives, but complex *O*- or *C*-methylation is rare. It should be noted that biflavones are not found either in the Pinaceae (Sawada, 1958; Harborne, 1967a) or in the three families of the Gnetales (Sawada, 1958; Cooper-Driver and Swain, 1974). The two new classes of compound which are found in the gymnosperms, albeit rarely, are the anthocyanins proper (18a–c) and two isoflavones, genistein (19) and podospicatin which were discovered in *Podocarpus spicatus* (Briggs *et al.*, 1959).

In the angiosperms, flavonoid evolution reaches its zenith (see Chapter 19) and leads to the formation of several new classes of flavonoids (neoflavonoids, e.g. 20, see chapter 15) aurones (e.g. 21, see Chapter 9) and the chromanocoumarans, like pisatin (22). In

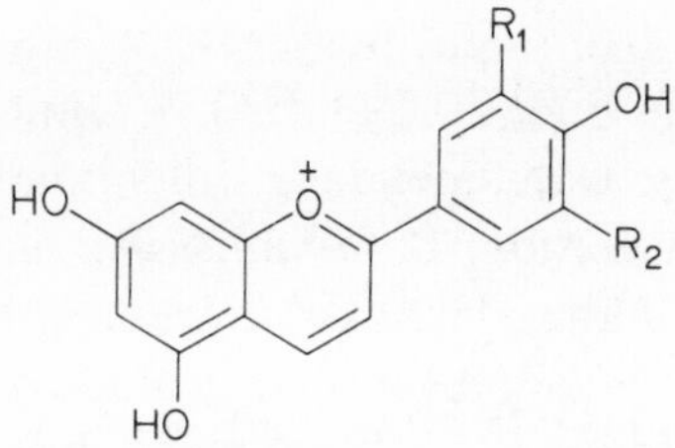

18 (a, R_1 = OH R_2 = H Cyanidin)
   (b, $R_1 R_2$ = H Pelargonidin)
   (c, $R_1 R_2$ = OH Delphinidin)

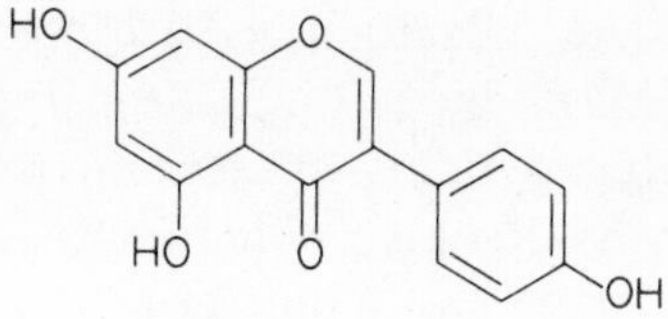

(19, Genistein)

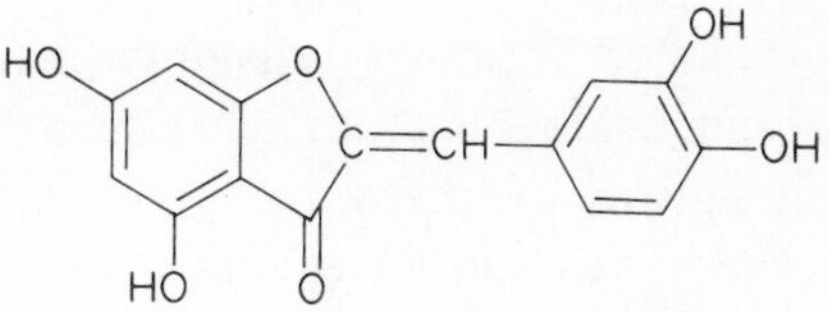

(21, Aureusidin)

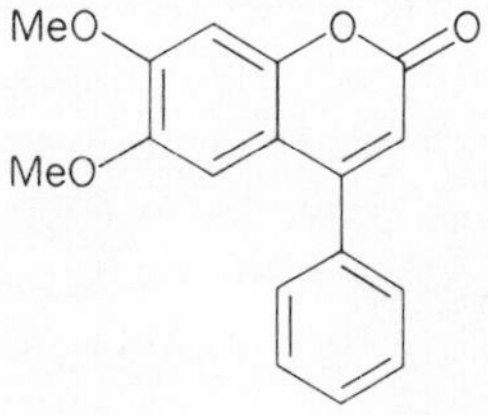

(20, 6, 7–dimethoxy–4–phenylcoumarin)

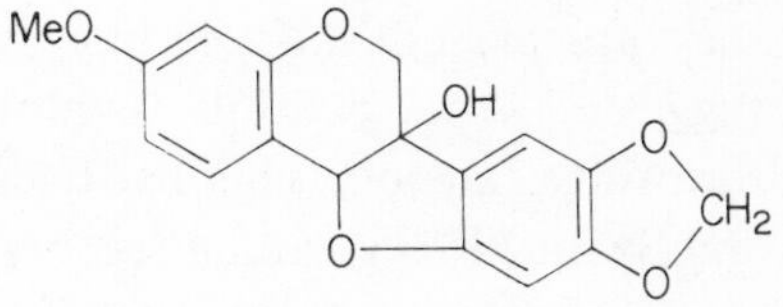

(22, Pisatin)

addition, complex hydroxylation and glycoside formation, multiple O-methylation and -prenylation are found as regular features in all classes of flavonoids. In spite of this complexity of the overall pattern, there are certain regularities which stand out (Harborne, 1972b).

One of the main generalizations to emerge from the study of flavonoid distribution in the angiosperms is the change following the transformation from the primitive woody to the more advanced herbaceous habit. This produces three typical changes in leaf flavonoids: overall loss of proanthocyanidins (15a–c), loss of B-ring trihydroxylation (e.g. 11e and 15a) and replacement of flavonols (e.g. quercetin, 11b) by flavones (e.g. luteolin, 6e) (Bate-Smith 1962, 1968; Harborne, 1972b). In relation to the first two features, Bate-Smith has drawn attention to the need to consider also the distribution of the ellagitannins. These compounds, which are esters of glucose, yield on hydrolysis either ellagic acid (23), one of its O-methyl ethers or a related compound. It may be noted that these compounds are absent from all lower classes of the Plant Kingdom.

The ellagitannins are virtually absent from all members of the Magnoliidae (and the Caryophyllidae) (see Figs. 20.1 and 20.2) while being richly present in the more woody Hamamelidae, Dilleniidae and Rosidae. On the basis of such information, Bate-Smith (1972, 1974) and Kubitski (1969) proposed that the Magnoliidae were not ancestral to the other three subdivisions (Fig. 20.1) as suggested by Cronquist (1968) and should be regarded as at least parallel in evolution (cf. Fig. 20.2). Bate-Smith has also postulated that in more advanced families ellagitannins are O-methylated and that the C-glucoside bergenin (24) may well represent a final stage in the evolution of these compounds, since it is distributed in a closely related fashion. A second generalization which can be made about flavonoid evolution in the angiosperms is with regard to the position of C-glycosylflavones; Harborne (1972b) has suggested that their distribution supports the idea that they are primitive relicts, since they are present more in woody families than in herbaceous ones. He suggested the possible evolutionary series flavonol → C-glycosyl-flavone → flavone. This has been described as 'switching' (Bate-Smith and Swain, 1965), and a further example of this is the substitution of C-glycosylflavones by mangiferin (25) in several cases. It may be noted that mangiferin, like hispidin (4), is formed by the condensation of a C_9 moiety with only two, rather than three, acetate units. The evolution of glycoflavones is not necessarily all in one direction. In certain advanced members of the monocotyledons, for example,

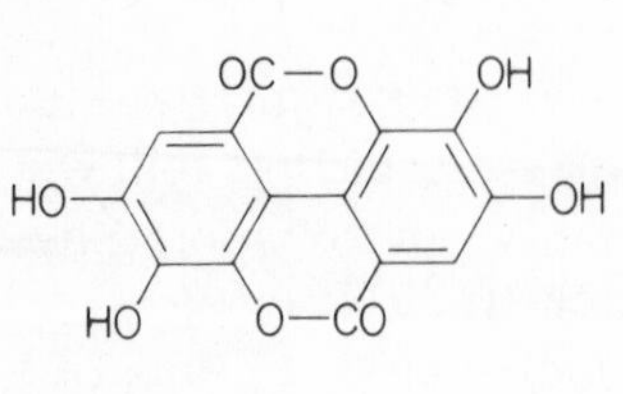

(23, Ellagic acid)

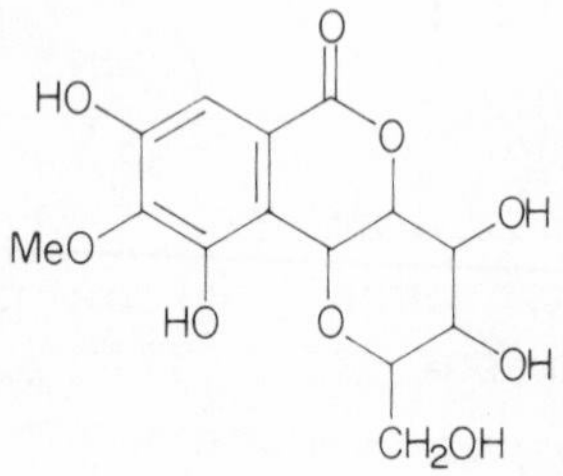

(24, Bergenin)

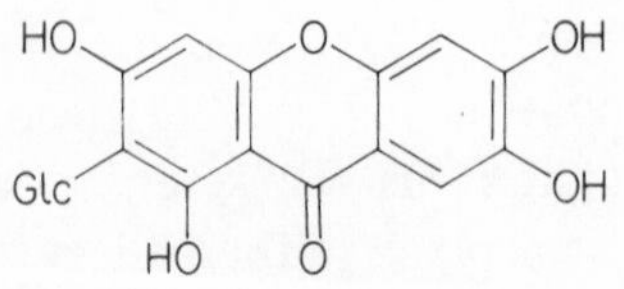

(25, Mangiferin)

the Gramineae, glycoflavones are extremely common. This supports the contention made earlier that evolution does not always lead to advanced flavonoid characters. Like all living processes, there is the element of chance, and since not all characters evolve at the same time, one obtains a reticulate evolutionary pattern.

One further example of flavonoid evolution in angiosperms should be noted and that is the distribution of 6- and 8-hydroxylation patterns in flavonoids (Harborne, 1972b). The 8-hydroxy derivatives of kaempferol (herbacetin, 11a) and quercetin (gossypetin, 11f) are regularly found in the Magnoliidae, Rosidae and Dilleniidae (cf. *Equisetum*) whereas the corresponding 6-hydroxyquercetin (quercetagetin 11f) 6-hydroxyluteolin (see 6e) and 6-methoxyapigenin (see 6c) are more widely distributed in the highly advanced Asteridae (Harborne, 1972b; Carman *et al.*, 1972). Again the distribution of 8-hydroxy derivatives points to a development of the subclasses in which they occur from a common early ancestor.

Very recently four groups of workers have reported the occurrence of flavonoids in a variety of angiosperm chloroplasts, apparently associated with the energy transfer systems (Monties, 1969; Weissenboeck *et al.*, 1971; Saunders and McClure, 1972, 1973; Oettmeier and Heupel, 1972). It is not certain, however, whether they are part of the electron transport chains and may, instead, again be acting as a light screen. The compounds involved include glycoflavones in barley, flavonols in *Impatiens,* and an alkylated quercetagetin in spinach, which can be regarded as a progression from primitive to advanced flavonoids respectively. Thus we cannot draw any conclusions about the possible evolution of their function in these organelles. However, these reports are exciting and, if one accepts the symbiont theory of the origin of the chloroplast from a blue-green ancestor (Margulis, 1971), they point to flavonoid evolution being older than presently believed and on par with lunularic acid. The data are too scanty to do more than note them here and suggest that an examination of the chloroplasts of lower algae for flavonoids may yield rich results.

20.5 Flavonoid evolution

Many of the generalizations made in the previous section, although based on adequate observations, are not easy to explain on a basis of evolutionary advance. That is, it is not clear what is the advantage in advanced angiosperms of, for example, a flavone glycoside over a flavonol glycoside. However, the loss of tannins (both ellagitannins and, more especially, proanthocyanins) from woody plants and their replacement in herbaceous angiosperms by more advanced flavonoid antifungal agents (e.g. pisatin, 22) which are only produced when infection starts, is more readily understood. Nevertheless, it will not always be possible to give such clear examples in discussing flavonoid evolution, and we will need to rely in our explanations as much on the changes in complexity of different routes of biosynthesis (including cellular control) as on potential ecological advantage.

It is obvious that since members of all algal divisions, including the procaryotic blue-greens, contain lunularic acid (7), they must have the ability to couple a C_9 unit (presumably p-coumaric acid, 1a) with three acetate units to form a stilbene precursor (Pryce, 1971; Hillis and Isui, 1965). This precursor, a cinnamoyltriacetic acid (Hillis and Yazaki, 1971) is presumed to be common to the biosynthesis of both stilbenes and flavonoids, the difference lying merely in the atoms of the polyketide chain which are involved in cyclization (Birch, 1962). It would appear, therefore, that with the exception of the Charophyceae which contain glycoflavones, the algae lack the specific cyclase which directs the conversion of this C_{15} precursor into a chalcone (see Chapter 16). Alternatively, since hydrangenol (26) is such a good precursor of lunularic acid (Pryce, 1971) it is possible that biosynthesis of the dihydrostilbene initially involves hydroxylation of the double bond of the cinnamoyl moiety to give a β-hydroxyphenylacrylic acid prior to condensation with the triketide. In any case, the formulation of lunularic acid, with its reduced central double bond and carboxylic acid group (7) perhaps shows that some different mechanism to that giving normal stilbenes, which do not have these features, may operate in the algae (Hillis and Yazaki, 1971).

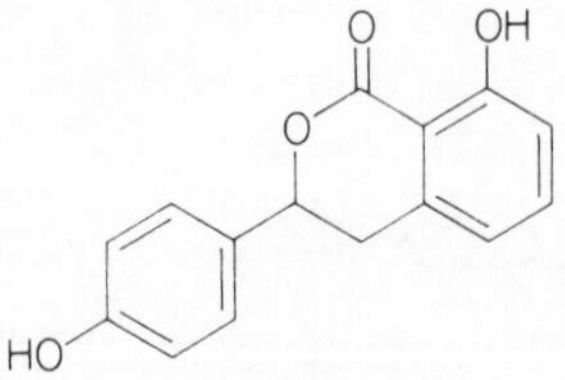

(26, Hydrangenol)

Leaving aside arguments concerning the relative biosynthetic ability of *Chara* and other algae, what possible benefit could the glycoflavones have to these thallophytes? It is unlikely that the selective advantage lies in the roles assigned to flavonoids in angiosperms (Harborne 1972a,b; see also Chapter 18). It has been proposed instead (Swain, 1970) that the glycoflavones with their high absorptivity at 257–270 and 335–349 nm (Mabry *et al.,* 1970) act as good UV screens (cf. Caldwell, 1971, see also Chapter 18, p. 1017). Their presence would thus prevent both mutagenesis or cellular death by dimerization of thymine units in the DNA which shows a maximum at 260 nm (Setlow, 1967) and possible photo-destruction of coenzymes NAD or NADP which have a maximum about 340 nm (Webb and Tai, 1969) and would obviously benefit those algae which were emerging from an aquatic environment onto the land. Many of the more primitive aquatic algae secrete calcareous material for such purposes (Banks, 1970) which obviously restricts their further evolution into complex multicellular forms. It is noticeable that flavonoids are predominant in pollen walls (Pratriel-Sosa and Percheron, 1972) and that tropical and high altitude plants contain a higher proportion of flavonoids in their immature growing tissue than do temperate ones (Caldwell, 1971). It should also be noted that the exposed scale leaves of *Dacrydium laxifolium* (Podocarpaceae) are coloured by cyanidin 3-glucoside while the shade leaves are not, and anthocyanins are reported in young leaves of several other gymnosperms (Lowry, 1972; Santamour, 1967). It may be noted that the hydroxy-benzoic and -cinnamic acids which have been found to the exclusion of flavonols in fern spore walls (Cooper-Driver and Swain, 1974) and even simple phenols can, in part, also fill this role, although they have much lower molar absorptivities and, in general, do not show absorption maxima in exactly the most effective regions. One might, for example, consider why lunularic acid (7) had not been so used. It should be noted that this compound acts as a potent growth inhibitor and it can be presumed that the algae could not tolerate the concentrations required for it to act as a UV screen. Certainly its concentration in the algae examined was only 1 μg per g fresh weight (Pryce, 1972) which would be too low to be effective.

Another possible advantage that flavonoids could have bestowed to early land plants is due to the fact that they possess anti-oxidant and metal chelatibility (Swain, 1962; Van Fleet, 1969). This would have reduced adventitious photo-oxidation of several sensitive compounds which was likely to be increased in the higher light intensity of the land environment.

With regard to the first hypothesis, it has recently been shown that the deleterious effect of UV light (264 nm) on the growth and pigmentation of *Alternaria* spp. can be completely prevented by screening a culture with a 0.5 cm layer of an aqueous solution of vitexin at 10 μg ml^{-1} (Woodhead *et al.*, 1974). An examination of the distribution of glycoflavones in *Chara* plants revealed, however, that there was virtually no difference in their concentration (as judged by UV absorption spectra) in the top or the bottom halves of 0.4 m high algae grown in Lake Windermere, England. These results do not, therefore, support the hypothesis although one should remember that there is probably little difference in the overall biochemistry of any of the cells in *Chara*.

Regardless of the exact role we assign to the glycoflavones in the algae, is there any reason why the compounds should be *C*-glycosides rather than the *O*-glycosides found in the vast majority of plants? Pridham (1964) showed that none of the algae which he examined had the ability to form phenolic *O*-glycosides when either quinol or resorcinol was fed to them. All higher plant taxa tested, including angiosperms, gymnosperms, ferns, mosses and liverworts possessed this activity, although in a few individual species, mainly mosses, and some aquatic plants, it was weak or absent. Since unglycosylated flavonoids are very insoluble in water and often toxic, it would presumably be necessary for the Charophyceae to solubilize in some way the compounds which they synthesized. *C*-Glycosylation seems to have been the answer. *O*-Glycosylation reactions are, of course, by far the most common feature in modern living organisms, but it is possible that the formation of such glycosides involving attachment of sugars to the acidic phenolic moiety is an advanced step and that a direct enzyme-catalysed electrophilic addition of a sugar to the nucleophilic *meta*-substituted A-ring of the flavonoids was the first to evolve. The carbanion reactivity of phloroglucinol-like rings is high, although it is reduced in the A-ring of normal flavonoids by the adjacent 4-carbonyl group. Presumably the orientation of the flavonoids on the enzyme involved in *C*-glycosylation reactions overcame this deactivation. One should note that only the simple flavones, apigenin and luteolin, are involved.

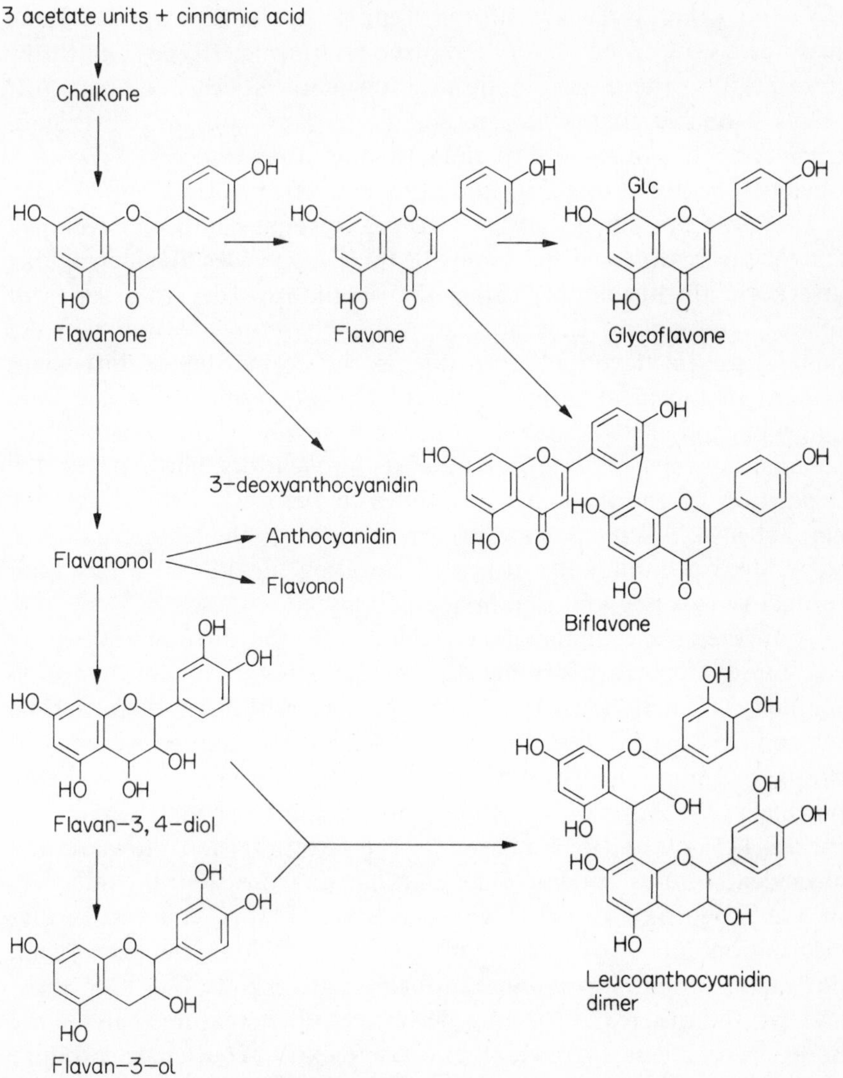

3 acetate units + cinnamic acid

Chalkone

Flavanone

Flavone

Glycoflavone

3–deoxyanthocyanidin

Flavanonol

Anthocyanidin

Flavonol

Biflavone

Flavan–3, 4–diol

Flavan–3–ol

Leucoanthocyanidin dimer

Figure 20.3 Biosynthesis of flavonoids.

The mosses took normal flavonoid biosynthesis one step further by developing enzymes to catalyse the oxidation of the flavanone stage directly to the 3-deoxyanthocyanidins (Fig. 20.3). The occurrence of *O*-glycosylation, *O*-methylation, 8-hydroxylation and B-ring trihydroxylation in flavones and the formation of flavonols in certain mosses and liverworts, however, is curious to say the least. The same is true of the unusual biluteolin in *Dicranium scoparium* since no other biflavone of this type is so far known. All these features are regarded as very advanced (Harborne, 1972b) and throw into question the phylogenetic position of the species concerned. It seems probable that some present day mosses and liverworts have advanced more in these directions than in others and further work is obviously required to clarify their phylogenetic position.

It can be presumed that once enzyme systems had been developed for the formation of C–C bonds in the C-glycosides by electrophilic attack on the flavonoid A-ring, they could give rise to a series of related compounds. It is not surprising, therefore, to find that the next stage in flavonoid evolution is the formation of biflavones (12a,b) in the most primitive vascular plants, *Psilotum*. Again only apigenin and its O-methyl ethers are involved. Presumably the vascular system of *Psilotum* and its progenitors allowed for the deposition of the insoluble biflavones in their cell wall where they can act as UV screens and also, on account of their more complex structure, can inhibit the attack by invading fungi as shown by their protective role in gymnosperm heartwoods.

A further stage in flavonoid evolution is the introduction of an enzyme system for forming the vicinal trihydroxy (or hydroxy-methoxy) function in the C_9 precursor. Although there are no flavonoids having this feature in *Selaginella* the occurrence of syringyl lignin (and of syringic acid in Lycopodales) presumably portends the formation of proanthocyanins having trihydroxy-B-rings (15a) found in Equisetales. The latter division shows several advanced features in flavonoid evolution: the consistent introduction of the 3-hydroxy group to give flavonols (11a,b), and the further elaboration of A-ring substitution to give, on the one hand, polymeric proanthocyanidins (probably with mainly C-4–C-8 links) and, on the other, C_8-hydroxy derivatives. Such features parallel the necessary increase in biosynthetic complexity required to produce these compounds (Fig. 20.3). This catalogue of advance more or less exhausts the evolution of common flavonoid features.

The occurrence of the polymeric proanthocyanidins in *Equisetum* and more especially in the ferns is probably due to the fact that these more advanced vascular plants required better means to prevent the ingress of fungal hyphae into their tracheids. These flavonoid polymers are known to be potent enzyme precipitants (Swain, 1965) and can readily inhibit the extracellular hydrolases secreted by fungal pathogens. The proanthocyanidins can thus be regarded as primitive broad-spectrum antibiotics. It should be noted that the introduction of the 3-hydroxylation reaction in the biosynthetic pathway was perhaps first exploited in the production of these compounds, for although corresponding 3-deoxy polymers are known (Roux, 1972), they are rare and probably less effective. As a bonus, the introduction of the 3-hydroxy group and, later, the 8-hydroxy group leading to compounds having longer wavelengths of absorption (360–400 nm) presumably gave the more advanced plants the ability to screen more light-sensitive compounds (e.g. FAD). At the same time many important light sensitive reactions might also be controlled, especially those concerned in growth and development, many aspects of which involve blue light (Wilkins, 1970). Their utilization as insect pollinator guides in flowers was not exploited until we reach the flowering plants.

There is little advancement in the flavonoids in ferns. Admittedly, the use of extracellular chalkones (3c,d) (presumably as a result of the evolution of some control which allowed these to accumulate instead of being metabolized in the normal flavonoid biosynthetic pathway) in the protection of spore cases is novel, but the other sporadic advances show nothing that is unexpected. The formation of 6,8-di-C-methyl substituted flavanones (17b,c) indicates a further development in the C–C bond formation A-ring substitution and foreshadows a wider use of this feature in the gymnosperms.

The development of flavonoids in the gymnosperms is also relatively minor. We have the first instances of anthocyanins proper. These compounds are prominent in the succulent receptacle or in the fleshy layer surrounding the seed of several species of Podocarpaceae (Lowry, 1972) and give the appearance of an angiosperm fruit. The anthocyanins obviously render the structures more readily detectable and aid in animal dispersal of the seed. The fruits of Gnetaceae and Taxaceae are pigmented by carotenoids which serve a similar purpose. Anthocyanins also appear in the young cones of many Pinaceae and in *Chamaecyparis* (Cupressaceae) as well as in the spring foliage of several other species.

The absence of the otherwise characteristic biflavones from the Pinaceae and Gnetales is worthy of comment. It seems possible that their fungistatic role in the wood of the Pinaceae may be taken over by C-methylflavonoid derivatives or, perhaps, the stilbenes or resins, while their absence from the Gnetales indicates the need to further re-examine the relationships of this atypical group to other gymnosperms.

The main generalizations about the distribution of flavonoids in the angiosperms have been outlined above and further examples have recently been given by Harborne (1967a, 1972b). The rich and complex variety of compounds present makes it difficult to discover many other generalizations. The formation of 6-hydroxy-flavones and -flavonols, like that of 6-C-glycosides, expectably follows later than 8-substitution and again, this follows from the known reactivity of the A-ring. The loss of the 3-hydroxy compounds (flavonols and proanthocyanidins) in more advanced angiosperm families is perhaps a reflection of both the instability of such compounds and the ability of the taxa concerned to produce the same sophisticated colourations given by highly hydroxylated flavonols by the synthesis of the more stable chalkones and aurones and polymethoxylated flavones. The replacement of proanthocyanidins by iso- and neo-flavones has obviously been due to the fact that the latter compounds are 'on-demand' anti-fungal agents which are needed in lower concentration than the polymeric proanthocyanidins or ellagitannins which, of course, may inhibit the cellular activities of the plant itself when damage occurs and thus impose a restraint on evolutionary expansion. On these grounds it would appear that the evolution of angiosperms is more likely to be as shown in Fig. 20.2 rather than that outlined in Fig. 20.1, for it is unlikely that any group patristically related to the main bulk of the angiosperms could have lacked proanthocyanins as do the Magnoliidae. A summary of the overall evolution of flavonoids, on the basis of the information and arguments presented above, is shown in Fig. 20.4. This must not be regarded as the final answer; but it is hoped that it will serve as a target for future experiment and speculation.

20.6 Conclusions

As was mentioned at the beginning of this chapter, we still know far too little about the distribution of biochemical pathways or even of their end products to be able to set down any definitive schemes dealing with the evolution of any class of compounds in the Plant Kingdom. But, in order to advance, we must try to erect on present

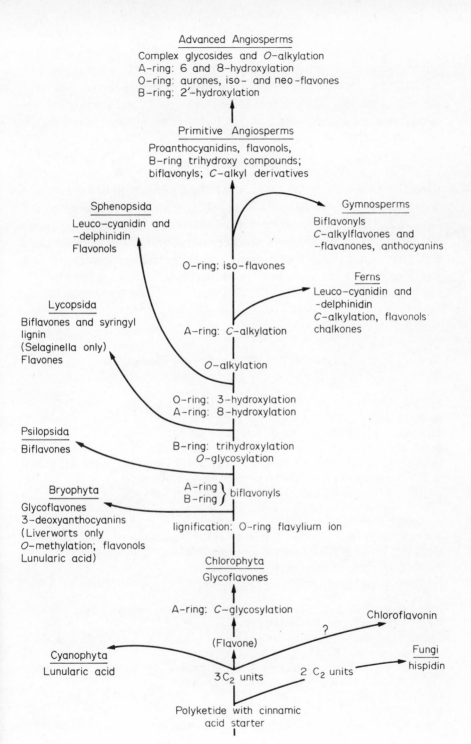

Figure 20.4 Main aspects of the evolution of flavonoids in higher plants.

information some house of cards which can be knocked down and reshuffled before we can rebuild another, hopefully, better one. It is my wish that this chapter will serve this purpose and stimulate more overall interest in evolution, and in flavonoid evolution in particular.

Acknowledgments

I wish to thank Dr P. W. D. Barnard for helpful comments on plant evolution and permission to use his data in the construction of Fig. 20.2; Dr J. B. Harborne for his always stimulating discussions; Mrs Gillian A. Cooper-Driver for her cogent and useful criticism; Drs E. C. Bate-Smith and J. McClure for their perceptive comments and Mrs Shirley Saunders for help in preparation of this manuscript for publication.

References

Akabori, Y. and Hasegawa, M. (1970), *Bot. Mag. (Tokyo)* **82**, 294.
Banks, H. P. (1970), *Evolution and Plants of the Past*, Wadsworth, Belmont, California.
Barnard, P. W. D. (1973), Private communication.
Bate-Smith, E. C. (1962), *J. Linn. Soc. (Bot.)* **58**, 95.
Bate-Smith, E. C. (1968), *J. Linn. Soc. (Bot.)* **60**, 325.
Bate-Smith, E. C. (1972), *Nature* **235**, 157.
Bate-Smith, E. C. (1974). In *Chemistry in Botanical Classification* (G. Bendz and J. Santesson, eds.) p. 93. Nobel Foundation, Stockholm.
Bate-Smith, E. C. and Swain, T. (1965), *Lloydia* **28**, 313.
Bell, P. and Woodcock, C. (1972), *The Diversity of Green Plants*, 2nd Ed. Arnold, London.
Bendz, G., Martensson, O. and Terenius, L. (1962), *Acta. chem. scand.* **16**, 1183.
Bierhorst, D. W., (1971), *Morphology of Vascular Plants*, Macmillan, New York.
Birch, A. J. (1962), *Proc. chem. Soc.* 3.
Bland, D. E., Logan, A., Menshun, M. and Sternhell, S. (1968), *Phytochemistry* **7**, 1373.
Briggs, L. H., Cain, B. F. and Cebalo, T. P. (1959), *Tetrahedron* **7**, 262.
Bu'Lock, J. and Smith, H. (1961), *Experimentia* **17**, 553.
Caldwell, M. M. (1971), *Photophysiology* (A. C. Glese, ed.), Vol. 6, p. 131. Academic Press, New York.
Carman, N. J., Watson, T., Bierner, M. W., Averett, J., Sanderson, S., Seaman, F. C. and Mabry, T. J. (1972), *Phytochemistry* **11**, 3271.
Cooper-Driver, G. A. and Swain, T. (1973), Suppl. No. 1 to *Bot. J. Linn. Soc.*, **67**, 111.
Cooper-Driver, G. A. and Swain, T. (1974), In preparation.

Copeland, E. B. (1947), Genera Filicum, Chronica Botanica, Waltham, Mass.

Cronquist, A. (1968), *The Evolution and Classification of Flowering Plants,* Houghton-Mifflin, Boston.

Cronquist, A. (1971), *Introducing Botany,* 2nd Ed. Harper Row, New York.

Dayhoff, M. O. (ed.) (1969), *Atlas of Protein Sequence and Structure,* Vol. 4. National Biomedical Research Foundation, Silver Springs, Maryland.

Echlin, P. (1970), *Phytochemical Phylogeny,* (J. B. Harborne, ed.), p. 1. Academic Press, London.

Emberger, L. (1960), *Traite de Botanique Systematique* (M. Chadefaud and L. Emberger, eds.), Vol. II(1), p. 754. Masson et Cie., Paris.

Fox, S. F. (1973), *Pure* and *appl. Chem.* **34,**

Fox, R. H., Scaplehorn, A. W. and Tonge, G. M. (1973), *J. Forensic Sci.,* in press.

Fritsch, F. E. (1945), *Ann. Bot.* (N.S.) **9,** 1.

Gass, I. G., Smith, P. J. and Wilson, R. C. L. (1971), *Understanding the Earth,* Artemis Press, Sussex.

Grambast, L. (1964), *Naturalia Monseliensia ser Bot.* **16,** 71.

Grisebach, H. (1972), Personal communication.

Harborne, J. B. (1966), *Phytochemistry* **5,** 589.

Harborne, J. B. (1967a), *Comparative Biochemistry of the Flavonoids,* Academic Press, London.

Harborne, J. B. (1967b), *Phytochemistry* **6,** 1643.

Harborne, J. B. (ed.) (1972a), *Phytochemical Ecology,* Academic Press, London.

Harborne, J. B. (1972b), In *Recent Advances in Phytochemistry* (V. C. Runeckles and J. E. Watkin, eds.), Vol. 4, p. 107. Appleton Century Crofts, New York.

Haug, A. and Larsen, B. (1958), *Acta chem. scand.* **12,** 650.

Hillis, W. E. and Isui, K. (1965), *Phytochemistry* **4,** 905.

Hillis, W. E. and Yazaki, Y. (1971), *Phytochemistry* **10,** 105.

Hudson, A. T. and Bentley, R. (1969), *Chem. Commun.* **830.**

Klohs, M. W. (1967), *Ethnopharmacologic search for psychoactive drugs* (D. H. Efron, ed.). U.S. Govt. Printing Office, Washington.

Kubitski, K. (1969), *Taxon* **18,** 360.

Lindberg, G., Osterdahl, B.-G., and Nilsson, E. (1974), *Chemica Scripta* **4,** 140.

List, P. H. and Freund, B. (1968), *Planta med.* 123.

Lowry, J. B. (1972), *Phytochemistry* **11,** 725.

Mabry, T. J., Markham, K. R. and Thomas, M. B. (1970), *Systematic Identification of the Flavonoids,* Springer-Verlag, Heidelberg.

Marchelli, R. and Vining, L. C. (1973), *Chem. Comm.* 555.

Margulis, L. (1971), *Evolution* **25,** 242.

Markham, K. R. (1972), *Phytochemistry* **11,** 2047.

Markham, K. R., Mabry, T. J. and Averett, J. E. (1972), *Phytochemistry* **11,** 2875.

Markham, K. R. and Porter, L. (1969), *Phytochemistry* **12,** 1777.

Markham, K. R. and Porter, L. (1973), *Phytochemistry* **12,** 2007.

Markham, K. R., Porter, L. and Brehm, B. G. (1969), *Phytochemistry* **8,** 2193.

McAlester, A. L. (1968), *The History of Life,* Prentice Hall, Englewood Cliffs, New Jersey.

McClure, J. and Miller, H. A. (1967), *Nova Hedwigia* **14**, 111.

Melchert, T. E. and Alston, R. E. (1965), *Science,* **150**, 1170.

Milborrow, B. V. (ed.), (1973), *Biosynthesis and its Control in Plants,* Academic Press, London.

Monties, P. (1969), *Bull. Soc. franc. Physiol. Veget.* **15**, 29.

Nilsson, E. (1967), *Acta. chem. scand.* **21**, 1942.

Nilsson, E. (1969), *Acta. chem. scand.* **23**, 2910.

Nilsson, E. (1973), *Phytochemistry* **12**, 722.

Nilsson, E. and Bendz, G. (1974), In *Chemistry in Botanical Classification* (G. Bendz and J. Santesson, eds.) p. 117. Nobel Foundation, Stockholm.

Oettmeier, W. and Heupel, A. (1972), *Z. Naturf.* **27B**, 177.

Parihar, N. S. (1962), *Pteridophytes,* Central Book Depot, Allahabad.

Pichi-Sermolli, R. E. G. (1959), *Vistas in Botany* (W. B. Turril, ed.) p. 421. Pergamon Press, London.

Ponnamperuma, C. (1972), *The Origins of Life,* Thames and Hudson, London.

Pratriel-Sosa, F. and Percheron, F. (1972), *Phytochemistry* **10**, 1809.

Pridham, J. (1964), *Phytochemistry* **3**, 493.

Pryce, R. J. (1971), *Phytochemistry* **10**, 2679.

Pryce, R. J. (1972), *Phytochemistry* **11**, 1759.

Reznik, H. and Wierman, R. (1966), *Naturwissenshaften* **53**, 230.

Richards, M., Bird, A. E. and Munden, J. E. (1969), *J. Antibiotics* (Tokyo) **22**, 388.

Roux, D. G. (1972), *Phytochemistry* **11**, 1219.

Saleh, N. A. M., Majak, W. and Towers, G. H. N. (1972), *Phytochemistry* **11**, 1095.

Santamour, F. S. (1967), *Morris Arb. Bull.,* **18**, 41.

Saunders, J. A. and McClure, J. W. (1972), *Am. J. Bot.* **59**(5), 673.

Saunders, J. A. and McClure, J. W. (1973), *Pl. Physiol.* **51**, 407.

Sawada, T. (1958), *J. pharm. Soc. Japan* **78**, 1023.

Schopf, J. W. (1970), *Biol. Rev.* **45**, 319.

Setlow, R. B. (1967), *Regulation of Nucleic Acid and Protein Biosynthesis* **10**, 51.

Siegel, S. M. (1968), *Comprehensive Biochemistry* (M. Florkin and E. Stotz, eds.) Vol. 26A, p. 1. Elsevier, Amsterdam.

Siegel, S. M. (1969), *Am. J. Bot.* **56**, 175.

Smith, I. (1968), *Chemotaxonomy and Serotaxonomy,* (J. G. Hawkes, ed.). Academic Press, London.

Sporne, K. R. (1965), *The Morphology of the Gymnosperms,* Hutchinson, London.

Sporne, K. R. (1970), *The Morphology of Pteridophytes,* Hutchinson, London.

Stebbins, G. L. (1966), *Processes of Organic Evolution,* Prentice Hall, Englewood Cliffs, New Jersey.

Swain, T. (1962), *The Chemistry of Flavonoid Compounds,* (T. A. Geissman, ed.) McMillan, New York.

Swain, T. (1965), *Plant Biochemistry,* (J. Bonner and J. E. Varner, eds.). Academic Press, New York.

Swain, T. (1970), *Biochemical Evolution and the Origin of Life,* (E. Schoffeniels, ed.). North Holland, Amsterdam.

Swain, T. (1974), *Comprehensive Biochemistry*, (M. Florkin and E. H. Stotz, eds.), Vol. 29A, Elsevier, p. 125.

Takhtajan, A. (1969), *Flowering Plants,* Oliver and Boyd, Edinburgh.

Towers, G. H. N. and Gibbs, R. D. (1953), *Nature* **172**, 25.

Towers, G. H. N. and Maas, W. S. G. (1965), *Phytochemistry* **4**, 57.

Ueno, A., Oguri, N., Hori, K., Saiki, Y. and Harada, T. (1963), *Yakugaku Zasshi* **83**, 420.

Van Fleet, D. S. (1969), *Advancing Frontiers of Plant Science,* **23**, 65. Impex, New Delhi.

Voirin, B. (1970), These Docteur-Science, L'Universite de Lyon.

Voirin, B. (1972). *Phytochemistry* **11**, 257.

Voirin, B. and Lebreton, P. (1971), *Boisseira* **19**, 259.

Watson, E. V. (1967), *The Structure and Life of Bryophytes* 2nd Ed. Hutchinson, London.

Webb, S. J. and Tai, C. E. (1969), *Nature* **224**, 1123.

Weinberg, E. D. (1970), *Adv. Microbiol. Physiol.* **4**, 1.

Weissenboeck, G., Tevini, M. and Reznik, H. (1971), *Z. Pflanzenphysiol.* **64**, 274.

White, E. and Towers, G. H. N. (1967), *Phytochemistry* **6**, 663.

Whittaker, R. H. and Feeny, P. P. (1971), *Science* **171**, 757.

Wilkins, M. B. (ed.) (1970). *Physiology of Plant Growth and Development,* McGraw Hill, Maidenhead, England.

Woodhead, S., Cooper-Driver, G. A. and Swain, T. (1974), In preparation.

Addendum

These addenda cover some of the more important references which have appeared since the book went to press. They mainly refer to reports of new flavonoids but also include a few corrections to earlier incorrect structures. Addenda are provided for Chapters 3–6, 8, 9, 11, 12 and 14 and are keyed in with page references to the main text.

Chapter 3

Chemical Ionization Mass Spectrometry (see pp. 83–118)

Kingston and Fales (1973) and Clark-Lewis *et al.* (1973) have recently described the chemical ionization mass spectrometry (CI-MS) of flavonoids. Clark-Lewis, using isobutane and hydrogen as reagent gases, found this technique useful in the structural investigation of flavans, flavan-3,4-diols, flavanones and dihydro-flavonols. In contrast to electron impact MS, simple fragmentation patterns and abundant metastable ions were produced. Kingston and Fales examined a wider range of flavonoids, but found when using methane as reagent gas, that flavones and flavonols showed no significant fragmentation in the CI mode. By contrast, flavanones and dihydroflavonols fragmented well and the CI mass spectra contained useful information for structure determination.

One of us (unpublished work of T. J. Mabry and H. M. Fales) has applied CI-MS to perdeuteriomethylated flavonoid glycosides. As with the aglycones, the glycosides show much less fragmentation than observed with EI-MS. Using isobutane at 175°, for example, it was found that rutin gave two base peaks, one due to $[M + H]^+$ and the other due to $[aglycone + H]^+$. Other peaks were observed for the sequential loss of the sugars. Myricetin under the same conditions gave an $[M + H]^+$ peak of 20% relative intensity while $[aglycone + H]^+$ was the base peak. Naringin, which opened to the chalcone on derivatization, gave a 100% $[M + H]^+$ peak with the only notable fragments being again due to the sequential loss of the sugars.

High Resolution Field Desorption Mass Spectrometry (see pp. 121–123)

Schulten and Games (1974) have recently applied field desorption mass spectrometry (FD-MS) to a number of flavonoid *O*-glycosides. Using this technique, the mass spectra of *underivatized* flavonoid glycosides were determined. In all cases molecular weights and elemental composition were established and fragments relating to both aglycone and saccharide moieties were observed. For example, with underivatized rutin, the electron impact mass spectrum (EI-MS) is similar to that of quercetin, whereas the FD-MS exhibits a molecular ion $[M + H]^+$ of 25% relative intensity and other ions for the sequential cleavage of the sugars. Thus, $[M\text{-rhamnose}]^+$ and $[M\text{-rhamnose-glucose} + H]^+$ are present at 20 and 100% relative intensity, respectively. Other flavonoid glycosides studied were naringin and hesperidin and these exhibited equally useful fragmentation patterns. The presence of additional bound salts in these two examples was also detected; for example, $[M + Na]^+$ was the base peak in the hesperidin FD-MS. (We suspect that these are salts of the chalcones derived from the flavones during the isolation.)

FD-MS because of its applicability to *underivatized* flavonoid glycosides and because of the informative fragmentations produced appears to be of considerable potential value for the structure determination of glycosides, generally. This technique offers a welcome alternative to EI-MS which is only of use with the often difficult to prepare permethylated and perdeuteriomethylated glycosides.

Chapter 4

Synthesis of isoflavonoids (see table 4.8, p. 186)

The following isoflavones have now been synthesized: alpinumisoflavone, osajin, scandenone (Jain and Sharma, 1973), dalpanol (Crombie *et al.* 1973), dalpatin, 7,2'-dimethoxy-6-hydroxy-4',5'-methylenedioxyisoflavone, 3',4'-methylenedioxy-5,6,7-trimethoxyisoflavone (Farkas *et al.*, unpublished), fujikinetin (Jain *et al.* 1973), glycitein (Naim *et al.* 1973), 6,7,2',3',4'- and 6,7,2',4',5'-pentamethoxyisoflavone (Farkas and Wolfner, 1974).

Synthesis of flavonol glycosides (see p. 181)

By a combination of selective benzylations and transacylations, the natural 3,3'-, 3,4'-, 3,7- and 7,4'-di-O-β-D-glucopyranosides of quercetin have been synthesized (Farkas *et al.*, 1974).

Synthesis of tachrosin (see p. 189)

The thallium nitrate method has proved to be useful (Antus *et al.*, 1974) for the construction of the novel dihydrofuranyl moiety of tachrosin, one of several related structures isolated from *Tephrosia* by Smalberger *et al.* (1973).

Chapter 5

The desirability of employing mild extraction conditions for isolation of labile anthocyanins (see p. 219) has been further emphasized by the finding of 3,5-diglucosides acylated with acetic acid during the re-investigation of the skin pigments of Royalty grapes (Fong *et al.*, 1974). The same authors and others (Hrazdina and Franzese, 1974a) have also confirmed the position of acylation of *p*-coumaric acid at the C-6 hydroxyl group of the 3-*O*-glucose moiety in 3-mono- and 3,5-diglucosides of grapes. Oxidation of the latter pigments with hydrogen peroxide gave acylated *o*-benzoyloxyphenylacetic acid esters under acid conditions and 3-(6-*O*-*p*-coumaryl)-glucosyl-5-*O*-glucosyl-7-hydroxycoumarin under neutral conditions, apparently *via* a Bayer-Villiger type oxidation (Hrazdina and Franzese, 1974b). The latter reaction has structural applications and was used by Thakur and Ibrahim (1974) to provide additional evidence for the occurrence of hirsutidin glycosides in flax hypocotyls (*Linum usitatissimum* Linaceae) (see p. 239), previously reported only in Primulaceae and Apocynaceae. However, the occurrence of these rare glycosides in *Linum* is in doubt since attempts to repeat their isolation from flax hypocotyls, using the same variety as Thakur and Ibrahim (1974), failed and cyanidin 3-rutinoside was the only pigment that could be found (J. B. Harborne, unpublished results).

Phenolic acid acylating groups have been further implicated in flower pigmentation (see p. 224), in the absence of metals and co-pigments. Thus, caffeic acid appears essential for the stable blue colour of the cineraria flower pigment and may act by association between its *o*-hydroxyl groups and those of the delphinidin B-ring (Yoshitama and Hayashi, 1974).

Anthocyanins representing new glycosidic types recently reported include: cyanidin 3-neohesperidoside from *Podocarpus lawrencii* (Crowden, 1974) and cyanidin 3-arabinosylsambubioside from *Viburnum trilobum* (Du *et al.* 1974).

Chapter 6

New flavones recently reported

The bark of *Morus alba* (see p. 283) contains mulberranol, a derivative of mulberrin (24) in which the 6-prenyl group has cyclised to a hydroxy-isopropyl-dihydrofuran; it is the first natural flavone of this type, of which there are several representatives among coumarins and rotenoids (Wakharkar, 1974).

Integrin and cyclointegrin, isolated from the heartwood of *Artocarpus integer* (see p. 278), an Indonesian species, are the first natural phloroglucinol-derived flavones with prenyl substituents in the 3-position (cf. the biosynthesis indicated in structures 13, 14, 15 and 16). Integrin is 5, 2',4'-trihydroxy-7-methoxy-3-prenylflavone; in cyclointegrin the 3-prenyl and 2'-hydroxyl groups have cyclised to a dihydro-2,2-dimethyl-2*H*-oxocin (Pendse, R., Rama Rao, A. V. and Venkataraman, K., unpublished work).

Semiglabrinol from *Tephrosia semiglabra* (see p. 284) is the first natural flavone with a bisfurano moiety, probably derived from an 8-prenyl precursor by secondary modifications (Smalberger *et al.*, 1973). Tephrostachin in *T. polystachyoides* is chrysin dimethyl ether substituted by $-CH=CH-CMe_2-OH$ in the 8-position (Vleggaar *et al.*, 1973).

Ugonin A and B, isolated from *Helminthostachys zeylanica* rhizome, are derivatives of luteolin and its 3'-methyl ether, in which a geranyl substituent in the 8-position has cyclised to a 2-benzoxepine ring (see p. 282) (Murakami *et al.*, 1973).

The stuctures of gardenin C and E (Table 6.6, p. 275) have been confirmed by synthesis (Kalra *et al.*, 1973).

Chapter 8

New flavone glycosides

Apigenin derivatives (see p. 403). Apigenin 4'-glucuronide has been found in *Chrysanthemum cinerarifolium* (Compositae) (Rao *et al.* 1973), the 7-rutinoside-4'-glucoside in *Galium mollugo* (Rubiaceae) (Borisov, 1974) and the 7-glucuronide-4'-glucoside in *Centaurea cyanus* flowers (Compositae) (Asen and Horowitz, 1974). The 5-xylosylglucoside of scutellarein 7,4'-dimethyl ether has been reported in leaf of *Ovida pillopillo* (Thymelaceae) by Nunez-Alarcon *et al.* (1973).

Luteolin derivatives (see p. 407). A range of luteolin glucuronides have been found in liverworts: the 3'-glucuronide and 3'4'-diglucuronide in *Lunularia cruciata*; and the 7,3'- and 7,4'-diglucuronides and the 7,3',4'-triglucuronide in *Marchantia polymorpha* (Markham and Porter, 1974). The 7-glucosylarabinoside-4'-glucoside of luteolin has been reported in *Galium mollugo* (Rubiaceae) by Borisov (1974). Chrysoeriol and diosmetin 7-bisulphates are two new conjugates from leaf of *Zostera marina* (Zosteraceae) and luteolin 7,3'-bis-bisulphate occurs in the same source (J. B. Harborne and C. A. Williams, unpublished results). Diosmetin 7-(xylosyl-(1 → 2)-glucoside) has been reported in *Galium mollugo* (Borisov, 1974). The 7-arabinoside and 7-arabinoside-4'-rhamnoside of 6-hydroxyluteolin have been discovered in *Lippia nodiflora* (Verbenaceae) by Nair *et al.* (1973).

New flavonol glycosides

Kaempferol derivatives (see p. 413). 8-O-Methylherbacetin 3-glucoside has been isolated from the inflorescence of *Sorbus aucuparia* (Rosaceae) by Jerzmanowska and Kamechi (1973). The 7-bisulphate and 7-bisulphate-3-glucuronide of kaempferol occur in leaf of *Frankenia pulverulenta* (Harborne, 1975).

Quercetin derivatives (see p. 418). The 3-(malonylglucoside) of quercetin has been reported in leaf of *Cichorium endiva* (Compositae) (Woeldecke and Herrmann, 1974). The 7-bisulphate of isorhamnetin and the 7-bisulphate-3-glucuronides of quercetin and isorhamnetin occur in leaf of *Frankenia pulverulenta* (Harborne, 1975). The 3-[rhamnosyl-(1 → 4)-rhamnosyl-(1 → 6)-galactosides] of rhamnetin and of rhamnazin have been identified in the fruits of *Rhamnus petiolaris* (Rhamnaceae) (Wagner *et al.* 1974).

A comprehensive review of the chemistry of flavonoid glycosides has just been published (Wagner, 1974).

Chapter 9

New chalcones and dihydrochalcones

Volsteedt *et al.* (1973) have found both the *cis-* and *trans-*isomers of α,2′,4′,6′,4-pentahydroxychalcone in *Berchemia zeyheri*. Montero and Winternitz (1973) have demonstrated that rubranine (19), originally reported to occur naturally in *Aniba rosaeodora* (see p. 453), is an artefact formed through condensation of pinocembrin and citral in the presence of anibine. Bhakuni *et al.* (1973) have found α,2′,4′,6′,4-pentahydroxydihydrochalcone in *Podocarpus nubigena* (Poducarpaceae). Shukla *et al.* (1973) have shown that the closely related α,2′,6′,4-tetrahydroxy-4′-methoxydihydrochalcone exists, along with its 2′-O-glucoside, in *Lyonia formosa* (Ericaceae). Hellyer and Pinhey (1966) report 1-(β-phenylpropionyl)-2-hydroxy-3,3,5,5-tetramethyl-4,6-diketocyclohex-1-ene, known as grandiflorone, in *Leptospermum flavescens*. This compound bears an interesting relationship to ceroptin (18) (see p. 453), a chalcone from *Pityrogramma* which also has *gem*-dimethyl substitution on the A-ring.

The role of α-hydroxychalcones as possible intermediates in flavonoid synthesis has been discussed in a review (Roux and Ferreira, 1974).

Chapter 11

Flavanones and dihydroflavonols

Murakami *et al.* (1973) have described a novel A-ring flavanone from *Helminthostachys*; 'ugonin-D' is based upon naringenin and has a heterocyclic ring between the C-7 oxygen and C-8 forming a 2,3,3-trimethyldihydrofuran system. Further reports have also appeared describing the complex flavonoid chemistry of *Sophora subprostrata* (Kyogoku *et al.* 1973; Komatsu *et al.* 1973).

Obara *et al.* (1973) synthesized 5,6,7,4′- and 5,7,8,4′-tetrahydroxyflavanone and showed that they compared with carthamidin and isocarthamidin, respectively. This is opposite to the assignment of structure to isocarthamin (the 5-O-glucoside) as given in structure (40) (see p. 575).

Birch and Thompson (1972) commented on the co-occurrence of the flavanones farrerol (46) and protofarrerol (49) with the chromone leptorumol (same substitution pattern) (see p. 577). They suggest that chromones may be formed by phenol oxidation of 4′-hydroxyflavanones to yield products of the general form of protofarrerol. Fission of these compounds would yield the chromone plus hydroquinone. In this respect it is interesting to note the co-occurrence of eriodictyol (5,7,3′,4′-tetrahydroxyflavanone) and 5,7-di-hydroxychromone in *Arachis hypogaea* (Pendse *et al.* 1973).

Hauteville *et al.* (1973) have described synthetic experiments confirming the existence of 2-hydroxyflavanones (18c, 18e) (p. 572) in *Populus nigra*. A correction in the structure of phellamurin has been published by Sakai and Hasegawa (1974) who believe it to be the 7-O-glucoside of 5,7,4′-trihydroxy-8-(γ,γ-dimethylallyl)dihydroflavonol. Earlier work (see p. 590) suggested that the compound was a tertiary alcohol with structure (95).

Chapter 12

New naturally occurring *C*-glycosylflavonoids (see p. 641)

2″-*O*-acetyl-7-*O*-methylvitexin from *Brackenridgea zanguebarica* (Ochnaceae): Bombardelli *et al.* (1974); 3′-*O*-glucosyliso-orientin from *Gentiana nivalis* (Gentianaceae): Hostettmann and Jacot-Guillarmod (1974); Potassium bisulphate salts of vitexin, isovitexin orientin, iso-orientin: Williams *et al.* (1973); Isoschaftoside (6-*C*-α-L-arabinopyranosyl-8-*C*-β-D-glucopyranosylapigenin) from *Flourensia cernua* (Compositae): Dillon, M. and Mabry, T. J. (unpub.); Biol *et al.* (1974); 6,8,di-*C*-pentosylapigenin from *Melilotus alba* (Leguminosae): Specht, J. and Gorz, H. J. (unpublished); 6-*C* and 8-*C*-β-neohesperidosylacacetin from *Fortunella margarita* (Rutaceae): Horowitz, R. M. *et al.* (unpublished).

Identification of *C*-glycosylflavonoids (see Table 12.4, p. 651)

Adonivernith = 2″-*O*-β-D-xylopyranosylorientin: Wagner, H. *et al.* (unpublished); Schaftoside = 6-*C*-β-D-glucopyranosyl   8-*C*-α-L-arabinopyranosylapigenin: Chopin, J., Bouillant, M. L., Wagner, H. and Galle, K. (unpublished); '6,8,di,*C*-β-D-glucopyranosylacacetin' from *Trigonella corniculata* = *6,8-C*-(pentosyl, hexosyl)acacetin: Bouillant, M. L. (unpublished).

Mass spectrometry (see p. 674)

Mono-*C*- and di-*C*-glycosylflavonoid permethyl ethers: Bouillant *et al.* (1974).

Circular dichroism (see p. 677)

6-*C*-β-D-xylopyranosyl,-6-*C*-β-D-galactopyranosyl and 6-*C*-α-L-arabinopyranosyl apigenins show a positive Cotton effect at 250–275 nm; 6-*C*-α-L-rhamnopyranosyl apigenin a negative one: Gaffield, W. *et al.* (unpublished).

Isomerization of *C*-glycosylflavonoids (see p. 680)

Enzymic conversion of isovitexin to vitexin by leaf extracts from *Briza media*: Murray and Williams (1973).

Chapter 14

ISOFLAVONES (see p. 747). New constituents of *Cordyla africana* heartwood include: 6-hydroxy-7,2′-dimethoxy-4′,5′-methylenedioxy-5,6,7-trimethoxy-3′,4′-methylenedioxy- and 5,6,7,8-tetramethoxy-3′,4′-methylenedioxyisoflavone (Campbell and Tannock, 1973). A phytoalexin from sugar beet leaves (*Beta vulgaris*; Chenopodiaceae) has been identified as 2′-methyltlatlancuayin (Geigert *et al.*, 1973). Occurring in *Iris germanica* (Iridaceae) (Pailer and Franke, 1973; Dhar and Kalla, 1973) are: 5,7,3′-trihydroxy-6,4′dimethoxy-, 5,4′-dihydroxy-6,7-methylenedioxy-(irilone),   5,3′,4′-trimethoxy-6,7-methylenedioxy-   and 5,3′,4′,5′-tetramethoxy-6,7-methylenedioxyisoflavone. The last compound (irisflorentin) also occurs in *I. florentina* (Morita *et al.*, 1973), as do 5,4′-dihydroxy-

3'-methoxy-6,7-methylenedioxyisoflavone-4'-O-β-D-glucoside (iritloside, iriflogenin 4'-glucoside) (Arisawa *et al.*, 1973), and irilone 4'-glucoside and irisolone 4'-bioside (Tsukida *et al.*, 1973). Neobaisoflavone, from the seeds of *Psoralea coryliflora*, is 3'-C-isopentenyldaidzein (Bajwa *et al.*, 1973), and licoricone, from licorice roots, is 7,6'-dihydroxy-2',4'-dimethoxy-3'-allylisoflavone (Kaneda *et al.*, 1973).

ISOFLAVANONES (see p. 760) Two new racemic 2'-methoxyisoflavanones, 2'-methylsophoral and 7-hydroxy-2',4'-dimethoxyisoflavanone have been isolated from *Dalbergia stevensonii*, together with the known pterocarpans and coumestans having the same substitution patterns (Donnelly *et al.*, 1973b).

ROTENOIDS (see p. 763) The co-occurrence of 12a-hydroxy- and 12a-methoxy-rotenoids has been reported for the first time (Oberholzer *et al.*, 1974). Extracts of roots of *Neorautanenia amboensis* yielded the 12a-hydroxy and 12a-methoxy derivatives oi rotenone and dolineone, in addition to rotenone, dehydrorotenone and dehydrodolineone. Three novel 6a,12a-dehydrorotenoid compounds with oxygenation at position C_6 of ring B have also recently been reported to occur in the roots of the Thai medicinal plant, *Stemona collinsae* (Shiengthong *et al.*, 1974). Stemonal is 6,11-dihydroxydehydromunduserone; stemonone is the corresponding 6-oxo derivative and stemonacetal is the 6-ethyl ether of stemonal.

PTEROCARPANS (see p. 768) Edulenol, a new pterocarpan from *Neorautanenia edulis*, is 3-hydroxy-1,9-dimethoxy-2-isopentenylpterocarpan (Rall *et al.*, 1972), and tuberosin from *Pueraria tuberosa* (Joshi and Kamat, 1973) is isomeric with 6a-hydroxyphaseollin.

ISOFLAVANS (see p. 776) A second anti-fungal isoflavan from diseased *Phaseolus vulgaris* tissues has been identified as the 2'-methyl ether of phaseollinisoflavan (Etten, 1973) and the induced isoflavan from leaves of *Medicago sativa* is (−)-2'-methylvestitol (Ingham and Millar, 1973). This compound (sativan) and (−)-vestitol also function as phytoalexins in *Lotus corniculatus* (Bonde *et al.*, 1973). (3R)-7,8,3'-Trihydroxy-2',4'-dimethoxyisoflavan occurs in *Dalbergia ecastophyllum* (Donnelly *et al.*, 1973a). (−)-Neorauflavane and neorauflavene are novel isoflavonoids from *Neorautanenia edulis* (Brink *et al.*, 1974), with the latter being the first natural isoflavene reported.

References

Antus, S., Farkas, L., Nogradi, M., and Sohar, P. (1974), *Chem. Comm.*, 799.

Arisawa, M., Morita, N., Kondo, Y. and Takemoto, T. (1973), *Chem. pharm. Bull.* 21, 2323.

Asen, S. and Horowitz, R. M. (1974), *Phytochemistry* 13, 1219.

Bajwa, B. S., Kharma, P. L. and Seshadri, T. R. (1972), *Curr. Sci.* 41, 882.

Bhakuni, D., Bittner, M., Silva, M. and Sammes, P. G. (1973), *Phytochemistry* 12, 2777.

Biol, M. C., Bouillant, M. L., Planche, G. and Chopin, J. (1974), *C. r. Acad. Sci., Ser. C* 279, 409.

Birch, A. J. and Thompson, D. J. (1972), *Australian J. Chem.* **25**, 2731.

Bombardelli, E., Bonati, A., Gabetta, B. and Mustich, G. (1974), *Phytochemistry* **13**, 295.

Bonde, M. R., Millar, R. L. and Ingham, J. L. (1973), *Phytochemistry* **12**, 2957.

Borisov, M. I. (1974), *Rast Resur.* **10**, 66.

Bouillant, M. L., Favre-Bonvin, J. and Chopin, J. (1974), *C. r. Acad. Sci., Ser. D* **279**, 295.

Brink, A. J., Rall, G. J. H. and Engelbrecht, J. P. (1974), *Tetrahedron* **30**, 311.

Campbell, R. V. P. and Tannock, J. (1973), *J. chem. Soc. Perkin I*, 2222.

Clark-Lewis, J. W., Harwood, C. N., Lacey, M. J. and Shannon, J. S. (1973), *Aust. J. Chem.* **26**, 1577.

Crombie, L., Freeman, P. W., and Whiting, D. A. (1973), *J. chem. Soc. Perkin I*, 1277.

Crowden, R. K. (1974), *Phytochemistry* **13**, 2877.

Dhar, K. L. and Kalla, A. K. (1973), *Phytochemistry* **12**, 734.

Donnelly, D. M. X., Keenan, P. J. and Prendergast, J. P. (1973a), *Phytochemistry* **12**, 1157.

Donnelly, D. M. X., Thompson, J. C., Whalley, W. B. and Ahmad, S. (1973b), *J. chem. Soc. Perkin I*, 1737.

Du, C. T., Wang, P. L. and Francis, F. J. (1974), *Phytochemistry* **13**, 1998.

Etten, H. D. V. (1973), *Phytochemistry* **12**, 1791.

Farkas, L. and Wolfner, A. (1974), *Chem. Ber.* **107**, in press.

Farkas, L., Vermes, B. and Nogradi, M. (1974), *Chem. Ber.* **107**, 1518.

Fong, R. A., Webb, A. G. and Kepner, R. E. (1974), *Phytochemistry* **13**, 1001.

Geigert, J., Stermitz, F. R., Johnson, G., Maag, D. D. and Johnson, D. K. (1973), *Tetrahedron* **29**, 2703.

Harborne, J. B. (1975), *Phytochemistry* **14**, in press.

Hauteville, M., Chadenson, M. and Chopin, J. (1973), *Bull. Soc. Chim. Fr.*, 1784.

Hellyer, R. O. and Pinhey, J. T. (1966), *J. Chem. Soc.*, 1496.

Hostettmann, K. and Jacot-Guillarmod, A. (1974), *Helv. Chim. Acta* **57**, 204.

Hrazdina, G. and Franzese, A. J. (1974a), *Phytochemistry* **13**, 225.

Hrazdina, G. and Franzese, A. J. (1974b), *Phytochemistry* **13**, 231.

Ingham, J. L. and Millar, R. L. (1973), *Nature* **242**, 125.

Jain, A. C. and Sharma, B. N. (1973), *Chem. Letters*, 1323.

Jain, A. C., Rohtagi, V. K. and Seshadri, T. R. (1973), *Indian J. Chem.* **11**, 98.

Jerzmanowska, Z. and Kamechi, J. (1973), *Rocz. Chem.* **47**, 1629.

Joshi, B. S. and Kamat, V. N. (1973), *J. chem. Soc. Perkin I* 907.

Kalra, A. J., Krishnamurti, M. and Seshadri, T. R. (1973), *Indian J. Chem.* **11**, 96, 1092.

Kaneda, M., Saitoh, T., Iitaka, Y. and Shibata, S. (1973), *Chem. pharm. Bull.* **21**, 1338.

Kingston, D. G. I. and Fales, H. M. (1973), *Tetrahedron* **29**, 4083.

Komatsu, M., Tomimori, T. Hatakeyamo, K. and Makiguchi, Y. (1973), *Japan. Patent* 73 11,931; *Chem. Abstr.* **79**: 63637q.

Kyogoku, K., Hatayama, K., Suzuki, K., Yokomori, S., Maejima, K., and Komatsu, M. (1973), *Chem. Pharm. Bull.* **21**, 1436, 1777, 1192, 2733.

Markham, K. R. and Porter, L. J. (1974), *Phytochemistry* **13**, 1553, 1937.

Montero, J. L. and Winternitz, F. (1973), *Tetrahedron* **29**, 1243.

Morita, N., Arisawa, M., Kondo, Y. and Takemoto, T. (1973), *Chem. pharm. Bull.* **21**, 600.

Murakami, T., Hagiwara, M., Tanaka, K. and Chen, C-M (1973), *Chem. pharm. Bull.* **21**, 1849, 1851.

Murray, B. G. and Williams, C. A. (1973), *Nature* **243**, 87.

Naim, M., Gestetner, B., Kirson, I., Birk, Y. and Bondi, A. (1973), *Phytochemistry* **12**, 169.

Nair, A. G., Ramachandran, R. P., Nagarajan, S. and Subramanian, S. S. (1973), *Ind. J. Chem.* **11**, 1316.

Nunez-Alarcon, J., Rodriguez, E., Schmid, R. D. and Mabry, T. J. (1973), *Phytochemistry* **12**, 1451.

Obara, H., Onodera, J. and Yamamoto, F. (1973), *Chemical Letters*, (8), 915.

Oberholzer, M. E., Rall, G. J. H. and Roux, D. G. (1974), *Tetrahedron Letters* 2211.

Pailer, M. and Franke, F. (1973), *Monatsh. Chem.* **104**, 1394.

Pendse, R., Rama Rao, A. V., and Venkataraman, K. (1973), *Phytochemistry* **12**, 2033.

Rall, G. J. H., Brink, A. J. and Engelbrecht, J. P. (1972), *J. S. African chem. Inst.* **25**, 131.

Rao, P. R., Seshadri, T. R. and Sharma, P. (1973), *Curr. Sci.* **42**, 811.

Roux, D. G. and Ferreira, D. (1974), *Phytochemistry* **13**, 2039.

Sakai, S. and Hasegawa, M. (1974), *Phytochemistry* **13**, 303.

Schulten, H-R. and Games, D. E. (1974), *Biomed. Mass Spectrom.* **2**, 120.

Shiengthong, D., Donavanik, T., Uaprasert, V. and Roengsumran, S. (1974), *Tetrahedron Letters* 2015.

Shukla, Y. N., Tandon, J. S. and Dhar, M. M. (1973), *Indian J. Chem.* **11**, 720.

Smalberger, T. M., van den Berg, A. J. and Vleggaar, R. (1973), *Tetrahedron* **29**, 3099.

Thakur, M. L. and Ibrahim, R. K. (1974), *Z. Pflanzenphysiol.* **71**, 391.

Tsukida, K., Saiki, K. and Ito, M. (1973), *Phytochemistry* **12**, 2318.

Vleggaar, R., Smalberger, T. M. and De Waal, H. L. (1973), *J. S. African chem. Inst.* **26**, 71.

Volsteedt, F. du R., Rall, G. J. H. and Roux, D. G. (1973), *Tetrahedron Letters*, (12), 1001.

Wagner, H. (1974), *Fortschr. org. Naturst.* **31**, 153.

Wagner, H., Ertan, M. and Seligmann, O. (1974), *Phytochemistry* **13**, 857.

Wakharkar, P. V. (1974), Ph.D. Thesis, University of Poona.

Williams, C. A., Harborne, J. B. and Clifford, H. T. (1973), *Phytochemistry* **12**, 2417.

Woeldecke, M. and Herrmann, K. (1974), *Z. Naturforsch.* **29c**, 355.

Yoshitama, K. and Hayashi, K. (1974), *Bot. Mag. (Tokyo)* **87**, 33.

Author Index

Gifford, A. M. 1010, 1046
Gilbert, A. H. 199, 207
Gilbert, B. 203, 627
Gilbert, R. I. 249, 258
Gilchrist, D. G. 997, 1046
Gillett, J. M. 1094
Ginzburg, C. 1010, 1046
Gisvold, O. 567, 602, 624
Giuffra, S. E. 439
Givan, C. V. 1010, 1046
Glass, A D. M. 944, 964
Glassner, S. 42
Glennie, C. W. 126, 327, 334, 335, 364,
    373, 383, 387, 407, 414, 423, 432,
    433, 434, 440, 493–496, 498, 639,
    650, 653, 686, 1063, 1068, 1088,
    1092, 1093
Gloggengiesser, F. 687
Gloppe, K. E. 612, 627
Glotzbach, B. 638, 660, 686
Gluszek, A. 300, 367
Glyzin, V. I. 277, 291, 372, 396, 419,
    434, 439, 585, 590, 606, 624, 637,
    686
Glyzin, V. N. 441
Godin, P. J. 204, 766, 796, 797
Godley, E. J. 660, 661, 688
Goel, R. N. 203, 611, 624
Goh, J. 367
Goldfien, A. 1049
Goldschmid, O. 608, 609, 625
Goldstein, J. L. 508, 528, 532, 557, 558,
    990, 1046
Goll, L. 557
Gombkötö, G. 219, 235, 236, 258, 262
Gomez, P. 373
Goncalves, T. M. M. 799
Goncalves da Lima, O. 802, 843, 863, 864
Gonnet, J. F. 240, 258, 292, 325, 328,
    246, 347, 364
Gonnet, M. 374
Gonzales, C. 258
Gonzalez, A. G. 290, 335, 360, 361, 364,
    433, 622
Gonzalez, J. G. 42
Goodall, D. W. 297, 372
Goodchild, J. 362
Goodman, M. M. 22, 41
Gopinath, K. W. 749, 797
Gorbaleva, G. N. 265
Gordzinskii, D. M. 974, 975, 1044
Gore, K. G. 368
Goris, A. 493, 494, 496, 500
Gorissen, H. 126, 441
Görlitz, K. 213, 559
Gorlitzer, K. 41, 75, 476, 501
Gorunovic, M. 294, 689
Goto, Y. 404, 434
Gotsiridze, A. A. 1045
Gottlieb, O. R. 40, 95, 98, 124, 211, 308,
    335, 337, 344, 345, 357, 360, 450,
    463, 466, 481, 491, 498, 499, 500,
    502, 563, 586, 587, 592, 598, 599,

602, 604, 621, 623, 624, 627,
    747–749, 760, 768, 769, 795, 796,
    797, 798, 799, 813, 856, 857, 861,
    862, 863, 864, 909, 914, 1088, 1092
Gottsegen, A. 205, 206, 290, 362, 797,
Govaert, F. 456, 478, 500
Govindarajan, V. S. 523, 557
Govindachari, T. R. 6, 23, 40, 143, 181,
    207, 270, 274, 275, 286, 289, 291,
    413, 434, 450, 451, 480, 500, 739,
    781, 782, 797
Gowan, J. E. 295
Graf, E. 208, 365
Graham, D. Z. 796
Graham, H. M. 609, 624
Graham, H. N. 42
Graham, J. H. 796
Grambast, L. 1108, 1127
Grambow, H. J. 429, 434, 879, 881, 884,
    886, 887, 912
Grankina, V. P. 637, 688
Grant, K. A. 1017, 1046
Grant, M. S. 554, 558
Grant, R. R. 358
Grant, V. 1017, 1046
Greatbanks, D. 555
Green, C. L. 499, 623, 911, 963
Green, J. 202, 210
Greger, H. 333, 334, 354, 364, 1068,
    1072, 1077, 1088, 1093
Gregson, M. 813, 816, 849, 863
Greguss, P. 737, 739
Gribovskaya, I. F. 994, 1047
Griffith, J. Q. 1034, 1046
Griffiths, F. P. 41
Griffiths, L. A. 955–957, 964, 965
Grill, R. 979, 997, 1046, 1047
Grimmer, J. 202, 211
Grimshaw, J. 292, 585, 590, 606, 624
Grinkevich, N. I. 994, 1047
Gripenberg, J. 583, 608, 624
Grippa, C. 445, 452, 480, 500
Grisebach, H. 205, 429, 434, 548, 557,
    760, 782, 797, 800, 868, 869, 875,
    878, 879, 880, 881, 882, 884, 886,
    887, 892, 895, 896, 902, 903, 905,
    906, 909, 911, 912, 913, 914, 915,
    921, 928, 929, 930, 935–937, 943,
    945, 955–957, 962–965, 967, 968,
    1046, 1106, 1127
Grist, K. L. 967
Gritsenko, E. N. 269, 291, 406, 434
Gritzky, R. 213
Grochowska, M. J. 925, 965, 1014, 1046
Groenwald, E. G. 974, 985, 1046
Grois, G. 256
Grommeck, R. 1012, 1046
Gromova, A. S. 627
de Groot-Pfleiderer, W. 702, 713, 721,
    725, 729, 738, 739
Grosourdy, D. de 803, 863
Gross, D. 941, 965
Gross, G. G. 875, 912

Pruden, G., 18, 19, 41
Prum, N., 262
Pryce, R. J., 1108, 1110, 1118, 1119, 1128
Psenak, M., 1021, 1051
Puggio, L., 1094
Puri, B., 446, 447, 457, 459, 466, 468, 470, 478, 481, 482, 485, 490, 502, 566, 601, 611, 629, 927, 967
Puri, R. N., 799
Purushothaman, K. K., 769, 772, 799
Puski, G., 428, 439
Pusztai, R., 1032, 1043, 1051
Putman, 237, 266

Quarmby, C., 24, 32, 33, 42,
Quat-Hao-Nguyen., 332, 334, 370
Quesnel, V. C., 540, 558
Quijano, L., 73, 76, 332, 334, 370, 417, 425, 439
Quinn, C. J., 732, 741, 1084, 1094
Quirin, M., 417, 438
Quiot, 253, 261

Rabate, J., 408, 414, 419, 427, 439, 446, 457, 482, 498, 611, 622
Rabcewicz-Wojcicka, A., 348, 370
Rabe, A., 206
Rackham, D. M., 74, 75
Racusen, D., 995, 1051
Radford, T., 395, 439
Radhakrishnan, R. V., 26, 42, 125, 279, 293, 739
Radhakrishniah, M., 795
Radner, B. S., 1053
Rae, J., 925, 967, 1026, 1051, 1083, 1092
Rahman, A., 616, 620, 629
Rahman, W., 39, 42, 76, 211, 299, 370, 373, 422, 423, 434, 439, 564, 567, 580, 604, 611, 629, 688, 699, 722, 723, 731, 738—741, 742
Rai, H. S., 210, 292
Raizada, K. S., 350, 370
Rajadurai, S., 359
Rajagopal, S., 358, 359
Rajagopalan, S., 358, 370
Rajajurai, S., 623
Rajaratnam, J. A., 995, 1051
Raju, R. R., 307, 371
Rakhimkhanov, Z. B., 242, 262, 263
Rakhimov, A. A., 502
Rakosi, M., 371
Rakusa, R., 213
Rall, G. J. H., 212, 769, 796, 799, 800
Ram, H., 243, 263
Raman, R. V., 210, 369, 741,
Ramanathan, J. D., 209, 436, 795
Rama Rao, A. V., 42, 86, 88, 125, 211, 274, 275, 282, 287, 290, 292, 293
Ramier, W., 585, 588, 591, 601, 629
Raminez, D. A., 265
Ramirez, R. H., 362
Ramway, M. V. J., 799

Ramwell, R. M., 925, 967
Randerath, K., 23, 42
Rane, D. F., 141, 209, 272, 292, 739
Ranft, G., 40, 500
Rangaswami, S., 285, 293, 294, 326, 327, 371, 439, 586, 601, 629
Ranjeva, R., 911, 1008, 1042
Rao, C. B., 639, 643, 651, 689
Rao, C. H., 263
Rao, D. S., 689
Rao, J. R., 659, 683, 748, 749, 758—760, 795
Rao, K. V., 132, 211, 293, 354, 374, 502,
Rao, M. M., 813, 815, 843, 861, 864
Rao, N. S. P., 111, 125, 702, 711, 713—715, 727, 734, 741
Rao, N. V. R., 210, 371, 797, 944, 968
Rao, P. L. N., 629
Rao, P. R., 290, 622
Rao, P. S., 211, 424, 428, 439
Rao, V., 274, 294
Rao, V. S., 567, 603, 629
Rasper, V., 252, 263
Rathi, S. S., 125, 281, 293, 294
Rathmell, W. G., 883, 914, 927, 967
Rattenbury, J. A., 1084, 1094
Rau, W., 154, 212
Raulais, D., 287, 292
Raut, K. B., 371
Raven, P. H., 366, 1017, 1031, 1045, 1051
Ravindranath, B., 207, 500, 624
Ray, J. N., 864
Raynaud, J., 233, 262, 263, 273, 294, 417, 437, 639, 689
Razaq, S., 162, 203, 738
Read, J., 473, 474, 503
Redman, B. T., 211, 798, 863, 864
Redman, G. H., 612—619, 621, 1086, 1092
Redner, A., 1052
Reed, D. J., 874, 875, 914
Reed, F. P., 502
Reed, G. F., 862
Reed, R. I., 82, 84, 118, 125
Regan, T. H., 263
Rege, D. V., 287, 293
Rêgo de Sousa, 844, 847, 865
Rehder, A., 497, 502
Reichel, L., 221, 263, 371
Reichardt, A., 691
Reid, P. D., 913, 1001, 1051
Reinert, J. H. C., 974, 1051
Reinhard, E., 1002, 1043
Reinhardt, H., 739
Remaily, G. W., 265
Reisener, H. J., 968
Rendina, M. A. C., 420, 439
Renedo, J., 265, 623
Rennie, E. H., 494, 496, 502
Rentschler, H., 43
Repas, A., 15, 42
Reuter, E. W., 947, 967

Index of Plant Species

1178 THE FLAVONOIDS

Alnus 563, 570, 583
  firma 582, 600
  glutinosa 332, 333, 335, 1061
  japonica 270–272
  pendula 570, 600
  sieboldiana 272, 324, 332, 338, 582,
    600, 1061
Alphitonia 474
  excelsa 473
Alpinia 445, 563, 570, 582, 583
  chinensis 324, 607
  japonica 324–326, 582, 607
  katsumadai 452, 483, 607
  kumatake 325, 326
  officinarum 324, 326
  speciosa 483, 607
Alisma 639
Alismataceae 639
Alternanthera phylloxeroides 1061
Alternaria 1120
Althaea rosea 417
Amaranthaceae 746, 1061
Amaranthus 979
  caudatus 980
  lividus 979
Ambrosia
  dumosa 274
  grayi 145, 342
  hispida 271
Amentotaxus formosana 724, 733
Ammi visnaga 401
Amorpha fruticosa 404, 763, 881, 908, 939
Ampelopsis 587, 594
  mediaefolia 607
Amphimas pterocarpoides 748
Anacardiaceae 239, 477, 478, 484, 591,
    600, 618, 726, 735
Anagallis arvensis 246
Andira
  inermis 768, 769
  parviflora 748, 760
Andrographis
  echioides 181, 270, 413
  paniculata 6, 274
  serpyllifolia 275
  wrightiana 270, 274
Anemone
  alpina 431
  coronaria 233
  hepatica 233
Anethum sowa 423, 638
Angelica archangelica 575
Angophora 566
  lanceolata 605
Aniba 450, 563
  rosaeodora 453, 480, 602
Anodendron affine 270
Antennaria dioica 411
Anthemis 333, 334, 1077
  cotula 1063
Anthonomus grandis 1031
Anthurium 567, 578
  andraeanum 253

  binoti 607
Anthyllis vulneraria 325, 328, 346, 347
Antirrhinum 270, 468, 469, 471, 472,
    486–488, 564, 574
  majus 249, 404, 408, 409, 431, 607
Apium graveolens 396, 408, 409
Apocynaceae 247
Apocynum pictum 419
Apuleia leiocarpa 335, 337, 344, 345, 357,
    769
Aquifoliaceae 240
Aquilegia 1069
Araceae 253, 607, 640
Aralia
  cordata 243
  elata 243
Araliaceae 243, 600
Araucaria
  bidwillii 722
  columnaris 722
  cooki 723
  cunninghamii 722
Araucariaceae 722, 729, 736, 737, 1101,
    1102
Arctostaphylos uva-ursi 422
Argemone mexicana 422, 423
Argyranthemum frutescens 334
Argyrolobium 637, 1078
Ariocarpus retusus 330
Arisaema serratum 253
Aristolochiaceae 234
Armeria 246
Armoracia rusticana 414
Artedia 1068
Artemisia 586, 639
  absinthium 336
  arborescens 336
  cana 336
  herba-alba 273
  pygmaea 591, 601
  transiliensis 327, 348, 422, 653
Arthraxon hispidus 285
Arthrotaxis 732, 733
Artocarpus 271, 277, 281–283, 568, 581,
    586, 592, 618
  chaplasha 281
  gomezianus 281
  heterophyllus 271, 278, 281, 604
  hirsutus 271, 281, 604
  incisa 281
  integrifolia 278, 344, 604
  lakoocha 281
Arum 640
  italicum 660
Asarum asaroides 234
Asclepiadaceae 247
Ascochyta imperfecta 999, 1025
Aspalathus 494, 496, 636
  acuminatus 646, 648, 659
  linearis 495, 659
Asparagus
  officinalis 252
  ramosus 252

Subject Index